Safety Symbols		Hazard	Examples	Precaution	Remedy
Disposal		Special disposal procedures need to be followed.	Certain chemicals, living organisms	Do not dispose of these materials in the sink or trash can.	Dispose of wastes as directed by your teacher.
Biological		Organisms or other biological materials that might be harmful to humans	Bacteria, fungi, blood, unpreserved tissues, plant materials	Avoid skin contact with these materials. Wear mask and gloves.	Notify your teacher if you suspect contact with material. Wash hands thoroughly.
Extreme Temperature		Objects that can burn skin by being too cold or too hot	Boiling liquids, hot plates, dry ice, liquid nitrogen	Use proper protection when handling.	Go to your teacher for first aid.
Sharp Object		Use of tools or glassware that can easily puncture or slice skin	Razor blades, pins, scalpels, pointed tools, dissecting probes, broken glass	Practice common-sense behavior and follow guidelines for use of the tool.	Go to your teacher for first aid.
Fume		Possible danger to respiratory tract from fumes	Ammonia, acetone, nail polish remover, heated sulfur, moth balls	Be sure there is good ventilation. Never smell fumes directly. Wear a mask.	Leave foul area and notify your teacher immediately.
Electrical		Possible danger from electrical shock or burn	Improper grounding, liquid spills, short circuits, exposed wires	Double-check setup with teacher. Check condition of wires and apparatus.	Do not attempt to fix electrical problems. Notify your teacher immediately.
Irritant		Substances that can irritate the skin or mucous membranes of the respiratory tract	Pollen, moth balls, steel wool, fiberglass, potassium permanganate	Wear dust mask and gloves. Practice extra care when handling these materials.	Go to your teacher for first aid.
Chemical		Chemicals that can react with and destroy tissue and other materials	Bleaches such as hydrogen peroxide; acids such as sulfuric acid, hydrochloric acid; bases such as ammonia, sodium hydroxide	Wear goggles, gloves, and an apron	Immediately flush the affected area with water and notify your teacher.
Toxic		Substance may be poisonous if touched, inhaled, or swallowed.	Mercury, many metal compounds, iodine, poinsettia plant parts	Follow your teacher's instructions.	Always wash hands thoroughly after use. Go to your teacher for first aid.
Flammable		Flammable chemicals may be ignited by open flame, spark, or exposed heat.	Alcohol, kerosene, potassium permanganate	Avoid open flames and heat when using flammable chemicals.	Notify your teacher immediately. Use fire safety equipment if applicable.
Open Flame		Open flame in use, may cause fire.	Hair, clothing, paper, synthetic materials	Tie back hair and loose clothing. Follow teacher's instruction on lighting and extinguishing flames.	Notify your teacher immediately. Use fire safety equipment if applicable.

 Eye Safety
Proper eye protection should be worn at all times by anyone performing or observing science activities.

 Clothing Protection
This symbol appears when substances could stain or burn clothing.

 Radioactivity
This symbol appears when radioactive materials are used.

 Handwashing
After the lab, wash hands with soap and water before removing goggles.

GLENCOE

PHYSICS

PRINCIPLES & PROBLEMS

Mc
Graw
Hill
Education

COVER: Spacedreamer/Shutterstock.com

mheducation.com/prek-12

Send all inquiries to:
McGraw-Hill Education
STEM Science
8787 Orion Place
Columbus, OH 43240

ISBN: 978-0-07-677476-0
MHID: 0-07-677476-7

Printed in the United States of America.

3 4 5 6 7 8 9 LWI 21 20 19 18 17

CONTENTS IN BRIEF

Welcome to

GLENCOE PHYSICS
PRINCIPLES & PROBLEMS

We are your partner in learning by meeting your diverse 21st century needs. Designed for today's tech-savvy high school students, the McGraw-Hill Education's *Physics: Principles and Problems* program offers hands-on investigations, rigorous science content, and engaging, real-world applications to make science fun, exciting, and stimulating.

Quick Start Guide
Glencoe Physics | Student Center

Login information

1 Go to **connected.mcgraw-hill.com**

2 Enter your registered Username and Password.

3 For **new users** click here to create a new account

4 Get **ConnectED Help** for creating accounts, verifying master codes, and more.

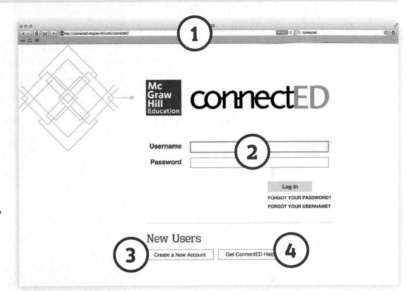

Your ConnectED Center

5 Scroll down to find the program from which you would like to work.

Quick Start Guide
Glencoe Physics | Student Center

1. The Menu allows you to easily jump to anywhere you need to be.

2. Click the **program icon** at the top left to **return to the main page** from any screen.

3. **Select a Chapter and Lesson** Use the drop down boxes to quickly jump to any lesson in any chapter.

4. Return to your **My Home** page for all your **ConnectED** content.

5. The **Help** icon will guide you to online help. It will also allow for a quick logout.

6. **Search Bar** allows you to search content by topic or standard.

7. **Access the eBook** Use the **Student Edition** to see content.

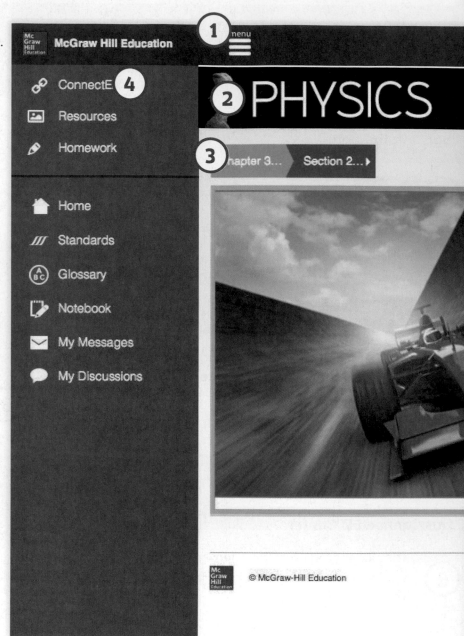

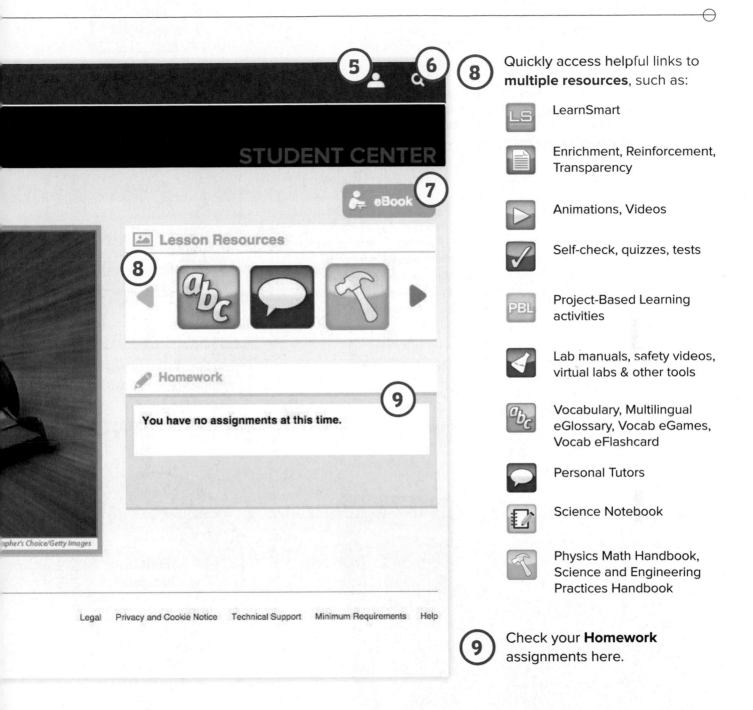

STUDENT CENTER

eBook ⑦

🖼 **Lesson Resources**

⑧

◄ *abc* 💬 🔨 ►

✏ **Homework**

⑨

You have no assignments at this time.

...apher's Choice/Getty Images

Legal   Privacy and Cookie Notice   Technical Support   Minimum Requirements   Help

⑤ ⑥ ⑧ Quickly access helpful links to **multiple resources**, such as:

LS — LearnSmart

📄 — Enrichment, Reinforcement, Transparency

▶ — Animations, Videos

✔ — Self-check, quizzes, tests

PBL — Project-Based Learning activities

🧪 — Lab manuals, safety videos, virtual labs & other tools

abc — Vocabulary, Multilingual eGlossary, Vocab eGames, Vocab eFlashcard

💬 — Personal Tutors

📓 — Science Notebook

🔨 — Physics Math Handbook, Science and Engineering Practices Handbook

⑨ Check your **Homework** assignments here.

connected.mcgraw-hill.com

Why A²?

Accuracy Assurance is central to McGraw-Hill Education's commitment to high-quality, learner-oriented, real-world, and error-free products. Also at the heart of our A² Development Process is a commitment to make the text **Accessible** and **Approachable** for both students and teachers. A collaboration among authors, content editors, academic advisors, and classroom teachers, the A² Development Process provides opportunities for continual improvement through customer feedback and thorough content review.

The A² Development Process begins with a review of the previous edition and a look forward to state and national standards. The authors for *Physics: Principles and Problems* combine expertise in teacher training and education with a mastery of chemistry content knowledge. As manuscript is created and edited, consultants review the accuracy of the content while our Teacher Advisory Board members examine the program from the points of view of both teacher and student. Student labs and teacher demonstrations are reviewed for both accuracy of content and safety. As design elements are applied, chapter content is again reviewed, as are photos and diagrams.

Throughout the life of the program, McGraw-Hill Education continues to troubleshoot and incorporate improvements. Our goal is to deliver to you a program that has been created, refined, tested, and validated as a successful tool for your continued **Academic Achievement.**

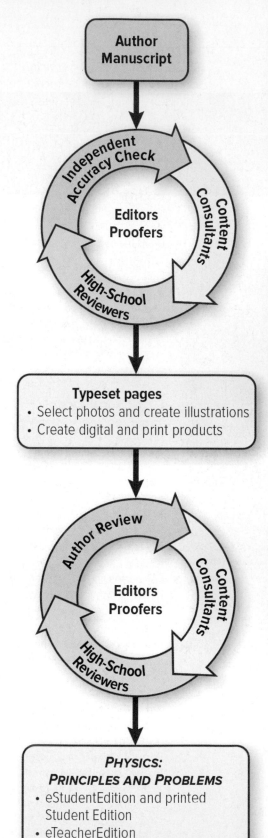

Author Manuscript

Independent Accuracy Check
Content Consultants
High-School Reviewers
Editors Proofers

Typeset pages
- Select photos and create illustrations
- Create digital and print products

Author Review
Content Consultants
High-School Reviewers
Editors Proofers

PHYSICS: PRINCIPLES AND PROBLEMS
- eStudentEdition and printed Student Edition
- eTeacherEdition
- Printed Teacher Essentials
- Student Worksheets
- eAssessment

©Roy Wylam/Alamy

Authors

The authors of *Physics: Principles and Problems* used their physics content knowledge and teaching expertise to craft manuscript that is accessible, accurate, and geared toward student achievement.

■ Paul W. Zitzewitz, lead author

was an emeritus professor of physics and of science education at the University of Michigan–Dearborn. He was awarded posthumously a Carleton College 2014 Alumni Association Distinguished Achievement Award. He received his BA from Carleton College and his MA and PhD from Harvard University, all in physics. Dr. Zitzewitz taught physics to undergraduates at the University of Michigan-Dearborn for 36 years and, as an active experimenter in the field of atomic physics, published more than 50 research papers. He was named a Fellow of the American Physical Society for his contributions to physics and science education for high school and middle school teachers and students. He is now treasurer of the American Association of Physics Teachers and has been the president of the Michigan Section of the American Association of Physics Teachers and chair of the American Physical Society's Forum on Education.

■ David G. Haase

is an Alumni Distinguished Undergraduate Professor of Physics at North Carolina State University. He earned a BA in physics and mathematics at Rice University and an MA and a PhD in physics at Duke University, where he was a J.B. Duke Fellow. He has been an active researcher in experimental low-temperature and nuclear physics. He teaches undergraduate and graduate physics courses and has worked many years in K–12 teacher training. He was the founding director of The Science House at NC State, a science-math learning center that leads teacher training and student programs across North Carolina. He has co-authored over 100 papers in experimental physics and in science education. He is a Fellow of the American Physical Society. He received the Alexander Holladay Medal for Excellence, NC State University; was awarded the Pegram Medal for Physics Teaching Excellence; and was chosen 1990 Professor of the Year in the state of North Carolina by the Council for the Advancement and Support of Education (CASE).

■ Kathleen A. Harper

is a senior lecturer in the Engineering Education Innovation Center at The Ohio State University. She received her MS in physics and BS in electrical engineering and applied physics from Case Western Reserve University and her PhD in physics from The Ohio State University. She has taught introductory physics, astronomy, and engineering courses to undergraduates for over 20 years and has been instrumental in offering Modeling Instruction workshops to in-service high school teachers, both in Ohio and nationwide. Her research interests include the teaching and learning of problem-solving skills and the development of alternative problem formats. She is active in AAPT, both on the local and national levels, often presenting talks and workshops on teaching problem solving. Additionally, she is co-editor of a selection of articles available through AAPT's Compadre portal entitled, "Getting Started in Physics Education Research."

Teacher Advisory Board

The Teacher Advisory Board gave the editorial staff and design team feedback on the content and design of both the Student Edition and the Teacher Essentials. They were instrumental in providing valuable input toward the development of *Physics: Principles and Problems.* We thank these teachers for their hard work and creative suggestions.

Janet Adams
Physics Teacher
Mars Area High School
Mars, PA

Craig Dowler
Physics Teacher
West Genesee High School
Camillus, NY

Chris Foust, M.S.
Physics Teacher
Hermitage High School
Richmond, VA

Ryan Hall
Physics Teacher
Palatine High School
Palatine, IL

Stan Hutto
Physics Teacher
Alamo Heights High School
San Antonio, TX

Richard A. Lines, Jr.
Physics Teacher
Garland High School
Garland, TX

Nikki Malatin
Physics Teacher
West Caldwell High School
Lenoir, NC

Jennifer McDonnell
Coordinator of Math and
Science
School District U-46
Elgin, IL

Jeremy Paschke
Physics Teacher
York High School
Elmhurst, IL

Don Pata
Physics Teacher
Grosse Pointe North High
School
Grosse Pointe Woods, MI

Charles Payne
Physics Teacher
Northern High School
Durham, NC

Content Consultants

Content consultants each reviewed selected chapters of *Physics: Principles and Problems* for content accuracy and clarity.

Dr. Solomon Bililign, PhD
Professor of Physics
Director: NOAA-ISET Center
North Carolina A&T State University
Greensboro, NC

Ruth Howes
Professor Emerita of Physics and
Astronomy
Ball State University
Muncie, IN

Dr. Keith H. Jackson
Professor and Chair of Department of
Physics
Morgan State University
Baltimore, MD

Kathleen Johnston
Associate Professor of Physics
Louisiana Tech University
Ruston, LA

Dr. Monika Kress
Associate Professor
Department of Physics & Astronomy
San Jose State University
San Jose, CA

Jorge Lopez
Professor
University of Texas at El Paso
El Paso, TX

Dr. Ramon E. Lopez
Department of Physics
University of Texas at Arlington
Arlington, TX

Albert J. Osei
Professor of Physics
Oakwood University
Huntsville, AL

Charles Ruggiero
Visiting Assistant Professor
Allegheny College
Meadeville, PA

Toni Sauncy
Associate Professor of Physics
Angelo State University
San Angelo, TX

Sally Seidel
Professor of Physics
University of New Mexico
Albuquerque, NM

Contributing Writers

Additional science writers added feature content, teacher materials, assessment, and laboratory investigations.

Molly Wetterschneider
Austin, TX

Steve Whitt
Columbus, OH

Safety Consultant

The safety consultant reviewed labs and lab materials for safety and implementation.

Kenneth Russell Roy, PhD
Director of Science and Safety
Glastonbury Public Schools
Glastonbury, CT

Teacher Reviewers

Each teacher reviewed selected chapters of *Physics: Principles and Problems* and provided feedback and suggestions regarding the effectiveness of the instruction.

Joseph S. Bonanno
Physics Teacher
Red Creek Central School
Red Creek, NY

Beverly Trina Cannon
Physics Teacher
Highland Park High School
Dallas TX

Katharine Chole
Physics Teacher
Villa Duchesne School
St. Louis, MO

Cheryl Rawlins Cowley
Lead Physics Teacher
Sherman High School
Sherman, TX

B. Wayne Davis
Physics Teacher (retired)
Henrico County Public Schools
Richmond, VA

Nina Morley Daye
Physics Teacher
Orange High School
Hillsborough, NC

David Eberst
Physics Teacher
Bishop Watterson High School
Columbus, OH

Terry Elmer
Physics Teacher
Red Creek Central School
Red Creek, NY

Michael Fetsko
Physics Teacher
Mills E. Godwin High School
Henrico, VA

Chris Foust, M.S.
Physics Teacher
Hermitage High School
Richmond, VA

Elaine Gwinn
Physics Teacher
Shenandoah High School
Middletown, IN

Janie Head
Physics Teacher
Foster High School
Richmond, TX

Stan Hutto
Physics Teacher
Alamo Heights High School
San Antonio, TX

Emily James
Physics Teacher
Brewster Academy
Wolfeboro, NH

Dr. Christopher D. Jones
School of Applied and Engineering Physics
Cornell University
Ithaca, NY

Dr. Mike Papadimitriou
Headmaster
Academy for Science and Health Professions
Conroe, TX

Julia Quaintance
Physics Teacher
Morgan Local High School
McConnelsville, OH

Stephen Rea
University of Michigan, Dearborn
Plymouth, MI

Patricia Rollison
Physics Teacher
St. Gertrude High School
Richmond, VA

Patrick Slattery
Physics Teacher
South Elgin High School
South Elgin, IL

James Stankevitz
Physics Teacher
Wheaton Warrenville South High School
Wheaton, IL

Jason Sterlace
Physics Teacher
J.R. Tucker High School
Henrico, VA

Christopher White
Physics Teacher
Seneca High School
Seneca, SC

Michael Young, MS
ACT, Inc.
Iowa City, IA

Tom Young
Physics Teacher
Whitehouse High School
Whitehouse, TX

PHYSICS STUDY TOOLS

BE THE SCIENTIST!
BE THE ENGINEER!

ConnectED is your one-stop online resource to explore real-world challenges that deepen understanding of core ideas and cross-cutting concepts!

- Physics ebook
- Science and Engineering Practices Handbook
- Applying Practices activities
- PBLs
- Design-your-own Labs
- Guided investigations

Get the LearnSmart© advantage! Improve your performance by using this interactive and adaptive study tool. Your personalized learning path will help you practice and master key physics concepts.

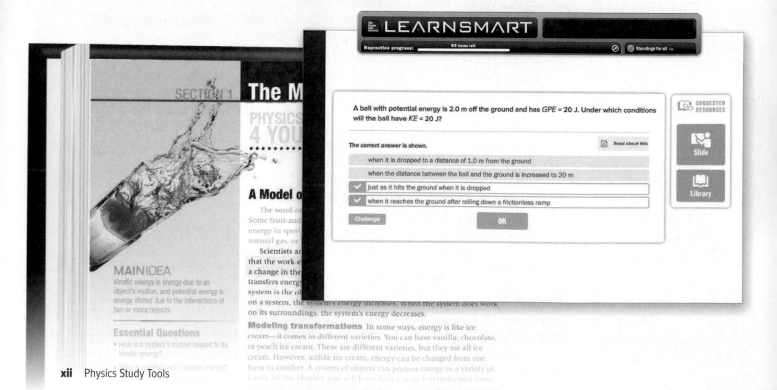

Go online!

66 It's **EASY** to get my assignments online and **QUICK** to find everything I need. 99

CHAPTER 11

Energy and Its Conservation

BIGIDEA Within a closed, isolated system, energy can change form, but the total energy is constant.

SECTIONS

1 **The Many Forms of Energy**

2 **Conservation of Energy**

LaunchLAB

ENERGY OF A BOUNCING BALL
What determines how high a basketball bounces?

WATCH THIS!

Video

FUEL EFFICIENCY
The phrase *fuel-efficient* is often used as a selling point when describing the features of a car. But what does that phrase really mean? How is fuel-efficiency calculated? And how does it relate to the conservation of energy?

PHYSICS T.V.

Video

Your introduction to the chapter concepts begins with **PHYSICS T.V.** videos, connecting physics to your life.

(l)McGraw-Hill Education; (r)Tadao Yanamoto/amana images/Getty Images

©Fancy Photography/Veer

PHYSICS STUDY TOOLS

Example Problems throughout each chapter give you step-by-step instructions and hints on how to solve each type of physics problem. Each Example Problem is followed by a number of **Practice Problems** on the same topic, allowing you to practice the skill.

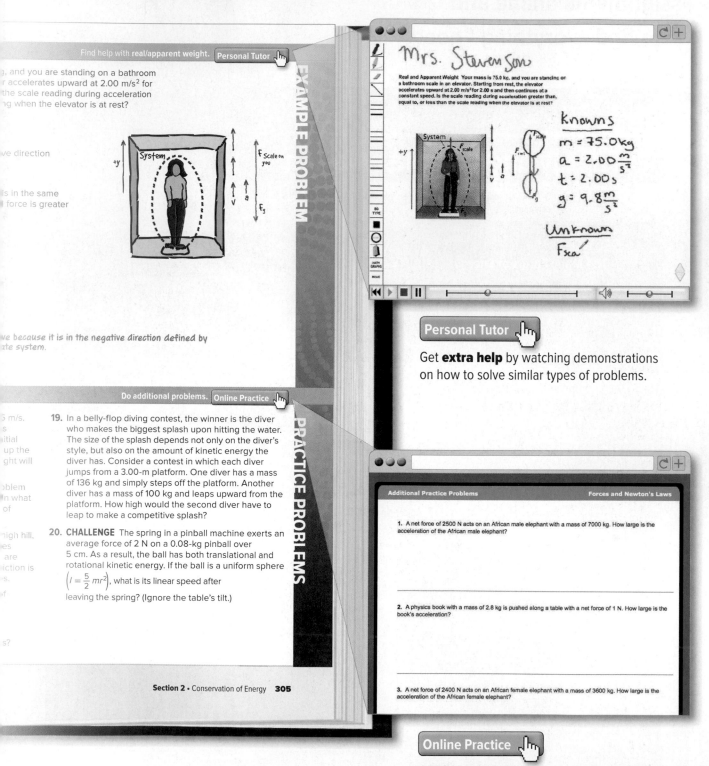

Find help with real/apparent weight. Personal Tutor

, and you are standing on a bathroom
accelerates upward at 2.00 m/s² for
the scale reading during acceleration
g when the elevator is at rest?

ve direction

s in the same
force is greater

ve because it is in the negative direction defined by
ate system.

Do additional problems. Online Practice

5 m/s.
s
itial
up the
ght will

oblem
n what
of

igh hill,
es
are
iction is
s.
of

s?

19. In a belly-flop diving contest, the winner is the diver who makes the biggest splash upon hitting the water. The size of the splash depends not only on the diver's style, but also on the amount of kinetic energy the diver has. Consider a contest in which each diver jumps from a 3.00-m platform. One diver has a mass of 136 kg and simply steps off the platform. Another diver has a mass of 100 kg and leaps upward from the platform. How high would the second diver have to leap to make a competitive splash?

20. **CHALLENGE** The spring in a pinball machine exerts an average force of 2 N on a 0.08-kg pinball over 5 cm. As a result, the ball has both translational and rotational kinetic energy. If the ball is a uniform sphere $\left(I = \frac{5}{2} mr^2\right)$, what is its linear speed after leaving the spring? (Ignore the table's tilt.)

Section 2 • Conservation of Energy **305**

Personal Tutor

Get **extra help** by watching demonstrations on how to solve similar types of problems.

Online Practice

Additional problems provide opportunities to practice the skills and concepts associated with each Example Problem.

" VIDEOS, ANIMATIONS, and tools to help me LEARN. "

Go **online!**

Virtual Investigations

Take labs to the next level, launching rockets, firing cannons, exploring electrons, and more in an **online laboratory.**

40 N | 40 N
Force exerted by first dog on sled | Force exerted by second dog on sled

Newton's Second Law

Figure 7 shows two dogs pulling a sled 40 N. From the cart and spring-scale sled accelerates as a result of the unba acceleration change if instead of two there was one bigger, stronger dog e sled? When considering forces and a sum of all forces, called the net force

Newton's second law states that proportional to the net force and in object being accelerated. This law is affect masses and is represented by th

NEWTON'S SECOND LAW

The acceleration of an object is equal to the on the object divided by the mass of the obje

$$a =$$

Solving problems using Newton's important steps in correctly apply the net force acting on the obj object, so you must add the f

Figure 7 The net force acting on an obj is the vector sum of all the forces acting that object.

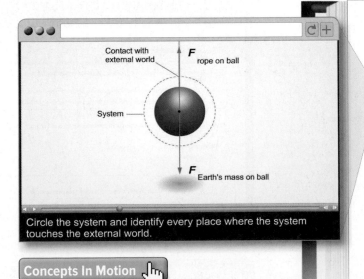

Circle the system and identify every place where the system touches the external world.

Concepts In Motion

See physics content come to life in **animated figures** and moving diagrams.

Figure 3 The drawings of the ball in the hand and the ball hanging from the string are both pictorial models. The free-body diagram for each situation is shown next to each pictorial model.

View an **animation of free body diagrams.**
Concepts In Motion

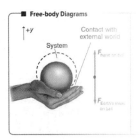

Free-body diagrams Just as pictoria diagrams are useful in solving problems tions will help you analyze how forces sketch the situation, as shown in Figure every place where the system touches places that an agent exerts a contact on the system. This gives you the pic

A **free-body diagram** is a physica forces acting on a system. Follow th body diagram:
- The free-body diagram is drawn problem situation.
- Apply the particle model, and re
- Represent each force with an force is applied. Always draw

PHYSICS STUDY TOOLS

Quizzes, assessments, and study tools provide opportunities for self-assessment, review, and additional practice.

net force acting on the above, then the object does not experience a change in speed or direction and is in equilibrium. As Figure 9 indicates, at least in terms of net forces, there is no difference between sitting in a chair and falling at a constant velocity while skydiving–velocity isn't changing, so the net force is zero.

When analyzing forces and motion, it is important to keep in mind that the real world is full of forces that resist motion, called frictional forces. Newton's ideal, friction-free world is not easy to obtain. If you analyze a situation and find that the result is different from a similar experience that you have had, ask yourself whether this is because of the presence of frictional forces. For example, if you shove a textbook across a table, it will quickly come to a stop. At first thought, it might seem Newton's first law is violated in this case because the book's velocity changes even though there is no apparent force acting on the book. The net force acting on the book, however, is a frictional force between the table and the book in the direction opposite motion.

SECTION 1 REVIEW

Section Self-Check

12. MAINIDEA Identify each of the following as either **a, b, or c:** mass, inertia, the push of a hand, friction, air resistance, spring force, gravity, and acceleration.

   a. contact force

   b. a field force

   c. not a force

13. Free-Body Diagram Draw a free-body diagram of a bag of sugar being lifted by your hand at an increasing speed. Specifically identify the system. Use subscripts to label all forces with their agents. Remember to make the arrows the correct lengths.

14. Free-Body Diagram Draw a free-body water bucket being lifted by a [...] speed. Specifically identify the [...] with their agents and make the [...] lengths.

15. Critical Thinking A force of 1 N is [...] force exerted on a block, and the [...] tion of the block is measured. W[...] tal force is the only force exert[...] the horizontal acceleration is th[...] What can you conclude about the[...] blocks?

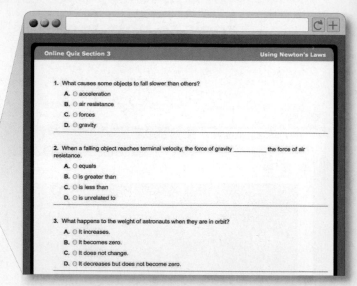

Section Self-Check

Review questions and problems for each section help you identify concepts that require additional study.

CHAPTER 4 ASSESSMENT

Chapter Self-Check
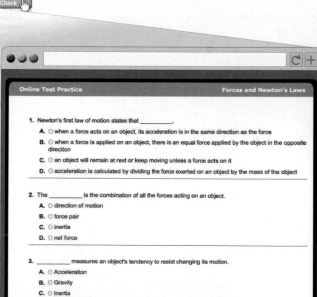

SECTION 1 Force and Motion

Mastering Concepts

39. BIGIDEA You kick a soccer ball across a field. It slows down and comes to a stop. You ask your younger brother to explain what happened to the ball. He says, "The force of your foot was transferred to the ball, which made it move. When that force ran out, the ball stopped." Would Newton agree with that explanation? If not, explain how Newton's laws would describe it.

40. Cycling Imagine riding a single-speed bicycle. Why do you have to push harder on the pedals to start the bicycle moving than to keep it moving at a constant velocity?

Mastering Problems

41. What is the net force acting on a 1.0-kg ball moving at a constant velocity?

42. Skating Joyce and Efua are skating. Joyce pushes Efua, whose mass is 40.0 kg, with a force of 5.0 N. What is Efua's resulting acceleration?

43. A 2300-kg car slows down at a rate of 3.0 m/s² when approaching a stop sign. What is the magnitude of the net force causing it to slow down?

44. Breaking the Wishbone After Thanksgiving, Kevin and Gamal use the turkey's wishbone to make a wish. If Kevin pulls on it with a force 0.17 N larger than the force Gamal pulls with in the opposite direction and the wishbone has a mass of 13 g, what is the wishbone's initial acceleration?

48. Before a skydiver opens her parachute, she [...] falling at a velocity higher than the termina[...] that she will have after the parachute deplo[...]

   a. Describe what happens to her velocity as [...] opens the parachute.

   b. Describe the sky diver's velocity from wh[...] parachute has been open for a time until [...] about to land.

49. Three objects are dropped simultaneously f[...] top of a tall building: a shot put, an air-fille[...] and a basketball.

   a. Rank the objects in the order in which t[...] reach terminal velocity, from first to last[...]

   b. Rank the objects according to the order [...] they will reach the ground, from first to [...]

   c. What is the relationship between your a[...] parts a and b?

Mastering Problems

50. What is your weight in newtons? Show you[...]

51. A rescue helicopter lifts two people using a[...] and a rescue ring as shown in **Figure 21.**

   a. The winch is capable of exerting a 2000 [...] What is the maximum mass it can lift?

   b. If the winch applies a force of 1200 N, w[...] rescuer's and victim's acceleration? Draw [...] body diagram for the people being lifted[...]

   c. Using the acceleration from part b, how [...] it take to pull the people up to the helic[...] Assume the people are initially at rest.

Chapter Self-Check

Chapter-level **study tools** provide more opportunities for review and practice.

Go online!

I have the online STUDY TOOLS I need to help me SUCCEED.

- gravitational field (p. 100)
- apparent weight (p. 102)
- weightlessness (p. 102)
- drag force (p. 104)
- terminal velocity (p. 105)

MAINIDEA

- The object's weight (F_g) depends on object's location.

- An object's apparent weight can be no apparent weight experiences w

- A falling object reaches a constant ve The constant velocity is called the te by the object's weight, size, and

VOCABULARY

- interaction pair (p. 106)
- Newton's third law (p. 106)
- tension (p. 109)
- normal force (p. 111)

SECTION 3 Newton's Th

MAINIDEA   All force occur in

- Newton's third law states that the for equal in magnitude, but opposite in pair, $F_{A\ on\ B}$ does not cause $F_{B\ on}$

- The normal force is a support for always perpendicular to the plane of

Games and Multilingual eGlossary

Vocabulary Practice

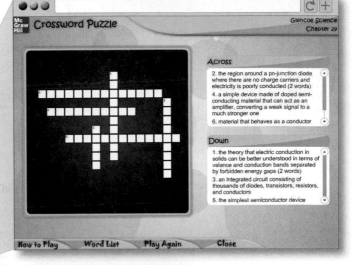

Vocabulary Practice

The **multilingual e-Glossary** and vocabulary study tools drive home important concepts.

---

new kid started in 2.5 m/s². Assume the sprint
starts from rest.

A. 13 m   C. 80 m

B. 20 m   D. 90 m

3. If a motorcycle starts from rest and maintains a constant acceleration of 3 m/s², what will its velocity be after 10 s?

A. 10 m/s   C. 90 m/s

B. 30 m/s   D. 100 m/s

4. How does an object's acceleration change if the net force on the object is doubled?

A. The acceleration is cut in half.

B. The acceleration does not change.

C. The acceleration is doubled.

D. The acceleration is multiplied by four.

5. What is the weight of a 225-kg space probe on the Moon? The gravitational field on the Moon is 1.62 N/kg.

A. 139 N   C. 1.35×10^3 N

B. 364 N   D. 2.21×10^3 N

Position v. Time Graph

Time (s)

Online Test Practice

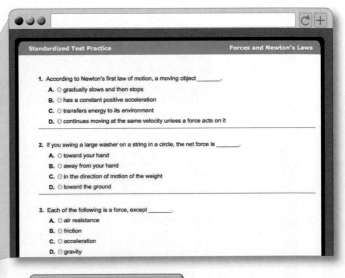

Online Test Practice

Practice taking **standardized tests** as you also review chapter content.

REAL-WORLD STEM

Physics: Principles and Problems makes physics real. Throughout the text, find personal physics connections, surprising examples of physics in careers, and how physicists are engaged in cutting-edge science research.

SECTION 1 Force and Motion

PHYSICS 4 YOU
To start a skateboard moving on a level surface, you must push on the ground with your foot. If the skateboard is on a ramp, however, gravity will pull you down the slope. In both cases unbalanced forces change the skateboard's motion.

Force

Consider a textbook resting on a table. To cause it to move, you could either push or pull on it. In physics, a push or a pull is called a **force**. If you push or pull harder on an object, you exert a greater force on the object. In other words, you increase the magnitude of the applied force. The direction in which the force is exerted also matters—if you push the resting book to the right, the book will start moving to the right. If you push the book to the left, it will start moving to the left. Because forces have both magnitude and direction, forces are vectors. The symbol *F* is vector notation that represents the size and direction of a force, while *F* represents only the magnitude. The magnitude of a force is measured in units called newtons (N).

Unbalanced forces change motion Recall that motion diagrams describe the positions of an object at equal time intervals. For example, the motion diagram for the book in **Figure 1** shows the distance between the dots increasing. This means the speed of the book is increasing. At *t* = 0, it is at rest, but after 2 seconds it is moving at 1.5 m/s. This change in speed means it is accelerating. What is the cause of this acceleration? The book was at rest until you pushed it, so the cause of the acceleration is the force exerted by your hand. In fact, all accelerations are the result of an unbalanced force acting on an object. What is the relationship between force and acceleration? By the end of this section, you will learn

MAINIDEA
A force is a push or a pull.

Essential Questions
- What is a force?
- What is the relationship between force and acceleration?
- How does motion change when the net force is zero?

Review Vocabulary

acceleration the rate at which the velocity of an object changes

New Vocabulary

PHYSICS 4 YOU

at the beginning of each section tells you how the physics you are about to learn relates to your life.

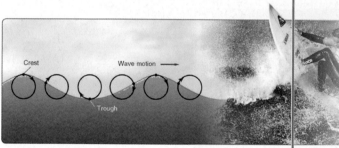

Crest   Wave motion ⟶

Trough

Surface waves Waves that are deep in a lake or an ocean are longitudinal. In a **surface wave**, however, the medium's particles follow a circular path that is at times parallel to the direction of travel and at other times perpendicular to the direction of wave travel, as shown in **Figure 8**. Surface waves set particles in the medium, in this case water, moving in a circular pattern. At the top and bottom of the circular path, particles are moving parallel to the direction of the wave's travel. This is similar to a longitudinal wave. At the left and right sides of each circle, particles are moving up or down. This up-and-down motion is perpendicular to the wave's direction, similar to a transverse wave.

Wave Properties

Many types of waves share a common set of wave properties. Some wave properties depend on how the wave is produced, whereas others depend on the medium through which the wave travels.

Amplitude How does the pulse generated by gently shaking a rope differ from the pulse produced by a violent shake? The difference is similar to the difference between a ripple in a pond and an ocean breaker—they have different amplitudes. You read earlier that the amplitude of periodic motion is the greatest distance from equilibrium. Similarly, as shown in **Figure 9**, a transverse wave's amplitude is the maximum distance of the wave from equilibrium. Since amplitude is a distance, it is

Figure 8 Surface waves in water cause movement both parallel and perpendicular to the direction of wave motion. When these waves interact with the shore, the regular, circular motion is disrupted and the waves break on the beach.

REAL-WORLD PHYSICS

Tsunamis On March 11, 2011, a wall of water estimated to be ten meters high hit areas on the East coast of Japan —tsunami! A tsunami is a series of ocean waves that can have wavelengths over 100 km, periods of one hour, and wave speeds of 500–1000 km/h.

Throughout the book,
REAL-WORLD PHYSICS
demonstrates how the physics you are learning applies to the world around you.

End-of-chapter features highlight physics in careers, how it connects to the real world, and what today's physicists are doing to learn more about our universe.

PHYSICS THAT'S ENTERTAINMENT!

How does physics apply to entertainment? You might be surprised! Explore the physics of special effects, 3-D movies, theater acoustics, and more!

A CLOSER LOOK

Take "A Closer Look" at a number of physics topics and discover the story behind some of the most interesting physics applications!

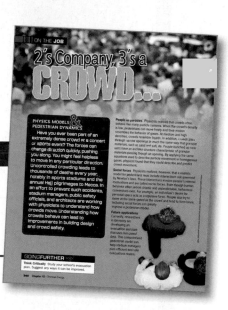

ON THE JOB

You might be surprised to discover the physics that can be found in many different careers. Explore jobs that unexpectedly rely on an understanding of physics.

FRONTIERS IN PHYSICS

What is being discovered in today's physics research? Explore the work being done by today's physicists.

HOW IT WORKS

Explore the physics of everyday objects or natural phenomena by discovering how they "work."

UNDERSTANDING PHYSICS

At the start of each chapter, you will see the **BIGIDEA** that will help you understand how what you are about to investigate fits into the big picture of science.

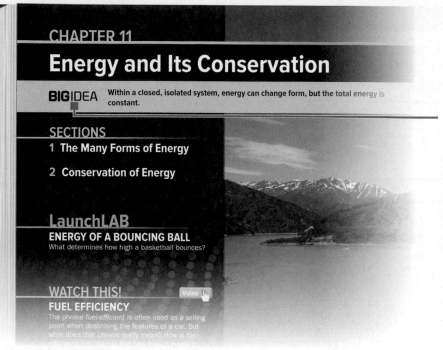

CHAPTER 11

Energy and Its Conservation

BIGIDEA   Within a closed, isolated system, energy can change form, but the total energy is constant.

SECTIONS

1 The Many Forms of Energy

2 Conservation of Energy

LaunchLAB

ENERGY OF A BOUNCING BALL
What determines how high a basketball bounces?

WATCH THIS!   Video

FUEL EFFICIENCY
The phrase *fuel-efficient* is often used as a selling point when describing the features of a car. But what does that phrase really mean? How is fuel-

The **BIGIDEA** is the focus of the chapter. By reading the text, doing labs and answering Practice Problems, Section Reviews, and Chapter Assessments, you will build an in-depth understanding of this idea.

CHAPTER 11   **ASSESSMENT**

Chapter Self-Ch...

SECTION 1 **The Many Forms of Energy**
Mastering Concepts

Unless otherwise noted, air resistance does no work.

31. Explain how work and energy change are related.

32. Explain how force and energy change are related.

33. What form of energy does a system that contains a wound-up spring toy have? What form of energy does the toy have when it is going? When the toy runs down, what has happened to the energy?

34. A ball is dropped from the top of a building. You choose the top of the building to be the reference level, while your friend chooses the bottom. Explain whether the energy calculated using these two reference levels is the same or different for the following situations:

   a. the ball-Earth system's potential energy

   b. the change in the system's potential energy as a result of the fall

   c. the kinetic energy of the system at any point

35. Can a baseball's kinetic energy ever be negative?

36. Can a baseball-Earth system ever have a negative gravitational potential energy? Explain without using a formula.

37. If a sprinter's velocity increases to three times the original velocity, by what factor does the kinetic energy increase?

38. What energy transformations take place when an athlete is pole-vaulting?

39. The sport of pole-vaulting was drastically changed when the stiff wooden poles were replaced by

43. A racing car has a mass of 1525 kg. What is its kinetic energy if it has a speed of 108 km/h?

44. Katia and Angela each have a mass of 45 kg and are moving together with a speed of 10.0 m/s.

   a. What is their combined kinetic energy?

   b. What is the ratio of their combined mass to Katia's mass?

   c. What is the ratio of their combined kinetic energy to Katia's kinetic energy? Explain how this ratio relates to the ratio of their masses.

45. **Train** In the 1950s an experimental train with a mass of 2.50×10^4 kg was powered along 509 m of track by a jet engine that produced a thrust of 5.00×10^5 N. Assume friction is negligible.

   a. Find the work done on the train by the jet engine.

   b. Find the change in kinetic energy.

   c. Find the final kinetic energy of the train if it started from rest.

   d. Find the final speed of the train.

46. **Car Brakes** The driver of the car in **Figure 15** suddenly applies the brakes, and the car slides to a stop. The average force between the tires and the road is 7100 N. How far will the car slide after the brakes are applied?

Before (initial)   After (final)
$v = 25$ m/s $v = 0.0$ m/s

Figure 15 $m = 14,700$ N

In the Chapter Assessment, you will find a question or problem that will help you evaluate your understanding of the

BIGIDEA.

The **MAIN**IDEA is the core concept covered in the section. Together, the Main Ideas from all the sections in the chapter support the chapter's Big Idea.

Essential Questions
reflect the important objectives of the section. Together, an understanding of these questions will lead toward understanding the section's Main Idea.

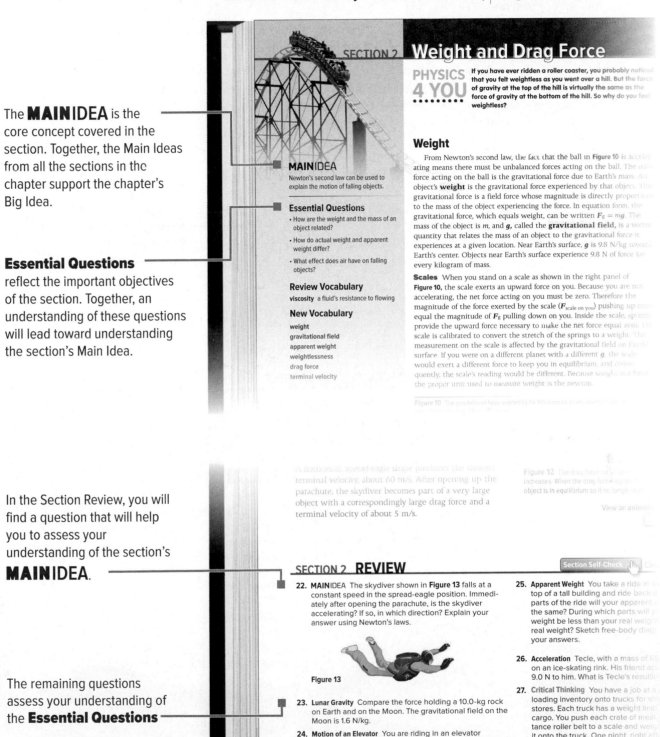

SECTION 2 | **Weight and Drag Force**

PHYSICS 4 YOU If you have ever ridden a roller coaster, you probably noticed that you felt weightless as you went over a hill. But the force of gravity at the top of the hill is virtually the same as the force of gravity at the bottom of the hill. So why do you feel weightless?

MAINIDEA
Newton's second law can be used to explain the motion of falling objects.

Essential Questions
• How are the weight and the mass of an object related?
• How do actual weight and apparent weight differ?
• What effect does air have on falling objects?

Review Vocabulary
viscosity a fluid's resistance to flowing

New Vocabulary
weight
gravitational field
apparent weight
weightlessness
drag force
terminal velocity

Weight

From Newton's second law, the fact that the ball in **Figure 10** is accelerating means there must be unbalanced forces acting on the ball. The only force acting on the ball is the gravitational force due to Earth's mass. An object's **weight** is the gravitational force experienced by that object. This gravitational force is a field force whose magnitude is directly proportional to the mass of the object experiencing the force. In equation form, the gravitational force, which equals weight, can be written $F_g = mg$. The mass of the object is m, and g, called the **gravitational field**, is a vector quantity that relates the mass of an object to the gravitational force it experiences at a given location. Near Earth's surface, g is 9.8 N/kg toward Earth's center. Objects near Earth's surface experience 9.8 N of force for every kilogram of mass.

Scales When you stand on a scale as shown in the right panel of **Figure 10,** the scale exerts an upward force on you. Because you are not accelerating, the net force acting on you must be zero. Therefore the magnitude of the force exerted by the scale ($F_{\text{scale on you}}$) pushing up must equal the magnitude of F_g pulling down on you. Inside the scale, springs provide the upward force necessary to make the net force equal zero. The scale is calibrated to convert the stretch of the springs to a weight. The measurement on the scale is affected by the gravitational field on Earth's surface. If you were on a different planet with a different g, the scale would exert a different force to keep you in equilibrium, and consequently, the scale's reading would be different. Because weight is a force, the proper unit used to measure weight is the newton.

Figure 10 The gravitational force exerted by Earth's mass on an object

A horizontal, spread-eagle shape produces the slowest terminal velocity, about 60 m/s. After opening up the parachute, the skydiver becomes part of a very large object with a correspondingly large drag force and a terminal velocity of about 5 m/s.

Figure 12 The drag force increases. When the drag force equals ... object is in equilibrium so it no longer ...

View an animation

In the Section Review, you will find a question that will help you to assess your understanding of the section's **MAIN**IDEA.

SECTION 2 **REVIEW**

Section Self-Check

22. MAINIDEA The skydiver shown in **Figure 13** falls at a constant speed in the spread-eagle position. Immediately after opening the parachute, is the skydiver accelerating? If so, in which direction? Explain your answer using Newton's laws.

Figure 13

23. Lunar Gravity Compare the force holding a 10.0-kg rock on Earth and on the Moon. The gravitational field on the Moon is 1.6 N/kg.

24. Motion of an Elevator You are riding in an elevator holding a spring scale with a 1-kg mass suspended from it. You look at the scale and see that it reads 9.3 N. What, if anything, can you conclude about the elevator's motion at that time?

25. Apparent Weight You take a ride in ... top of a tall building and ride back d... parts of the ride will your apparent ... the same? During which parts will y... weight be less than your real weigh... real weight? Sketch free-body diag... your answers.

26. Acceleration Tecle, with a mass of 6... on an ice-skating rink. His friend ap... 9.0 N to him. What is Tecle's resultin...

27. Critical Thinking You have a job at a ... loading inventory onto trucks for sh... stores. Each truck has a weight limit... cargo. You push each crate of meat ... tance roller belt to a scale and weig... it onto the truck. One night, right aft... 1000-N crate, the scale breaks. Desc... which you could apply Newton's law... the masses of the remaining crates...

Section 2 • Weight...

The remaining questions assess your understanding of the **Essential Questions**

TABLE OF CONTENTS

Your eStudentEdition, found at **connectED.mcgraw-hill.com**, includes interactive features to improve your understanding of the physics content presented in each chapter.

Concepts in Motion include animations and interactive diagrams to help explain topics and eliminate misconceptions. BrainPOP **Videos** connect physics to your world. **Virtual Investigations** are online laboratory experiences, and **Personal Tutors** provide video demonstrations of solutions to various example problems.

Throughout each chapter, you will see references to laboratory experiences. **Go Online**, to find all your online resources for lab activities and worksheets.

Each chapter begins with a **LaunchLAB,** an introductory laboratory investigation designed to introduce the concepts in that chapter. **MiniLABs** are short investigations that can improve your understanding of physics content. You will also find one or more **PhysicsLABs** in each chapter, providing opportunities for more in-depth investigations.

Concepts in Motion	Precision and Accuracy
	Graphing Data
BrainPOPs	Scientific Methods
	Measuring Matter
LaunchLAB	Mass and Falling Objects
MiniLABs	Measuring Change
	How far around?
PhysicsLABs	It's in the Blood
	Mass and Volume
	Exploring Objects in Motion

Go online!

Colin Anderson/Photographer's Choice/Getty Images

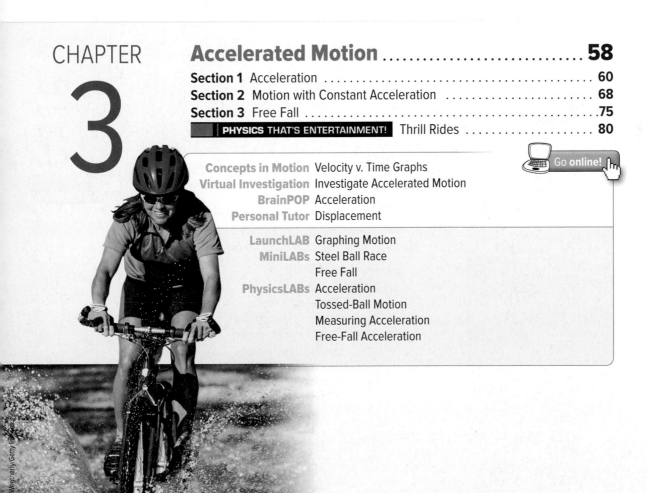

Mechanics

Mechanics

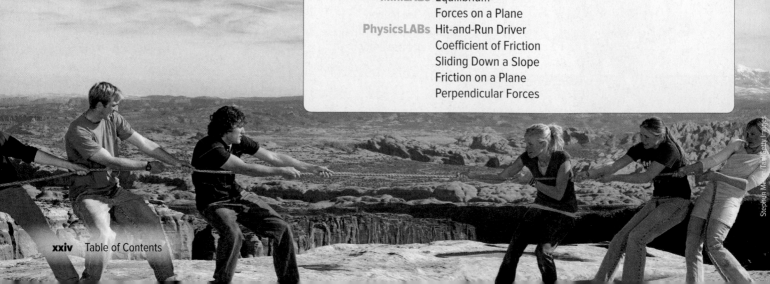

Stephan Mark Truid/Getty Images

Mechanics

Mechanics

Pete Gardner/Getty Images

Go online!

Energy

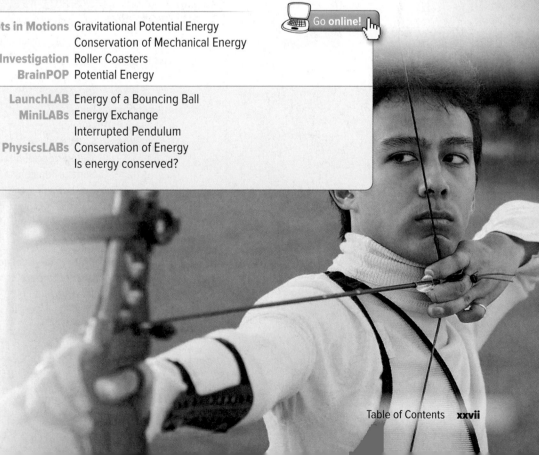

image100/Jupiterimages

TABLE OF CONTENTS

Energy

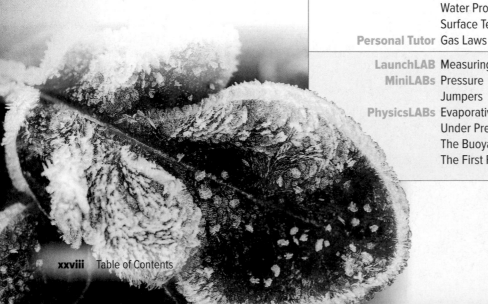

Ingram Publishing

Waves and Light

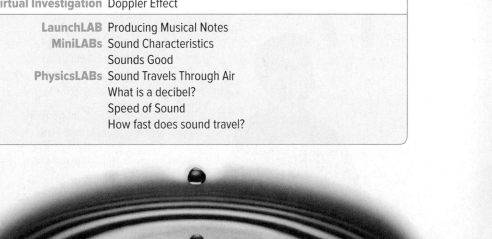

TABLE OF CONTENTS

Go online!

Waves and Light

Electricity and Magnetism

Go online!

Electricity and Magnetism

Photodisc Collection/Getty Images

TABLE OF CONTENTS

Go online!

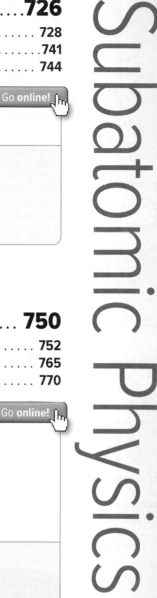
Subatomic Physics

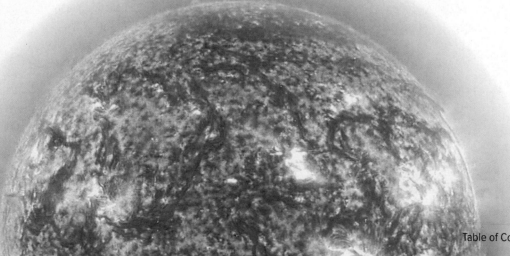

SOHO (ESA & NASA)

Subatomic Physics

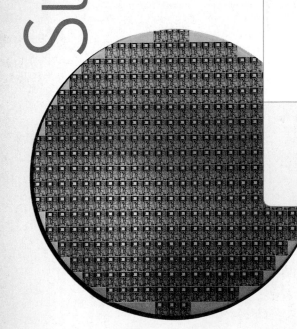

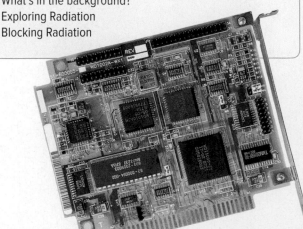

(l)©Clark Dunbar/Corbis; (r)©Comstock Images/Alamy

STUDENT RESOURCES

CHAPTER 1

A Physics Toolkit

BIGIDEA   Physicists use scientific methods to investigate energy and matter.

SECTIONS

1 **Methods of Science**

2 **Mathematics and Physics**

3 **Measurement**

4 **Graphing Data**

LaunchLAB

MASS AND FALLING OBJECTS

Does mass affect the rate at which an object falls?

WATCH THIS!

Video

ROCK STACKING

How do you get a stack of rocks to remain upright and balanced? It's all about understanding the physics! Explore the science behind the art of rock stacking.

PHYSICS T.V.

(l)Jose Luis Stephens/Getty Images; (r)Brand Simons/Stringer/Getty Images News/Getty Images

Methods of Science

PHYSICS 4 YOU

Think about what the world would be like if we still thought Earth was flat or if we didn't have indoor plumbing or electricity. Science helps us learn about the natural world and improve our lives.

MAIN IDEA

Scientific investigations do not always proceed with identical steps but do contain similar methods.

Essential Questions

- What are the characteristics of scientific methods?
- Why do scientists use models?
- What is the difference between a scientific theory and a scientific law?
- What are some limitations of science?

Review Vocabulary

control the standard by which test results in an experiment can be compared

New Vocabulary

physics
scientific methods
hypothesis
model
scientific theory
scientific law

What is physics?

Science is not just a subject in school. It is a method for studying the natural world. After all, science comes from the Latin word *scientia*, which means "knowledge." Science is a process based on inquiry that helps develop explanations about events in nature. **Physics** is a branch of science that involves the study of the physical world: energy, matter, and how they are related.

When you see the word *physics* you might picture a chalkboard full of formulas and mathematics: $E = mc^2$, $I = \dfrac{V}{R}$, $x = \left(\dfrac{1}{2}\right)at^2 + v_0t + x_0$. Maybe you picture scientists in white lab coats or well-known figures such as Marie Curie and Albert Einstein. Alternatively, you might think of the many modern technologies created with physics, such as weather satellites, laptop computers, or lasers. Physicists investigate the motions of electrons and rockets, the energy in sound waves and electric circuits, the structure of the proton and of the universe. The goal of this course is to help you better understand the physical world.

People who study physics go on to many different careers. Some become scientists at universities and colleges, at industries, or in research institutes. Others go into related fields, such as engineering, computer science, teaching, medicine, or astronomy, as shown in **Figure 1.** Still others use the problem-solving skills of physics to work in finance, construction, or other very different disciplines. In the last 50 years, research in the field of physics has led to many new technologies, including satellite-based communications and high-speed microscanners used to detect disease.

Figure 1 Physicists may choose from a variety of careers.

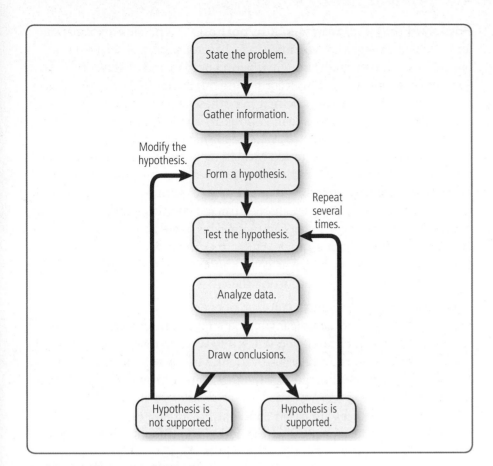

Figure 2 The series of procedures shown here is one way to use scientific methods to solve a problem.

Scientific Methods

Although physicists do not always follow a rigid set of steps, investigations often follow similar patterns. These patterns of investigation procedures are called **scientific methods.** Common steps found in scientific methods are shown in **Figure 2.** Depending on the particular investigation, a scientist might add new steps, repeat some steps, or skip steps altogether.

State the problem When you begin an investigation, you should state what you are going to investigate. Many investigations begin when someone observes an event in nature and wonders why or how it occurs. The question of "why" or "how" is the problem.

Scientists once posed questions about why objects fall to Earth, what causes day and night, and how to generate electricity for daily use. Many times a statement of a problem arises when an investigation is complete and its results lead to new questions. For example, once scientists understood why we experience day and night, they wanted to know why Earth rotates.

Sometimes a new question is posed during the course of an investigation. In the 1940s, researcher Percy Spencer was trying to answer the question of how to mass-produce the magnetron tubes used in radar systems. When he stood in front of an operating magnetron, which produces microwaves, a candy bar in his pocket melted. The new question of how the magnetron was cooking food was then asked.

Research and gather information Before beginning an investigation, it is useful to research what is already known about the problem. Making and examining observations and interpretations from reliable sources fine-tune the question and form it into a hypothesis.

View a **BrainPOP video on scientific methods.**

MEASURING CHANGE

How does increasing mass affect the length of a spring?

Form and test a hypothesis A **hypothesis** is a possible explanation for a problem using what you know and have observed. A scientific hypothesis can be tested through experimentation and observation. Sometimes scientists must wait for new technologies before a hypothesis can be tested. For example, the first hypotheses about the existence of atoms were developed more than 2300 years ago, but the technologies to test these hypotheses were not available for many centuries.

Some hypotheses can be tested by making observations. Others can be tested by building a model and relating it to real-life situations. One common way to test a hypothesis is to perform an experiment. An experiment tests the effect of one thing on another, using a control. Sometimes it is not possible to perform experiments; in these cases, investigations become descriptive in nature. For example, physicists cannot conduct experiments in deep space. They can, however, collect and analyze valuable data to help us learn more about events occurring there.

Analyze the data An important part of every investigation includes recording observations and organizing data into easy-to-read tables and graphs. Later in this chapter, you will study ways to display data. When you are making and recording observations, you should include all results, even unexpected ones. Many important discoveries have been made from unexpected results.

Scientific inferences are based on scientific observations. All possible scientific explanations must be considered. If the data are not organized in a logical manner, incorrect conclusions can be drawn. When a scientist communicates and shares data, other scientists will examine those data, how the data were analyzed, and compare the data to the work of others. Scientists, such as the physicist in **Figure 3,** share their data and analyses through reports and conferences.

Draw conclusions Based on the analysis of the data, the next step is to decide whether the hypothesis is supported. For the hypothesis to be considered valid and widely accepted, the results of the experiment must be the same every time it is repeated. If the experiment does not support the hypothesis, the hypothesis must be reconsidered. Perhaps the hypothesis needs to be revised, or maybe the experimenter's procedure needs to be refined.

Figure 3 An important part of scientific methods is to share data and results with other scientists. This physicist is giving a presentation at the World Science Festival.

Peer review Before it is made public, science-based information is reviewed by scientists' peers—scientists who are in the same field of study. Peer review is a process by which the procedures and results of an experiment are evaluated by peer scientists of those who conducted the research. Reviewing other scientists' work is a responsibility that many scientists have.

Being objective One also should be careful to reduce bias in scientific investigations. Bias can occur when the scientist's expectations affect how the results are analyzed or the conclusions are made. This might cause a scientist to select a result from one trial over those from other trials. Bias might also be found if the advantages of a product being tested are used in a promotion and the drawbacks are not presented. Scientists can lessen bias by running as many trials as possible and by keeping accurate notes of each observation made.

Models

Sometimes, scientists cannot see everything they are testing. They might be observing an object that is too large or too small, a process that takes too much time to see completely, or a material that is hazardous. In these cases, scientists use models. A **model** is a representation of an idea, event, structure, or object that helps people better understand it.

Models in history Models have been used throughout history. In the early 1900s, British physicist J.J. Thomson created a model of the atom that consisted of electrons embedded in a ball of positive charge. Several years later, physicist Ernest Rutherford created a model of the atom based on new research. Later in the twentieth century, scientists discovered the nucleus is not a solid ball but is made of protons and neutrons. The present-day model of the atom is a nucleus made of protons and neutrons surrounded by an electron cloud. All three of these models are shown in **Figure 4.** Scientists use models of atoms to represent their current understanding because of the small size of an atom.

Figure 4 Throughout history, scientists have made models of the atom.

Infer Why have models of the atom changed over the years?

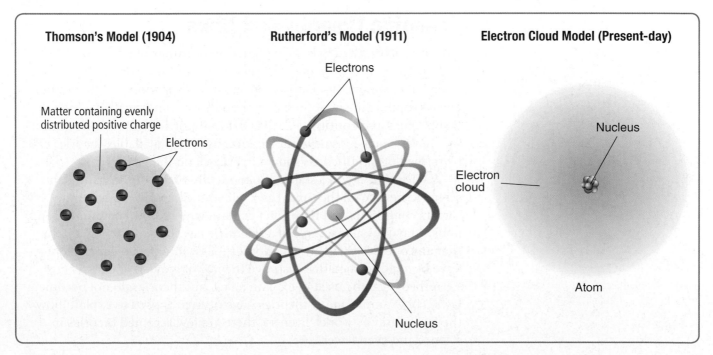

| Thomson's Model (1904) | Rutherford's Model (1911) | Electron Cloud Model (Present-day) |

Matter containing evenly distributed positive charge

Electrons

Electrons

Nucleus

Nucleus

Electron cloud

Nucleus

Atom

Figure 5 This is a computer simulation of an aircraft landing on a runway. The image on the screen in front of the pilot mimics what he would see if he were landing a real plane.

Identify other models around your classroom.

High-tech models Scientific models are not always something you can touch. Another type of model is a computer simulation. A computer simulation uses a computer to test a process or procedure and to collect data. Computer software is designed to mimic the processes under study. For instance, it is not possible for astronomers to observe how our solar system was formed, but when models of the process are proposed, they can be tested with computers.

Computer simulations also enable pilots, such as the ones shown in **Figure 5,** to practice all aspects of flight without ever leaving the ground. In addition, the computer simulation can simulate harsh weather conditions or other potentially dangerous in-flight challenges that pilots might face.

☑ **READING CHECK** **Identify** two advantages of using computer simulations.

Scientific Theories and Laws

A **scientific theory** is an explanation of things or events based on knowledge gained from many observations and investigations.

It is not a guess. If scientists repeat an investigation and the results always support the hypothesis, the hypothesis can be called a theory. Just because a scientific theory has data supporting it does not mean it will never change. As new information becomes available, theories can be refined or modified, as shown in **Figure 6** on the next page.

A **scientific law** is a statement about what happens in nature and seems to be true all the time. Laws tell you what will happen under certain conditions, but they don't explain why or how something happens. Gravity is an example of a scientific law. The law of gravity states that any one mass will attract another mass. To date, no experiments have been performed that disprove the law of gravity.

A theory can be used to explain a law, but theories do not become laws. For example, many theories have been proposed to explain how the law of gravity works. Even so, there are few accepted theories in science and even fewer laws.

Greek philosophers proposed that objects fall because they seek their natural places. The more massive the object, the faster it falls.

Revision

Galileo showed that the speed at which an object falls depends on the amount of time for which that object has fallen and not on the object's mass.

Revision

Newton provided an explanation for why objects fall. Newton proposed that objects fall because the object and Earth are attracted by a force. Newton also stated that there is a force of attraction between any two objects with mass.

Revision

Einstein suggested that the force of attraction between two objects is due to mass causing the space around it to curve.

Figure 6 If experiments provide new insight and evidence about a theory, the theory is modified accordingly. The theory describing the behavior of falling objects has undergone many revisions based on new evidence.

The Limitations of Science

Science can help you explain many things about the world, but science cannot explain or solve everything. Although it is the scientist's job to make guesses, the scientist also has to make sure his or her guesses can be tested and verified.

Questions about opinions, values, or emotions are not scientific because they cannot be tested. For example, some people may find a particular piece of art beautiful while others do not. Some people may think that certain foods, such as pizza, taste delicious while others do not. Or, some people might think that the best color is blue, while others think it is green. You might take a survey to gather opinions about such questions, but that would not prove the opinions are true for everyone.

SECTION 1 REVIEW

Section Self-Check Check your understanding.

1. **MAIN**IDEA Summarize the steps you might use to carry out an investigation using scientific methods.

2. **Define** the term hypothesis and identify three ways in which a hypothesis can be tested.

3. **Describe** why it is important for scientists to avoid bias.

4. **Explain** why scientists use models. Give an example of a scientific model not mentioned in this section.

5. **Explain** why a scientific theory cannot become a scientific law.

6. **Analyze** Your friend conducts a survey, asking students in your school about lunches provided by the cafeteria. She finds that 90 percent of students surveyed like pizza. She concludes that this scientifically proves that everyone likes pizza. How would you respond to her conclusion?

7. **Critical Thinking** An accepted value for free-fall acceleration is 9.8 m/s^2. In an experiment with pendulums, you calculate that the value is 9.4 m/s^2. Should the accepted value be tossed out to accommodate your new finding? Explain.

Mathematics and Physics

PHYSICS 4 YOU

If you were to toss a tennis ball straight up into the air, how could you determine how far the ball would rise or how long it would stay in the air? How could you determine the velocity of the skydiver in the photo? Physicists use mathematics to help find the answers to these and other questions about motion, forces, energy, and matter.

MAIN IDEA

We use math to express concepts in physics.

Essential Questions

• Why do scientists use the metric system?

• How can dimensional analysis help evaluate answers?

• What are significant figures?

Review Vocabulary

SI Système International d'Unités—the improved, universally accepted version of the metric system that is based on multiples of ten; also called the International System of Units

New Vocabulary

dimensional analysis
significant figures

Mathematics in Physics

Physicists often use the language of mathematics. In physics, equations are important tools for modeling observations and for making predictions. Equations are one way of representing relationships between measurements. Physicists rely on theories and experiments with numerical results to support their conclusions. For example, you can predict that if you drop a penny, it will fall, but can you predict how fast it will be going when it strikes the ground below? Different models of falling objects give different answers to how the speed of the object changes as it falls or on what the speed depends. By measuring how an object falls, you can compare the experimental data with the results predicted by different models. This tests the models, allowing you to pick the best one or to develop a new model.

SI Units

To communicate results, it is helpful to use units that everyone understands. The worldwide scientific community currently uses an adaptation of the metric system for measurements. **Table 1** shows that the Système International d'Unités, or SI, uses seven base quantities. Other units, called derived units, are created by combining the base units in various ways. Velocity is measured in meters per second (m/s). Often, derived units are given special names. For example, electric charge is measured in ampere-seconds (A·s), which are also called coulombs (C).

The base quantities were originally defined in terms of direct measurements. Scientific institutions have since been created to define and regulate measurements. SI is regulated by the International Bureau of Weights and Measures in Sèvres, France.

Table 1 SI Base Units		
Base Quantity	**Base Unit**	**Symbol**
Length	meter	m
Mass	kilogram	kg
Time	second	s
Temperature	kelvin	K
Amount of a substance	mole	mol
Electric current	ampere	A
Luminous intensity	candela	cd

This bureau and the National Institute of Science and Technology (NIST) in Gaithersburg, Maryland, keep the standards of length, time, and mass against which our metersticks, clocks, and balances are calibrated. The standard for a kilogram is shown in **Figure 7.**

You probably learned in math class that it is much easier to convert meters to kilometers than feet to miles. The ease of switching between units is another feature of SI. To convert between units, multiply or divide by the appropriate power of 10. Prefixes are used to change SI base units by powers of 10, as shown in **Table 2.** You often will encounter these prefixes in daily life, as in, for example, milligrams, nanoseconds, and centimeters.

Figure 7 The International Prototype Kilogram, the standard for the mass of a kilogram, is a mixture of platinum and iridium. It is kept in a vacuum so it does not lose mass. Scientists are working to redefine the standard for a kilogram, using a perfect sphere made of silicon.

Describe Why is it important to have standards for measurements?

Table 2 Prefixes Used with SI Units				
Prefix	**Symbol**	**Multiplier**	**Scientific Notation**	**Example**
femto–	f	0.000000000000001	10^{-15}	femtosecond (fs)
pico–	p	0.000000000001	10^{-12}	picometer (pm)
nano–	n	0.000000001	10^{-9}	nanometer (nm)
micro–	μ	0.000001	10^{-6}	microgram (μg)
milli–	m	0.001	10^{-3}	milliamps (mA)
centi–	c	0.01	10^{-2}	centimeter (cm)
deci–	d	0.1	10^{-1}	deciliter (dL)
kilo–	k	1000	10^{3}	kilometer (km)
mega–	M	1,000,000	10^{6}	megagram (Mg)
giga–	G	1,000,000,000	10^{9}	gigameter (Gm)
tera–	T	1,000,000,000,000	10^{12}	terahertz (THz)

☑ **READING CHECK Identify** the prefix that would be used to express 2,000,000,000 bytes of computer memory.

Dimensional Analysis

You can use units to check your work. You often will need to manipulate a formula, or use a string of formulas, to solve a physics problem. One way to check whether you have set up a problem correctly is to write out the equation or set of equations you plan to use. Before performing calculations, check that the answer will be in the expected units. For example, if you are finding a car's speed and you see that your answer will be measured in s/m or m/s^2, you have made an error in setting up the problem. This method of treating the units as algebraic quantities that can be cancelled is called **dimensional analysis.** Knowing that your answer will be in the correct units is not a guarantee that your answer is right, but if you find that your answer will have the wrong units, you can be sure that you have made an error. Dimensional analysis also is used in choosing conversion factors. A conversion factor is a multiplier equal to 1.

Find help with **dimensional analysis.**

For example, because 1 kg = 1000 g, you can construct the following conversion factors:

$$1 = \frac{1 \text{ kg}}{1000 \text{ g}} \qquad 1 = \frac{1000 \text{ g}}{1 \text{ kg}}$$

Choose a conversion factor that will make the initial units cancel, leaving the answer in the desired units. For example, to convert 1.34 kg of iron ore to grams, do as shown below.

$$1.34 \cancel{\text{kg}} \left(\frac{1000 \text{ g}}{1 \cancel{\text{kg}}} \right) = 1340 \text{ g}$$

You also might need to do a series of conversions. To convert 43 km/h to m/s, do the following:

$$\left(\frac{43 \cancel{\text{km}}}{1 \cancel{\text{h}}} \right) \left(\frac{1000 \text{ m}}{1 \cancel{\text{km}}} \right) \left(\frac{1 \cancel{\text{h}}}{60 \cancel{\text{min}}} \right) \left(\frac{1 \cancel{\text{min}}}{60 \text{ s}} \right) = 12 \text{ m/s}$$

Significant Figures

Suppose you use a metric ruler to measure a pen and you find that the end of the pen is just past 138 mm, as shown in **Figure 8.** You estimate that the pen is one-tenth of a millimeter past the last tic mark on the ruler and record the pen as being 138.1 mm long. This measurement has four valid digits: the first three digits are certain, and the last one is uncertain. The valid digits in a measurement are referred to as **significant figures.** The last digit given for any measurement is the uncertain digit. All nonzero digits in a measurement are significant.

Are all zeros significant? No. For example, in the measurement 0.0860 m, the first two zeros serve only to locate the decimal point and are not significant. The last zero, however, is the estimated digit and is significant. The measurement 172,000 m could have 3, 4, 5, or 6 significant figures. This ambiguity is one reason to use scientific notation. It is clear that the measurement 1.7200×10^5 m has five significant figures.

Arithmetic with significant figures When you perform any arithmetic operation, it is important to remember that the result never can be more precise than the least-precise measurement.

To add or subtract measurements, first perform the operation, then round off the result to correspond to the least-precise value involved. For example, 3.86 m + 2.4 m = 6.3 m because the least-precise measure is to one-tenth of a meter.

To multiply or divide measurements, perform the calculation and then round to the same number of significant figures as the least-precise measurement. For example, $\frac{409.2 \text{ km}}{11.4 \text{ L}} = 35.9$ km/L, because the least-precise measurement has three significant figures. Some calculators display several additional digits, while others round at different points. Be sure to record your answers with the correct number of digits.

Solving Problems

As you continue this course, you will complete practice problems. Most problems will be complex and require a strategy to solve. This textbook includes many example problems, each of which is solved using a three-step process. Read Example Problem 1 and follow the steps to calculate a car's average speed using distance and time.

Figure 8 The student measuring this pen recorded the length as 138.1 mm.

Infer Why is the last digit in this measurement uncertain?

Find help with **significant figures** and **scientific notation.**

Math Handbook

EXAMPLE PROBLEM 1

Get help with **example problems.** [Personal Tutor] 🖐

USING DISTANCE AND TIME TO FIND SPEED When a car travels 434 km in 4.5 h, what is the car's average speed?

1 ANALYZE AND SKETCH THE PROBLEM

The car's speed is unknown. The known values include the distance the car traveled and time. Use the relationship among speed, distance, and time to solve for the car's speed.

KNOWN

distance = 434 km
time = 4.5 h

UNKNOWN

speed = ? km/h

2 SOLVE FOR THE UNKNOWN

distance = **speed** × time   ◀ *State the relationship as an equation.*

$$speed = \frac{distance}{time}$$ ◀ *Solve the equation for speed.*

$$speed = \frac{434 \text{ km}}{4.5 \text{ h}}$$ ◀ *Substitute distance = 434 km and time = 4.5 h.*

speed = 96.4 km/h   ◀ *Divide, and calculate units.*

3 EVALUATE THE ANSWER

Check your answer by using it to calculate the distance the car traveled.

distance = speed × time = 96.4 km/h̶ × 4.5 h̶ = 434 km

The calculated distance matches the distance stated in the problem. This means the average speed is correct.

THE PROBLEM

1. Read the problem carefully.
2. Be sure you understand what is being asked.

ANALYZE AND SKETCH THE PROBLEM

1. Read the problem again.
2. Identify what you are given, and list the known data. If needed, gather information from graphs, tables, or figures.
3. Identify and list the unknowns.
4. Determine whether you need a sketch to help solve the problem.
5. Plan the steps you will follow to find the answer.

SOLVE FOR THE UNKNOWN

1. If the solution is mathematical, write the equation and isolate the unknown factor.
2. Substitute the known quantities into the equation.
3. Solve the equation.
4. Continue the solution process until you solve the problem.

EVALUATE THE ANSWER

1. Reread the problem. Is the answer reasonable?
2. Check your math. Are the units and the significant figures correct?

SECTION 2 REVIEW

[Section Self-Check] 🖐 Check your understanding.

8. **MAIN**IDEA Why are concepts in physics described with formulas?

9. **SI Units** What is one advantage to using SI Units in science?

10. **Dimensional Analysis** How many kilohertz are 750 megahertz?

11. **Dimensional Analysis** How many seconds are in a leap year?

12. **Significant Figures** Solve the following problems, using the correct number of significant figures each time.

  a. 10.8 g − 8.264 g

  b. 4.75 m − 0.4168 m

  c. 139 cm × 2.3 cm

  d. 13.78 g / 11.3 mL

  e. 6.201 cm + 7.4 cm + 0.68 m + 12.0 cm

  f. 1.6 km + 1.62 m + 1200 cm

13. **Solving Problems** Rewrite $F = Bqv$ to find v in terms of F, q, and B.

14. **Critical Thinking** Using values given in a problem and the equation of distance = speed × time, you calculate a car's speed to be 290 km/h. Is this a reasonable answer? Why or why not? Under what circumstances might this be a reasonable answer?

Measurement

There are many devices that you often use to make measurements. Clocks measure time, rulers measure distance, and speedometers measure speed. What other measuring devices have you used?

MAIN IDEA

Making careful measurements allows scientists to repeat experiments and compare results.

Essential Questions

- Why are the results of measurements often reported with an uncertainty?
- What is the difference between precision and accuracy?
- What is a common source of error when making a measurement?

Review Vocabulary

parallax the apparent shift in the position of an object when it is viewed from different angles

New Vocabulary

measurement
precision
accuracy

What is measurement?

When you visit the doctor for a checkup, many measurements are taken: your height, weight, blood pressure, and heart rate. Even your vision is measured and assigned numbers. Blood might be drawn so measurements can be made of blood cells or cholesterol levels. Measurements quantify our observations: a person's blood pressure isn't just "pretty good," it's $\frac{110}{60}$, the low end of the good range.

A **measurement** is a comparison between an unknown quantity and a standard. For example, if you measure the mass of a rolling cart used in an experiment, the unknown quantity is the mass of the cart and the standard is the gram, as defined by the balance or the spring scale you use. In the MiniLab in Section 1, the length of the spring was the unknown and the centimeter was the standard.

Comparing Results

As you learned in Section 1, scientists share their results. Before new data are fully accepted, other scientists examine the experiment, look for possible sources of error, and try to reproduce the results. Results often are reported with an uncertainty. A new measurement that is within the margin of uncertainty is in agreement with the old measurement.

For example, archaeologists use radiocarbon dating to find the age of cave paintings, such as those from the Lascaux cave, in **Figure 9,** and the Chauvet cave. Each radiocarbon date is reported with an uncertainty. Three radiocarbon ages from a panel in the Chauvet cave are 30,940 ± 610 years, 30,790 ± 600 years, and 30,230 ± 530 years. While none of the measurements exactly matches, the uncertainties in all three overlap, and the measurements agree with each other.

Figure 9 These drawings are from the Lascaux cave in France. Scientists estimate that the drawings were made 17,000 years ago.

Suppose three students performed the MiniLab from Section 1 several times, starting with springs of the same length. With two washers on the spring, student 1 made repeated measurements, which ranged from 14.4 cm to 14.8 cm. The average of student 1's measurements was 14.6 cm, as shown in **Figure 10.** This result was reported as (14.6 ± 0.2) cm. Student 2 reported finding the spring's length to be (14.8 ± 0.3) cm. Student 3 reported a length of (14.0 ± 0.1) cm.

Could you conclude that the three measurements are in agreement? Is student 1's result reproducible? The ranges of the results of students 1 and 2 overlap between 14.5 cm and 14.8 cm. However, there is no overlap and, therefore, no agreement, between their results and the result of student 3.

Precision Versus Accuracy

Both precision and accuracy are characteristics of measured values as shown in **Figure 11.** How precise and accurate are the measurements of the three students above? The degree of exactness of a measurement is called its **precision.** In the example above, student 3's measurements are the most precise, within ± 0.1 cm. Both the measurements of student 1 and student 2 are less precise because they have a larger uncertainty (student 1 = ± 0.2 cm, student 2 = ± 0.3 cm).

Precision depends on the instrument and technique used to make the measurement. Generally, the device that has the finest division on its scale produces the most precise measurement. The precision of a measurement is one-half the smallest division of the instrument. For example, suppose a graduated cylinder has divisions of 1 mL. You could measure an object to within 0.5 mL with this device. However, if the smallest division on a beaker is 50 mL, how precise would your measurements be compared to those taken with the graduated cylinder?

The significant figures in an answer show its precision. A measure of 67.100 g is precise to the nearest thousandth of a gram. Recall from Section 2 the rules for performing operations with measurements given to different levels of precision. If you add 1.2 mL of acid to a beaker containing 2.4×10^2 mL of water, you cannot say you now have 2.412×10^2 mL of fluid because the volume of water was not measured to the nearest tenth of a milliliter, but to the nearest 10 mL.

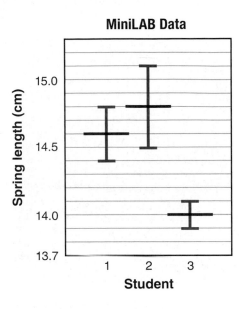

MiniLAB Data

Figure 10 Three students took multiple measurements. The red bars show the uncertainty of each measurement.

Explain Are the measurements in agreement? Is student 3's result reproducible? Why or why not?

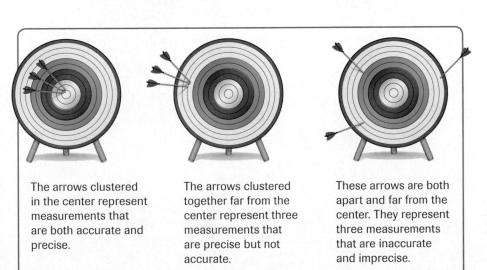

The arrows clustered in the center represent measurements that are both accurate and precise.

The arrows clustered together far from the center represent three measurements that are precise but not accurate.

These arrows are both apart and far from the center. They represent three measurements that are inaccurate and imprecise.

Figure 11 The yellow area in the center of each target represents an accepted value for a particular measurement. The arrows represent measurements taken by a scientist during an experiment.

View an **animation of precision and accuracy.**

Concepts In Motion

Figure 12 Accuracy is checked by zeroing an instrument before measuring.

Infer Is this instrument accurate? Why or why not?

Accuracy describes how well the results of a measurement agree with the "real" value; that is, the accepted value as measured by competent experimenters, as shown in **Figure 11.** If the length of the spring that the three students measured had been 14.8 cm, then student 2 would have been most accurate and student 3 least accurate. What might have led someone to make inaccurate measurements? How could you check the accuracy of measurements?

A common method for checking the accuracy of an instrument is called the two-point calibration. First, does the instrument read zero when it should, as shown in **Figure 12?** Second, does it give the correct reading when it is measuring an accepted standard? Regular checks for accuracy are performed on critical measuring instruments, such as the radiation output of the machines used to treat cancer.

☑ **READING CHECK** **Compare and contrast** precision and accuracy.

Techniques of Good Measurement

To assure accuracy and precision, instruments also have to be used correctly. Measurements have to be made carefully if they are to be as precise as the instrument allows. One common source of error comes from the angle at which an instrument is read. Scales should be read with one's eye directly in front of the measure, as shown in **Figure 13.** If the scale is read from an angle, also shown in **Figure 13,** a different, less accurate, value will be obtained. The difference in the readings is caused by parallax, which is the apparent shift in the position of an object when it is viewed from different angles. To experiment with parallax, place your pen on a ruler and read the scale with your eye directly over the tip, then read the scale with your head shifted far to one side.

View a **BrainPOP video on measuring matter.**

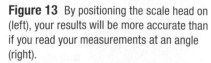

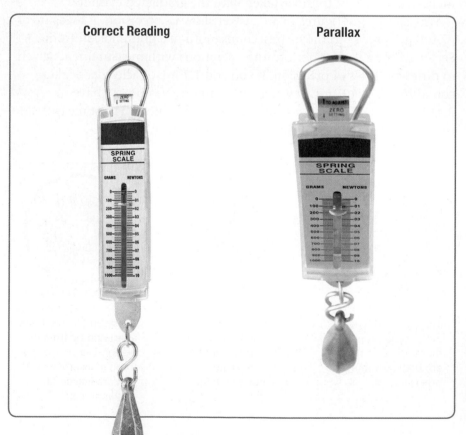

Figure 13 By positioning the scale head on (left), your results will be more accurate than if you read your measurements at an angle (right).

Identify How far did parallax shift the measurement on the right?

(t)McGraw-Hill Education; (b)Hutchings Photography/Digital Light Source

GPS The Global Positioning System, or GPS, offers an illustration of accuracy and precision in measurement. The GPS consists of 24 satellites with transmitters in orbit and numerous receivers on Earth. The satellites send signals with the time, measured by highly accurate atomic clocks. The receiver uses the information from at least four satellites to determine latitude, longitude, and elevation. (The clocks in the receivers are not as accurate as those on the satellites.)

Receivers have different levels of precision. A device in an automobile might give your position to within a few meters. Devices used by geophysicists, as in **Figure 14,** can measure movements of millimeters in Earth's crust.

The GPS was developed by the United States Department of Defense. It uses atomic clocks, which were developed to test Einstein's theories of relativity and gravity. The GPS eventually was made available for civilian use. GPS signals now are provided worldwide free of charge and are used in navigation on land, at sea, and in the air, for mapping and surveying, by telecommunications and satellite networks, and for scientific research into earthquakes and plate tectonics.

Figure 14 This scientist is setting up a highly accurate GPS receiver in order to record and analyze the movements of continental plates.

PhysicsLAB

MASS AND VOLUME
How does mass depend on volume?

SECTION 3 REVIEW

Section Self-Check   Check your understanding.

15. **MAIN**IDEA You find a micrometer (a tool used to measure objects to the nearest 0.001 mm) that has been badly bent. How would it compare to a new, high-quality meterstick in terms of its precision? Its accuracy?

16. **Accuracy** Some wooden rulers do not start with 0 at the edge, but have it set in a few millimeters. How could this improve the accuracy of the ruler?

17. **Parallax** Does parallax affect the precision of a measurement that you make? Explain.

18. **Uncertainty** Your friend tells you that his height is 182 cm. In your own words, explain the range of heights implied by this statement.

19. **Precision** A box has a length of 18.1 cm and a width of 19.2 cm, and it is 20.3 cm tall.

   a. What is its volume?

   b. How precise is the measurement of length? Of volume?

   c. How tall is a stack of 12 of these boxes?

   d. How precise is the measurement of the height of one box? Of 12 boxes?

20. **Critical Thinking** Your friend states in a report that the average time required for a car to circle a 1.5-mi track was 65.414 s. This was measured by timing 7 laps using a clock with a precision of 0.1 s. How much confidence do you have in the results of the report? Explain.

Graphing Data

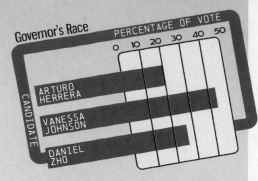

Governor's Race
PERCENTAGE OF VOTE
0 10 20 30 40 50
CANDIDATE
ARTURO HERRERA
VANESSA JOHNSON
DANIEL ZHO

PHYSICS 4 YOU

Graphs are often used in news stories after elections. Bar and circle graphs are used to show the number or percentage of votes various candidates received. Other graphs are used to show increases and decreases in population or resources over years.

MAINIDEA

Graphs make it easier to interpret data, identify trends, and show relationships among a set of variables.

Essential Questions

- What can be learned from graphs?
- What are some common relationships in graphs?
- How do scientists make predictions?

Review Vocabulary

slope on a graph, the ratio of vertical change to horizontal change

New Vocabulary

independent variable
dependent variable
line of best fit
linear relationship
quadratic relationship
inverse relationship

Identifying Variables

When you perform an experiment, it is important to change only one factor at a time. For example, **Table 3** gives the length of a spring with different masses attached. Only the mass varies; if different masses were hung from different types of springs, you wouldn't know how much of the difference between two data pairs was due to the different masses and how much was due to the different springs.

Independent and dependent variables A variable is any factor that might affect the behavior of an experimental setup. The factor that is manipulated during an investigation is the **independent variable.** In this investigation, the mass was the independent variable. The factor that depends on the independent variable is the **dependent variable.** In this investigation, the amount the spring stretched depended on the mass, so the amount of stretch was the dependent variable. A scientist might also look at how radiation varies with time or how the strength of a magnetic field depends on the distance from a magnet.

Line of best fit A line graph shows how the dependent variable changes with the independent variable. The data from **Table 3** are graphed in **Figure 15.** The line in blue, drawn as close to all the data points as possible, is called a **line of best fit.** The line of best fit is a better model for predictions than any one point along the line. **Figure 15** gives detailed instructions on how to construct a graph, plot data, and sketch a line of best fit.

A well-designed graph allows patterns that are not immediately evident in a list of numbers to be seen quickly and simply. The graph in **Figure 15** shows that the length of the spring increases as the mass suspended from the spring increases.

Table 3 Length of a Spring for Different Masses	
Mass Attached to Spring (g)	**Length of Spring (cm)**
0	13.7
5	14.1
10	14.5
15	14.9
20	15.3
25	15.7
30	16.0
35	16.4

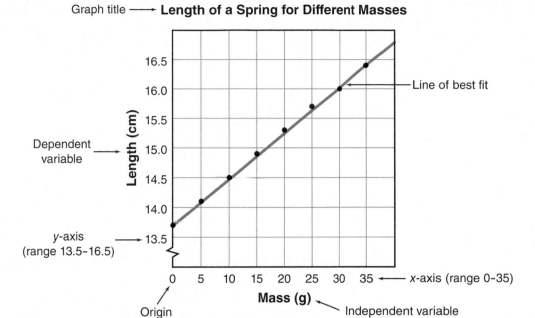

Graph title ⟶ **Length of a Spring for Different Masses**

Dependent variable ⟶

Line of best fit

y-axis (range 13.5–16.5) ⟶

x-axis (range 0–35)

Origin

Mass (g)

Independent variable

1. Identify the independent variable and dependent variable in your data. In this example, the independent variable is mass (g) and the dependent variable is length (cm). The independent variable is plotted on the horizontal axis, the x-axis. The dependent variable is plotted on the vertical axis, the y-axis.

2. Determine the range of the independent variable to be plotted. In this case the range is 0–35.

3. Decide whether the origin (0,0) is a valid data point.

4. Spread the data out as much as possible. Let each division on the graph paper stand for a convenient unit. This usually means units that are multiples of 2, 5, or 10.

5. Number and label the horizontal axis. The label should include the units, such as Mass (g).

6. Repeat steps 2–5 for the dependent variable.

7. Plot the data points on the graph.

8. Draw the best-fit straight line or smooth curve that passes through as many data points as possible. This is sometimes called *eyeballing.* Do not use a series of straight-line segments that connect the dots. The line that looks like the best fit to you may not be exactly the same as someone else's. There is a formal procedure, which many graphing calculators use, called the least-squares technique, that produces a unique best-fit line, but that is beyond the scope of this textbook.

9. Give the graph a title that clearly tells what the graph represents.

Figure 15 Use the steps above to plot line graphs from data tables.

View an **animation of graphing data.**

Concepts In Motion

Figure 16 In a linear relationship, the dependent variable—in this case, length—varies linearly with the independent variable. The independent variable in this experiment is mass.

Describe What happens to the length of the spring as mass decreases?

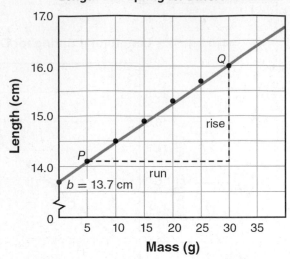

Length of a Spring for Different Masses

Linear Relationships

Scatter plots of data take many different shapes, suggesting different relationships. Three of the most common relationships include linear relationships, quadratic relationships, and inverse relationships. You probably are familiar with them from math class.

When the line of best fit is a straight line, as in **Figure 15,** there is a linear relationship between the variables. In a **linear relationship,** the dependent variable varies linearly with the independent variable. The relationship can be written as the following equation.

LINEAR RELATIONSHIP BETWEEN TWO VARIABLES $y = mx + b$

Find the y-intercept (b) and the slope (m) as illustrated in **Figure 16.** Use points on the line–they may or may not be data points. The slope is the ratio of the vertical change to the horizontal change. To find the slope, select two points, P and Q, far apart on the line. The vertical change, or rise (Δy), is the difference between the vertical values of P and Q. The horizontal change, or run (Δx), is the difference between the horizontal values of P and Q.

SLOPE
The slope of a line is equal to the rise divided by the run, which also can be expressed as the vertical change divided by the horizontal change.

$$m = \frac{rise}{run} = \frac{\Delta y}{\Delta x}$$

In **Figure 16:** $m = \dfrac{(16.0 \text{ cm} - 14.1 \text{ cm})}{(30 \text{ g} - 5 \text{ g})} = 0.08 \text{ cm/g}$

If y gets smaller as x gets larger, then $\dfrac{\Delta y}{\Delta x}$ is negative, and the line slopes downward from left to right. The y-intercept (b) is the point at which the line crosses the vertical axis, or the y-value when the value of x is zero. In this example, $b = 13.7$ cm. This means that when no mass is suspended by the spring, it has a length of 13.7 cm. When $b = 0$, or $y = mx$, the quantity y is said to vary directly with x. In physics, the slope of the line and the y-intercept always contain information about the physical system that is described by the graph.

MiniLAB

HOW FAR AROUND?
What is the relationship between circumference and diameter?

Get help with **determining slope.**

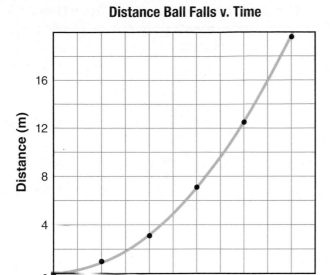

Distance Ball Falls v. Time

Figure 17 The quadratic, or parabolic, relationship shown here is an example of a nonlinear relationship.

Nonlinear Relationships

Figure 17 shows the distance a brass ball falls versus time. Note that the graph is not a straight line, meaning the relationship is not linear. There are many types of nonlinear relationships in science. Two of the most common are the quadratic and inverse relationships.

Quadratic relationship The graph in **Figure 17** is a quadratic relationship, represented by the equation below. A **quadratic relationship** exists when one variable depends on the square of another.

QUADRATIC RELATIONSHIP BETWEEN TWO VARIABLES

$$y = ax^2 + bx + c$$

A computer program or graphing calculator easily can find the values of the constants a, b, and c in this equation. In **Figure 17**, the equation is $d = 5t^2$. See the Math Handbook in the back of this book or online for more on making and using line graphs.

Get help with **quadratic graphs** and **quadratic equations**.

☑ **READING CHECK Explain** how two variables related to each other in a quadratic relationship.

PHYSICS CHALLENGE

An object is suspended from spring 1, and the spring's elongation (the distance it stretches) is x_1. Then the same object is removed from the first spring and suspended from a second spring. The elongation of spring 2 is x_2. x_2 is greater than x_1.

1. On the same axes, sketch the graphs of the mass versus elongation for both springs.

2. Should the origin be included in the graph? Why or why not?

3. Which slope is steeper?

4. At a given mass, $x_2 = 1.6x_1$. If $x_2 = 5.3$ cm, what is x_1?

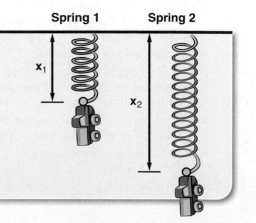

Spring 1   Spring 2

Figure 18 This graph shows the inverse relationship between speed and travel time.

Describe How does travel time change as speed increases?

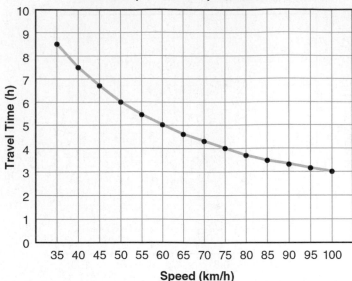

Relationship Between Speed and Travel Time

Inverse relationship The graph in **Figure 18** shows how the time it takes to travel 300 km varies as a car's speed increases. This is an example of an inverse relationship, represented by the equation below. An **inverse relationship** is a hyperbolic relationship in which one variable depends on the inverse of the other variable.

INVERSE RELATIONSHIP BETWEEN TWO VARIABLES $y = \dfrac{a}{x}$

The three relationships you have learned about are a sample of the relations you will most likely investigate in this course. Many other mathematical models are used. Important examples include sinusoids, used to model cyclical phenomena, and exponential growth and decay, used to study radioactivity. Combinations of different mathematical models represent even more complex phenomena.

☑ **READING CHECK Explain** how two variables are related to each other in an inverse relationship.

PhysicsLAB

IT'S IN THE BLOOD
FORENSICS LAB How can blood spatter provide clues?

PRACTICE PROBLEMS

Do additional problems. **Online Practice** 🖱

21. The mass values of specified volumes of pure gold nuggets are given in **Table 4.**

   a. Plot mass versus volume from the values given in the table and draw the curve that best fits all points.

   b. Describe the resulting curve.

   c. According to the graph, what type of relationship exists between the mass of the pure gold nuggets and their volume?

   d. What is the value of the slope of this graph? Include the proper units.

   e. Write the equation showing mass as a function of volume for gold.

   f. Write a word interpretation for the slope of the line.

Table 4 Mass of Pure Gold Nuggets

Volume (cm³)	Mass (g)
1.0	19.4
2.0	38.6
3.0	58.1
4.0	77.4
5.0	96.5

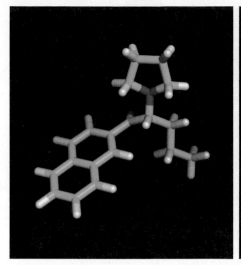
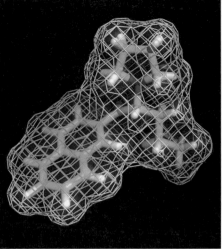
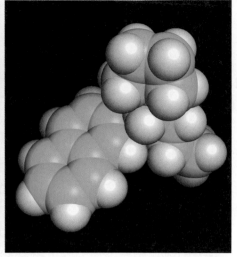

Predicting Values

When scientists discover relationships like the ones shown in the graphs in this section, they use them to make predictions. For example, the equation for the linear graph in **Figure 16** is as follows:

$$y = (0.08 \text{ cm/g})x + 13.7 \text{ cm}$$

Relationships, either learned as formulas or developed from graphs, can be used to predict values you haven't measured directly. How far would the spring in **Table 3** stretch with 49 g of mass?

$$y = (0.08 \text{ cm/g})(49 \text{ g}) + 13.7 \text{ cm}$$
$$= 18 \text{ cm}$$

It is important to decide how far you can extrapolate from the data you have. For example, 90 g is a value far outside the ones measured, and the spring might break rather than stretch that far.

Physicists use models to accurately predict how systems will behave: what circumstances might lead to a solar flare (an immense outburst of material from the Sun's surface into space), how changes to a grandfather clock's pendulum will change its ability to keep accurate time, or how magnetic fields will affect a medical instrument. People in all walks of life use models in many ways. One example is shown in **Figure 19**. With the tools you have learned in this chapter, you can answer questions and produce models for the physics questions you will encounter in the rest of this textbook.

Figure 19 In order to create a realistic animation, computer animators use mathematical models of the real world to create a convincing fictional world. This computer model of a naphyrone molecule is in development on an animator's computer.

PhysicsLAB

EXPLORING OBJECTS IN MOTION

INTERNET LAB How can you determine the speed of a vehicle?

SECTION 4 REVIEW

Section Self-Check Check your understanding.

22. **MAIN IDEA** Graph the following data. Time is the independent variable.

Time (s)	0	5	10	15	20	25	30	35
Speed (m/s)	12	10	8	6	4	2	2	2

23. **Interpret a Graph** What would be the meaning of a nonzero y-intercept in a graph of total mass versus volume?

24. **Predict** Use the relationship illustrated in **Figure 16** to determine the mass required to stretch the spring 15 cm.

25. **Predict** Use the relationship shown in **Figure 18** to predict the travel time when speed is 110 km/h.

26. **Critical Thinking** Look again at the graph in **Figure 16**. In your own words, explain how the spring would be different if the line in the graph were shallower or had a smaller slope.

Coming to Life

The Physics Behind Animation

If you were asked to name careers that use physics, animator would probably not be the first that comes to your mind. Three-dimensional (3-D) computer animation has replaced traditional two-dimensional, hand-drawn animation as the preferred medium for big-screen animated features. Knowing the physics involved in movement and light interaction is important for would-be animators who aim to create physically accurate models.

Modeling movement Initially, 3-D models are either sculpted by hand or modeled directly in the computer. Internal control points are connected to a larger grid with fewer external control points called a cage, shown in **Figure 1.** Linear geometric equations linking the cage to animation variables allow animators to produce complex, physically accurate movement without needing to move each individual control point.

Computer power The computer power required to render all of these equations is substantial. For example, the rendering equation needed for global illumination—the simulation of light bouncing around an environment—typically involves 10 million points, each with its own equation. Each frame of the animation, representing 0.04 s of screen time, generally takes about six hours to render.

Realistic characters In the past, proponents of math-based animation avoided using complicated characters, such as human beings, who appeared jarringly unrealistic compared to their nonhuman counterparts. In these cases, many animation studios preferred the technique of motion capture. Improvements in the last decade have led to increasingly complex virtual environments, however, such as oceans, and more compelling "purely animated" human characters.

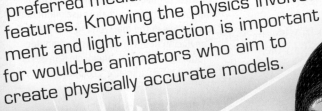

FIGURE 1 Each point on the numerous triangles that make up the character grid are linked by geometric equations.

GOING**FURTHER** >>>

Research There is a debate that motion capture is a technique that takes the art out of animation. Compare the benefits and drawbacks of math-based animation with those of motion-capture animation.

STUDY GUIDE

BIGIDEA   Physicists use scientific methods to investigate energy and matter.

VOCABULARY

- **physics** *(p. 4)*
- **scientific methods** *(p. 5)*
- **hypothesis** *(p. 6)*
- **model** *(p. 7)*
- **scientific theory** *(p. 8)*
- **scientific law** *(p. 8)*

SECTION 1 **Methods of Science**

MAINIDEA   Scientific investigations do not always proceed with identical steps but do contain similar methods.

- Scientific methods include making observations and asking questions about the natural world.
- Scientists use models to represent things that may be too small or too large, processes that take too much time to see completely, or a material that is hazardous.
- A scientific theory is an explanation of things or events based on knowledge gained from observations and investigations. A scientific law is a statement about what happens in nature, which seems to be true all the time.
- Science can't explain or solve everything. Questions about opinions or values can't be tested.

VOCABULARY

- **dimensional analysis** *(p. 11)*
- **significant figures** *(p. 12)*

SECTION 2 **Mathematics and Physics**

MAINIDEA   We use math to express concepts in physics.

- Using the metric system helps scientists around the world communicate more easily.
- Dimensional analysis is used to check that an answer will be in the correct units.
- Significant figures are the valid digits in a measurement.

VOCABULARY

- **measurement** *(p. 14)*
- **precision** *(p. 15)*
- **accuracy** *(p. 16)*

SECTION 3 **Measurement**

MAINIDEA   Making careful measurements allows scientists to repeat experiments and compare results.

- Measurements are reported with uncertainty because a new measurement that is within the margin of uncertainty confirms the old measurement.
- Precision is the degree of exactness with which a quantity is measured. Accuracy is the extent to which a measurement matches the true value.
- A common source of error that occurs when making a measurement is the angle at which an instrument is read. If the scale of an instrument is read at an angle, as opposed to at eye level, the measurement will be less accurate.

VOCABULARY

- **independent variable** *(p. 18)*
- **dependent variable** *(p. 18)*
- **line of best fit** *(p. 18)*
- **linear relationship** *(p. 20)*
- **quadratic relationship** *(p. 21)*
- **inverse relationship** *(p. 22)*

SECTION 4 **Graphing Data**

MAINIDEA   Graphs make it easier to interpret data, identify trends, and show relationships among a set of variables.

- Graphs contain information about the relationships among variables. Patterns that are not immediately evident in a list of numbers are seen more easily when the data are graphed.
- Common relationships shown in graphs include linear relationships, quadratic relationships, and inverse relationships. In a linear relationship the dependent variable varies linearly with the independent variable. A quadratic relationship occurs when one variable depends on the square of another. In an inverse relationship, one variable depends on the inverse of the other variable.
- Scientists use models and relationships between variables to make predictions.

Games and Multilingual eGlossary

Vocabulary Practice

Chapter Self-Check

SECTION 1 Methods of Science

Mastering Concepts

27. Describe a scientific method.

28. Explain why scientists might use each of the models listed below.

    a. physical model of the solar system

    b. computer model of airplane aerodynamics

    c. mathematical model of the force of attraction between two objects

SECTION 2 Mathematics and Physics

Mastering Concepts

29. Why is mathematics important to science?

30. What is the SI system?

31. How are base units and derived units related?

32. Suppose your lab partner recorded a measurement as 100 g.

    a. Why is it difficult to tell the number of significant figures in this measurement?

    b. How can the number of significant figures in such a number be made clear?

33. Give the name for each of the following multiples of the meter.

    a. $\frac{1}{100}$ m

    b. $\frac{1}{1000}$ m

    c. 1000 m

34. To convert 1.8 h to minutes, by what conversion factor should you multiply?

35. Solve each problem. Give the correct number of significant figures in the answers.

    a. 4.667×10^4 g $+ 3.02 \times 10^5$ g

    b. $(1.70 \times 10^2$ J$) \div (5.922 \times 10^{-4}$ cm$^3)$

Mastering Problems

36. Convert each of the following measurements to meters.

    a. 42.3 cm

    b. 6.2 pm

    c. 21 km

    d. 0.023 mm

    e. 214 μm

    f. 57 nm

37. Add or subtract as indicated.

    a. 5.80×10^9 s $+ 3.20 \times 10^8$ s

    b. 4.87×10^{-6} m $- 1.93 \times 10^{-6}$ m

    c. 3.14×10^{-5} kg $+ 9.36 \times 10^{-5}$ kg

    d. 8.12×10^7 g $- 6.20 \times 10^6$ g

38. **Ranking Task** Rank the following numbers according to the number of significant figures they have, from most to least: 1.234, 0.13, 0.250, 7.603, 0.08. Specifically indicate any ties.

39. State the number of significant figures in each of the following measurements.

    a. 0.00003 m

    b. 64.01 fm

    c. 80.001 m

    d. 6×10^8 kg

    e. 4.07×10^{16} m

40. Add or subtract as indicated.

    a. 16.2 m $+$ 5.008 m $+$ 13.48 m

    b. 5.006 m $+$ 12.0077 m $+$ 8.0084 m

    c. 78.05 cm^2 $-$ 32.046 cm^2

    d. 15.07 kg $-$ 12.0 kg

41. Multiply or divide as indicated.

    a. $(6.2 \times 10^{18}$ m$)(4.7 \times 10^{-10}$ m$)$

    b. $\dfrac{(5.6 \times 10^{-7}\text{ m})}{(2.8 \times 10^{-12}\text{ s})}$

    c. $(8.1 \times 10^{-4}$ km$)(1.6 \times 10^{-3}$ km$)$

    d. $\dfrac{(6.5 \times 10^5\text{ kg})}{(3.4 \times 10^3\text{ m}^3)}$

42. **Gravity** The force due to gravity is $F = mg$ where $g = 9.8$ N/kg.

    a. Find the force due to gravity on a 41.63-kg object.

    b. The force due to gravity on an object is 632 N. What is its mass?

43. **Dimensional Analysis** Pressure is measured in pascals, where 1 Pa $= 1$ kg/(m·s^2). Will the following expression give a pressure in the correct units?

$$\frac{(0.55\text{ kg})(2.1\text{ m/s})}{9.8\text{ m/s}^2}$$

SECTION 3 Measurement

Mastering Concepts

44. What determines the precision of a measurement?

45. How does the last digit differ from the other digits in a measurement?

Mastering Problems

46. A water tank has a mass of 3.64 kg when it is empty and a mass of 51.8 kg when it is filled to a certain level. What is the mass of the water in the tank?

47. The length of a room is 16.40 m, its width is 4.5 m, and its height is 3.26 m. What volume does the room enclose?

48. The sides of a quadrangular plot of land are 132.68 m, 48.3 m, 132.736 m, and 48.37 m. What is the perimeter of the plot?

49. How precise a measurement could you make with the scale shown in **Figure 20?**

Figure 20

50. Give the measurement shown on the meter in **Figure 21** as precisely as you can. Include the uncertainty in your answer.

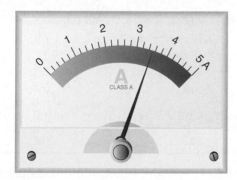

Figure 21

51. Estimate the height of the nearest door frame in centimeters. Then measure it. How accurate was your estimate? How precise was your estimate? How precise was your measurement? Why are the two precisions different?

52. Temperature The temperature drops linearly from 24°C to 10°C in 12 hours.

  a. Find the average temperature change per hour.

  b. Predict the temperature in 2 more hours if the trend continues.

  c. Could you accurately predict the temperature in 24 hours? Explain why or why not.

SECTION 4 Graphing Data

Mastering Concepts

53. How do you find the slope of a linear graph?

54. When driving, the distance traveled between seeing a stoplight and stepping on the brakes is called the reaction distance. Reaction distance for a given driver and vehicle depends linearly on speed.

  a. Would the graph of reaction distance versus speed have a positive or a negative slope?

  b. A driver who is distracted takes a longer time to step on the brake than a driver who is not. Would the graph of reaction distance versus speed for a distracted driver have a larger or smaller slope than for a normal driver? Explain.

55. During a laboratory experiment, the temperature of the gas in a balloon is varied and the volume of the balloon is measured. Identify the independent variable and the dependent variable.

56. What type of relationship is shown in **Figure 22?** Give the general equation for this type of relation.

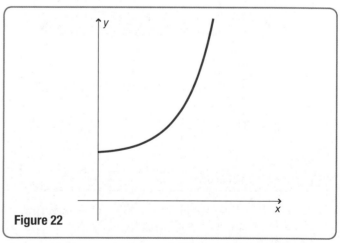

Figure 22

57. Given the equation $F = \dfrac{mv^2}{R}$, what kind of relationship exists between each of the following?

  a. F and R

  b. F and m

  c. F and v

Mastering Problems

58. Figure 23 shows the masses of three substances for volumes between 0 and 60 cm³.

a. What is the mass of 30 cm³ of each substance?

b. If you had 100 g of each substance, what would be each of their volumes?

c. In one or two sentences, describe the meaning of the slopes of the lines in this graph.

d. Explain the meaning of each line's y-intercept.

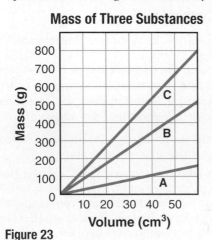

Mass of Three Substances

Figure 23

59. Suppose a mass is placed on a horizontal table that is nearly frictionless. Various horizontal forces are applied to the mass. The distance the mass traveled in 5 seconds for each force applied is measured. The results of the experiment are shown in **Table 5**.

Table 5 Distance Traveled with Different Forces

Force (N)	Distance (cm)
5.0	24
10.0	49
15.0	75
20.0	99
25.0	120
30.0	145

a. Plot the values given in the table and draw the curve that best fits all points.

b. Describe the resulting curve.

c. Use the graph to write an equation relating the distance to the force.

d. What is the constant in the equation? Find its units.

e. Predict the distance traveled when a 22.0-N force is exerted on the object for 5 s.

60. Suppose the procedure from the previous problem changed. The mass was varied while the force was kept constant. Time and distance were measured, and the acceleration of each mass was calculated. The results of the experiment are shown in **Table 6**.

Table 6 Acceleration of Different Masses

Mass (kg)	Acceleration (m/s²)
1.0	12.0
2.0	5.9
3.0	4.1
4.0	3.0
5.0	2.5
6.0	2.0

a. Plot the values given in the table and draw the curve that best fits all points.

b. Describe the resulting curve.

c. Write the equation relating acceleration to mass given by the data in the graph.

d. Find the units of the constant in the equation.

e. Predict the acceleration of an 8.0-kg mass.

61. During an experiment, a student measured the mass of 10.0 cm³ of alcohol. The student then measured the mass of 20.0 cm³ of alcohol. In this way, the data in **Table 7** were collected.

Table 7
The Mass Values of Specific Volumes of Alcohol

Volume (cm³)	Mass (g)
10.0	7.9
20.0	15.8
30.0	23.7
40.0	31.6
50.0	39.6

a. Plot the values given in the table and draw the curve that best fits all the points.

b. Describe the resulting curve.

c. Use the graph to write an equation relating the volume to the mass of the alcohol.

d. Find the units of the slope of the graph. What is the name given to this quantity?

e. What is the mass of 32.5 cm³ of alcohol?

Applying Concepts

62. Is a scientific method one set of clearly defined steps? Support your answer.

63. Explain the difference between a scientific theory and a scientific law.

64. **Figure 24** gives the height above the ground of a ball that is thrown upward from the roof of a building, for the first 1.5 s of its trajectory. What is the ball's height at $t = 0$? Predict the ball's height at $t = 2$ s and at $t = 5$ s.

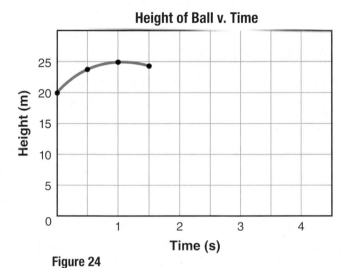

Figure 24

65. **Density** The density of a substance is its mass divided by its volume.

  a. Give the metric unit for density.

  b. Is the unit for density a base unit or a derived unit?

66. What metric unit would you use to measure each of the following?

  a. the width of your hand

  b. the thickness of a book cover

  c. the height of your classroom

  d. the distance from your home to your classroom

67. **Size** Make a chart of sizes of objects. Lengths should range from less than 1 mm to several kilometers. Samples might include the size of a cell, the distance light travels in 1 s, and the height of a room.

68. **Time** Make a chart of time intervals. Sample intervals might include the time between heartbeats, the time between presidential elections, the average lifetime of a human, and the age of the United States. In your chart, include several examples of very short and very long time intervals.

69. **Speed of Light** Two scientists measure the speed of light. One obtains $(3.001 \pm 0.001) \times 10^8$ m/s; the other obtains $(2.999 \pm 0.006) \times 10^8$ m/s.

  a. Which is more precise?

  b. Which is more accurate? (You can find the speed of light in the back of this textbook.)

70. You measure the dimensions of a desk as 132 cm, 83 cm, and 76 cm. The sum of these measures is 291 cm, while the product is 8.3×10^5 cm^3. Explain how the significant figures were determined in each case.

71. **Money** Suppose you receive $15.00 at the beginning of a week and spend $2.50 each day for lunch. You prepare a graph of the amount you have left at the end of each day for one week. Would the slope of this graph be positive, zero, or negative? Why?

72. Data are plotted on a graph, and the value on the y-axis is the same for each value of the independent variable. What is the slope? Why? How does y depend on x?

73. **Driving** The graph of braking distance versus car speed is part of a parabola. Thus, the equation is written $d = av^2 + bv + c$. The distance (d) has units in meters, and velocity (v) has units in meters/second. How could you find the units of a, b, and c? What would they be?

74. How long is the leaf in **Figure 25?** Include the uncertainty in your measurement.

Figure 25

75. Explain the difference between a hypothesis and a scientific theory.

76. Give an example of a scientific law.

77. What reason might the ancient Greeks have had not to question the (incorrect) hypothesis that heavier objects fall faster than lighter objects? *Hint: Did you ever question which falls faster?*

78. A graduated cylinder is marked every mL. How precise a measurement can you make with this instrument?

79. Reverse Problem Write a problem with real-life objects for which the graph in **Figure 26** could be part of the solution.

Number of People in a Room over Time

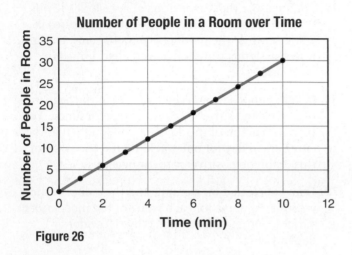

Figure 26

Mixed Review

80. Arrange the following numbers from most precise to least precise: 0.0034 m, 45.6 m, 1234 m.

81. Figure 27 shows an engine of a jet plane. Explain why a width of 80 m would be an unreasonable value for the diameter of the engine. What would be a reasonable value?

Figure 27

82. You are cracking a code and have discovered the following conversion factors: 1.23 longs = 23.0 mediums, and 74.5 mediums = 645 shorts. How many shorts are equal to one long?

83. You are given the following measurements of a rectangular bar: length = 2.347 m, thickness = 3.452 cm, height = 2.31 mm, mass = 1659 g. Determine the volume, in cubic meters, and density, in g/cm^3, of the beam.

84. A drop of water contains 1.7×10^{21} molecules. If the water evaporated at the rate of one million molecules per second, how many years would it take for the drop to completely evaporate?

85. A 17.6-gram sample of metal is placed in a graduated cylinder containing $10.0 \ cm^3$ of water. If the water level rises to $12.20 \ cm^3$, what is the density of the metal?

Thinking Critically

86. Apply Concepts It has been said that fools can ask more questions than the wise can answer. In science, it is frequently the case that one wise person is needed to ask the right question rather than to answer it. Explain.

87. Apply Concepts Find the approximate mass of water in kilograms needed to fill a container that is 1.40 m long and 0.600 m wide to a depth of 34.0 cm. Report your result to one significant figure. (Use a reference source to find the density of water.)

88. Analyze and Conclude A container of gas with a pressure of 101 kPa has a volume of $324 \ cm^3$ and a mass of 4.00 g. If the pressure is increased to 404 kPa, what is the density of the gas? Pressure and volume are inversely proportional.

89. BIGIDEA Design an Experiment How high can you throw a ball? What variables might affect the answer to this question?

90. Problem Posing Complete this problem so that the final answer will have 3 significant figures: "A home remedy used to prevent swimmer's ear calls for equal parts vinegar and rubbing alcohol. You measure 45.62 mL of vinegar . . . ."

Writing in Physics

91. Research and describe a topic in the history of physics. Explain how ideas about the topic changed over time. Be sure to include the contributions of scientists and to evaluate the impact of their contributions on scientific thought and the world outside the laboratory.

92. Explain how improved precision in measuring time would have led to more accurate predictions about how an object falls.

MULTIPLE CHOICE

1. Two laboratories use radiocarbon dating to measure the age of two wooden spear handles found in the same grave. Lab A finds an age of 2250 ± 40 years for the first object; lab B finds an age of 2215 ± 50 years for the second object. Which is true?

 A. Lab A's reading is more accurate than lab B's.

 B. Lab A's reading is less accurate than lab B's.

 C. Lab A's reading is more precise than lab B's.

 D. Lab A's reading is less precise than lab B's.

2. Which of the following is equal to 86.2 cm?

 A. 8.62 m

 B. 0.862 mm

 C. 8.62×10^{-4} km

 D. 862 dm

3. Jario has a problem to do involving time, distance, and velocity, but he has forgotten the formula. The question asks him for a measurement in seconds, and the numbers that are given have units of m/s and km. What could Jario do to get the answer in seconds?

 A. Multiply the km by the m/s, then multiply by 1000.

 B. Divide the km by the m/s, then multiply by 1000.

 C. Divide the km by the m/s, then divide by 1000.

 D. Multiply the km by the m/s, then divide by 1000.

4. What is the slope of the graph?

 A. 0.25 m/s^2

 B. 0.4 m/s^2

 C. 2.5 m/s^2

 D. 4.0 m/s^2

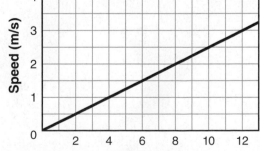

Stopping Distance

5. Which formula is equivalent to $D = \frac{m}{V}$?

 A. $V = \frac{m}{D}$

 B. $V = Dm$

 C. $V = \frac{mD}{V}$

 D. $V = \frac{D}{m}$

6. A computer simulation is an example of what?

 A. a hypothesis

 B. a model

 C. a scientific law

 D. a scientific theory

FREE RESPONSE

7. You want to calculate an acceleration, in units of m/s^2, given a force, in N, and the mass, in g, on which the force acts. $(1\ \text{N} = 1\ \text{kg·m/s}^2)$

 a. Rewrite the equation $F = ma$ so a is in terms of m and F.

 b. What conversion factor will you need to multiply by to convert grams to kilograms?

 c. A force of 2.7 N acts on a 350-g mass. Write the equation you will use, including the conversion factor, to find the acceleration.

8. Find an equation for a line of best fit for the data shown below.

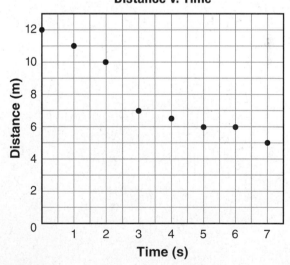

Distance v. Time

NEED EXTRA HELP?

If You Missed Question	1	2	3	4	5	6	7	8
Review Section	3	2	2	4	2	1	2	4

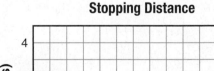
Online Test Practice

Representing Motion

BIGIDEA   You can use displacement and velocity to describe an object's motion.

SECTIONS

LaunchLAB

TOY CAR RACE

What factors determine an object's speed?

WATCH THIS!

Video

MEASURING SPEED

Have you ever been passed by another car on the freeway? If you know a few important details, it's possible to determine how fast that car is going. It's physics in action on the freeway.

PHYSICS T.V.

(l)Steve Allen/Brand X Pictures; (r)John Giustina/Photodisc/Getty Images

Picturing Motion

Look at this multiple-exposure photograph of a bird's movement. Physicists can use these photographs to evaluate changes in position and velocity.

MAIN IDEA

You can use motion diagrams to show how an object's position changes over time.

Essential Questions

- How do motion diagrams represent motion?

- How can you use a particle model to represent a moving object?

Review Vocabulary

model a representation of an idea, event, structure, or object to help people better understand it

New Vocabulary

motion diagram
particle model

All Kinds of Motion

You have learned about scientific processes that will be useful in your study of physics. You will now begin to use these tools to analyze motion. In subsequent chapters, you will apply these processes to many kinds of motion. You will use words, sketches, diagrams, graphs, and equations. These concepts will help you determine how fast and how far an object moves, in which direction that object is moving, whether that object is speeding up or slowing down, and whether that object is standing still or moving at a constant speed.

Changes in position What comes to your mind when you hear the word *motion*? A spinning ride at an amusement park? A baseball soaring over a fence for a home run? Motion is all around you—from fast trains and speedy skiers to slow breezes and lazy clouds. Objects move in many different ways, such as the straight-line path of a bowling ball in a bowling lane's gutter, the curved path of a car rounding a turn, the spiral of a falling kite, and swirls of water circling a drain. When an object is in motion, such as the subway train in **Figure 1,** its position changes.

Some types of motion are more complicated than others. When beginning a new area of study, it is generally a good idea to begin with the least complicated situation, learn as much as possible about it, and then gradually add more complexity to that simple model. In the case of motion, you will begin your study with movement along a straight line.

Figure 1 The subway train appears blurry in the photograph because its position changed during the time the camera shutter was open.

Describe how the picture would be different if the train were sitting still.

Movement along a straight line In general, an object can move along many different kinds of paths, but straight-line motion follows a path directly between two points without turning left or right. For example, you might describe an object's motion as forward and backward, up and down, or north and south. In each of these cases, the object moves along a straight line.

Suppose you are reading this textbook at home. As you start to read, you glance over at your pet hamster and see that it is sitting in a corner of the cage. Sometime later you look over again, and you see that it now is sitting next to the food dish in the opposite corner of the cage. You can infer that your hamster has moved from one place to another in the time between your observations. What factors helped you make this inference about the hamster's movement?

The description of motion is a description of place and time. You must answer the questions of where an object is located and when it is at that position in order to clearly describe its motion. Next, you will look at some tools that help determine when an object is at a particular place.

☑ **READING CHECK Identify** two factors you must know in order to describe the motion of an object along a straight line.

Motion Diagrams

Consider the following example of straight-line motion: a runner jogs along a straight path. One way of representing the runner's motion is to create a series of images showing the runner's position at equal time intervals. You can do this by photographing the runner in motion to obtain a sequence of pictures. Each photograph will show the runner at a point that is farther along the straight path.

Consecutive images Suppose you point a camera in a direction and a runner crosses the camera's field of view. Then you take a series of photographs of the runner at equal time intervals, without moving the camera. **Figure 2** shows what a series of consecutive images for a runner might look like. Notice that the runner is in a different position in each image, but everything in the background remains in the same position. This indicates that, relative to the camera and the ground, only the runner is in motion.

☑ **READING CHECK Decide** whether the spaces between a moving object's position must be equal if photographs are taken of the object at equal time intervals. Explain.

Figure 2 You can tell that the jogger is in motion because her position changes relative to the tree and the ground.

Figure 3 Combining the images from **Figure 2** produces this motion diagram of the jogger's movement. The series of dots at the bottom of the figure is a particle model that corresponds to the motion diagram.

Explain how the particle model shows that the jogger's speed is not changing.

View an **animation of motion diagrams v. particle motion.**

Concepts In Motion

PhysicsLAB

MOTION DIAGRAMS

How do the motion diagrams of a fast toy car and a slow toy car differ?

Combining images Suppose that you layered the four images of the runner from **Figure 2** one on top of the other. **Figure 3** shows what such a layered image might look like. You see more than one image of the moving runner, but you see only a single image of the tree and other motionless objects in the background. A series of images showing the positions of a moving object at equal time intervals is called a **motion diagram.**

Particle Models

Keeping track of the runner's motion is easier if you disregard the movement of her arms and her legs and instead concentrate on a single point at the center of her body. In effect, you can disregard the fact that the runner has size and imagine that she is a very small object located precisely at that central point. In a **particle model,** you replace the object or objects of interest with single points. Use of the particle model is common throughout the study of physics.

To use the particle model, the object's size must be much less than the distance it moves. The object's internal motions, such as the waving of the runner's arms or the movement of her legs, are ignored in the particle model. In the photographic motion diagram, you could identify one central point on the runner, such as a point centered at her waistline, and draw a dot at its position at different times. The bottom of **Figure 3** shows the particle model for the runner's motion. In the next section, you will learn how to create and use a motion diagram that shows how far an object moved and how much time it took to move that far.

SECTION 1 REVIEW

Section Self-Check Check your understanding.

1. **MAIN**IDEA How does a motion diagram represent an object's motion?

2. **Motion Diagram of a Bike Rider** Draw a particle model motion diagram for a bike rider moving at a constant pace along a straight path.

3. **Motion Diagram of a Car** Draw a particle model motion diagram corresponding to the motion diagram in **Figure 4** for a car coming to a stop at a stop sign. What point on the car did you use to represent the car?

Figure 4

4. **Motion Diagram of a Bird** Draw a particle model motion diagram corresponding to the motion diagram in **Figure 5** for a flying bird. What point on the bird did you choose to represent the bird?

Figure 5

5. **Critical Thinking** Draw particle model motion diagrams for two runners during a race in which the first runner crosses the finish line as the other runner is three-fourths of the way to the finish line.

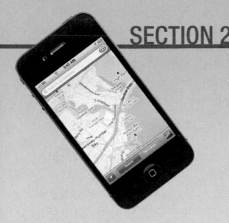

Where and When?

PHYSICS 4 YOU Have you ever used an electronic map for directions? These useful devices display the distances and directions you need to go. Many even show the time for different parts of the trip. To find your way to a place, you need clear directions for getting there.

MAIN IDEA

A coordinate system is helpful when you are describing motion.

Essential Questions

• What is a coordinate system?

• How does the chosen coordinate system affect the sign of objects' positions?

• How are time intervals measured?

• What is displacement?

• How are motion diagrams helpful in answering questions about an object's position or displacement?

Review Vocabulary

dimension extension in a given direction; one dimension is along a straight line; three dimensions are height, width, and length

New Vocabulary

coordinate system
origin
position
distance
magnitude
vector
scalar
time interval
displacement
resultant

Coordinate Systems

Is it possible to measure distance and time on a motion diagram? Before photographing a runner, you could place a long measuring tape on the ground to show where the runner is in each image. A stopwatch within the camera's view could show the time. But where should you place the end of the measuring tape? When should you start the stopwatch?

Position and distance It is useful to identify a system in which you have chosen where to place the zero point of the measuring tape and when to start the stopwatch. A **coordinate system** gives the location of the zero point of the variable you are studying and the direction in which the values of the variable increase. The **origin** is the point at which all variables in a coordinate system have the value zero. In the example of the runner, the origin, which is the zero point of the measuring tape, could be 6 m to the left of the tree. Because the motion is in a straight line, your measuring tape should lie along this line. The straight line is an axis of the coordinate system.

You can indicate how far the runner in **Figure 6** is from the origin at a certain time on the motion diagram by drawing an arrow from the origin to the point that represents the runner, shown at the bottom of the figure. This arrow represents the runner's **position,** the distance and direction from the origin to the object. In general, **distance** is the entire length of an object's path, even if the object moves in many directions. Because the motion in **Figure 6** is in one direction, the arrow lengths represent distance.

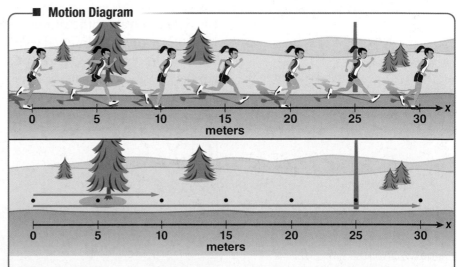

Figure 6 A simplified motion diagram uses dots to represent a moving object and arrows to indicate positions.

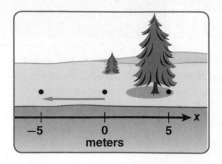

Figure 7 The green arrow indicates a negative position of −5 m if the direction right of the origin is chosen as positive.

Infer What position would the arrow indicate if you chose the direction left of the origin as positive?

Negative position Is there such a thing as a negative position? Suppose you chose the coordinate system just described but this time placed the origin 4 m left of the tree with the *x*-axis extending in a positive direction to the right. A position 9 m left of the tree, or 5 m left of the origin, would be a negative position, as shown in **Figure 7.**

Vectors and Scalars

Many quantities in physics have both size, also called **magnitude,** and direction. A quantity that has both magnitude and direction is called a **vector.** You can represent a vector with an arrow. The length of the arrow represents the magnitude of the vector, and the direction of the arrow represents the direction of the vector. A quantity that is just a number without any direction, such as distance, time, or temperature, is called a **scalar.** In this textbook, we will use boldface letters to represent vector quantities and regular letters to represent scalars.

Time intervals are scalars. When analyzing the runner's motion, you might want to know how long it took her to travel from the tree to the lamppost. You can obtain this value by finding the difference between the stopwatch reading at the tree and the stopwatch reading at the lamppost. **Figure 8** shows these stopwatch readings. The difference between two times is called a **time interval.**

A common symbol for a time interval is Δt, where the Greek letter delta (Δ) is used to represent a change in a quantity. Let t_i represent the initial (starting) time, when the runner was at the tree. Let t_f represent the final (ending) time of the interval, when the runner was at the lamppost. We define a time interval mathematically as follows.

TIME INTERVAL

The time interval is equal to the change in time from the initial time to the final time.

$$\Delta t = t_f - t_i$$

The subscripts i and f represent the initial and final times, but they can be the initial and final times of any time interval you choose. In the example of the runner, the time it takes for her to go from the tree to the lamppost is $t_f - t_i = 5.0 \text{ s} - 1.0 \text{ s} = 4.0 \text{ s}$. You could instead describe the time interval for the runner to go from the origin to the lamppost. In this case the time interval would be $t_f - t_i = 5.0 \text{ s} - 0.0 \text{ s} = 5.0 \text{ s}$. The time interval is a scalar because it has no direction. What about the runner's position? Is it also a scalar?

Figure 8 You can use the clocks in the figure to calculate the time interval (Δt) for the runner's movement from one position to another.

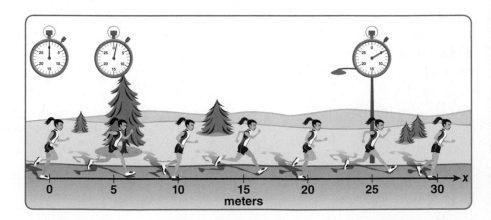

Positions and displacements are vectors. You have already seen how a position can be described as negative or positive in order to indicate whether that position is to the left or the right of a coordinate system's origin. This suggests that position is a vector because position has direction–either right or left in this case.

Figure 9 shows the position of the runner at both the tree and the lamppost. Notice that you can draw an arrow from the origin to the location of the runner in each case. These arrows have magnitude and direction. In common speech, a position refers to a certain place, but in physics, the definition of a position is more precise. A position is a vector with the arrow's tail at the origin of a coordinate system and the arrow's tip at the place.

You can use the symbol x to represent position vectors mathematically. In **Figure 9,** the symbol x_i represents the position at the tree, and the symbol x_f represents the position at the lamppost. The symbol Δx represents the change in position from the tree to the lamppost. Because a change in position is described and analyzed so often in physics, it has a special name. In physics, a change in position is called a **displacement.** Because displacement has direction, it is a vector.

☑ **READING CHECK** **Contrast** the distance an object moves and the object's displacement for straight-line motion.

What was the runner's displacement when she ran from the tree to the lamppost? By looking at **Figure 9,** you can see that this displacement is 20 m to the right. Notice also, that the displacement from the tree to the lamppost (Δx) equals the position at the lamppost (x_f) minus the position at the tree (x_i). This is true in general; displacement equals final position minus initial position.

DISPLACEMENT

Displacement is the change in position from initial position to final position.

$$\Delta x = x_f - x_i$$

Remember that the initial and final positions are the start and the end of any interval you choose. Although position is a vector, sometimes the magnitude of a position is described without the boldface. In this case, a plus or minus sign might be used to indicate direction.

☑ **READING CHECK** **Describe** what the direction and length of a displacement arrow indicate.

MiniLAB

VECTOR MODELS
How can you model vector addition using construction toys?

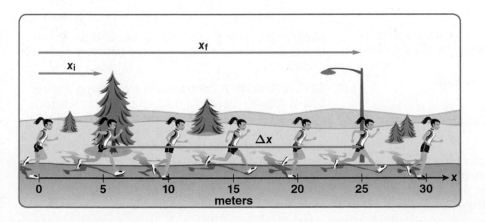

Figure 9 The vectors x_i and x_f represent positions. The vector Δx represents displacement from x_i to x_f.

Describe the displacement from the lamppost to the tree.

Vector addition and subtraction You will learn about many different types of vectors in physics, including velocity, acceleration, and momentum. Often, you will need to find the sum of two vectors or the difference between two vectors. A vector that represents the sum of two other vectors is called a **resultant. Figure 10** shows how to add and subtract vectors in one dimension. In a later chapter, you will learn how to add and subtract vectors in two dimensions.

Figure 10 You can use a diagram or an equation to combine vectors.

Analyze What is the sum of a vector 12 m north and a vector 8 m north?

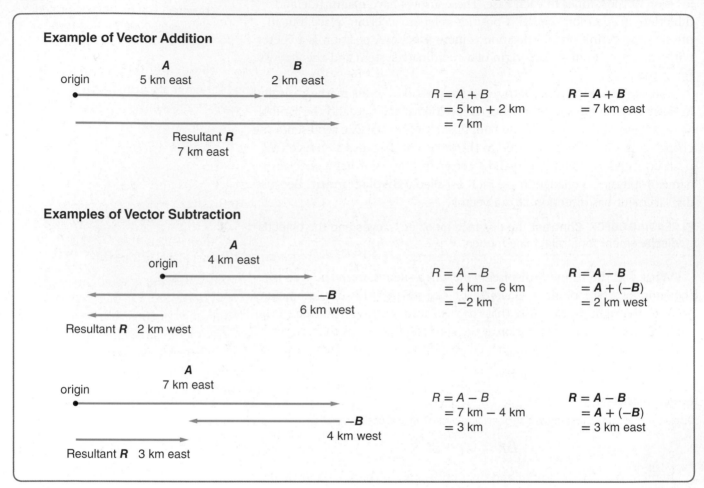

Example of Vector Addition

origin   **A**   **B**
           5 km east   2 km east

$R = A + B$ $R = A + B$
$= 5\ \text{km} + 2\ \text{km}$ $= 7\ \text{km east}$
$= 7\ \text{km}$

Resultant **R**
7 km east

Examples of Vector Subtraction

           A
origin   4 km east

                    −B
                    6 km west

$R = A - B$ $R = A - B$
$= 4\ \text{km} - 6\ \text{km}$ $= A + (-B)$
$= -2\ \text{km}$ $= 2\ \text{km west}$

Resultant **R**   2 km west

           A
origin   7 km east

                    −B
                    4 km west

$R = A - B$ $R = A - B$
$= 7\ \text{km} - 4\ \text{km}$ $= A + (-B)$
$= 3\ \text{km}$ $= 3\ \text{km east}$

Resultant **R**   3 km east

SECTION 2 REVIEW

Section Self-Check   Check your understanding.

6. **MAIN**IDEA Identify a coordinate system you could use to describe the motion of a girl swimming across a rectangular pool.

7. **Displacement** The motion diagram for a car traveling on an interstate highway is shown below. The starting and ending points are indicated.

   Start • • • • • End

   Make a copy of the diagram. Draw a vector to represent the car's displacement from the starting time to the end of the third time interval.

8. **Position** Two students added a vector for a moving object's position at $t = 2$ s to a motion diagram. When they compared their diagrams, they found that their vectors did not point in the same direction. Explain.

9. **Displacement** The motion diagram for a boy walking to school is shown below.

   Home • • • • • • • • • • School

   Make a copy of this motion diagram, and draw vectors to represent the displacement between each pair of dots.

10. **Critical Thinking** A car travels straight along a street from a grocery store to a post office. To represent its motion, you use a coordinate system with its origin at the grocery store and the direction the car is moving as the positive direction. Your friend uses a coordinate system with its origin at the post office and the opposite direction as the positive direction. Would the two of you agree on the car's position? Displacement? Distance? The time interval the trip took? Explain.

Position-Time Graphs

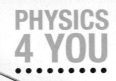

PHYSICS 4 YOU

Many graphs show trends over time. For example, a graph might show the price of gasoline through the course of several months or years. Similarly, a position-time graph can show how a rower's position changes through time. Rowers can use graphs to analyze their performances.

MAINIDEA

You can use position-time graphs to determine an object's position at a certain time.

Essential Questions

- What information do position-time graphs provide?

- How can you use a position-time graph to interpret an object's position or displacement?

- What are the purposes of equivalent representations of an object's motion?

Review Vocabulary

intersection a point where lines meet and cross

New Vocabulary

position-time graph
instantaneous position

Finding Positions

When analyzing complex motion, it often is useful to represent the motion in a variety of ways. A motion diagram contains information about an object's position at various times. Tables and graphs can also show this same information. Review the motion diagrams in **Figure 8** and **Figure 9.** You can use these diagrams to organize the times and corresponding positions of the runner, as in **Table 1.**

Plotting data The data listed in **Table 1** can be presented on a **position-time graph,** in which the time data is plotted on a horizontal axis and the position data is plotted on a vertical axis. The graph of the runner's motion is shown in **Figure 11.** To draw this graph, first plot the runner's positions. Then, draw a line that best fits the points.

Estimating time and position Notice that the graph is not a picture of the runner's path—the graphed line is sloped, but the runner's path was horizontal. Instead, the line represents the most likely positions of the runner at the times between the recorded data points. Even though there is no data point exactly when the runner was 12.0 m beyond her starting point or where she was at $t = 4.5$ s, you can use the graph to estimate the time or her position. The example problem on the next page shows how.

Table 1 Position v. Time	
Time (s)	Position (m)
0.0	0.0
1.0	5.0
2.0	10.0
3.0	15.0
4.0	20.0
5.0	25.0

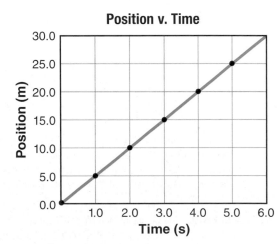

Figure 11 You can create a position-time graph by plotting the positions and times from the table. By drawing a best-fit line, you can estimate other times and positions.

Explain Why is the line on the graph sloped even though it describes motion along a flat path?

View an **animation of position-time graphs.**

Concepts In Motion

ANALYZE A POSITION-TIME GRAPH When did the runner whose motion is described in **Figure 11** reach 12.0 m beyond the starting point? Where was she after 4.5 s?

1 ANALYZE THE PROBLEM

Restate the questions.

Question 1: At what time was the magnitude of the runner's position (x) equal to 12.0 m?

Question 2: What was the runner's position at time $t = 4.5$ s?

2 SOLVE FOR THE UNKNOWN

Question 1

Examine the graph to find the intersection of the best-fit line with a horizontal line at the 12.0 m mark. Next, find where a vertical line from that point crosses the time axis. The value of t there is 2.4 s.

Question 2

Find the intersection of the graph with a vertical line at 4.5 s (halfway between 4.0 s and 5.0 s on this graph). Next, find where a horizontal line from that point crosses the position axis. The value of x is approximately 22.5 m.

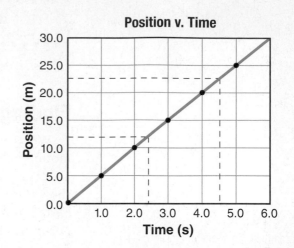

Position v. Time

PRACTICE PROBLEMS

Do additional problems. [Online Practice]

*For problems 11–13, refer to **Figure 12**.*

11. The graph in **Figure 12** represents the motion of a car moving along a straight highway. Describe in words the car's motion.

12. Draw a particle model motion diagram that corresponds to the graph.

13. Answer the following questions about the car's motion. Assume that the positive x-direction is east of the origin and the negative x-direction is west of the origin.

   a. At what time was the car's position 25.0 m east of the origin?

   b. Where was the car at time $t = 1.0$ s?

   c. What was the displacement of the car between times $t = 1.0$ s and $t = 3.0$ s?

14. The graph in **Figure 13** represents the motion of two pedestrians who are walking along a straight sidewalk in a city. Describe in words the motion of the pedestrians. Assume that the positive direction is east of the origin.

15. CHALLENGE Ari walked down the hall at school from the cafeteria to the band room, a distance of 100.0 m. A class of physics students recorded and graphed his position every 2.0 s, noting that he moved 2.6 m every 2.0 s. When was Ari at the following positions?

   a. 25.0 m from the cafeteria

   b. 25.0 m from the band room

   c. Create a graph showing Ari's motion.

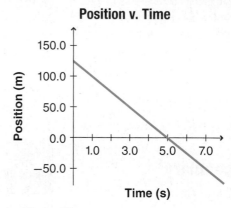

Position v. Time

Figure 12

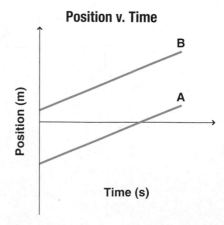

Position v. Time

Figure 13

Table 1 Position v. Time	
Time (s)	Position (m)
0.0	0.0
1.0	5.0
2.0	10.0
3.0	15.0
4.0	20.0
5.0	25.0

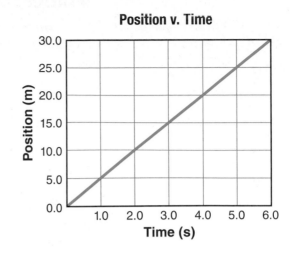

Position v. Time

Motion Diagram

Begin ● ● ● ● ● ● ● End

Instantaneous position How long did the runner spend at any location? Each position has been linked to a time, but how long did that time last? You could say "an instant," but how long is that? If an instant lasts for any finite amount of time, then the runner would have stayed at the same position during that time, and she would not have been moving. An instant is not a finite period of time, however. It lasts zero seconds. The symbol x represents the runner's **instantaneous position**—the position at a particular instant. Instantaneous position is usually simply called position.

☑ **READING CHECK Explain** what is meant by the instantaneous position of a runner.

Equivalent representations As shown in **Figure 14,** you now have several different ways to describe motion. You might describe motion using words, pictures (or pictorial representations), motion diagrams, data tables, or position-time graphs. All of these representations contain the same information about the runner's motion. However, depending on what you want to learn about an object's motion, some types of representations will be more useful than others. In the pages that follow, you will practice constructing these equivalent representations and learn which ones are most useful for solving different kinds of problems.

Figure 14 You can describe the runner's motion using the data table, the motion diagram, and the graph.

Identify one benefit the table has over the graph.

PHYSICS CHALLENGE

POSITION-TIME GRAPHS Natana, Olivia, and Phil all enjoy exercising and often go to a path along the river for this purpose. Natana bicycles at a very consistent 40.25 km/h, Olivia runs south at a constant speed of 16.0 km/h, and Phil walks south at a brisk 6.5 km/h. Natana starts biking north at noon from the waterfalls. Olivia and Phil both start at 11:30 A.M. at the canoe dock, 20.0 km north of the falls.

1. Draw position-time graphs for each person.
2. At what time will the three exercise enthusiasts be located within the smallest distance interval from each other?
3. What is the length of that distance interval?

Karl Weatherly/Getty Images

Multiple Objects on a Position-Time Graph

A position-time graph for two different runners is shown in Example Problem 2 below. Notice that runner A is ahead of runner B at time $t = 0$, but the motion of each runner is different. When and where does one runner pass the other? First, you should restate this question in physics terms: At what time are the two runners at the same position? What is their position at this time? You can evaluate these questions by identifying the point on the position-time graph at which the lines representing the two runners' motions intersect.

The intersection of two lines on a position-time graph tells you when objects have the same position, but does this mean that they will collide? Not necessarily. For example, if the two objects are runners and if they are in different lanes, they will not collide, even though they might be the same distance from the starting point.

☑ **READING CHECK** **Explain** what the intersection of two lines on a position-time graph means.

What else can you learn from a position-time graph? Notice in Example Problem 2 that the lines on the graph have different slopes. What does the slope of the line on a position-time graph tell you? In the next section, you will use the slope of a line on a position-time graph to determine the velocity of an object. When you study accelerated motion, you will draw other motion graphs and learn to interpret the areas under the plotted lines. In later studies, you will continue to refine your skills with creating and interpreting different types of motion graphs.

EXAMPLE PROBLEM 2

Find help with **interpolating and extrapolating.** **Math Handbook**

INTERPRETING A GRAPH The graph to the right describes the motion of two runners moving along a straight path. The lines representing their motion are labeled A and B. When and where does runner B pass runner A?

1 ANALYZE THE PROBLEM

Restate the questions.

Question 1: At what time are runner A and runner B at the same position?

Question 2: What is the position of runner A and runner B at this time?

2 SOLVE FOR THE UNKNOWN

Question 1

Examine the graph to find the intersection of the line representing the motion of runner A with the line representing the motion of runner B. These lines intersect at time 45 s.

Question 2

Examine the graph to determine the position when the lines representing the motion of the runners intersect. The position of both runners is about 190 m from the origin.

Runner B passes runner A about 190 m beyond the origin, 45 s after A has passed the origin.

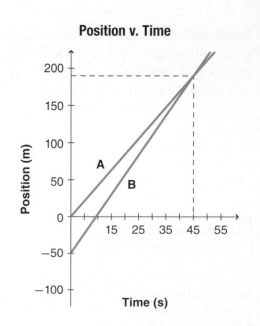

Position v. Time

For problems 16–19, refer to the figure in Example Problem 2 on the previous page.

16. Where was runner A located at $t = 0$ s?

17. Which runner was ahead at $t = 48.0$ s?

18. When runner A was at 0.0 m, where was runner B?

19. How far apart were runners A and B at $t = 20.0$ s?

20. CHALLENGE Juanita goes for a walk. Later her friend Heather starts to walk after her. Their motions are represented by the position-time graph in **Figure 15.**

    a. How long had Juanita been walking when Heather started her walk?

    b. Will Heather catch up to Juanita? How can you tell?

    c. What was Juanita's position at $t = 0.2$ h?

    d. At what time was Heather 5.0 km from the start?

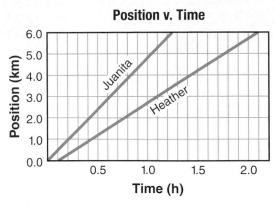

Position v. Time

Figure 15

SECTION 3 REVIEW

Section Self-Check 🖐 Check your understanding.

21. MAINIDEA Using the particle model motion diagram in **Figure 16** of a baby crawling across a kitchen floor, plot a position-time graph to represent the baby's motion. The time interval between successive dots on the diagram is 1 s.

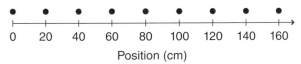

Figure 16

*For problems 22–25, refer to **Figure 17.***

22. Particle Model Create a particle model motion diagram from the position-time graph in **Figure 17** of a hockey puck gliding across a frozen pond.

23. Time Use the hockey puck's position-time graph to determine the time when the puck was 10.0 m beyond the origin.

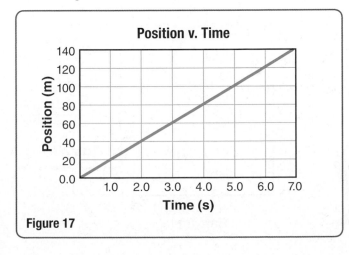

Position v. Time

Figure 17

24. Distance Use the position-time graph in **Figure 17** to determine how far the hockey puck moved between times 0.0 s and 5.0 s.

25. Time Interval Use the position-time graph for the hockey puck to determine how much time it took for the puck to go from 40 m beyond the origin to 80 m beyond the origin.

26. Critical Thinking Look at the particle model diagram and the position-time graph shown in **Figure 18.** Do they describe the same motion? How do you know? Do not confuse the position coordinate system in the particle model with the horizontal axis in the position-time graph. The time intervals in the particle model diagram are 2 s.

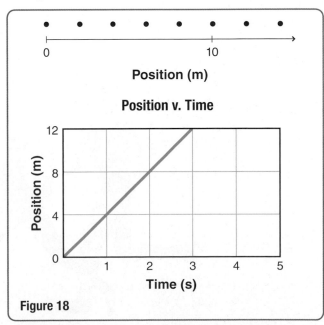

Position v. Time

Figure 18

How Fast?

Snails move much slower than cheetahs. You can see this by observing how far the animals travel during a given time period. For example, a cheetah can travel 30 m in a second, but a snail might move only 1 cm in that time interval.

MAIN IDEA

An object's velocity is the rate of change in its position.

Essential Questions

• What is velocity?

• What is the difference between speed and velocity?

• How can you determine an object's average velocity from a position-time graph?

• How can you represent motion with pictorial, physical, and mathematical models?

Review Vocabulary

absolute value magnitude of a number, regardless of sign

New Vocabulary

average velocity
average speed
instantaneous velocity

Velocity and Speed

Suppose you recorded the motion of two joggers on one diagram, as shown by the graph in **Figure 19.** The position of the jogger wearing red changes more than that of the jogger wearing blue. For a fixed time interval, the magnitude of the displacement (Δx) is greater for the jogger in red because she is moving faster. Now, suppose that each jogger travels 100 m. The time interval (Δt) for the 100 m would be smaller for the jogger in red than for the one in blue.

Slope on a position-time graph Compare the lines representing the joggers in the graph in **Figure 19.** The slope of the red jogger's line is steeper, indicating a greater change in position during each time interval. Recall that you find the slope of a line by first choosing two points on the line. Next, you subtract the vertical coordinate (x in this case) of the first point from the vertical coordinate of the second point to obtain the rise of the line. After that, you subtract the horizontal coordinate (t in this case) of the first point from the horizontal coordinate of the second point to obtain the run. The rise divided by the run is the slope.

Figure 19 A greater slope shows that the red jogger traveled faster.

Analyze How much farther did the red jogger travel than the blue jogger in the 3 s interval described by the graph?

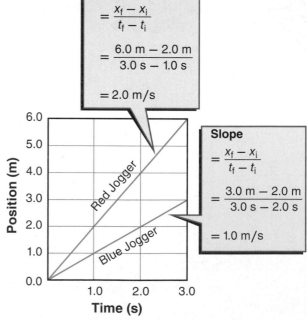

$$\text{Slope} = \frac{x_f - x_i}{t_f - t_i}$$

$$= \frac{6.0 \text{ m} - 2.0 \text{ m}}{3.0 \text{ s} - 1.0 \text{ s}}$$

$$= 2.0 \text{ m/s}$$

$$\text{Slope} = \frac{x_f - x_i}{t_f - t_i}$$

$$= \frac{3.0 \text{ m} - 2.0 \text{ m}}{3.0 \text{ s} - 2.0 \text{ s}}$$

$$= 1.0 \text{ m/s}$$

Red Jogger

Blue Jogger

Position (m)

Time (s)

IT Stock Free/Alamy

Average velocity Notice that the slope of the faster runner's line in **Figure 19** is a greater number. A greater slope indicates a faster speed. Also notice that the slope's units are meters per second. Looking at how the slope is calculated, you can see that slope is the change in the magnitude of the position divided by the time interval during which that change took place: $\frac{x_f - x_i}{t_f - t_i}$, or $\frac{\Delta x}{\Delta t}$. When Δx gets larger, the slope gets larger; when Δt gets larger, the slope gets smaller. This agrees with the interpretation given on the previous page of the speeds of the red and blue joggers. **Average velocity** is the ratio of an object's change in position to the time interval during which the change occurred. If the object is in uniform motion, so that its speed does not change, then its average velocity is the slope of its position-time graph.

AVERAGE VELOCITY

Average velocity is defined as the change in position divided by the time during which the change occurred.

$$\overline{v} \equiv \frac{\Delta x}{\Delta t} = \frac{x_f - x_i}{t_f - t_i}$$

The symbol $\equiv$ means that the left-hand side of the equation is defined by the right-hand side.

Interpreting slope The position-time graph's slope in **Figure 20** is −5.0 m/s. Notice that the slope of the graph indicates both magnitude and direction. By calculating the slope from the rise divided by the run between two points, you find that the object whose motion is represented by the graph has an average velocity of −5.0 m/s. The object started out at a positive position and moves toward the origin. After 4 s, it passes the origin and continues moving in the negative direction at a rate of 5.0 m/s.

> ☑ **READING CHECK Explain** the meaning of a position-time graph slope that is upward or downward, and above or below the x-axis.

Average speed The slope's absolute value is the object's **average speed,** 5.0 m/s, which is the distance traveled divided by the time taken to travel that distance. For uniform motion, average speed is the absolute value of the slope of the object's position-time graph. The combination of an object's average speed ($\overline{v}$) and the direction in which it is moving is the average velocity ($\overline{v}$). Remember that if an object moves in the negative direction, its change in position is negative. This means that an object's displacement and velocity are both always in the same direction.

PhysicsLABs

CONSTANT SPEED
How can you determine average speed by measuring distance and time?

MEASURE VELOCITY
PROBEWARE LAB How can you measure velocity with a motion detector?

Figure 20 The downward slope of this position-time graph shows that the motion is in the negative direction.

Analyze What would the graph look like if the motion were at the same speed, but in the positive direction?

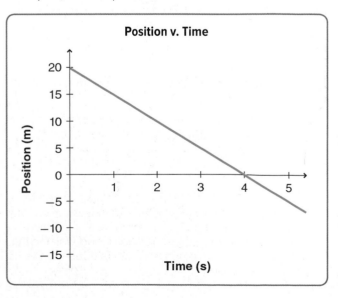

Get help with **average velocity and average speed.** | Personal Tutor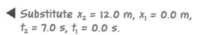

AVERAGE VELOCITY The graph at the right describes the straight-line motion of a student riding her skateboard along a smooth, pedestrian-free sidewalk. What is her average velocity? What is her average speed?

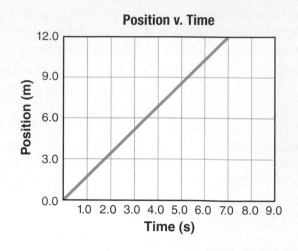

Position v. Time

1 **ANALYZE AND SKETCH THE PROBLEM**

Identify the graph's coordinate system.

UNKNOWN

$\bar{v} = ?$ $\bar{v} = ?$

2 **SOLVE FOR THE UNKNOWN**

Find the average velocity using two points on the line.

$$\bar{v} = \frac{\Delta x}{\Delta t}$$

$$= \frac{x_f - x_i}{t_f - t_i}$$

$$= \frac{12.0 \text{ m} - 0.0 \text{ m}}{7.0 \text{ s} - 0.0 \text{ s}}$$

◀ Substitute $x_2 = 12.0$ m, $x_1 = 0.0$ m, $t_2 = 7.0$ s, $t_1 = 0.0$ s.

$\bar{v} = 1.7$ m/s in the positive direction

The average speed ($\bar{v}$) is the absolute value of the average velocity, or 1.7 m/s.

3 **EVALUATE THE ANSWER**

• **Are the units correct?** The units for both velocity and speed are meters per second.

• **Do the signs make sense?** The positive sign for the velocity agrees with the coordinate system. No direction is associated with speed.

PRACTICE PROBLEMS

Do additional problems. | Online Practice

27. The graph in **Figure 21** describes the motion of a cruise ship drifting slowly through calm waters. The positive *x*-direction (along the vertical axis) is defined to be south.

  a. What is the ship's average speed?

  b. What is its average velocity?

28. Describe, in words, the cruise ship's motion in the previous problem.

29. What is the average velocity of an object that moves from 6.5 cm to 3.7 cm relative to the origin in 2.3 s?

30. The graph in **Figure 22** represents the motion of a bicycle.

  a. What is the bicycle's average speed?

  b. What is its average velocity?

31. Describe, in words, the bicycle's motion in the previous problem.

32. CHALLENGE When Marshall takes his pet dog for a walk, the dog walks at a very consistent pace of 0.55 m/s. Draw a motion diagram and a position-time graph to represent Marshall's dog walking the 19.8-m distance from in front of his house to the nearest stop sign.

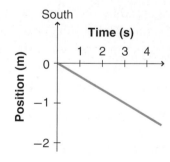

Figure 21

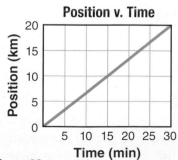

Figure 22

Instantaneous velocity Why do we call the quantity $\frac{\Delta x}{\Delta t}$ average velocity? Why don't we just call it velocity? A motion diagram shows the position of a moving object at the beginning and end of a time interval. It does not, however, indicate what happened within that time interval. During the time interval, the object's speed could have remained the same, increased, or decreased. The object may have stopped or even changed direction. You can find the average velocity for each time interval in the motion diagram, but you cannot find the speed and the direction of the object at any specific instant. The speed and the direction of an object at a particular instant is called the **instantaneous velocity.** In this textbook, the term velocity will refer to instantaneous velocity, represented by the symbol $\boldsymbol{v}$.

☑ **READING CHECK** **Explain** how average velocity is different from velocity.

Average velocity on motion diagrams When an object moves between two points, its average velocity is in the same direction as its displacement. The two quantities are also proportional—when displacement is greater during a given time interval, so is average velocity. A motion diagram indicates the average velocity's direction and magnitude.

Imagine two cars driving down the road at different speeds. A video camera records the motion of the cars at the rate of one frame every second. Imagine that each car has a paintbrush attached to it that automatically descends and paints a red line on the ground for half a second every second. The faster car would paint a longer line on the ground. The vectors you draw on a motion diagram to represent the velocity are like the lines that the paintbrushes make on the ground below the cars. In this book, we use red to indicate velocity vectors on motion diagrams. **Figure 23** shows motion diagrams with velocity vectors for two cars. One is moving to the right, and the other is moving to the left.

☑ **READING CHECK** **Identify** what the lengths of velocity vectors mean.

Equation of Motion

Often it is more efficient to use an equation, rather than a graph, to solve problems. Any time you graph a straight line, you can find an equation to describe it. Take another look at the graph in **Figure 20** for the object moving with a constant velocity of −5.0 m/s. Recall that you can represent any straight line with the equation $y = mx + b$, where y is the quantity plotted on the vertical axis, m is the line's slope, x is the quantity plotted on the horizontal axis, and b is the line's y-intercept.

For the graph in **Figure 20,** the quantity plotted on the vertical axis is position, represented by the variable $\boldsymbol{x}$. The line's slope is −5.0 m/s, which is the object's average velocity ($\overline{\boldsymbol{v}}$). The quantity plotted on the horizontal axis is time (t). The y-intercept is 20.0 m. What does this 20.0 m represent? This shows that the object was at a position of 20.0 m when $t = 0.0$ s. This is called the initial position of the object and it is designated $\boldsymbol{x}_i$.

REAL-WORLD PHYSICS

● ● ● ● ● ● ● ● ● ● ● ●

SPEED RECORDS The world record for the men's 100-m dash is 9.58 s, established in 2009 by Usain Bolt. The world record for the women's 100-m dash is 10.49 s, established in 1988 by Florence Griffith-Joyner.

MiniLAB

VELOCITY VECTORS
How can velocity vectors represent the motion of a mass on a string?

Figure 23 The length of each velocity vector is proportional to the magnitude of the velocity that it represents.

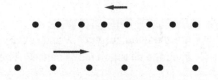

Lines and Graphs Symbols used in the point-slope equation of a line relate to symbols used for motion variables on a position-time graph.

General Variable	Specific Motion Variable	Value in Figure 20
y	x	
m	$\bar{v}$	−5.0 m/s
x	t	
b	x_i	20.0 m

A summary is given to the left of how the general variables in the straight-line formula are changed to the specific variables you have been using to describe motion. The table also shows the numerical values for the average velocity and initial position. Consider the graph shown in **Figure 20.** The mathematical equation for the line graphed is as follows:

$$y = (-5.0 \text{ m/s})x + 20.0 \text{ m}$$

You can rewrite this equation, using **x** for position and t for time.

$$x = (-5.0 \text{ m/s})t + 20.0 \text{ m}$$

It might be confusing to use y and x in math but use **x** and t in physics. You do this because there are many types of graphs in physics, including position v. time graphs, velocity v. time graphs, and force v. position graphs. For a position v. time graph, the math equation $y = mx + b$ can be rewritten as follows:

POSITION

An object's position is equal to the average velocity multiplied by time plus the initial position.

$$\boldsymbol{x} = \boldsymbol{\bar{v}}t + \boldsymbol{x_i}$$

This equation gives you another way to represent motion. Note that a graph of x v. t would be a straight line.

EXAMPLE PROBLEM 4

Find help with **solving equations.** Math Handbook

POSITION The figure shows a motorcyclist traveling east along a straight road. After passing point **B,** the cyclist continues to travel at an average velocity of 12 m/s east and arrives at point **C** 3.0 s later. What is the position of point **C?**

1 ANALYZE THE PROBLEM

Choose a coordinate system with the origin at **A.**

KNOWN

$\bar{v}$ = 12 m/s east
x_i = 46 m east
t = 3.0 s

UNKNOWN

x = ?

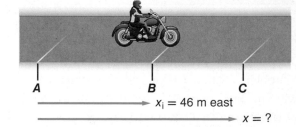

2 SOLVE FOR THE UNKNOWN

$x = \bar{v}t + x_i$ ◀ Use magnitudes for the calculations.
$\quad = (12 \text{ m/s})(3.0 \text{ s}) + 46 \text{ m}$ ◀ Substitute $\bar{v}$ = 12 m/s, t = 3.0 s, and x_i = 46 m.
$\quad = 82 \text{ m}$
$x = 82 \text{ m east}$

3 EVALUATE THE ANSWER

• **Are the units correct?** Position is measured in meters.
• **Does the direction make sense?** The motorcyclist is traveling east the entire time.

EXAMPLE PROBLEM

Do additional problems. Online Practice

*For problems 33–36, refer to **Figure 24**.*

33. The diagram at the right shows the path of a ship that sails at a constant velocity of 42 km/h east. What is the ship's position when it reaches point *C*, relative to the starting point, *A*, if it sails from point *B* to point *C* in exactly 1.5 h?

34. Another ship starts at the same time from point *B*, but its average velocity is 58 km/h east. What is its position, relative to *A*, after 1.5 h?

35. What would a ship's position be if that ship started at point *B* and traveled at an average velocity of 35 km/h west to point *D* in a time period of 1.2 h?

36. CHALLENGE Suppose two ships start from point *B* and travel west. One ship travels at an average velocity of 35 km/h for 2.2 h. Another ship travels at an average velocity of 26 km/h for 2.5 h. What is the final position of each ship?

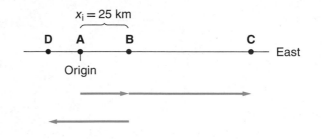

Figure 24

SECTION 4 REVIEW

Section Self-Check   Check your understanding.

37. MAINIDEA How is an object's velocity related to its position?

*For problems 38–40, refer to **Figure 25**.*

38. Ranking Task Rank the position-time graphs according to the average speed, from greatest average speed to least average speed. Specifically indicate any ties.

39. Contrast Average Velocities Describe differences in the average velocities shown on the graph for objects A and B. Describe differences in the average velocities shown on the graph for objects C and D.

40. Ranking Task Rank the graphs in **Figure 25** according to each object's initial position, from most positive position to most negative position. Specifically indicate any ties. Would your ranking be different if you ranked according to initial distance from the origin?

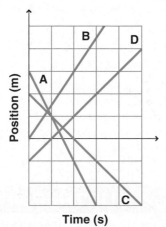

Figure 25

41. Average Speed and Average Velocity Explain how average speed and average velocity are related to each other for an object in uniform motion.

42. Position Two cars are traveling along a straight road, as shown in **Figure 26**. They pass each other at point B and then continue in opposite directions. The red car travels for 0.25 h from point B to point C at a constant velocity of 32 km/h east. The blue car travels for 0.25 h from point B to point D at a constant velocity of 48 km/h west. How far has each car traveled from point B? What is the position of each car relative to the origin, point A?

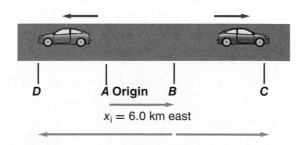

Figure 26

43. Position A car travels north along a straight highway at an average speed of 85 km/h. After driving 2.0 km, the car passes a gas station and continues along the highway. What is the car's position relative to the start of its trip 0.25 h after it passes the gas station?

44. Critical Thinking In solving a physics problem, why is it important to create pictorial and physical models before trying to solve an equation?

Got the time?

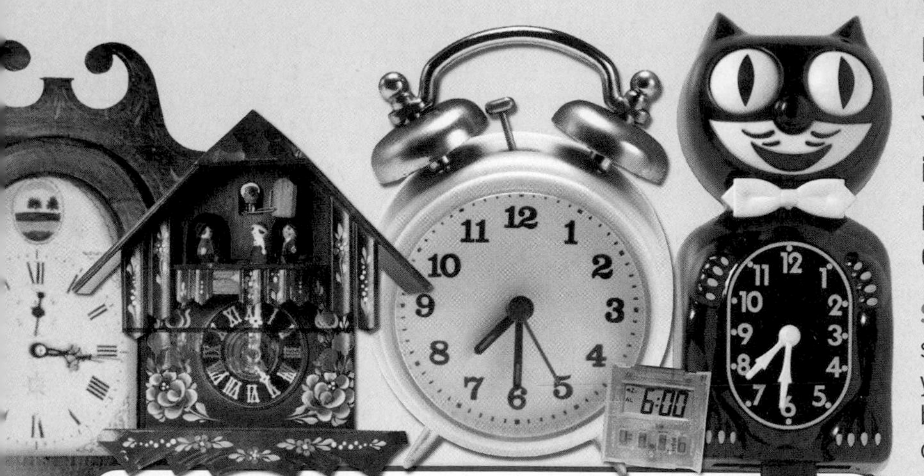

What is time? If one hour of time passes for you, does one hour of time also pass for your friend? You might think that the answer is yes, but it is actually no. Time passes at different rates depending on your point of view.

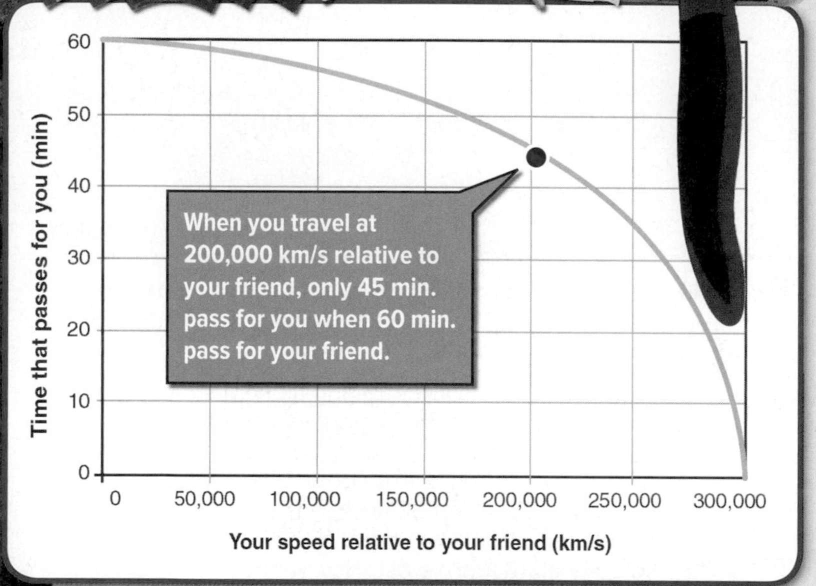

Speed and time Think about how wrong that last sentence seems. For example, suppose that you tell your friend to meet you at the mall in one hour. You both assume that when one hour passes for you, one hour also passes for your friend.

This is because you and your friend move very slowly relative to each other. At slow speeds, one hour for you is almost exactly the same as one hour for your friend. As you move faster relative to your friend, however, the difference between your time and your friend's time increases.

How fast? You would need to travel very fast relative to your friend in order for any difference to be noticeable. If you travel at 100,000 km/s, then only 57 minutes passes for you when one hour passes for your friend. At 200,000 km/s, only 45 minutes passes for you during your friend's hour. **Figure 1** shows how your time compares to one hour of your friend's time as you travel faster and faster relative to your friend.

Real–World Application All of this might seem rather pointless. After all, even the fastest spacecraft travel at less than 100 km/s. Have you ever used a GPS receiver, such as the one shown in **Figure 2?** At 4 km/s, a GPS satellite travels fast enough for time differences to affect the accuracy of the GPS receiver. The effect is small—approximately 10 μs in one day. It is enough, however, that the GPS would become completely useless within one month if engineers did not account for it.

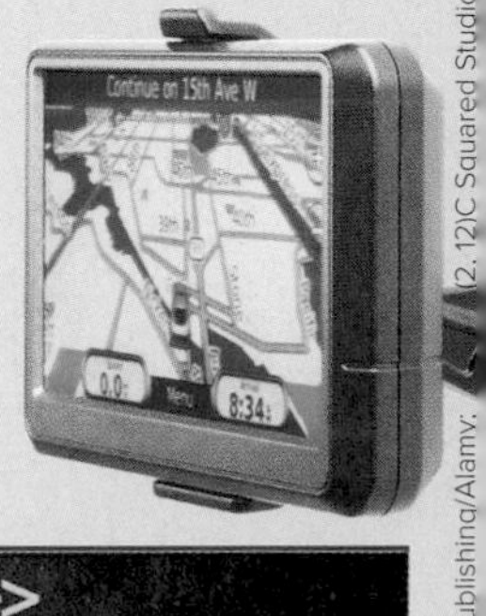

When you travel at 200,000 km/s relative to your friend, only 45 min. pass for you when 60 min. pass for your friend.

FIGURE 1 In this graph, 60 minutes always passes for your friend, but other amounts of time pass for you.

FIGURE 2 This GPS receiver would be completely inaccurate if the designers of the Global Positioning System did not understand the relativity of time.

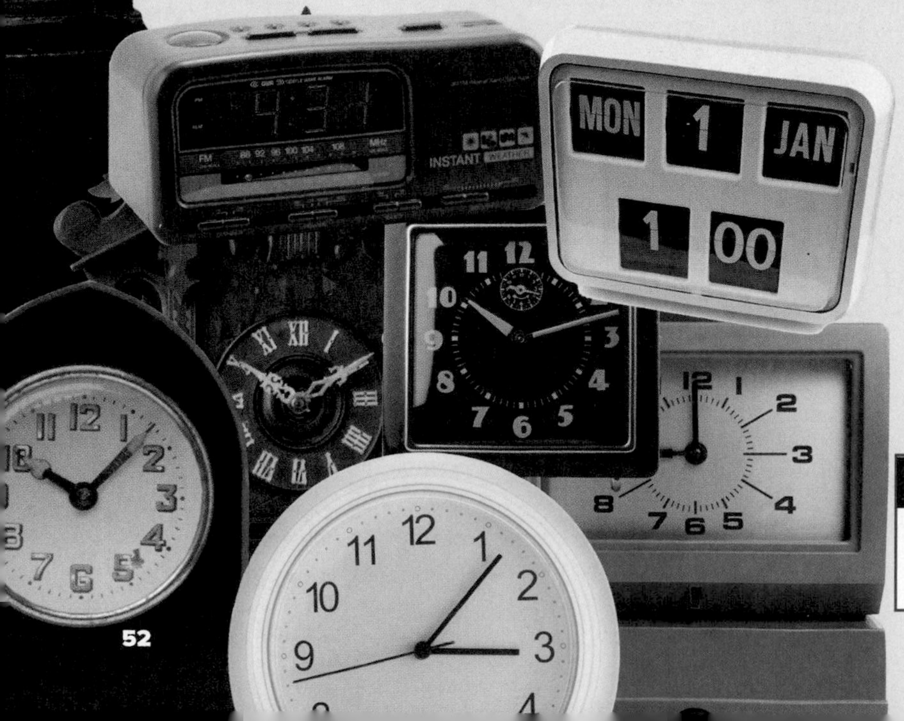

GOING FURTHER >>>

Research Gravity also affects time. Research how gravity affects time on Earth and on a GPS satellite.

STUDY GUIDE

BIGIDEA   You can use displacement and velocity to describe an object's motion.

SECTION 1 **Picturing Motion**

MAINIDEA   You can use motion diagrams to show how an object's position changes over time.

- A motion diagram shows the position of an object at successive equal time intervals.

- In a particle model motion diagram, an object's position at successive times is represented by a series of dots. The spacing between dots indicates whether the object is moving faster or slower.

SECTION 2 **Where and When?**

MAINIDEA   A coordinate system is helpful when you are describing motion.

- A coordinate system gives the location of the zero point of the variable you are studying and the direction in which the values of the variable increase.

- A vector drawn from the origin of a coordinate system to an object indicates the object's position in that coordinate system. The directions chosen as positive and negative on the coordinate system determine whether the objects' positions are positive or negative in the coordinate system.

- A time interval is the difference between two times.

$$\Delta t = t_\mathrm{f} - t_\mathrm{i}$$

- Change in position is displacement, which has both magnitude and direction.

$$\Delta \boldsymbol{x} = \boldsymbol{x}_\mathrm{f} - \boldsymbol{x}_\mathrm{i}$$

- On a motion diagram, the displacement vector's length represents how far the object was displaced. The vector points in the direction of the displacement, from $\boldsymbol{x}_\mathrm{i}$ to $\boldsymbol{x}_\mathrm{f}$.

SECTION 3 **Position-Time Graphs**

MAINIDEA   You can use a position-time graph to determine an object's position at a certain time.

- Position-time graphs provide information about the motion of objects. They also might indicate where and when two objects meet.

- The line on a position-time graph describes an object's position at each time.

- Motion can be described using words, motion diagrams, data tables, or graphs.

SECTION 4 **How Fast?**

MAINIDEA   An object's velocity is the rate of change in its position.

- An object's velocity tells how fast it is moving and in what direction it is moving.

- Speed is the magnitude of velocity.

- Slope on a position-time graph describes the average velocity of the object.

$$\overline{\boldsymbol{v}} \equiv \frac{\Delta \boldsymbol{x}}{\Delta t} = \frac{\boldsymbol{x}_\mathrm{f} - \boldsymbol{x}_\mathrm{i}}{t_\mathrm{f} - t_\mathrm{i}}$$

- You can represent motion with pictures and physical models. A simple equation relates an object's initial position ($\boldsymbol{x}_\mathrm{i}$), its constant average velocity ($\overline{\boldsymbol{v}}$), its position ($\boldsymbol{x}$), and the time ($t$) since the object was at its initial position.

$$\boldsymbol{x} = \overline{\boldsymbol{v}}t + \boldsymbol{x}_\mathrm{i}$$

Games and Multilingual eGlossary

Vocabulary Practice

Chapter Self-Check

SECTION 1 **Picturing Motion**

Mastering Concepts

45. What is the purpose of drawing a motion diagram?

46. Under what circumstances is it legitimate to treat an object as a particle when solving motion problems?

SECTION 2 **Where and When?**

Mastering Concepts

47. The following quantities describe location or its change: position, distance, and displacement. Briefly describe the differences among them.

48. How can you use a clock to find a time interval?

SECTION 3 **Position-Time Graphs**

Mastering Concepts

49. **In-line Skating** How can you use the position-time graphs for two in-line skaters to determine if and when one in-line skater will pass the other one?

SECTION 4 **How Fast?**

Mastering Concepts

50. **BIGIDEA** Which equation describes how the average velocity of a moving object relates to its displacement?

51. **Walking Versus Running** A walker and a runner leave your front door at the same time. They move in the same direction at different constant velocities. Describe the position-time graphs of each.

52. What does the slope of a position-time graph measure?

53. If you know the time it took an object to travel between two points and the positions of the object at the points, can you determine the object's instantaneous velocity? Its average velocity? Explain.

Mastering Problems

54. You ride a bike at a constant speed of 4.0 m/s for 5.0 s. How far do you travel?

55. **Astronomy** Light from the Sun reaches Earth in about 8.3 min. The speed of light is 3.00×10^8 m/s. What is the distance from the Sun to Earth?

56. **Problem Posing** Complete this problem so that someone must solve it using the concept of average speed: "A butterfly travels 15 m from one flower to another ...."

57. Nora jogs several times a week and always keeps track of how much time she runs each time she goes out. One day she forgets to take her stopwatch with her and wonders if there is a way she can still have some idea of her time. As she passes a particular bank building, she remembers that it is 4.3 km from her house. She knows from her previous training that she has a consistent pace of 4.0 m/s. How long has Nora been jogging when she reaches the bank?

58. **Driving** You and a friend each drive 50.0 km. You travel at 90.0 km/h; your friend travels at 95.0 km/h. How much sooner will your friend finish the trip?

Applying Concepts

59. **Ranking Task** The position-time graph in **Figure 27** shows the motion of four cows walking from the pasture back to the barn. Rank the cows according to their average velocity, from slowest to fastest.

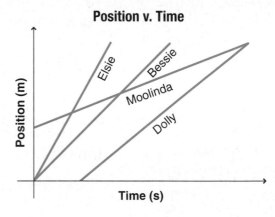

Figure 27

60. **Figure 28** is a position-time graph for a rabbit running away from a dog. How would the graph differ if the rabbit ran twice as fast? How would it differ if the rabbit ran in the opposite direction?

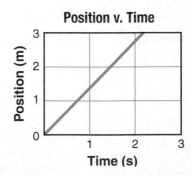

Figure 28

61. Test the following combinations and explain why each does not have the properties needed to describe the concept of velocity: $\Delta x + \Delta t$, $\Delta x - \Delta t$, $\Delta x \times \Delta t$, $\frac{\Delta t}{\Delta x}$.

62. Football When solving physics problems, what must be true about the motion of a football in order for you to treat the football as if it were a particle?

63. Figure 29 is a graph of two people running.

   a. Describe the position of runner A relative to runner B at the *y*-intercept.

   b. Which runner is faster?

   c. What occurs at point P and beyond?

Position v. Time

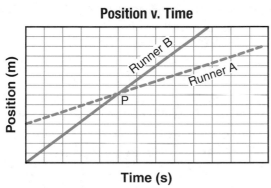

Figure 29

Mixed Review

64. Cycling A cyclist traveling along a straight path maintains a constant velocity of 5.0 m/s west. At time $t = 0.0$ s, the cyclist is 250 m west of point A.

   a. Plot a position-time graph of the cyclist's location from point A at 10.0-s intervals for a total time of 60.0 s.

   b. What is the cyclist's position from point A at 60.0 s?

   c. What is the displacement from the starting position at 60.0 s?

65. Figure 30 is a particle model diagram for a chicken casually walking across a road. Draw the corresponding position-time graph, and write an equation to describe the chicken's motion.

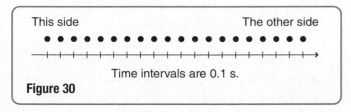

Figure 30

66. Figure 31 shows position-time graphs for Joszi and Heike paddling canoes in a local river.

   a. At what time(s) are Joszi and Heike in the same place?

   b. How long does Joszi paddle before passing Heike?

   c. Where on the river does it appear that there might be a swift current?

Position v. Time

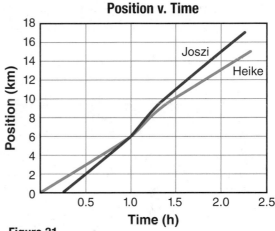

Figure 31

67. Driving Both car A and car B leave school when a stopwatch reads zero. Car A travels at a constant 75 km/h, and car B travels at a constant 85 km/h.

   a. Draw a position-time graph showing the motion of both cars over 3 hours. How far are the two cars from school when the stopwatch reads 2.0 h? Calculate the distances and show them on your graph.

   b. Both cars passed a gas station 120 km from the school. When did each car pass the gas station? Calculate the times and show them on your graph.

68. Draw a position-time graph for two cars traveling to a beach that is 50 km from school. At noon, car A leaves a store that is 10 km closer to the beach than the school is and moves at 40 km/h. Car B starts from school at 12:30 P.M. and moves at 100 km/h. When does each car get to the beach?

69. Two cars travel along a straight road. When a stopwatch reads $t = 0.00$ h, car A is at $x_A = 48.0$ km moving at a constant speed of 36.0 km/h. Later, when the watch reads $t = 0.50$ h, car B is at $x_B = 0.00$ km moving at 48.0 km/h. Answer the following questions, first graphically by creating a position-time graph and then algebraically by writing equations for the positions x_A and x_B as a function of the stopwatch time (t).

   a. What will the watch read when car B passes car A?

   b. At what position will car B pass car A?

   c. When the cars pass, how long will it have been since car A was at the reference point?

70. The graph in **Figure 32** depicts Jim's movement along a straight path. The origin is at one end of the path.

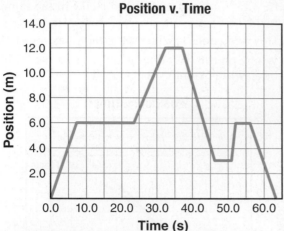

Position v. Time

Figure 32

a. Reverse Problem Write a story describing Jim's movements along the path that would correspond to the motion represented by the graph.

b. When is Jim 6.0 m from the origin?

c. How much time passes between when Jim starts moving and when he is 12.0 m from the origin?

d. What is Jim's average velocity between 37.0 s and 46.0 s?

Thinking Critically

71. Apply Calculators Members of a physics class stood 25 m apart and used stopwatches to measure the time at which a car traveling on the highway passed each person. **Table 2** shows their data.

Table 2 Position v. Time	
Time (s)	**Position (m)**
0.0	0.0
1.3	25.0
2.7	50.0
3.6	75.0
5.1	100.0
5.9	125.0
7.0	150.0
8.6	175.0
10.3	200.0

Use a graphing calculator to fit a line to a position-time graph of the data and to plot this line. Be sure to set the display range of the graph so that all the data fit on it. Find the line's slope. What was the car's speed?

72. Apply Concepts You want to average 90 km/h on a car trip. You cover the first half of the distance at an average speed of 48 km/h. What average speed must you have for the second half of the trip to meet your goal? Is this reasonable? Note that the velocities are based on half the distance, not half the time.

73. Design an Experiment Every time someone drives a particular red motorcycle past your friend's home, his father becomes angry. He thinks the motorcycle is going too fast for the posted 25 mph (40 km/h) speed limit. Describe a simple experiment you could do to determine whether the motorcycle is speeding the next time it passes your friend's house.

74. Interpret Graphs Is it possible for an object's position-time graph to be a horizontal line? A vertical line? If you answer yes to either situation, describe the associated motion in words.

Writing in Physics

75. Physicists have determined that the speed of light is 3.00×10^8 m/s. How did they arrive at this number? Read about some of the experiments scientists have performed to determine light's speed. Describe how the experimental techniques improved to make the experiments' results more accurate.

76. Some species of animals have good endurance, while others have the ability to move very quickly, but only for a short amount of time. Use reference sources to find two examples of each quality, and describe how it is helpful to that animal.

Cumulative Review

77. Convert each of the following time measurements to its equivalent in seconds:

a. 58 ns   **c.** 9270 ms

b. 0.046 Gs   **d.** 12.3 ks

78. State the number of significant figures in the following measurements:

a. 3218 kg   **c.** 801 kg

b. 60.080 kg   **d.** 0.000534 kg

79. Using a calculator, Chris obtained the following results. Rewrite each answer using the correct number of significant figures.

a. 5.32 mm + 2.1 mm = 7.4200000 mm

b. 13.597 m × 3.65 m = 49.62905 m²

c. 83.2 kg − 12.804 kg = 70.3960000 kg

MULTIPLE CHOICE

1. Which statement would be true about the particle model motion diagram for an airplane flying at a constant speed of 850 km/h?

  A. The dots would start close together and get farther apart as the plane moved away from the airport.

  B. The dots would be far apart at the beginning and get closer together as the plane moved away from the airport.

  C. The dots would form an evenly spaced pattern.

  D. The dots would start close together, get farther apart, and then get close together again as the airplane traveled away from the airport.

2. Which statement about drawing vectors is true?

  A. The vector's length should be proportional to its magnitude.

  B. You need a vector diagram to solve all physics problems properly.

  C. A vector is a quantity that has a magnitude but no direction.

  D. All quantities in physics are vectors.

3. The figure below shows a simplified graph of a bicyclist's motion. (Speeding up and slowing down motion is ignored.) When is the person's velocity greatest?

  A. section I   **C.** point D

  B. section III   **D.** point B

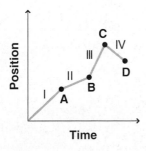

4. What is the average velocity of a train moving along a straight track if its displacement is 192 m east during a time period of 8.0 s?

  A. 12 m/s east   **C.** 48 m/s east

  B. 24 m/s east   **D.** 96 m/s east

5. A squirrel descends an 8-m tree at a constant speed in 1.5 min. It remains still at the base of the tree for 2.3 min. A loud noise then causes the squirrel to scamper back up the tree in 0.1 min to the exact position on the branch from which it started. Ignoring speeding up and slowing down motion, which graph most closely represents the squirrel's vertical displacement from the base of the tree?

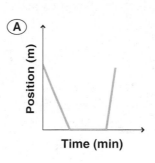
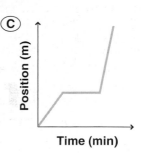
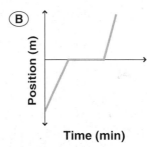
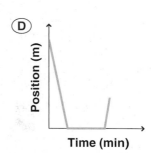

FREE RESPONSE

6. A rat is moving along a straight path. Find the rat's position relative to its starting point if it moves 12.8 cm/s north for 3.10 s.

NEED EXTRA HELP?

If You Missed Question	1	2	3	4	5	6
Review Section	1	2	3	4	4	4

Online Test Practice

CHAPTER 3

Accelerated Motion

BIGIDEA   Acceleration is the rate of change in an object's velocity.

SECTIONS

1 **Acceleration**

2 **Motion with Constant Acceleration**

3 **Free Fall**

LaunchLAB

GRAPHING MOTION

How does a graph showing constant speed compare to a graph of an object that is accelerating?

WATCH THIS!

SKATEBOARD PHYSICS

How does a trip to your local skate park involve physics? You might be surprised! Explore acceleration as skateboarders show off their best moves.

Acceleration

PHYSICS 4 YOU

As an airplane takes off, its speed changes from 5 m/s on the runway to nearly 300 m/s once it's in the air. If you've ever ridden on an airplane, you've felt the seat push against your back as the plane rapidly accelerates.

MAINIDEA

An object accelerates when its velocity changes—that is, when it speeds up, slows down, or changes direction.

Essential Questions

- What is acceleration?
- How is acceleration different from velocity?
- What information can you learn from velocity-time graphs?

Review Vocabulary

vector a quantity that has magnitude and direction

New Vocabulary

acceleration
velocity-time graph
average acceleration
instantaneous acceleration

Nonuniform Motion Diagrams

An object in uniform motion moves along a straight line with an unchanging velocity, but few objects move this way all the time. More common is nonuniform motion, in which velocity is changing. In this chapter, you will study nonuniform motion along a straight line. Examples include balls rolling down hills, cars braking to a stop, and falling objects. In later chapters you will analyze nonuniform motion that is not confined to a straight line, such as motion along a circular path and the motion of thrown objects, such as baseballs.

Describing nonuniform motion You can feel a difference between uniform and nonuniform motion. Uniform motion feels smooth. If you close your eyes, it feels as if you are not moving at all. In contrast, when you move around a curve or up and down a roller coaster hill, you feel pushed or pulled.

How would you describe the motion of the person in **Figure 1**? In the first diagram, the person is motionless, but in the others, her position is changing in different ways. What information do the diagrams contain that could be used to distinguish the different types of motion? Notice the distances between successive positions. Because there is only one image of the person in the first diagram, you can conclude that she is at rest. The distances between images in the second diagram are the same because the jogger is in uniform motion; she moves at a constant velocity. In the remaining two diagrams, the distance between successive positions changes. The change in distance increases if the jogger speeds up. The change decreases if the jogger slows down.

■ Motion Diagram

a. The person is motionless.

b. Equally spaced images show her moving at a constant speed.

c. She is speeding up.

d. She is slowing down.

Figure 1 The distance the jogger moves in each time interval indicates the type of motion.

Particle model diagram What does a particle model motion diagram look like for an object with changing velocity? **Figure 2** shows particle model motion diagrams below the motion diagrams of the jogger when she is speeding up and slowing down. There are two major indicators of the change in velocity in this form of the motion diagram. The change in the spacing of the dots and the differences in the lengths of the velocity vectors indicate the changes in velocity. If an object speeds up, each subsequent velocity vector is longer, and the spacing between dots increases. If the object slows down, each vector is shorter than the previous one, and the spacing between dots decreases. Both types of motion diagrams indicate how an object's velocity is changing.

☑ **READING CHECK Analyze** What do increasing and decreasing lengths of velocity vectors indicate on a motion diagram?

Displaying acceleration on a motion diagram For a motion diagram to give a full picture of an object's movement, it should contain information about the rate at which the object's velocity is changing. The rate at which an object's velocity changes is called the **acceleration** of the object. By including acceleration vectors on a motion diagram, you can indicate the rate of change for the velocity.

Figure 3 shows a particle motion diagram for an object with increasing velocity. Notice that the lengths of the red velocity vectors get longer from left to right along the diagram. The figure also describes how to use the diagram to draw an acceleration vector for the motion. The acceleration vector that describes the increasing velocity is shown in violet on the diagram.

Notice in the figure that if the object's acceleration is constant, you can determine the length and direction of an acceleration vector by subtracting two consecutive velocity vectors and dividing by the time interval. That is, first find the change in velocity, $\Delta v = v_f - v_i = v_f + (-v_i)$, where v_i and v_f refer to the velocities at the beginning and the end of the chosen time interval. Then divide by the time interval (Δt). The time interval between each dot in **Figure 3** is 1 s. You can draw the acceleration vector from the tail of the final velocity vector to the tip of the initial velocity vector.

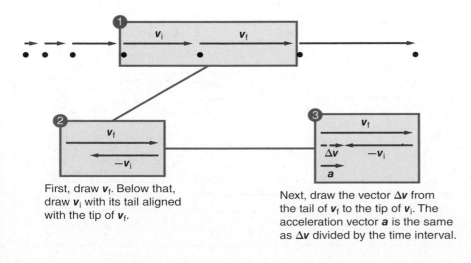

Finding Acceleration Vectors

First, draw v_f. Below that, draw v_i with its tail aligned with the tip of v_f.

Next, draw the vector Δv from the tail of v_f to the tip of v_i. The acceleration vector a is the same as Δv divided by the time interval.

Figure 2 The change in length of the velocity vectors on these motion diagrams indicates whether the jogger is speeding up or slowing down.

Figure 3 For constant acceleration, an acceleration vector on a particle model diagram is the difference in the two velocity vectors divided by the time interval: $a = \dfrac{\Delta v}{\Delta t}$.

Analyze Can you draw an acceleration vector for two successive velocity vectors that are the same length and direction? Explain.

COLOR CONVENTION

acceleration ◄──────► **violet**

velocity ◄──────► **red**

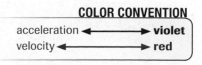

Direction of Acceleration

Consider the four situations shown in **Figure 4** in which an object can accelerate by changing speed. The first motion diagram shows the car moving in the positive direction and speeding up. The second motion diagram shows the car moving in the positive direction and slowing down. The third shows the car speeding up in the negative direction, and the fourth shows the car slowing down as it moves in the negative direction. The figure also shows the velocity vectors for the second time interval of each diagram, along with the corresponding acceleration vectors. Note that Δt is equal to 1 s.

In the first and third situations, when the car is speeding up, the velocity and acceleration vectors point in the same direction. In the other two situations, in which the acceleration vector is in the opposite direction from the velocity vectors, the car is slowing down. In other words, when the car's acceleration is in the same direction as its velocity, the car's speed increases. When they are in opposite directions, the speed of the car decreases.

Both the direction of an object's velocity and its direction of acceleration are needed to determine whether it is speeding up or slowing down. An object has a positive acceleration when the acceleration vector points in the positive direction and a negative acceleration when the acceleration vector points in the negative direction. It is important to notice that the sign of acceleration alone does not indicate whether the object is speeding up or slowing down.

☑ **READING CHECK Describe** the motion of an object if its velocity and acceleration vectors have opposite signs.

Investigate **accelerated motion.**

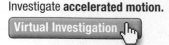
Virtual Investigation

Figure 4 You need to know the direction of both the velocity and acceleration vectors in order to determine whether an object is speeding up or slowing down.

■ **Velocity and Motion Diagram**

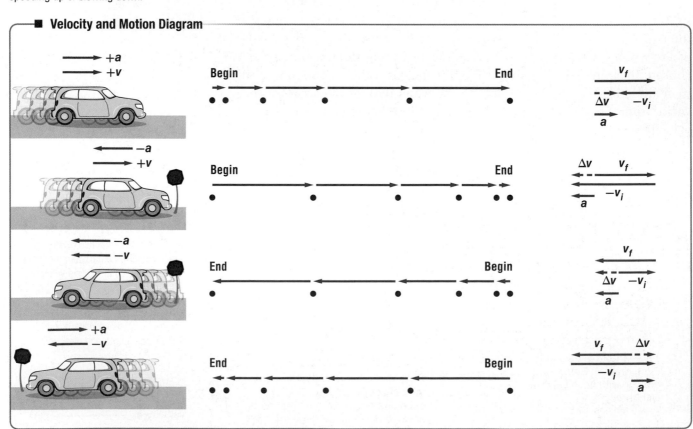

Velocity-Time Graphs

Just as it was useful to graph position versus time, it also is useful to plot velocity versus time. On a **velocity-time graph,** or *v-t* graph, velocity is plotted on the vertical axis and time is plotted on the horizontal axis.

Slope The velocity-time graph for a car that started at rest and sped up along a straight stretch of road is shown in **Figure 5.** The positive direction has been chosen to be the same as that of the car's motion. Notice that the graph is a straight line. This means the car sped up at a constant rate. The rate at which the car's velocity changed can be found by calculating the slope of the velocity-time graph.

The graph shows that the slope is 5.00 (m/s)/s, which is commonly written as 5 m/s². Consider the time interval between 4.00 s and 5.00 s. At 4.00 s, the car's velocity was 20.0 m/s in the positive direction. At 5.00 s, the car was traveling at 25.0 m/s in the same direction. Thus, in 1.00 s, the car's velocity increased by 5.0 m/s in the positive direction. When the velocity of an object changes at a constant rate, it has a constant acceleration.

Reading velocity-time graphs The motions of five runners are shown in **Figure 6.** Assume that the positive direction is east. The slopes of Graphs A and E are zero. Thus, the accelerations are zero. Both graphs show motion at a constant velocity—Graph A to the east and Graph E to the west. Graph B shows motion with a positive velocity eastward. Its slope indicates a constant, positive acceleration. You can infer that the speed increases because velocity and acceleration are positive. Graph C has a negative slope. It shows motion that begins with a positive velocity, slows down, and then stops. This means the acceleration and the velocity are in opposite directions. The point at which Graphs C and B cross shows that the runners' velocities are equal at that time. It does not, however, identify their positions.

Graph D indicates motion that starts out toward the west, slows down, for an instant has zero velocity, and then moves east with increasing speed. The slope of Graph D is positive. Because velocity and acceleration are initially in opposite directions, the speed decreases to zero at the time the graph crosses the *x*-axis. After that time, velocity and acceleration are in the same direction, and the speed increases.

☑ **READING CHECK Describe** the meaning of a line crossing the *x*-axis in a velocity-time graph.

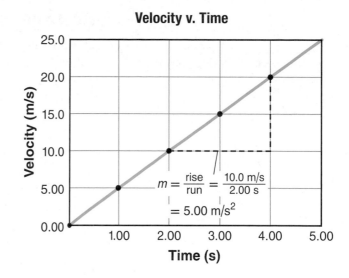

Velocity v. Time

Figure 5 You can determine acceleration from a velocity-time graph by calculating the slope of the data. The slope is the rise divided by the run using any two points on the line.

View an **animation on velocity v. time graphs.**

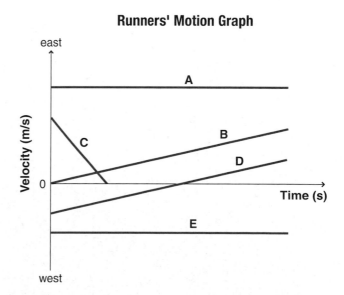

Runners' Motion Graph

Figure 6 Because east is chosen as the positive direction on the graph, velocity is positive if the line is above the horizontal axis and negative if the line is below it. Acceleration is positive if the line is slanted upward on the graph. Acceleration is negative if the line is slanted downward on the graph. A horizontal line indicates constant velocity and zero acceleration.

View a **BrainPOP video on acceleration.**

Video

MiniLAB

STEEL BALL RACE
Does the height of a ramp affect the motion of a ball rolling down it?

PhysicsLABs

ACCELERATION
How can you use motion measurements to calculate the acceleration of a rolling ball?

TOSSED-BALL MOTION
PROBEWARE LAB What does the graph of ball tossed upward look like?

Average and Instantaneous Acceleration

How does it feel differently if the car you ride in accelerates a little or if it accelerates a lot? As with velocity, the acceleration of most moving objects continually changes. If you want to describe an object's acceleration, it is often more convenient to describe the overall change in velocity during a certain time interval rather than describing the continual change.

The **average acceleration** of an object is its change in velocity during some measurable time interval divided by that time interval. Average acceleration is measured in meters per second per second (m/s/s), or simply meters per second squared (m/s²). A car might accelerate quickly at times and more slowly at times. Just as average velocity depends only on the starting and ending displacement, average acceleration depends only on the starting and ending velocity during a time interval. **Figure 7** shows a graph of motion in which the acceleration is changing. The average acceleration during a certain time interval is determined just as it is in **Figure 5** for constant acceleration. Notice, however, that because the line is curved, the average acceleration in this graph varies depending on the time interval that you choose.

The change in an object's velocity at an instant of time is called **instantaneous acceleration.** You can determine the instantaneous acceleration of an object by drawing a tangent line on the velocity-time graph at the point of time in which you are interested. The slope of this line is equal to the instantaneous acceleration. Most of the situations considered in this textbook assume an ideal case of constant acceleration. When the acceleration is the same at all points during a time interval, the average acceleration and the instantaneous accelerations are equal.

✓ **READING CHECK** **Contrast** How is instantaneous acceleration different from average acceleration?

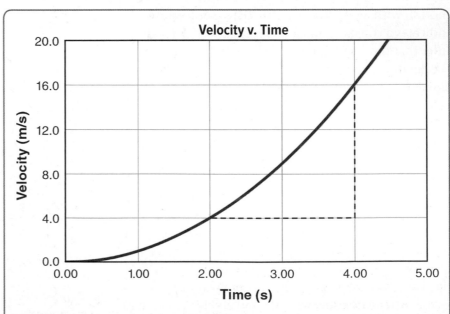

Figure 7 A curved line on a velocity-time graph shows that the acceleration is changing. The slope indicates the average acceleration during a time interval that you choose.

Calculate How large is the average acceleration between 0.00 s and 2.00 s?

Calculating Acceleration

How can you describe the acceleration of an object mathematically? Recall that the acceleration of an object is the slope of that object's velocity v. time graph. On a velocity v. time graph, slope equals $\Delta v/\Delta t$.

AVERAGE ACCELERATION

Average acceleration is defined as the change in velocity divided by the time it takes to make that change.

$$\overline{a} \equiv \frac{\Delta v}{\Delta t} = \frac{v_f - v_i}{t_f - t_i}$$

Suppose you run wind sprints back and forth across the gym. You first run at a speed of 4.0 m/s toward the wall. Then, 10.0 s later, your speed is 4.0 m/s as you run away from the wall. What is your average acceleration if the positive direction is toward the wall?

$$\overline{a} = \frac{\Delta v}{\Delta t} = \frac{v_f - v_i}{t_f - t_i}$$

$$= \frac{-4.0 \text{ m/s} - 4.0 \text{ m/s}}{10.0 \text{ s}} = -0.80 \text{ m/s}^2$$

EXAMPLE PROBLEM 1

Find help with **slope**. Math Handbook

VELOCITY AND ACCELERATION How would you describe the sprinter's velocity and acceleration as shown on the graph?

1 ANALYZE AND SKETCH THE PROBLEM

From the graph, note that the magnitude of the sprinter's velocity starts at zero, increases rapidly for the first few seconds, and then, after reaching about 10.0 m/s, remains almost constant.

KNOWN	UNKNOWN
v = varies	a = ?

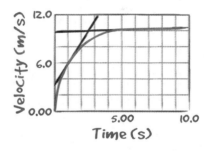

2 SOLVE FOR THE UNKNOWN

Draw tangents to the curve at two points. Choose t = 1.00 s and t = 5.00 s.
Solve for magnitude of the acceleration at 1.00 s:

$a = \dfrac{\text{rise}}{\text{run}}$ ◄ *The slope of the line at 1.00 s is equal to the acceleration at that time.*

$= \dfrac{10.0 \text{ m/s} - 6.0 \text{ m/s}}{2.4 \text{ s} - 1.00 \text{ s}}$

$= 2.9 \text{ m/s/s} = 2.9 \text{ m/s}^2$

Solve for the magnitude of the instantaneous acceleration at 5.0 s:

$a = \dfrac{\text{rise}}{\text{run}}$ ◄ *The slope of the line at 5.0 s is equal to the acceleration at that time.*

$= \dfrac{10.3 \text{ m/s} - 10.0 \text{ m/s}}{10.0 \text{ s} - 0.00 \text{ s}}$

$= 0.030 \text{ m/s/s} = 0.030 \text{ m/s}^2$

The acceleration is not constant because its magnitude changes from 2.9 m/s² at 1.0 s to 0.030 m/s² at 5.0 s.

The acceleration is in the direction chosen to be positive because both values are positive.

3 EVALUATE THE ANSWER

Are the units correct? Acceleration is measured in m/s².

PRACTICE PROBLEMS

1. The velocity-time graph in **Figure 8** describes Steven's motion as he walks along the midway at the state fair. Sketch the corresponding motion diagram. Include velocity vectors in your diagram.

2. Use the *v-t* graph of the toy train in **Figure 9** to answer these questions.

   a. When is the train's speed constant?

   b. During which time interval is the train's acceleration positive?

   c. When is the train's acceleration most negative?

3. Refer to **Figure 9** to find the average acceleration of the train during the following time intervals.

   a. 0.0 s to 5.0 s   **b.** 15.0 s to 20.0 s   **c.** 0.0 s to 40.0 s

4. **CHALLENGE** Plot a *v-t* graph representing the following motion: An elevator starts at rest from the ground floor of a three-story shopping mall. It accelerates upward for 2.0 s at a rate of 0.5 m/s², continues up at a constant velocity of 1.0 m/s for 12.0 s, and then slows down with a constant downward acceleration of 0.25 m/s² for 4.0 s as it reaches the third floor.

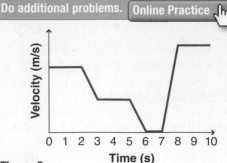

Figure 8

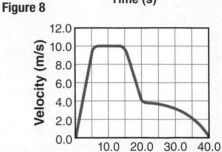

Figure 9

EXAMPLE PROBLEM 2

ACCELERATION Describe a ball's motion as it rolls up a slanted driveway. It starts at 2.50 m/s, slows down for 5.00 s, stops for an instant, and then rolls back down. The positive direction is chosen to be up the driveway. The origin is where the motion begins. What are the sign and the magnitude of the ball's acceleration as it rolls up the driveway?

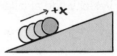

1 ANALYZE AND SKETCH THE PROBLEM

- Sketch the situation.

- Draw the coordinate system based on the motion diagram.

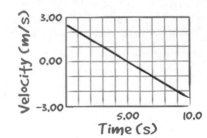

KNOWN

$v_i = +2.50$ m/s

$v_f = 0.00$ m/s at $t = 5.00$ s

UNKNOWN

$a = ?$

2 SOLVE FOR THE UNKNOWN

Find the acceleration from the slope of the graph.

Solve for the change in velocity and the time taken to make that change.

$$\Delta v = v_f - v_i$$

$$= 0.00 \text{ m/s} - 2.50 \text{ m/s} = -2.50 \text{ m/s}$$ ◀ Substitute $v_f = 0.00$ m/s at $t_f = 5.00$ s, $v_i = 2.50$ m/s at $t_i = 0.00$ s.

$$\Delta t = t_f - t_i$$

$$= 5.00 \text{ s} - 0.00 \text{ s} = 5.00 \text{ s}$$ ◀ Substitute $t_f = 5.00$ s, $t_i = 0.00$ s.

Solve for the acceleration.

$$\overline{a} \equiv \frac{\Delta v}{\Delta t} = (-2.50 \text{ m/s}) / 5.00 \text{ s}$$ ◀ Substitute $\Delta v = -2.50$ m/s, $\Delta t = 5.00$ s.

$$= -0.500 \text{ m/s}^2 \text{ or } 0.500 \text{ m/s}^2 \text{ down the driveway}$$

3 EVALUATE THE ANSWER

- **Are the units correct?** Acceleration is measured in m/s².

- **Do the directions make sense?** As the ball slows down, the direction of acceleration is opposite that of velocity.

5. A race car's forward velocity increases from 4.0 m/s to 36 m/s over a 4.0-s time interval. What is its average acceleration?

6. The race car in the previous problem slows from 36 m/s to 15 m/s over 3.0 s. What is its average acceleration?

7. A bus is moving west at 25 m/s when the driver steps on the brakes and brings the bus to a stop in 3.0 s.

   a. What is the average acceleration of the bus while braking?

   b. If the bus took twice as long to stop, how would the acceleration compare with what you found in part **a?**

8. A car is coasting backward downhill at a speed of 3.0 m/s when the driver gets the engine started. After 2.5 s, the car is moving uphill at 4.5 m/s. If uphill is chosen as the positive direction, what is the car's average acceleration?

9. Rohith has been jogging east toward the bus stop at 3.5 m/s when he looks at his watch and sees that he has plenty of time before the bus arrives. Over the next 10.0 s, he slows his pace to a leisurely 0.75 m/s. What was his average acceleration during this 10.0 s?

10. **CHALLENGE** If the rate of continental drift were to abruptly slow from 1.0 cm/y to 0.5 cm/y over the time interval of a year, what would be the average acceleration?

Acceleration with Constant Speed

Think again about running wind sprints across the gym. Notice that your speed is the same as you move toward the wall of the gym and as you move away from it. In both cases, you are running at a speed of 4.0 m/s. How is it possible for you to be accelerating?

Acceleration can occur even when speed is constant. The average acceleration for the entire trip you make toward the wall of the gym and back again is -0.80 m/s^2. The negative sign indicates that the direction of your acceleration is away from the wall because the positive direction was chosen as toward the wall. The velocity changes from positive to negative when the direction of motion changes. A change in velocity results in acceleration. Thus, acceleration can also be associated with a change in the direction of motion.

☑ **READING CHECK** **Explain** how it is possible for an object to accelerate when the object is traveling at a constant speed.

SECTION 1 REVIEW

[Section Self-Check 🖑] Check your understanding.

11. **MAIN**IDEA What are three ways an object can accelerate?

12. **Position-Time and Velocity-Time Graphs** Two joggers run at a constant velocity of 7.5 m/s east. **Figure 10** shows the positions of both joggers at time $t = 0$.

   a. What would be the difference(s) in the position-time graphs of their motion?

   b. What would be the difference(s) in their velocity-time graphs?

13. **Velocity-Time Graph** Sketch a velocity-time graph for a car that goes east at 25 m/s for 100 s, then west at 25 m/s for another 100 s.

14. **Average Velocity and Average Acceleration** A canoeist paddles upstream at a velocity of 2.0 m/s for 4.0 s and then floats downstream at 4.0 m/s for 4.0 s.

   a. What is the average velocity of the canoe during the 8.0-s time interval?

   b. What is the average acceleration of the canoe during the 8.0-s time interval?

15. **Critical Thinking** A police officer clocked a driver going 32 km/h over the speed limit just as the driver passed a slower car. When the officer stopped the car, the driver argued that the other driver should get a ticket as well. The driver said that the cars must have been going the same speed because they were observed next to each other. Is the driver correct? Explain with a sketch and a motion diagram.

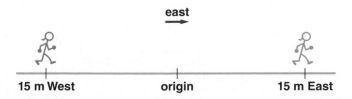

east
→

15 m West   origin   15 m East

Figure 10

Motion with Constant Acceleration

Suppose a car is moving along a road and suddenly the driver sees a fallen tree blocking the way ahead. Will the driver be able to stop in time? It all depends on how effectively the car's brakes can cause the car to accelerate in the direction opposite its motion.

Position with Constant Acceleration

If an object experiences constant acceleration, its velocity changes at a constant rate. How does its position change? The positions at different times of a car with constant acceleration are graphed in **Figure 11.** The graph shows that the car's motion is not uniform. The displacements for equal time intervals on the graph get larger and larger. As a result, the slope of the line in **Figure 11** gets steeper as time goes on. For an object with constant acceleration, the position-time graph is a parabola.

The slopes from the position-time graph in **Figure 11** have been used to create the velocity-time graph on the left in **Figure 12.** For an object with constant acceleration, the velocity-time graph is a straight line.

A unique position-time graph cannot be created using a velocity-time graph because it does not contain information about position. It does, however, contain information about displacement. Recall that for an object moving at a constant velocity, the velocity is the displacement divided by the time interval. The displacement is then the product of the velocity and the time interval. On the right graph in **Figure 12** on the next page, v is the height of the plotted line above the horizontal axis, and Δt is the width of the shaded triangle. The area is $\left(\frac{1}{2}\right)v\Delta t$, or Δx. Thus, the area under the v-t graph equals the displacement.

☑ **READING CHECK Identify** What is the shape of a position–time graph of an object traveling with constant acceleration?

MAINIDEA

For an object with constant acceleration, the relationships among position, velocity, acceleration, and time can be described by graphs and equations.

Essential Questions

• What do a position-time graph and a velocity-time graph look like for motion with constant acceleration?

• How can you determine the displacement of a moving object from its velocity-time graph?

• What are the relationships among position, velocity, acceleration, and time?

Review Vocabulary

displacement change in position having both magnitude and direction; it is equal to the final position minus the initial position

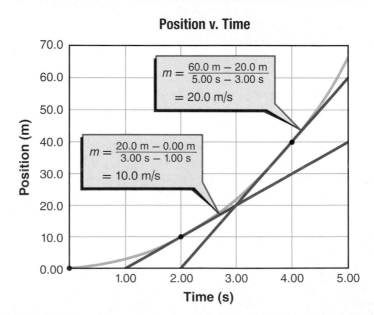

Position v. Time

$$m = \frac{60.0\ m - 20.0\ m}{5.00\ s - 3.00\ s}$$
$$= 20.0\ m/s$$

$$m = \frac{20.0\ m - 0.00\ m}{3.00\ s - 1.00\ s}$$
$$= 10.0\ m/s$$

Figure 11 The slope of a position-time graph changes with time for an object with constant acceleration.

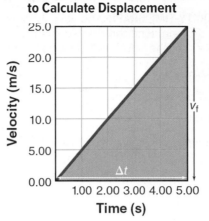

Slope Used to Calculate Acceleration

$$m = \frac{20.0\ \text{m/s} - 15.0\ \text{m/s}}{4.00\ \text{s} - 3.00\ \text{s}}$$

$$= 5.00\ \text{m/s}^2$$

Area Under the Graph Used to Calculate Displacement

Figure 12 The slopes of the position-time graph in **Figure 11** are shown in these velocity-time graphs. The rise divided by the run gives the acceleration on the left. The area under the curve gives the displacement on the right.

Calculate What is the slope of the velocity-time graph on the left between $t = 2.00$ s and $t = 5.00$ s?

Velocity with Average Acceleration

You have read that the equation for average velocity can be algebraically rearranged to show the new position after a period of time, given the initial position and the average velocity. The definition of average acceleration can be manipulated similarly to show the new velocity after a period of time, given the initial velocity and the average acceleration.

If you know an object's average acceleration during a time interval, you can use it to determine how much the velocity changed during that time. You can rewrite the definition of average acceleration $\left(\bar{a} \equiv \frac{\Delta v}{\Delta t}\right)$ as follows:

$$\Delta v = \bar{a}\Delta t$$
$$v_f - v_i = \bar{a}\Delta t$$

The equation for final velocity with average acceleration can be written:

FINAL VELOCITY WITH AVERAGE ACCELERATION
The final velocity is equal to the initial velocity plus the product of the average acceleration and the time interval.

$$v_f = v_i + \bar{a}\Delta t$$

In cases when the acceleration is constant, the average acceleration ($\bar{a}$) is the same as the instantaneous acceleration (a). This equation can be rearranged to find the time at which an object with constant acceleration has a given velocity. You can also use it to calculate the initial velocity of an object when both a velocity and the time at which it occurred are given.

PhysicsLAB

MEASURING ACCELERATION
PROBEWARE LAB What does a graph show about the motion of a cart?

PRACTICE PROBLEMS — Do additional problems. | Online Practice

16. A golf ball rolls up a hill toward a miniature-golf hole. Assume the direction toward the hole is positive.

   a. If the golf ball starts with a speed of 2.0 m/s and slows at a constant rate of 0.50 m/s², what is its velocity after 2.0 s?

   b. What is the golf ball's velocity if the constant acceleration continues for 6.0 s?

   c. Describe the motion of the golf ball in words and with a motion diagram.

17. A bus traveling 30.0 km/h east has a constant increase in speed of 1.5 m/s². What is its velocity 6.8 s later?

18. If a car accelerates from rest at a constant rate of 5.5 m/s² north, how long will it take for the car to reach a velocity of 28 m/s north?

19. **CHALLENGE** A car slows from 22 m/s to 3.0 m/s at a constant rate of 2.1 m/s². How many seconds are required before the car is traveling at a forward velocity of 3.0 m/s?

FINDING DISPLACEMENT FROM A VELOCITY-TIME GRAPH The velocity-time graph at the right shows the motion of an airplane. Find the displacement of the airplane for $\Delta t = 1.0$ s and for $\Delta t = 2.0$ s. Let the positive direction be forward.

1 ANALYZE AND SKETCH THE PROBLEM

- The displacement is the area under the *v-t* graph.
- The time intervals begin at $t = 0.0$ s.

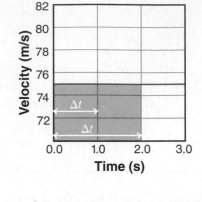

KNOWN	UNKNOWN
$v = +75$ m/s	$\Delta x = ?$
$\Delta t = 1.0$ s	
$\Delta t = 2.0$ s	

2 SOLVE FOR THE UNKNOWN

Use the relationship among displacement, velocity, and time interval to find Δx during $\Delta t = 1.0$ s.

$\Delta x = v\Delta t$

$\quad = (+75 \text{ m/s})(1.0 \text{ s})$ ◀ Substitute $v = +75$ m/s, $\Delta t = 1.0$ s.

$\quad = +75$ m

Use the same relationship to find Δx during $\Delta t = 2.0$ s.

$\Delta x = v\Delta t$

$\quad = (+75 \text{ m/s})(2.0 \text{ s})$ ◀ Substitute $v = +75$ m/s, $\Delta t = 2.0$ s.

$\quad = +150$ m

3 EVALUATE THE ANSWER

- **Are the units correct?** Displacement is measured in meters.
- **Do the signs make sense?** The positive sign agrees with the graph.
- **Is the magnitude realistic?** Moving a distance of about one football field per second is reasonable for an airplane.

PRACTICE PROBLEMS

Do additional problems. Online Practice

20. The graph in **Figure 13** describes the motion of two bicyclists, Akiko and Brian, who start from rest and travel north, increasing their speed with a constant acceleration. What was the total displacement of each bicyclist during the time shown for each?

    Hint: Use the area of a triangle: area = $\left(\frac{1}{2}\right)$(base)(height).

21. The motion of two people, Carlos and Diana, moving south along a straight path is described by the graph in **Figure 14.** What is the total displacement of each person during the 4.0-s interval shown on the graph?

22. **CHALLENGE** A car, just pulling onto a straight stretch of highway, has a constant acceleration from 0 m/s to 25 m/s west in 12 s.

    a. Draw a *v-t* graph of the car's motion.

    b. Use the graph to determine the car's displacement during the 12.0-s time interval.

    c. Another car is traveling along the same stretch of highway. It travels the same distance in the same time as the first car, but its velocity is constant. Draw a *v-t* graph for this car's motion.

    d. Explain how you knew this car's velocity.

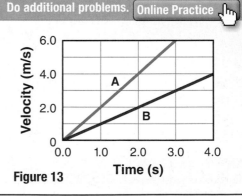

Figure 13

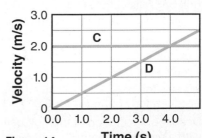

Figure 14

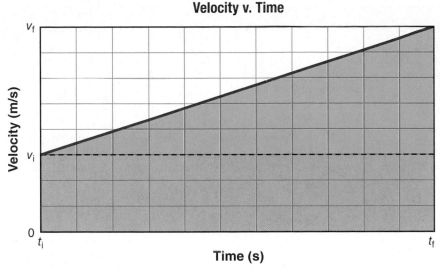

Velocity v. Time

Velocity (m/s)

v_f

v_i

0

t_i t_f

Time (s)

Figure 15 For motion with constant acceleration, if the initial velocity on a velocity-time graph is not zero, the area under the graph is the sum of a rectangular area and a triangular area.

Motion with an initial nonzero velocity The graph in **Figure 15** describes constant acceleration that started with an initial velocity of v_i. To determine the displacement, you can divide the area under the graph into a rectangle and a triangle. The total area is then:

$$\Delta x = \Delta x_{rectangle} + \Delta x_{triangle} = v_i(\Delta t) + \left(\frac{1}{2}\right)\Delta v \Delta t$$

Substituting $a\Delta t$ for the change in velocity in the equation yields:

$$\Delta x = \Delta x_{rectangle} + \Delta x_{triangle} = v_i(\Delta t) + \left(\frac{1}{2}\right)a(\Delta t)^2$$

When the initial or final position of the object is known, the equation can be written as follows:

$$x_f - x_i = v_i(\Delta t) + \left(\frac{1}{2}\right)a(\Delta t)^2 \text{ or } x_f = x_i + v_i(\Delta t) + \left(\frac{1}{2}\right)a(\Delta t)^2$$

If the initial time is $t_i = 0$, the equation then becomes the following.

POSITION WITH AVERAGE ACCELERATION
An object's final position is equal to the sum of its initial position, the product of the initial velocity and the final time, and half the product of the acceleration and the square of the final time.

$$x_f = x_i + v_i t_f + \left(\frac{1}{2}\right)at_f^2$$

REAL-WORLD PHYSICS
••••••••••••

DRAG RACING A dragster driver tries to obtain maximum acceleration over a course. The fastest U.S. National Hot Rod Association time on record for the 402-m course is 3.771 s. The highest final speed on record is 145.3 m/s (324.98 mph).

An Alternative Equation

Often, it is useful to relate position, velocity, and constant acceleration without including time. Rearrange the equation

$v_f = v_i + at_f$ to solve for time: $t_f = \dfrac{v_f - v_i}{a}$.

You can then rewrite the position with average acceleration equation by substituting t_f to obtain the following:

$$x_f = x_i + v_i\left(\frac{v_f - v_i}{a}\right) + \left(\frac{1}{2}\right)a\left(\frac{v_f - v_i}{a}\right)^2$$

This equation can be solved for the velocity (v_f) at any position (x_f).

VELOCITY WITH CONSTANT ACCELERATION
The square of the final velocity equals the sum of the square of the initial velocity and twice the product of the acceleration and the displacement since the initial time.

$$v_f^2 = v_i^2 + 2a(x_f - x_i)$$

EXAMPLE PROBLEM

DISPLACEMENT An automobile starts at rest and accelerates at 3.5 m/s² after a traffic light turns green. How far will it have gone when it is traveling at 25 m/s?

1 ANALYZE AND SKETCH THE PROBLEM

- Sketch the situation.
- Establish coordinate axes. Let the positive direction be to the right.
- Draw a motion diagram.

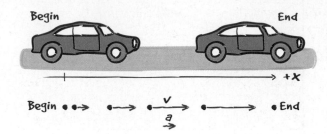

KNOWN	UNKNOWN
$x_i = 0.00$ m	$x_f = ?$
$v_i = 0.00$ m/s	
$v_f = +25$ m/s	
$\bar{a} = a = +3.5$ m/s²	

2 SOLVE FOR THE UNKNOWN

Use the relationship among velocity, acceleration, and displacement to find x_f.

$$v_f^2 = v_i^2 + 2a(x_f - x_i)$$

$$x_f = x_i + \frac{v_f^2 - v_i^2}{2a}$$

◀ **Substitute** $x_i = 0.00$ m, $v_f = +25$ m/s, $v_i = 0.00$ m/s, $a = +3.5$ m/s².

$$= 0.00 \text{ m} + \frac{(+25 \text{ m/s})^2 - (0.00 \text{ m/s})^2}{2(+3.5 \text{ m/s}^2)}$$

$$= +89 \text{ m}$$

3 EVALUATE THE ANSWER

- **Are the units correct?** Position is measured in meters.
- **Does the sign make sense?** The positive sign agrees with both the pictorial and physical models.
- **Is the magnitude realistic?** The displacement is almost the length of a football field. The result is reasonable because 25 m/s (about 55 mph) is fast.

PRACTICE PROBLEMS

Do additional problems. | **Online Practice**

PRACTICE PROBLEMS

23. A skateboarder is moving at a constant speed of 1.75 m/s when she starts up an incline that causes her to slow down with a constant acceleration of −0.20 m/s². How much time passes from when she begins to slow down until she begins to move back down the incline?

24. A race car travels on a straight racetrack with a forward velocity of 44 m/s and slows at a constant rate to a velocity of 22 m/s over 11 s. How far does it move during this time?

25. A car accelerates at a constant rate from 15 m/s to 25 m/s while it travels a distance of 125 m. How long does it take to achieve the final speed?

26. A bike rider pedals with constant acceleration to reach a velocity of 7.5 m/s north over a time of 4.5 s. During the period of acceleration, the bike's displacement is 19 m north. What was the initial velocity of the bike?

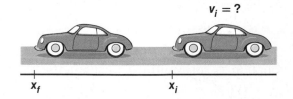

Figure 16

27. CHALLENGE The car in **Figure 16** travels west with a forward acceleration of 0.22 m/s². What was the car's velocity ($\mathbf{v}_i$) at point $\mathbf{x}_i$ if it travels a distance of 350 m in 18.4 s?

TWO-PART MOTION You are driving a car, traveling at a constant velocity of 25 m/s along a straight road, when you see a child suddenly run onto the road. It takes 0.45 s for you to react and apply the brakes. As a result, the car slows with a steady acceleration of 8.5 m/s^2 in the direction opposite your motion and comes to a stop. What is the total displacement of the car before it stops?

1 ANALYZE AND SKETCH THE PROBLEM

- Sketch the situation.
- Choose a coordinate system with the motion of the car in the positive direction.
- Draw the motion diagram, and label **v** and **a**.

KNOWN

$v_{reacting} = +25$ m/s
$t_{reacting} - 0.45$ s
$a_{braking} = -8.5$ m/s^2
$v_{i, braking} = +25$ m/s
$v_{f, braking} = 0.00$ m/s

UNKNOWN

$x_{reacting} = ?$
$x_{braking} = ?$
$x_{total} = ?$

2 SOLVE FOR THE UNKNOWN

Reacting:

Use the relationship among displacement, velocity, and time interval to find the displacement of the car as it travels at a constant speed.

$x_{reacting} = v_{reacting} t_{reacting}$
$x_{reacting} = (+25$ m/s$)(0.45$ s$)$ ◀ Substitute $v_{reacting} = +25$ m/s, $t_{reacting} = 0.45$ s.
$\quad\quad = +11$ m

Braking:

Use the relationship among velocity, acceleration, and displacement to find the displacement of the car while it is braking.

$v_{f, braking}^2 = v_{reacting}^2 + 2a_{braking}(x_{braking})$
Solve for $x_{braking}$.

$x_{braking} = \dfrac{v_{f, braking}^2 - v_{reacting}^2}{2a_{braking}}$

$\quad\quad = \dfrac{(0.00 \text{ m/s})^2 - (+25 \text{ m/s})^2}{2(-8.5 \text{ m/s}^2)}$ ◀ Substitute $v_{f, braking} = 0.00$ m/s, $v_{reacting} = +25$ m/s, $a_{braking} = -8.5$ m/s^2.

$\quad\quad = +37$ m

The total displacement is the sum of the reaction displacement and the braking displacement.
Solve for x_{total}.

$x_{total} = x_{reacting} + x_{braking}$
$\quad\quad = +11$ m $+ 37$ m ◀ Substitute $x_{reacting} = +11$ m, $x_{braking} = +37$ m.
$\quad\quad = +48$ m

3 EVALUATE THE ANSWER

- **Are the units correct?** Displacement is measured in meters.
- **Do the signs make sense?** Both $x_{reacting}$ and $x_{braking}$ are positive, as they should be.
- **Is the magnitude realistic?** The braking displacement is small because the magnitude of the acceleration is large.

PRACTICE PROBLEMS

28. A car with an initial velocity of 24.5 m/s east has an acceleration of 4.2 m/s² west. What is its displacement at the moment that its velocity is 18.3 m/s east?

29. A man runs along the path shown in **Figure 17**. From point A to point B, he runs at a forward velocity of 4.5 m/s for 15.0 min. From point B to point C, he runs up a hill. He slows down at a constant rate of 0.050 m/s² for 90.0 s and comes to a stop at point C. What was the total distance the man ran?

30. You start your bicycle ride at the top of a hill. You coast down the hill at a constant acceleration of 2.00 m/s². When you get to the bottom of the hill, you are moving at 18.0 m/s, and you pedal to maintain that speed. If you continue at this speed for 1.00 min, how far will you have gone from the time you left the hilltop?

31. Sunee is training for a 5.0-km race. She starts out her training run by moving at a constant pace of 4.3 m/s for 19 min. Then she accelerates at a constant rate until she crosses the finish line 19.4 s later. What is her acceleration during the last portion of the training run?

32. **CHALLENGE** Sekazi is learning to ride a bike without training wheels. His father pushes him with a constant acceleration of 0.50 m/s² east for 6.0 s. Sekazi then travels at 3.0 m/s east for another 6.0 s before falling. What is Sekazi's displacement? Solve this problem by constructing a velocity-time graph for Sekazi's motion and computing the area underneath the graphed line.

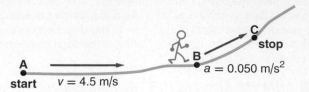

Figure 17

SECTION 2 REVIEW

33. **MAIN**IDEA If you were given initial and final velocities and the constant acceleration of an object, and you were asked to find the displacement, what mathematical relationship would you use?

34. **Acceleration** A woman driving west along a straight road at a speed of 23 m/s sees a deer on the road ahead. She applies the brakes when she is 210 m from the deer. If the deer does not move and the car stops right before it hits the deer, what is the acceleration provided by the car's brakes?

35. **Distance** The airplane in **Figure 18** starts from rest and accelerates east at a constant 3.00 m/s² for 30.0 s before leaving the ground.

    a. What was the plane's displacement (Δx)?

    b. How fast was the airplane going when it took off?

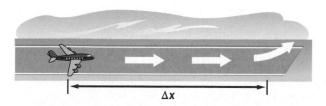

Figure 18

36. **Distance** An in-line skater first accelerates from 0.0 m/s to 5.0 m/s in 4.5 s, then continues at this constant speed for another 4.5 s. What is the total distance traveled by the in-line skater?

37. **Final Velocity** A plane travels a distance of 5.0×10^2 m north while being accelerated uniformly from rest at the rate of 5.0 m/s². What final velocity does it attain?

38. **Final Velocity** An airplane accelerated uniformly from rest at the rate of 5.0 m/s² south for 14 s. What final velocity did it attain?

39. **Graphs** A sprinter walks up to the starting blocks at a constant speed and positions herself for the start of the race. She waits until she hears the starting pistol go off and then accelerates rapidly until she attains a constant velocity. She maintains this velocity until she crosses the finish line, and then she slows to a walk, taking more time to slow down than she did to speed up at the beginning of the race. Sketch a velocity-time and a position-time graph to represent her motion. Draw them one above the other using the same time scale. Indicate on your position-time graph where the starting blocks and finish line are.

40. **Critical Thinking** Describe how you could calculate the acceleration of an automobile. Specify the measuring instruments and the procedures you would use.

SECTION 3 | Free Fall

PHYSICS 4 YOU

Before their parachutes open, skydivers sometimes join hands to form a ring as they fall toward Earth. What happens if the skydivers have different masses? Do they fall at the same rate or different rates?

MAINIDEA

The acceleration of an object in free fall is due to gravity alone.

Essential Questions

• What is free-fall acceleration?

• How do objects in free fall move?

Review Vocabulary

origin the point at which both variables in a coordinate system have the value zero

New Vocabulary

free fall
free-fall acceleration

Galileo's Discovery

Which falls with more acceleration, a piece of paper or your physics book? If you hold one in each hand and release them, the book hits the ground first. Do heavier objects accelerate more as they fall? Try dropping them again, but first place the paper flat on the book. Without air pushing against it, the paper falls as fast as the book. For a lightweight object such as paper, collisions with particles of air have a greater effect than they do on a heavy book.

To understand falling objects, first consider the case in which air does not have an appreciable effect on motion. Recall that gravity is an attraction between objects. **Free fall** is the motion of an object when gravity is the only significant force acting on it.

About 400 years ago, Galileo Galilei discovered that, neglecting the effect of the air, all objects in free fall have the same acceleration. It doesn't matter what they are made of or how much they weigh. The acceleration of an object due only to the effect of gravity is known as **free-fall acceleration. Figure 19** depicts the results of a 1971 free-fall experiment on the Moon in which astronauts verified Galileo's results.

Near Earth's surface, free-fall acceleration is about 9.8 m/s² downward (which is equal to about 22 mph/s downward). Think about the skydivers above. Each second the skydivers fall, their downward velocity increases by 9.8 m/s. When analyzing free fall, whether you treat the acceleration as positive or negative depends on the coordinate system you use. If you define upward as the positive direction, then the free-fall acceleration is negative. If you decide that downward is the positive direction, then free-fall acceleration is positive.

Figure 19 In 1971 astronaut David Scott dropped a hammer and a feather at the same time from the same height above the Moon's surface. The hammer's mass was greater, but both objects hit the ground at the same time because the Moon has gravity but no air.

Free-Fall Acceleration

Galileo's discovery explains why parachutists can form a ring in midair. Regardless of their masses, they fall with the same acceleration. To understand the acceleration that occurs during free fall, look at the multiflash photo of a dropped ball in **Figure 20.** The time interval between the images is 0.06 s. The distance between each pair of images increases, so the speed is increasing. If the upward direction is positive, then the velocity is becoming more and more negative.

Ball thrown upward Instead of a dropped ball, could this photo also illustrate a ball thrown upward? Suppose you throw a ball upward with a speed of 20.0 m/s. If you choose upward to be positive, then the ball starts at the bottom of the photo with a positive velocity. The acceleration is $a = -9.8$ m/s². Because velocity and acceleration are in opposite directions, the speed of the ball decreases. If you think of the bottom of the photo as the start, this agrees with the multiflash photo.

Rising and falling motion After 1 s, the ball's velocity is reduced by 9.8 m/s, so it now is traveling at +10.2 m/s. After 2 s, the velocity is +0.4 m/s, and the ball still is moving upward. What happens during the next second? The ball's velocity is reduced by another 9.8 m/s and equals −9.4 m/s. The ball now is moving downward. After 4 s, the velocity is −19.2 m/s, meaning the ball is falling even faster.

Velocity-time graph The v-t graph for the ball as it goes up and down is shown in **Figure 21.** The straight line sloping downward does not mean that the speed is always decreasing. The speed decreases as the ball rises and increases as it falls. At around 2 s, the velocity changes smoothly from positive to negative. As the ball falls, its speed increases in the negative direction. The figure also shows a closer view of the v-t graph. At an instant of time, near 2.04 s, the velocity is zero.

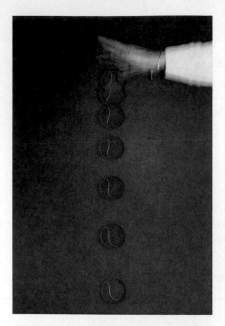

Figure 20 Because of free-fall acceleration, the speed of this falling ball increases 9.8 m/s each second.

MiniLAB

FREE FALL

How can you use the motion of a falling object to estimate free-fall acceleration?

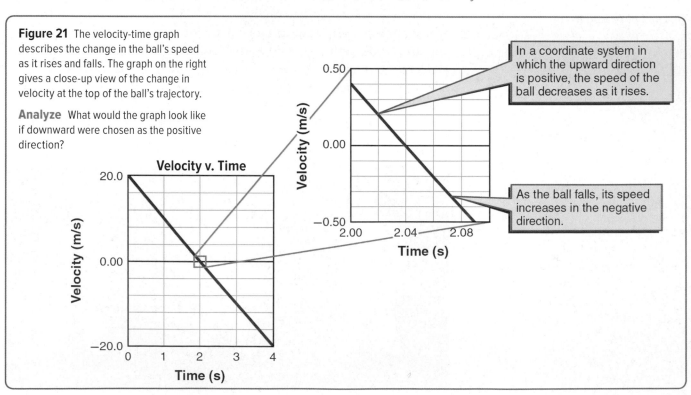

Figure 21 The velocity-time graph describes the change in the ball's speed as it rises and falls. The graph on the right gives a close-up view of the change in velocity at the top of the ball's trajectory.

Analyze What would the graph look like if downward were chosen as the positive direction?

In a coordinate system in which the upward direction is positive, the speed of the ball decreases as it rises.

As the ball falls, its speed increases in the negative direction.

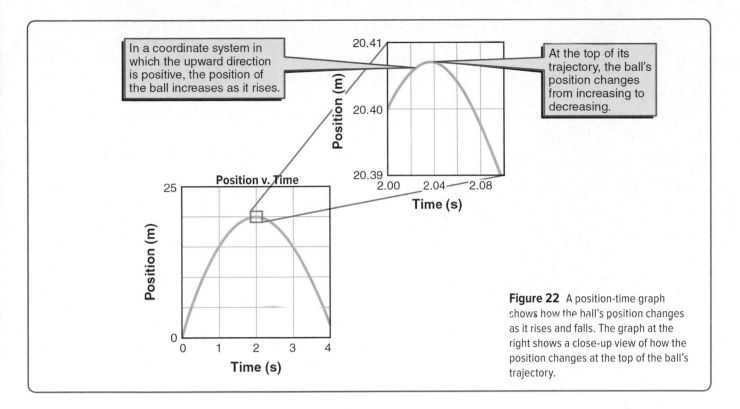

In a coordinate system in which the upward direction is positive, the position of the ball increases as it rises.

At the top of its trajectory, the ball's position changes from increasing to decreasing.

Figure 22 A position-time graph shows how the ball's position changes as it rises and falls. The graph at the right shows a close-up view of how the position changes at the top of the ball's trajectory.

Position-time graph Look at the position-time graphs in **Figure 22.** These graphs show how the ball's height changes as it rises and falls. If an object is moving with constant acceleration, its position-time graph forms a parabola. Because the ball is rising and falling, its graph is an inverted parabola. The shape of the graph shows the progression of time. It does not mean that the ball's path was in the shape of a parabola. The close-up graph on the right shows that at about 2.04 s, the ball reaches its maximum height.

Maximum height Compare the close-up graphs in **Figure 21** and **Figure 22.** Just before the ball reaches its maximum height, its velocity is decreasing in the negative direction. At the instant of time when its height is maximum, its velocity is zero. Just after it reaches its maximum height, the ball's velocity is increasing in the negative direction.

Acceleration The slope of the line on the velocity-time graph in **Figure 21** is constant at -9.8 m/s^2. This shows that the ball's free-fall acceleration is 9.8 m/s^2 in the downward direction the entire time the ball is rising and falling.

It may seem that the acceleration should be zero at the top of the trajectory, but this is not the case. At the top of the flight, the ball's velocity is 0 m/s. If its acceleration were also zero, the ball's velocity would not change and would remain at 0 m/s. The ball would not gain any downward velocity and would simply hover in the air. Have you ever seen that happen? Objects tossed in the air on Earth always fall, so you know the acceleration of an object at the top of its flight must not be zero. Further, because the object falls down, you know the acceleration must be downward.

☑ **READING CHECK Analyze** If you throw a ball straight up, what are its velocity and acceleration at the uppermost point of its path?

VOCABULARY
Science Usage v. Common Usage

Free fall
• **Science usage**
motion of a body when air resistance is negligible and the acceleration can be considered due to gravity alone
Acceleration during free fall is 9.8 m/s^2 downward.

• **Common usage**
a rapid and continuing drop or decline
The stock market's free fall in 1929 marked the beginning of the Great Depression.

PhysicsLAB

FREE-FALL ACCELERATION
INTERNET LAB How can you use motion data to calculate free-fall acceleration?

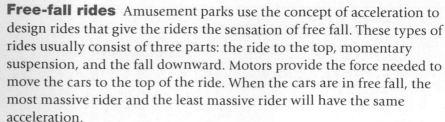

Free-fall rides Amusement parks use the concept of acceleration to design rides that give the riders the sensation of free fall. These types of rides usually consist of three parts: the ride to the top, momentary suspension, and the fall downward. Motors provide the force needed to move the cars to the top of the ride. When the cars are in free fall, the most massive rider and the least massive rider will have the same acceleration.

Suppose the free-fall ride shown in **Figure 23** starts from the top at rest and is in free fall for 1.5 s. What would be its velocity at the end of 1.5 s? Choose a coordinate system with a positive axis upward and the origin at the initial position of the car. Because the car starts at rest, v_i would be equal to 0.0 m/s. To calculate the final velocity, use the equation for velocity with constant acceleration.

$$v_f = v_i + \overline{a}t_f$$
$$= 0.0 \text{ m/s} + (-9.8 \text{ m/s}^2)(1.5 \text{ s})$$
$$= -15 \text{ m/s}$$

How far do people on the ride fall during this time? Use the equation for displacement when time and constant acceleration are known.

$$x_f = x_i + v_i t_f + \left(\frac{1}{2}\right)\overline{a}t_f^2$$
$$= 0.0 \text{ m} + (0.0 \text{ m/s})(1.5 \text{ s}) + \left(\frac{1}{2}\right)(-9.8 \text{ m/s}^2)(1.5 \text{ s})^2$$
$$= -11 \text{ m}$$

Figure 23 The people on this amusement-park ride experience free-fall acceleration.

PRACTICE PROBLEMS

Do additional problems. | Online Practice |

41. A construction worker accidentally drops a brick from a high scaffold.

   a. What is the velocity of the brick after 4.0 s?

   b. How far does the brick fall during this time?

42. Suppose for the previous problem you choose your coordinate system so that the opposite direction is positive.

   a. What is the brick's velocity after 4.0 s?

   b. How far does the brick fall during this time?

43. A student drops a ball from a window 3.5 m above the sidewalk. How fast is it moving when it hits the sidewalk?

44. A tennis ball is thrown straight up with an initial speed of 22.5 m/s. It is caught at the same distance above the ground.

   a. How high does the ball rise?

   b. How long does the ball remain in the air?
   Hint: The time it takes the ball to rise equals the time it takes to fall.

45. You decide to flip a coin to determine whether to do your physics or English homework first. The coin is flipped straight up.

   a. What are the velocity and acceleration of the coin at the top of its trajectory?

   b. If the coin reaches a high point of 0.25 m above where you released it, what was its initial speed?

   c. If you catch it at the same height as you released it, how much time did it spend in the air?

46. CHALLENGE A basketball player is holding a ball in her hands at a height of 1.5 m above the ground. She drops the ball, and it bounces several times. After the first bounce, the ball only returns to a height of 0.75 m. After the second bounce, the ball only returns to a height of 0.25 m.

   a. Suppose downward is the positive direction. What would the shape of a velocity-time graph look like for the first two bounces?

   b. What would be the shape of a position-time graph for the first two bounces?

Variations in Free Fall

When astronaut David Scott performed his free-fall experiment on the Moon, the hammer and the feather did not fall with an acceleration of magnitude 9.8 m/s². The value 9.8 m/s² is free-fall acceleration only near Earth's surface. The magnitude of free-fall acceleration on the Moon is approximately 1.6 m/s², which is about one-sixth its value on Earth.

When you study force and motion, you will learn about factors that affect the value of free-fall acceleration. One factor is the mass of the object, such as Earth or the Moon, that is responsible for the acceleration. Free-fall acceleration is not as great near the Moon as near Earth because the Moon has much less mass.

Free-fall acceleration also depends on the distance from the object responsible for it. The rings drawn around Earth in **Figure 24** show how free-fall acceleration decreases with distance from Earth. It is important to understand, however, that variations in free-fall acceleration at different locations on Earth's surface are very small, even with great variations in elevation. In New York City, for example, the magnitude of free-fall acceleration is about 9.81 m/s². In Denver, Colorado, it is about 9.79 m/s², despite a change in elevation of almost 1600 m greater. For calculations in this book, a value of 9.8 m/s² will be used for free-fall acceleration.

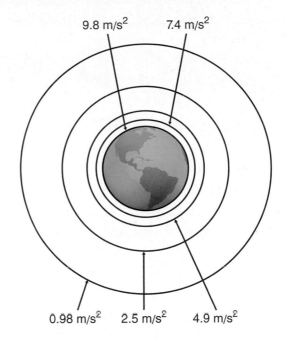

Figure 24 As the distance from Earth increases, the effect of free-fall acceleration decreases.

Analyze According to the diagram, what is the magnitude of free-fall acceleration a distance above Earth's surface equal to Earth's radius?

SECTION 3 REVIEW

Section Self-Check Check your understanding.

47. MAINIDEA Suppose you hold a book in one hand and a flat sheet of paper in another hand. You drop them both, and they fall to the ground. Explain why the falling book is a good example of free fall, but the paper is not.

48. Final Velocity Your sister drops your house keys down to you from the second-floor window, as shown in **Figure 25.** What is the velocity of the keys when you catch them?

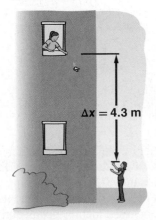

Δx = 4.3 m

Figure 25

49. Free-Fall Ride Suppose a free-fall ride at an amusement park starts at rest and is in free fall. What is the velocity of the ride after 2.3 s? How far do people on the ride fall during the 2.3-s time period?

50. Maximum Height and Flight Time The free-fall acceleration on Mars is about one-third that on Earth. Suppose you throw a ball upward with the same velocity on Mars as on Earth.

a. How would the ball's maximum height compare to that on Earth?

b. How would its flight time compare?

51. Velocity and Acceleration Suppose you throw a ball straight up into the air. Describe the changes in the velocity of the ball. Describe the changes in the acceleration of the ball.

52. Critical Thinking A ball thrown vertically upward continues upward until it reaches a certain position, and then falls downward. The ball's velocity is instantaneously zero at that highest point. Is the ball accelerating at that point? Devise an experiment to prove or disprove your answer.

Going Down?

Amusement–Park Thrill Rides

Your stomach jumps as you plummet 100 m downward. Just when you think you might smash into the ground, the breaks kick in and the amusement-park ride brings you to a slow and safe stop. "Let's go again!" cries your friend.

It's all about acceleration. Whether it's the 100-m plunge of a drop tower or the gentle up-and-down action of a carousel ride, the thrills of many amusement-park rides are based on the same principle—changes in velocity are exciting. Thrill rides take advantage of accelerations due to both changes in speed and changes in direction. A roller coaster is exciting because of accelerations produced by hills, loops, and banked turns, as shown in **Figure 1.**

Rides that use free-fall acceleration Many rides use free-fall acceleration to generate thrills. Drop-tower rides let passengers experience free fall and then carefully slow their descent just before they reach the ground. Pendulum rides, such as the one in **Figure 2,** act like giant swings. Passengers experience a stomach-churning moment as the pendulum reaches the top of its swing and begins to move downward. Passengers on a roller coaster experience a similar moment at the top of a hill.

FIGURE 2 Passengers sit in the pendulum of this pirate ship–themed ride, which swings back and forth like the pendulum on a grandfather clock.

FIGURE 1 Passengers accelerate as they move around a banked turn of a roller coaster. Acceleration is also part of what provides the excitement in a series of loops.

GOING**FURTHER** >>>

Research amusement-park ride design online. Design your own ride and present your design to the class. Identify points during the ride where the passengers accelerate.

BIGIDEA   Acceleration is the rate of change in an object's velocity.

SECTION 1 Acceleration

MAINIDEA   An object accelerates when its velocity changes—that is, when it speeds up, slows down, or changes direction.

- Acceleration is the rate at which an object's velocity changes.

- Velocity and acceleration are not the same thing. An object moving with constant velocity has zero acceleration. When the velocity and the acceleration of an object are in the same direction, the object speeds up; when they are in opposite directions, the object slows down.

- You can use a velocity-time graph to find the velocity and the acceleration of an object. The average acceleration of an object is the slope of its velocity-time graph.

$$\overline{a} \equiv \frac{\Delta v}{\Delta t} = \frac{v_f - v_i}{t_f - t_i}$$

SECTION 2 Motion with Constant Acceleration

MAINIDEA   For an object with constant acceleration, the relationships among position, velocity, acceleration, and time can be described by graphs and equations.

- If an object is moving with constant acceleration, its position-time graph is a parabola, and its velocity-time graph is a straight line.

- The area under an object's velocity-time graph is its displacement.

- In motion with constant acceleration, position, velocity, acceleration, and time are related:

$$v_f = v_i + \overline{a}\,\Delta t$$
$$x_f = x_i + v_i t_f + \frac{1}{2}\overline{a}t_f^2$$
$$v_f^2 = v_i^2 + 2\overline{a}(x_f - x_i)$$

SECTION 3 Free Fall

MAINIDEA   The acceleration of an object in free fall is due to gravity alone.

- Free-fall acceleration on Earth is about 9.8 m/s² downward. The sign associated with free-fall acceleration in equations depends on the choice of the coordinate system.

- When an object is in free fall, gravity is the only force acting on it. Equations for motion with constant acceleration can be used to solve problems involving objects in free fall.

Games and Multilingual eGlossary

Vocabulary Practice

SECTION 1 Acceleration

Mastering Concepts

53. BIGIDEA How are velocity and acceleration related?

54. Give an example of each of the following:

a. an object that is slowing down but has a positive acceleration

b. an object that is speeding up but has a negative acceleration

c. an object that is moving at a constant speed but has an acceleration

55. Figure 26 shows the velocity-time graph for an automobile on a test track. Describe how the velocity changes with time.

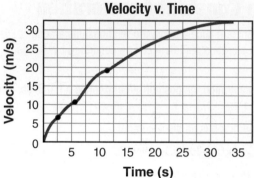

Velocity v. Time

Figure 26

56. If the velocity-time graph of an object moving on a straight path is a line parallel to the horizontal axis, what can you conclude about its acceleration?

Mastering Problems

57. Ranking Task Rank the following objects according to the magnitude of the acceleration, from least to greatest. Specifically indicate any ties.

A. A falling acorn accelerates from 0.50 m/s to 10.3 m/s in 1.0 s.

B. A car accelerates from 20 m/s to rest in 1.0 s.

C. A centipede accelerates from 0.40 cm/s to 2.0 cm/s in 0.50 s.

D. While being hit, a golf ball accelerates from rest to 4.3 m/s in 0.40 s.

E. A jogger accelerates from 2.0 m/s to 1.0 m/s in 8.3 s.

58. Problem Posing Complete this problem so that it can be solved using the concept listed: "Angela is playing basketball . . ."

a. acceleration

b. speed

59. The graph in **Figure 27** describes the motion of an object moving east along a straight path. Find the acceleration of the object at each of these times:

a. during the first 5.0 min of travel

b. between 5.0 min and 10.0 min

c. between 10.0 min and 15.0 min

d. between 20.0 min and 25.0 min

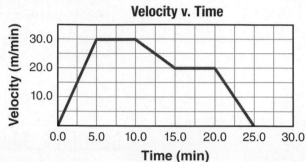

Velocity v. Time

Figure 27

60. Plot a velocity-time graph using the information in **Table 1,** and answer the following questions:

a. During what time interval is the object speeding up? Slowing down?

b. At what time does the object reverse direction?

c. How does the average acceleration of the object between 0.0 s and 2.0 s differ from the average acceleration between 7.0 s and 12.0 s?

Table 1 Velocity v. Time	
Time (s)	**Velocity (m/s)**
0.00	4.00
1.00	8.00
2.00	12.0
3.00	14.0
4.00	16.0
5.00	16.0
6.00	14.0
7.00	12.0
8.00	8.00
9.00	4.00
10.0	0.00
11.0	−4.00
12.0	−8.00

61. Determine the final velocity of a proton that has an initial forward velocity of 2.35×10^5 m/s and then is accelerated uniformly in an electric field at the rate of -1.10×10^{12} m/s² for 1.50×10^{-7} s.

62. Ranking Task Marco wants to buy the used sports car with the greatest acceleration. Car A can go from 0 m/s to 17.9 m/s in 4.0 s. Car B can accelerate from 0 m/s to 22.4 m/s in 3.5 s. Car C can go from 0 to 26.8 m/s in 6.0 s. Rank the three cars from greatest acceleration to least. Indicate if any are the same.

SECTION 2
Motion with Constant Acceleration
Mastering Concepts

63. What quantity does the area under a velocity-time graph represent?

64. Reverse Problem Write a physics problem with real-life objects for which the graph in **Figure 28** would be part of the solution.

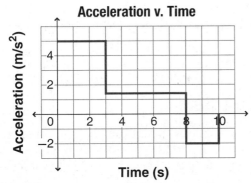

Figure 28

Mastering Problems

65. A car moves forward up a hill at 12 m/s with a uniform backward acceleration of 1.6 m/s².

  a. What is its displacement after 6.0 s?

  b. What is its displacement after 9.0 s?

66. Airplane Determine the displacement of a plane that experiences uniform acceleration from 66 m/s north to 88 m/s north in 12 s.

67. Race Car A race car is slowed with a constant acceleration of 11 m/s² opposite the direction of motion.

  a. If the car is going 55 m/s, how many meters will it travel before it stops?

  b. How many meters will it take to stop a car going twice as fast?

68. Refer to **Figure 29** to find the magnitude of the displacement during the following time intervals. Round answers to the nearest meter.

  a. $t = 5.0$ min and $t = 10.0$ min

  b. $t = 10.0$ min and $t = 15.0$ min

  c. $t = 25.0$ min and $t = 30.0$ min

  d. $t = 0.0$ min and $t = 25.0$ min

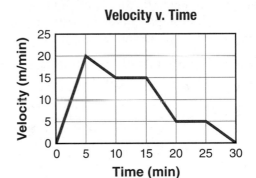

Figure 29

SECTION 3 **Free Fall**
Mastering Concepts

69. Explain why an aluminum ball and a steel ball of similar size and shape, dropped from the same height, reach the ground at the same time.

70. Give some examples of falling objects for which air resistance can and cannot be ignored.

Mastering Problems

71. Suppose an astronaut drops a feather from a height of 1.2 m above the surface of the Moon. If the free-fall acceleration on the Moon is 1.62 m/s² downward, how long does it take the feather to hit the Moon's surface?

72. A stone that starts at rest is in free fall for 8.0 s.

  a. Calculate the stone's velocity after 8.0 s.

  b. What is the stone's displacement during this time?

73. A bag is dropped from a hovering helicopter. The bag has fallen for 2.0 s. What is the bag's velocity? How far has the bag fallen? Ignore air resistance.

74. You throw a ball downward from a window at a speed of 2.0 m/s. How fast will it be moving when it hits the sidewalk 2.5 m below?

75. If you throw the ball in the previous problem up instead of down, how fast will it be moving when it hits the sidewalk?

76. Beanbag You throw a beanbag in the air and catch it 2.2 s later at the same place at which you threw it.

a. How high did it go?

b. What was its initial velocity?

Applying Concepts

77. Croquet A croquet ball, after being hit by a mallet, slows down and stops. Do the velocity and the acceleration of the ball have the same signs?

78. Explain how you would walk to produce each of the position-time graphs in **Figure 30.**

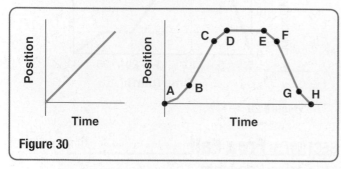

Figure 30

79. If you were given a table of velocities of an object at various times, how would you determine whether the acceleration was constant?

80. Look back at the graph in **Figure 26.** The three notches in the graph occur where the driver changed gears. Describe the changes in velocity and acceleration of the car while in first gear. Is the acceleration just before a gear change larger or smaller than the acceleration just after the change? Explain your answer.

81. An object shot straight up rises for 7.0 s before it reaches its maximum height. A second object falling from rest takes 7.0 s to reach the ground. Compare the displacements of the two objects during this time interval.

82. Draw a velocity-time graph for each of the graphs in **Figure 31.**

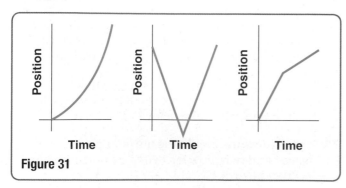

Figure 31

83. The Moon The value of free-fall acceleration on the Moon is about one-sixth of its value on Earth.

a. Would a ball dropped by an astronaut hit the surface of the Moon with a greater, equal, or lesser speed than that of a ball dropped from the same height to Earth?

b. Would it take the ball more, less, or equal time to fall?

84. Jupiter An object on the planet Jupiter has about three times the free-fall acceleration as on Earth. Suppose a ball could be thrown vertically upward with the same initial velocity on Earth and on Jupiter. Neglect the effects of Jupiter's atmospheric resistance, and assume that gravity is the only force on the ball.

a. How would the maximum height reached by the ball on Jupiter compare to the maximum height reached on Earth?

b. If the ball on Jupiter were thrown with an initial velocity that is three times greater, how would this affect your answer to part **a**?

85. Rock A is dropped from a cliff, and rock B is thrown upward from the same position.

a. When they reach the ground at the bottom of the cliff, which rock has a greater velocity?

b. Which has a greater acceleration?

c. Which arrives first?

Mixed Review

86. Suppose a spaceship far from any star or planet had a uniform forward acceleration from 65.0 m/s to 162.0 m/s in 10.0 s. How far would the spaceship move?

87. Figure 32 is a multiflash photo of a horizontally moving ball. What information about the photo would you need and what measurements would you make to estimate the acceleration?

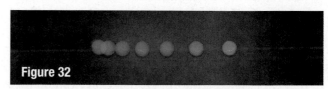

Figure 32

88. Bicycle A bicycle accelerates from 0.0 m/s to 4.0 m/s in 4.0 s. What distance does it travel?

89. A weather balloon is floating at a constant height above Earth when it releases a pack of instruments.

a. If the pack hits the ground with a downward velocity of 73.5 m/s, how far did the pack fall?

b. How long did it take for the pack to fall?

90. The total distance a steel ball rolls down an incline at various times is given in **Table 2.**

    a. Draw a position-time graph of the motion of the ball. When setting up the axes, use five divisions for each 10 m of travel on the *x*-axis. Use five divisions for 1 s of time on the *t*-axis.

    b. Calculate the distance the ball has rolled at the end of 2.2 s.

Table 2 Position v. Time	
Time (s)	Position (m)
0.0	0.0
1.0	2.0
2.0	8.0
3.0	18.0
4.0	32.0
5.0	50.0

91. Engineers are developing new types of guns that might someday be used to launch satellites as if they were bullets. One such gun can give a small object a forward velocity of 3.5 km/s while moving it through a distance of only 2.0 cm.

    a. What acceleration does the gun give this object?

    b. Over what time interval does the acceleration take place?

92. **Safety Barriers** Highway safety engineers build soft barriers, such as the one shown in **Figure 33,** so that cars hitting them will slow down at a safe rate. Suppose a car traveling at 110 km/h hits the barrier, and the barrier decreases the car's velocity at a rate of 32 m/s². What distance would the car travel along the barrier before coming to a stop?

Figure 33

93. **Baseball** A baseball pitcher throws a fastball at a speed of 44 m/s. The ball has constant acceleration as the pitcher holds it in his hand and moves it through an almost straight-line distance of 3.5 m. Calculate the acceleration. Compare this acceleration to the free-fall acceleration on Earth.

94. **Sleds** Rocket-powered sleds are used to test the responses of humans to acceleration. Starting from rest, one sled can reach a speed of 444 m/s in 1.80 s and can be brought to a stop again in 2.15 s.

    a. Calculate the acceleration of the sled when starting, and compare it to the magnitude of free-fall acceleration, 9.8 m/s².

    b. Find the acceleration of the sled as it is braking, and compare it to the magnitude of free-fall acceleration.

95. The forward velocity of a car changes over an 8.0-s time period, as shown in **Table 3.**

    a. Plot the velocity-time graph of the motion.

    b. What is the car's displacement in the first 2.0 s?

    c. What is the car's displacement in the first 4.0 s?

    d. What is the displacement of the car during the entire 8.0 s?

    e. Find the slope of the line between $t = 0.0$ s and $t = 4.0$ s. What does this slope represent?

    f. Find the slope of the line between $t = 5.0$ s and $t = 7.0$ s. What does this slope indicate?

Table 3 Velocity v. Time	
Time (s)	Velocity (m/s)
0.0	0.0
1.0	4.0
2.0	8.0
3.0	12.0
4.0	16.0
5.0	20.0
6.0	20.0
7.0	20.0
8.0	20.0

96. A truck is stopped at a stoplight. When the light turns green, the truck accelerates at 2.5 m/s². At the same instant, a car passes the truck going at a constant 15 m/s. Where and when does the truck catch up with the car?

97. **Karate** The position-time and velocity-time graphs of George's fist breaking a wooden board during karate practice are shown in **Figure 34.**

   a. Use the velocity-time graph to describe the motion of George's fist during the first 10 ms.

   b. Estimate the slope of the velocity-time graph to determine the acceleration of his fist when it suddenly stops.

   c. Express the acceleration as a multiple of the magnitude of free-fall acceleration, 9.8 m/s^2.

   d. Determine the area under the velocity-time curve to find the displacement of the fist in the first 6 ms. Compare this with the position-time graph.

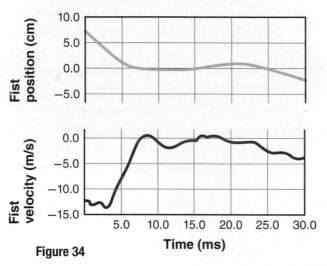

Figure 34

98. **Cargo** A helicopter is rising at 5.0 m/s when a bag of its cargo is dropped. The bag falls for 2.0 s.

   a. What is the bag's velocity?

   b. How far has the bag fallen?

   c. How far below the helicopter is the bag?

Thinking Critically

99. **Probeware** Design a probeware lab to measure the distance an accelerated object moves over time. Use equal time intervals so that you can plot velocity over time as well as distance. A pulley at the edge of a table with a mass attached is a good way to achieve uniform acceleration. Suggested materials include a motion detector, lab cart, string, pulley, C-clamp, and masses. Generate position-time and velocity-time graphs using different masses on the pulley. How does the change in mass affect your graphs?

100. **Analyze and Conclude** Which (if either) has the greater acceleration: a car that increases its speed from 50 km/h to 60 km/h or a bike that goes from 0 km/h to 10 km/h in the same time? Explain.

101. **Analyze and Conclude** An express train traveling at 36.0 m/s is accidentally sidetracked onto a local train track. The express engineer spots a local train exactly 1.00×10^2 m ahead on the same track and traveling in the same direction. The local engineer is unaware of the situation. The express engineer jams on the brakes and slows the express train at a constant rate of 3.00 m/s^2. If the speed of the local train is 11.0 m/s, will the express train be able to stop in time, or will there be a collision? To solve this problem, take the position of the express train when the engineer first sights the local train as a point of origin. Next, keeping in mind that the local train has exactly a 1.00×10^2 m lead, calculate how far each train is from the origin at the end of the 12.0 s it would take the express train to stop (accelerate at −3.00 m/s^2 from 36 m/s to 0 m/s).

   a. On the basis of your calculations, would you conclude that a collision will occur?

   b. To check the calculations from part **a** and to verify your conclusion, take the position of the express train when the engineer first sights the local train as the point of origin and calculate the position of each train at the end of each second after the sighting. Make a table showing the distance of each train from the origin at the end of each second. Plot these positions on the same graph and draw two lines. Compare your graph to your answer to part **a**.

Writing in Physics

102. Research and describe Galileo's contributions to physics.

103. Research the maximum acceleration a human body can withstand without blacking out. Discuss how this impacts the design of three common entertainment or transportation devices.

Cumulative Review

104. Solve the following problems. Express your answers in scientific notation.

   a. 6.2×10^{-4} m $+ 5.7 \times 10^{-3}$ m

   b. 8.7×10^8 km $- 3.4 \times 10^7$ km

   c. $(9.21 \times 10^{-5}$ cm$)(1.83 \times 10^8$ cm$)$

   d. $\dfrac{2.63 \times 10^{-6} \text{ m}}{4.08 \times 10^6 \text{ s}}$

105. The equation below describes an object's motion. Create the corresponding position-time graph and motion diagram. Then write a physics problem that could be solved using that equation. Be creative.

$$x = (35.0 \text{ m/s}) \, t - 5.0 \text{ m}$$

MULTIPLE CHOICE

Use the following information to answer the first two questions.

A ball rolls down a hill with a constant acceleration of 2.0 m/s². The ball starts at rest and travels for 4.0 s before it reaches the bottom of the hill.

1. How far did the ball travel during this time?
- **A.** 8.0 m
- **C.** 16 m
- **B.** 12 m
- **D.** 20 m

2. What was the ball's speed at the bottom of the hill?
- **A.** 2.0 m/s
- **C.** 12 m/s
- **B.** 8.0 m/s
- **D.** 16 m/s

3. A driver of a car enters a new 110-km/h speed zone on the highway. The driver begins to accelerate immediately and reaches 110 km/h after driving 500 m. If the original speed was 80 km/h, what was the driver's forward acceleration?
- **A.** 0.44 m/s²
- **C.** 8.4 m/s²
- **B.** 0.60 m/s²
- **D.** 9.8 m/s²

4. A flowerpot falls off a balcony 85 m above the street. How long does it take to hit the ground?
- **A.** 4.2 s
- **C.** 8.7 s
- **B.** 8.3 s
- **D.** 17 s

5. A rock climber's shoe loosens a rock, and her climbing buddy at the bottom of the cliff notices that the rock takes 3.20 s to fall to the ground. How high up the cliff is the rock climber?
- **A.** 15.0 m
- **C.** 50.0 m
- **B.** 31.0 m
- **D.** 1.00×10^2 m

6. A car traveling at 91.0 km/h approaches the turnoff for a restaurant 30.0 m aheadxd. If the driver slams on the brakes with an acceleration of -6.40 m/s², what will be her stopping distance?
- **A.** 14.0 m
- **C.** 50.0 m
- **B.** 29.0 m
- **D.** 100.0 m

7. What is the correct formula manipulation to find acceleration when using the equation $v_f^2 = v_i^2 + 2ax$?
- **A.** $\dfrac{v_f^2 - v_i^2}{x}$
- **C.** $\dfrac{(v_f + v_i)^2}{2x}$
- **B.** $\dfrac{v_f^2 + v_i^2}{2x}$
- **D.** $\dfrac{v_f^2 - v_i^2}{2x}$

Online Test Practice

8. The graph below shows the motion of a farmer's truck. What is the truck's total displacement? Assume north is the positive direction.
- **A.** 150 m south
- **C.** 300 m north
- **B.** 125 m north
- **D.** 600 m south

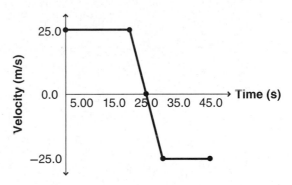

9. How can the instantaneous acceleration of an object with varying acceleration be calculated?
- **A.** by calculating the slope of the tangent on a position-time graph
- **B.** by calculating the area under the graph on a position-time graph
- **C.** by calculating the area under the graph on a velocity-time graph
- **D.** by calculating the slope of the tangent on a velocity-time graph

FREE RESPONSE

10. Graph the following data, and then show calculations for acceleration and displacement after 12.0 s on the graph.

Time (s)	Velocity (m/s)
0.00	8.10
6.00	36.9
9.00	51.3
12.00	65.7

NEED EXTRA HELP?

If you Missed Question	1	2	3	4	5	6	7	8	9	10
Review Section	2	2	2	3	3	2	2	2	1	2

Forces in One Dimension

BIGIDEA   Net forces cause changes in motion.

SECTIONS

1 **Force and Motion**

2 **Weight and Drag Force**

3 **Newton's Third Law**

LaunchLAB

FORCES IN OPPOSITE DIRECTIONS

What happens when more than one force acts on an object?

WATCH THIS!

APPARENT WEIGHT

Have you ever had the sensation of feeling weightless while on a freefall ride? An elevator acts in much the same way, giving the sensation of being lighter when you're accelerating down and heavier when you're moving up.

PHYSICS T.V.

(l)McGraw-Hill Education; (r)James Lauritz/Digital Vision/Getty Images

Force and Motion

PHYSICS
4 YOU

To start a skateboard moving on a level surface, you must push on the ground with your foot. If the skateboard is on a ramp, however, gravity will pull you down the slope. In both cases unbalanced forces change the skateboard's motion.

MAIN IDEA

A force is a push or a pull.

Essential Questions

• What is a force?

• What is the relationship between force and acceleration?

• How does motion change when the net force is zero?

Review Vocabulary

acceleration the rate at which the velocity of an object changes

New Vocabulary

force
system
free-body diagram
net force
Newton's second law
Newton's first law
inertia
equilibrium

Force

Consider a textbook resting on a table. To cause it to move, you could either push or pull on it. In physics, a push or a pull is called a **force.** If you push or pull harder on an object, you exert a greater force on the object. In other words, you increase the magnitude of the applied force. The direction in which the force is exerted also matters—if you push the resting book to the right, the book will start moving to the right. If you push the book to the left, it will start moving to the left. Because forces have both magnitude and direction, forces are vectors. The symbol F is vector notation that represents the size and direction of a force, while F represents only the magnitude. The magnitude of a force is measured in units called newtons (N).

Unbalanced forces change motion Recall that motion diagrams describe the positions of an object at equal time intervals. For example, the motion diagram for the book in **Figure 1** shows the distance between the dots increasing. This means the speed of the book is increasing. At $t = 0$, it is at rest, but after 2 seconds it is moving at 1.5 m/s. This change in speed means it is accelerating. What is the cause of this acceleration? The book was at rest until you pushed it, so the cause of the acceleration is the force exerted by your hand. In fact, all accelerations are the result of an unbalanced force acting on an object. What is the relationship between force and acceleration? By the end of this section, you will be able to answer that question as well as apply laws of motion to solve many different types of problems.

☑ **READING CHECK** **Identify** the cause of all accelerations.

Figure 1 The hand pushing on the book exerts a force that causes the book to accelerate in the direction of the unbalanced force.

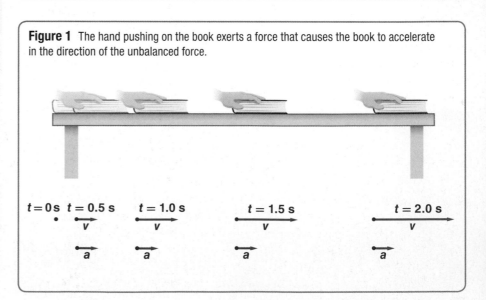

$t=0\text{ s}$ $t=0.5\text{ s}$ $t=1.0\text{ s}$ $t=1.5\text{ s}$ $t=2.0\text{ s}$

Fuse/Getty Images

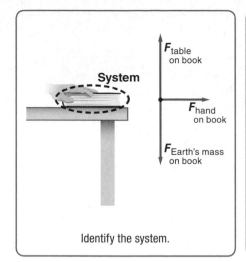

Identify the system.

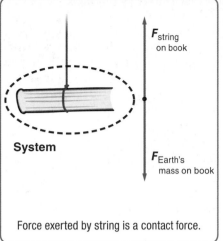

Force exerted by string is a contact force.

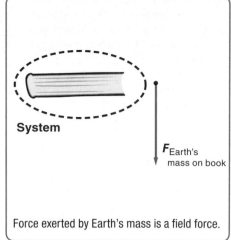

Force exerted by Earth's mass is a field force.

Systems and external world When considering how a force affects motion, it is important to identify the object or objects of interest, called the **system.** Everything around the system with which the system can interact is called the external world. In **Figure 2,** the book is the system. Your hand, Earth, string and the table are parts of the external world that interact with the book by pushing or pulling on it.

Contact forces Again, think about the different ways in which you could move a textbook. You could push or pull it by directly touching it, or you could tie a string around it and pull on the string. These are examples of contact forces. A contact force exists when an object from the external world touches a system, exerting a force on it. If you are holding this physics textbook right now, your hands are exerting a contact force on it. If you place the book on a table, you are no longer exerting a contact force on the book. The table, however, is exerting a contact force because the table and the book are in contact.

Field forces There are other ways in which the motion of the textbook can change. You could drop it, and as you learned in a previous chapter, it would accelerate as it falls to the ground. The gravitational force of Earth acting on the book causes this acceleration. This force affects the book whether or not Earth is actually touching it. Gravitational force is an example of a field force. Field forces are exerted without contact. Can you think of other kinds of field forces? If you have ever investigated magnets, you know that they exert forces without touching. You will investigate magnetism and other field forces in future chapters. For now, the only field force you need to consider is the gravitational force.

Agents Forces result from interactions; every contact and field force has a specific and identifiable cause, called the agent. You should be able to name the agent exerting each force as well as the system upon which the force is exerted. For example, when you push your textbook, your hand (the agent) exerts a force on the textbook (the system). If there are not both an agent and a system, a force does not exist. What about the gravitational force? The agent is the mass of Earth exerting a field force on the book. The labels on the forces in **Figure 2** are good examples of how to identify a force's agent and the system upon which the force acts.

Figure 2 The book is the system in each of these situations.

Classify each force in the first panel as either a contact force or a field force.

View an **animation on systems.**

Concepts In Motion

VOCABULARY
Science Use v. Common Use

Force
• Science usage
a push or a pull exerted on an object
The force of gravity exerted by the Sun on Earth pulls Earth into orbit around the Sun.

• Common usage
to compel by physical, moral, or intellectual means
Emi forced her younger brother to wash the dishes.

Figure 3 The drawings of the ball in the hand and the ball hanging from the string are both pictorial models. The free-body diagram for each situation is shown next to each pictorial model.

View an **animation of free body diagrams.**

Concepts In Motion 👆

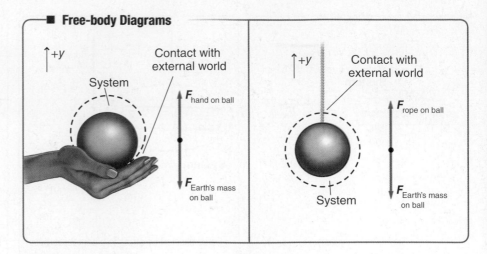

■ **Free-body Diagrams**

Free-body diagrams Just as pictorial representations and motion diagrams are useful in solving problems about motion, similar representations will help you analyze how forces affect motion. The first step is to sketch the situation, as shown in **Figure 3.** Circle the system, and identify every place where the system touches the external world. It is at these places that an agent exerts a contact force. Then identify any field forces on the system. This gives you the pictorial representation.

A **free-body diagram** is a physical representation that shows the forces acting on a system. Follow these guidelines when drawing a free-body diagram:

- The free-body diagram is drawn separately from the sketch of the problem situation.
- Apply the particle model, and represent the object with a dot.
- Represent each force with an arrow that points in the direction the force is applied. Always draw the force vectors pointing away from the particle, even when the force is a push.
- Make the length of each arrow proportional to the size of the force. Often you will draw these diagrams before you know the magnitudes of all the forces. In such cases, make your best estimate.
- Label each force. Use the symbol F with a subscript label to identify both the agent and the object on which the force is exerted.
- Choose a direction to be positive, and indicate this on the diagram.

Using free-body diagrams and motion diagrams Recall that all accelerations are the result of unbalanced forces. If a motion diagram shows that an object is accelerating, a free-body diagram of that object should have an unbalanced force in the same direction as the acceleration.

☑ **READING CHECK Compare** the direction of an object's acceleration with the direction of the unbalanced force exerted on the object.

Combining Forces

What happens if you and a friend each push a table and exert 100 N of force on it? When you push together in the same direction, you give the table twice the acceleration that it would have if just one of you applied 100 N of force. When you push on the table in opposite directions with the same amount of force, as in **Figure 4,** there is no unbalanced force, so the table does not accelerate but remains at rest.

$F_1 = 100$ N

$F_2 = 100$ N $F_1 = 100$ N

$F_2 = 100$ N

$F_2 = 200$ N $F_1 = 100$ N

$F_{net} = 0$ N

$F_{net} = 200$ N

$F_{net} = 100$ N

Equal forces
Opposite directions

Equal forces
Same direction

Unequal forces
Opposite directions

Net force The bottom portion of **Figure 4** shows free-body diagrams for these two situations. The third diagram in **Figure 4** shows the free-body diagram for a third situation in which your friend pushes on the table twice as hard as you in the opposite direction. Below each free-body diagram is a vector representing the resultant of the two forces. When the force vectors are in the same direction, they can be replaced by one vector with a length equal to their combined length. When the forces are in opposite directions, the resultant is the length of the difference between the two vectors. Another term for the vector sum of all the forces on an object is the **net force.**

You also can analyze the situation mathematically. Call the positive direction the direction in which you are pushing the table with a 100 N force. In the first case, your friend is pushing with a negative force of 100 N. Adding them together gives a total force of 0 N, which means there is no acceleration. In the second case, your friend's force is 100 N, so the total force is 200 N in the positive direction and the table accelerates in the positive direction. In the third case, your friend's force is −200 N, so the total force is −100 N and the table accelerates in the negative direction.

PRACTICE PROBLEMS

Do additional problems. **Online Practice**

For each of the following situations, specify the system and draw a motion diagram and a free-body diagram. Label all forces with their agents, and indicate the direction of the acceleration and of the net force. Draw vectors of appropriate lengths. Ignore air resistance unless otherwise indicated.

1. A skydiver falls downward through the air at constant velocity. (The air exerts an upward force on the person.)

2. You hold a softball in the palm of your hand and toss it up. Draw the diagrams while the ball is still touching your hand.

3. After the softball leaves your hand, it rises, slowing down.

4. After the softball reaches its maximum height, it falls down, speeding up.

5. **CHALLENGE** You catch the ball in your hand and bring it to rest.

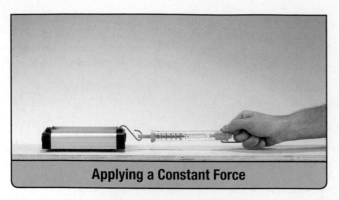

Applying a Constant Force

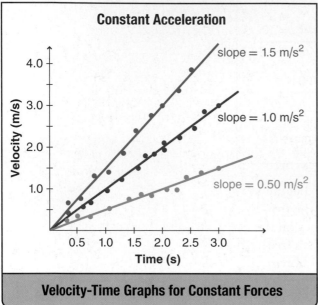

Velocity-Time Graphs for Constant Forces

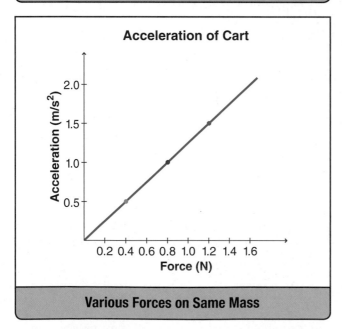

Various Forces on Same Mass

Figure 5 A spring scale exerts a constant unbalanced force on the cart. Repeating the investigation with different forces produces velocity-time graphs with different slopes.

View an **animation of force and acceleration.**

Acceleration and Force

To explore how forces affect an object's motion, think about doing a series of investigations. Consider the simple situation shown in the top photo of **Figure 5** in which we exert one force horizontally on an object. Starting with the horizontal direction is helpful because gravity does not act horizontally. To reduce complications resulting from the object rubbing against the surface, the investigations should be done on a smooth surface, such as a well-polished table. We'll also use a cart with wheels that spin easily.

Apply constant force How can you exert a constant unbalanced force? One way is to use a device called a spring scale. Inside the scale is a spring that stretches proportionally to the magnitude of the applied force. The front of the scale is calibrated to read the force in newtons. If you pull on the scale so that the reading on the front stays constant, the applied force is constant. The top photo in **Figure 5** shows a spring scale pulling a low-resistance cart with a constant unbalanced force.

If you perform this investigation and measure the cart's velocity for a period of time, you could construct a graph like the green line shown in the velocity-time graphs for constant forces in the middle panel of **Figure 5.** The constant slope of the line in the velocity-time graph indicates the cart's velocity increases at a constant rate. The constant rate of change of velocity means the acceleration is constant. This constant acceleration is a result of the constant unbalanced force applied by the spring scale to the cart.

How does the acceleration depend on the force? Repeat the investigation with a larger constant force. Then repeat it again with an even greater force. For each force, plot a velocity-time graph like the red and blue lines in the middle panel of **Figure 5.** Recall that the line's slope is the cart's acceleration. Calculate the slope of each line and plot the results for each force to make an acceleration-force graph, as shown in the bottom panel of **Figure 5.**

The graph indicates the relationship between force and acceleration is linear. Because the relationship is linear, you can apply the equation for a straight line:

$$y = kx + b$$

The y-intercept is 0, so the linear equation simplifies to $y = kx$. The y-variable is acceleration and the x-variable is force, so acceleration equals the slope of the line times the applied net force.

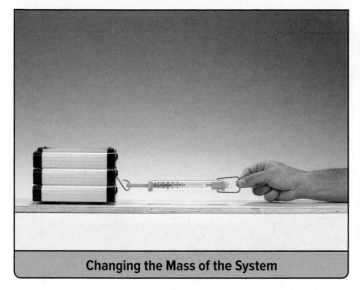

Changing the Mass of the System

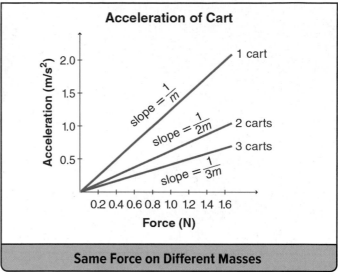

Acceleration of Cart

Same Force on Different Masses

Interpreting slope What is the physical meaning of the slope of the acceleration-force graph? Does it describe something about the object that is accelerating? To see, change the object. Suppose that a second, identical cart is placed on top of the first, and then a third cart is added as in **Figure 6.** The spring scale would be pulling two carts and then three. A plot of the force versus acceleration for one, two, and three carts is shown in the graph in **Figure 6.**

The graph shows that if the same force is applied in each case, the acceleration of two carts is $\frac{1}{2}$ the acceleration of one cart, and the acceleration of three carts is $\frac{1}{3}$ the acceleration of one cart. This means that as the number of carts increases, the acceleration decreases. In other words, a greater force is needed to produce the same acceleration. The slopes of the lines in **Figure 6** depend upon the number of carts; that is, the slope depends on the total mass of the carts. In fact, the slope is the reciprocal of the mass (slope $= \frac{1}{\text{mass}}$). Using this value for slope, the mathematical equation $y = kx$ becomes the physics equation $a = \frac{F_{\text{net}}}{m}$.

What information is contained in the equation $a = \frac{F_{\text{net}}}{m}$? It tells you that a net force applied to an object causes that object to experience a change in motion–the force causes the object to accelerate. It also tells you that for the same object, if you double the force, you will double the object's acceleration. Lastly, if you apply the same force to objects with different masses, the one with the most mass will have the smallest acceleration and the one with the least mass will have the greatest acceleration.

☑ **READING CHECK Determine** how the force exerted on an object must be changed to reduce the object's acceleration by half.

Recall that forces are measured in units called newtons. Because $F_{\text{net}} = ma$, a newton has the units of mass times the units of acceleration. So one newton is equal to one kg·m/s². To get an approximate idea of the size of 1 N, think about the downward force you feel when you hold an apple in your hand. The force exerted by the apple on your hand is approximately one newton. **Table 1** shows the magnitudes of some other common forces.

Figure 6 Changing an object's mass affects that object's acceleration.

Compare the acceleration of one cart to the acceleration of two carts for an applied force of 1 N.

⚙ APPLYING PRACTICES

Analyze Data Go to the resources tab in ConnectED to find the Applying Practices worksheet *Newton's Second Law.*

Table 1 Common Forces

Description	F (N)
Force of gravity on a coin (nickel)	0.05
Force of gravity on a 0.45-kg bag of sugar	4.5
Force of gravity on a 70-kg person	686
Force exerted by road on an accelerating car	3000
Force of a rocket engine	5,000,000

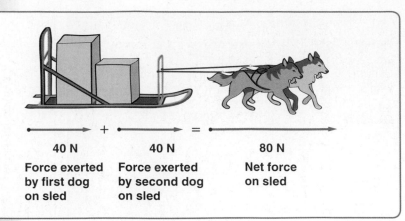

40 N
Force exerted
by first dog
on sled

40 N
Force exerted
by second dog
on sled

80 N
Net force
on sled

Newton's Second Law

Figure 7 shows two dogs pulling a sled. Each dog pulls with a force of 40 N. From the cart and spring-scale investigations, you know that the sled accelerates as a result of the unbalanced force acting it. Would the acceleration change if instead of two dogs each exerting a 40-N force, there was one bigger, stronger dog exerting a single 80-N force on the sled? When considering forces and acceleration, it is important to find the sum of all forces, called the net force, acting on a system.

Newton's second law states that the acceleration of an object is proportional to the net force and inversely proportional to the mass of the object being accelerated. This law is based on observations of how forces affect masses and is represented by the following equation.

NEWTON'S SECOND LAW

The acceleration of an object is equal to the sum of the forces acting on the object divided by the mass of the object.

$$a = \frac{F_{net}}{m}$$

Solving problems using Newton's second law One of the most important steps in correctly applying Newton's second law is determining the net force acting on the object. Often, more than one force acts on an object, so you must add the force vectors to determine the net force. Draw a free-body diagram showing the direction and relative strength of each force acting on the system. Then, add the force vectors to find the net force. Next, use Newton's second law to calculate the acceleration. Finally, if necessary, you can use what you know about accelerated motion to find the velocity or position of the object.

Figure 7 The net force acting on an object is the vector sum of all the forces acting on that object.

Investigate **Newton's second law.**

Virtual Investigation

PRACTICE PROBLEMS

Do additional problems. Online Practice

6. Two horizontal forces, 225 N and 165 N, are exerted on a canoe. If these forces are applied in the same direction, find the net horizontal force on the canoe.

7. If the same two forces as in the previous problem are exerted on the canoe in opposite directions, what is the net horizontal force on the canoe? Be sure to indicate the direction of the net force.

8. CHALLENGE Three confused sled dogs are trying to pull a sled across the Alaskan snow. Alutia pulls east with a force of 35 N, Seward also pulls east but with a force of 42 N, and big Kodiak pulls west with a force of 53 N. What is the net force on the sled?

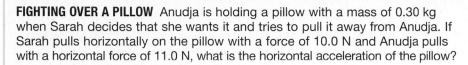
FIGHTING OVER A PILLOW Anudja is holding a pillow with a mass of 0.30 kg when Sarah decides that she wants it and tries to pull it away from Anudja. If Sarah pulls horizontally on the pillow with a force of 10.0 N and Anudja pulls with a horizontal force of 11.0 N, what is the horizontal acceleration of the pillow?

1 ANALYZE AND SKETCH THE PROBLEM

- Sketch the situation.
- Identify the pillow as the system, and the direction in which Anudja pulls as positive.
- Draw the free-body diagram. Label the forces.

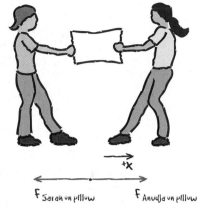

KNOWN	UNKNOWN
$m = 0.30$ kg	$a = ?$
$F_{\text{Anudja on pillow}} = 11.0$ N	
$F_{\text{Sarah on pillow}} = 10.0$ N	

2 SOLVE FOR THE UNKNOWN

$$F_{\text{net}} = F_{\text{Anudja on pillow}} + (-F_{\text{Sarah on pillow}})$$

Use Newton's second law.

$$a = \frac{F_{\text{net}}}{m}$$

$$= \frac{F_{\text{Anudja on pillow}} + (-F_{\text{Sarah on pillow}})}{m}$$

$$= \frac{11.0 \text{ N} - 10.0 \text{ N}}{0.30 \text{ kg}}$$ ◀ Substitute $F_{\text{Anudja on pillow}} = 11.0$ N, $F_{\text{Sarah on pillow}} = 10.0$ N, $m = 0.30$ kg.

$$= 3.3 \text{ m/s}^2$$

$$a = 3.3 \text{ m/s}^2 \text{ toward Anudja}$$

3 EVALUATE THE ANSWER

- **Are the units correct?** m/s^2 is the correct unit for acceleration.
- **Does the sign make sense?** The acceleration is toward Anudja because Anudja is pulling toward herself with a greater force than Sarah is pulling in the opposite direction.
- **Is the magnitude realistic?** The net force is 1 N and the mass is 0.3 kg, so the acceleration is realistic.

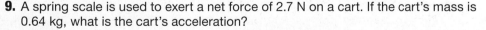
PRACTICE PROBLEMS   Do additional problems. Online Practice

9. A spring scale is used to exert a net force of 2.7 N on a cart. If the cart's mass is 0.64 kg, what is the cart's acceleration?

10. Kamaria is learning how to ice skate. She wants her mother to pull her along so that she has an acceleration of 0.80 m/s^2. If Kamaria's mass is 27.2 kg, with what force does her mother need to pull her? (Neglect any resistance between the ice and Kamaria's skates.)

11. **CHALLENGE** Two horizontal forces are exerted on a large crate. The first force is 317 N to the right. The second force is 173 N to the left.

   a. Draw a force diagram for the horizontal forces acting on the crate.

   b. What is the net force acting on the crate?

   c. The box is initially at rest. Five seconds later, its velocity is 6.5 m/s to the right. What is the crate's mass?

Newton's First Law

What is the motion of an object when the net force acting on it is zero? Newton's second law says that if $F_{net} = 0$, then acceleration equals zero. Recall that if acceleration equals zero, then velocity does not change. Thus a stationary object with no net force acting on it will remain at rest. What about a moving object, such as a ball rolling on a surface? How long will the ball continue to roll? It will depend on the surface. If the ball is rolled on a thick carpet that exerts a force on the ball, it will come to rest quickly. If it is rolled on a hard, smooth surface that exerts very little force, such as a bowling alley, the ball will roll for a long time with little change in velocity. Galileo did many experiments and he concluded that if he could remove all forces opposing motion, horizontal motion would never stop. Galileo was the first to recognize that the general principles of motion could be found by extrapolating experimental results to an ideal case.

In the absence of a net force, the velocity of the moving ball and the lack of motion of the stationary object do not change. Newton recognized this and generalized Galileo's results into a single statement. Newton's statement, "an object that is at rest will remain at rest, and an object that is moving will continue to move in a straight line with constant speed, if and only if the net force acting on that object is zero," is called **Newton's first law.**

Inertia Newton's first law is sometimes called the law of inertia because **inertia** is the tendency of an object to resist changes in velocity. The car and the red block in **Figure 8** demonstrate the law of inertia. In the left panel, both objects are moving to the right. In the right panel, the wooden box applies a force to the car, causing it to stop. The red block does not experience the force applied by the wooden box. It continues to move to the right with the same velocity as in the left panel.

Is inertia a force? No. Forces are results of interactions between two objects; they are not properties of single objects, so inertia cannot be a force. Remember that because velocity includes both the speed and direction of motion, a net force is required to change either the speed or the direction of motion. If the net force is zero, Newton's first law means the object will continue with the same speed and direction.

Figure 8 The car and the block approach the wooden box at the same speed. After the collision, the block continues on with the same horizontal speed.

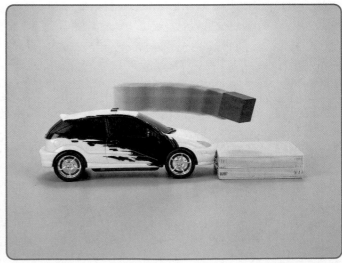

Equilibrium, $v = 0$, $a = 0$

Equilibrium, $v \neq 0$, $a = 0$

Figure 9 An object is in equilibrium if its velocity isn't changing. In both cases pictured here, velocity isn't changing, so the net force must be zero.

Equilibrium According to Newton's first law, a net force causes the velocity of an object to change. If the net force on an object is zero, then the object is in **equilibrium.** An object is in equilibrium if it is moving at a constant velocity. Note that being at rest is simply a special case of the state of constant velocity, $v = 0$. Newton's first law identifies a net force as something that disturbs a state of equilibrium. Thus, if there is no net force acting on the object, then the object does not experience a change in speed or direction and is in equilibrium. As **Figure 9** indicates, at least in terms of net forces, there is no difference between sitting in a chair and falling at a constant velocity while skydiving—velocity isn't changing, so the net force is zero.

When analyzing forces and motion, it is important to keep in mind that the real world is full of forces that resist motion, called frictional forces. Newton's ideal, friction-free world is not easy to obtain. If you analyze a situation and find that the result is different from a similar experience that you have had, ask yourself whether this is because of the presence of frictional forces. For example, if you shove a textbook across a table, it will quickly come to a stop. At first thought, it might seem Newton's first law is violated in this case because the book's velocity changes even though there is no apparent force acting on the book. The net force acting on the book, however, is a frictional force between the table and the book in the direction opposite motion.

SECTION 1 REVIEW

Section Self-Check Check your understanding.

12. **MAIN**IDEA Identify each of the following as either **a, b,** or **c**: mass, inertia, the push of a hand, friction, air resistance, spring force, gravity, and acceleration.

    a. contact force

    b. a field force

    c. not a force

13. **Free-Body Diagram** Draw a free-body diagram of a bag of sugar being lifted by your hand at an increasing speed. Specifically identify the system. Use subscripts to label all forces with their agents. Remember to make the arrows the correct lengths.

14. **Free-Body Diagram** Draw a free-body diagram of a water bucket being lifted by a rope at a decreasing speed. Specifically identify the system. Label all forces with their agents and make the arrows the correct lengths.

15. **Critical Thinking** A force of 1 N is the only horizontal force exerted on a block, and the horizontal acceleration of the block is measured. When the same horizontal force is the only force exerted on a second block, the horizontal acceleration is three times as large. What can you conclude about the masses of the two blocks?

Weight and Drag Force

If you have ever ridden a roller coaster, you probably noticed that you felt weightless as you went over a hill. But the force of gravity at the top of the hill is virtually the same as the force of gravity at the bottom of the hill. So why do you feel weightless?

MAIN IDEA

Newton's second law can be used to explain the motion of falling objects.

Essential Questions

• How are the weight and the mass of an object related?

• How do actual weight and apparent weight differ?

• What effect does air have on falling objects?

Review Vocabulary

viscosity a fluid's resistance to flowing

New Vocabulary

weight
gravitational field
apparent weight
weightlessness
drag force
terminal velocity

Weight

From Newton's second law, the fact that the ball in **Figure 10** is accelerating means there must be unbalanced forces acting on the ball. The only force acting on the ball is the gravitational force due to Earth's mass. An object's **weight** is the gravitational force experienced by that object. This gravitational force is a field force whose magnitude is directly proportional to the mass of the object experiencing the force. In equation form, the gravitational force, which equals weight, can be written $F_g = mg$. The mass of the object is m, and g, called the **gravitational field,** is a vector quantity that relates the mass of an object to the gravitational force it experiences at a given location. Near Earth's surface, g is 9.8 N/kg toward Earth's center. Objects near Earth's surface experience 9.8 N of force for every kilogram of mass.

Scales When you stand on a scale as shown in the right panel of **Figure 10,** the scale exerts an upward force on you. Because you are not accelerating, the net force acting on you must be zero. Therefore the magnitude of the force exerted by the scale ($F_{\text{scale on you}}$) pushing up must equal the magnitude of F_g pulling down on you. Inside the scale, springs provide the upward force necessary to make the net force equal zero. The scale is calibrated to convert the stretch of the springs to a weight. The measurement on the scale is affected by the gravitational field on Earth's surface. If you were on a different planet with a different g, the scale would exert a different force to keep you in equilibrium, and consequently, the scale's reading would be different. Because weight is a force, the proper unit used to measure weight is the newton.

Figure 10 The gravitational force exerted by Earth's mass on an object equals the object's mass times the gravitational field, ($F_g = mg$).

Identify the forces acting on you when you are in equilibrium while standing on a scale.

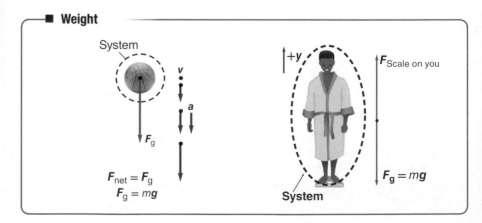

■ **Weight**

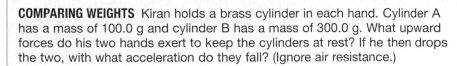

Find help with **operations with significant digits.** | Math Handbook |

COMPARING WEIGHTS Kiran holds a brass cylinder in each hand. Cylinder A has a mass of 100.0 g and cylinder B has a mass of 300.0 g. What upward forces do his two hands exert to keep the cylinders at rest? If he then drops the two, with what acceleration do they fall? (Ignore air resistance.)

1 ANALYZE AND SKETCH THE PROBLEM

- Sketch the situation.
- Identify the two cylinders as the systems, and choose the upward direction as positive.
- Draw the free-body diagrams. Label the forces.

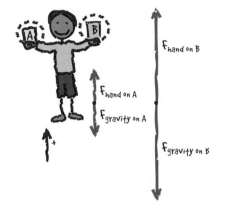

KNOWN	**UNKNOWNS**
$m_A = 0.1000$ kg	$F_{\text{Hand on A}} = ?$
$m_B = 0.3000$ kg	$F_{\text{Hand on B}} = ?$
$g = -9.8$ N/kg	$a_A = ?$ $a_B = ?$

2 SOLVE FOR THE UNKNOWNS

For cylinder A:

$$F_{\text{Net on A}} = F_{\text{Hand on A}} + F_{\text{Gravity on A}}$$
$$0 = F_{\text{Hand on A}} + F_{\text{Gravity on A}}$$
$$F_{\text{Hand on A}} = -F_{\text{Gravity on A}}$$
$$F_{\text{Hand on A}} = -m_A g$$
$$= -(0.1000 \text{ kg})(-9.8 \text{ N/kg})$$
$$= 0.98 \text{ N up}$$

For cylinder B:

$$F_{\text{Net on B}} = F_{\text{Hand on B}} + F_{\text{Gravity on B}}$$
$$0 = F_{\text{Hand on B}} + F_{\text{Gravity on B}}$$
$$F_{\text{Hand on B}} = -F_{\text{Gravity on B}}$$
$$F_{\text{Hand on B}} = -m_B g$$
$$= -(0.3000 \text{ kg })(-9.8 \text{ N/kg})$$
$$= 2.9 \text{ N up}$$

After the cylinders are dropped, the only force on each is the force of gravity. Use Newton's second law.

$$a_A = \frac{F_{\text{Net on A}}}{m_A} \qquad a_B = \frac{F_{\text{Net on B}}}{m_B}$$

$$a_A = \frac{m_A g}{m_A} = g \qquad a_B = \frac{m_B g}{m_B} = g \qquad \blacktriangleleft \text{ Substitute } F_{Net \text{ on } A} = m_A g \text{ and } F_{Net \text{ on } B} = m_B g.$$

$$= -9.8 \text{ m/s}^2 \qquad = -9.8 \text{ m/s}^2 \qquad \blacktriangleleft \text{ Substitute } g = -9.8 \text{ N/kg} = -9.8 \text{ m/s}^2.$$

3 EVALUATE THE ANSWER

- **Are the units correct?** N is the correct unit for force. m/s^2 is the correct unit for acceleration.
- **Does the sign make sense?** The direction of the fall is downward, the negative direction, and the object is speeding up, so the acceleration should be negative.
- **Is the magnitude realistic?** Forces are 1–5 N, typical of that exerted by objects that have a mass of one kg or less. The accelerations are both equal to free fall acceleration.

PRACTICE PROBLEMS

Do additional problems. | Online Practice |

16. You place a watermelon on a spring scale calibrated to measure in newtons. If the watermelon's mass is 4.0 kg, what is the scale's reading?

17. You place a 22.50-kg television on a spring scale. If the scale reads 235.2 N, what is the gravitational field at that location?

18. A 0.50-kg guinea pig is lifted up from the ground. What is the smallest force needed to lift it? Describe the particular motion resulting from this minimum force.

19. CHALLENGE A grocery sack can withstand a maximum of 230 N before it rips. Will a bag holding 15 kg of groceries that is lifted from the checkout counter at an acceleration of 7.0 m/s^2 hold?

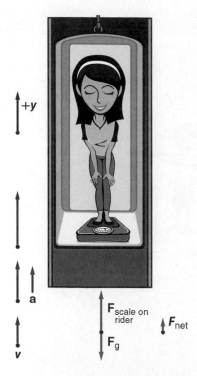

Figure 11 If you are accelerating upward, the net force acting on you must be upward. The scale must exert an upward force greater than the downward force of your weight.

View an **animation of apparent weight.**

Concepts In Motion

PhysicsLAB

FORCES IN AN ELEVATOR
INTERNET LAB How does apparent weight change when riding in an elevator?

MiniLAB

MASS AND WEIGHT
How are mass and weight related?

Apparent weight What is weight? Because the weight force is defined as $F_g = mg$, F_g changes when g varies. On or near the surface of Earth, g is approximately constant, so an object's weight does not change appreciably as it moves around near Earth's surface. If a bathroom scale provides the only upward force on you, then it reads your weight. What would it read if you stood with one foot on the scale and one foot on the floor? What if a friend pushed down on your shoulders or lifted up on your elbows? Then there would be other contact forces on you, and the scale would not read your weight.

What happens if you stand on a scale in an elevator? As long as you are not accelerating, the scale will read your weight. What would the scale read if the elevator accelerated upward? **Figure 11** shows the pictorial and physical representations for this situation. You are the system, and upward is the positive direction. Because the acceleration of the system is upward, the net force must be upward. The upward force of the scale must be greater than the downward force of your weight. Therefore, the scale reading is greater than your weight.

If you ride in an elevator accelerating upward, you feel as if you are heavier because the floor presses harder on your feet. On the other hand, if the acceleration is downward, then you feel lighter, and the scale reads less than your weight. The force exerted by the scale is an example of **apparent weight,** which is the support force exerted on an object.

☑ **READING CHECK Describe** the reading on the scale as the elevator accelerates upward from rest, reaches a constant speed, then comes to a stop.

Imagine that the cable holding the elevator breaks. What would the scale read then? The scale and you would both accelerate at $a = g$. According to this formula, the scale would read zero and your apparent weight would be zero. That is, you would be weightless. However, **weightlessness** does not mean that an object's weight is actually zero; rather, it means that there are no contact forces acting to support the object, and the object's *apparent weight* is zero. Similar to the falling elevator, astronauts experience weightlessness in orbit because they and their spacecraft are in free fall. You will study gravity and weightlessness in greater detail in a later chapter.

PROBLEM-SOLVING STRATEGIES

FORCE AND MOTION
When solving force and motion problems, use the following strategies.

1. Read the problem carefully, and sketch a pictorial model.
2. Circle the system and choose a coordinate system.
3. Determine which quantities are known and which are unknown.
4. Create a physical model by drawing a motion diagram showing the direction of the acceleration.
5. Create a free-body diagram showing all the forces acting on the object.
6. Use Newton's laws to link acceleration and net force.
7. Rearrange the equation to solve for the unknown quantity.
8. Substitute known quantities with their units into the equation and solve.
9. Check your results to see whether they are reasonable.

REAL AND APPARENT WEIGHT Your mass is 75.0 kg, and you are standing on a bathroom scale in an elevator. Starting from rest, the elevator accelerates upward at 2.00 m/s² for 2.00 s and then continues at a constant speed. Is the scale reading during acceleration greater than, equal to, or less than the scale reading when the elevator is at rest?

1 ANALYZE AND SKETCH THE PROBLEM

- Sketch the situation.
- Choose a coordinate system with the positive direction as upward.
- Draw the motion diagram. Label v and a.
- Draw the free-body diagram. The net force is in the same direction as the acceleration, so the upward force is greater than the downward force.

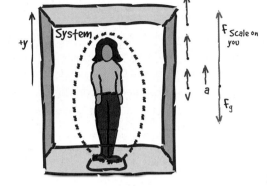

KNOWN

$m = 75.0$ kg
$a = 2.00$ m/s²
$t = 2.00$ s
$g = 9.8$ N/kg

UNKNOWN

$F_{scale} = ?$

2 SOLVE FOR THE UNKNOWN

$F_{net} = ma$

$F_{net} = F_{scale} + (-F_g)$ ◀ F_g is negative because it is in the negative direction defined by the coordinate system.

Solve for F_{scale}.

$F_{scale} = F_{net} + F_g$

Elevator at rest:

$F_{scale} = F_{net} + F_g$ ◀ The elevator is not accelerating. Thus, $F_{net} = 0.00$ N.

$\quad\ = F_g$ ◀ Substitute $F_{net} = 0.00$ N.

$\quad\ = mg$ ◀ Substitute $F_g = mg$.

$\quad\ = (75.0$ kg$)(9.8$ N/kg$)$ ◀ Substitute $m = 75.0$ kg, $g = 9.8$ N/kg.

$\quad\ = 735$ N

Elevator accelerating upward:

$F_{scale} = F_{net} + F_g$

$\quad\ = ma + mg$ ◀ Substitute $F_{net} = ma$, $F_g = mg$

$\quad\ = (75.0$ kg$)(2.00$ m/s²$) + (75.0$ kg$)(9.8$ N/kg$)$ ◀ Substitute $m = 75.0$ kg, $a = 2.00$ m/s², $g = 9.8$ N/kg

$\quad\ = 885$ N

The scale reading when the elevator is accelerating (885 N) is larger than the scale reading when the elevator is at rest (735 N).

3 EVALUATE THE ANSWER

- **Are the units correct?** kg·m/s² is the force unit, N.
- **Does the sign make sense?** The positive sign agrees with the coordinate system.
- **Is the magnitude realistic?** $F_{scale} = 885$ N is larger than it would be at rest when F_{scale} would be 735 N. The increase is 150 N, which is about 20 percent of the rest weight. The upward acceleration is about 20 percent of that due to gravity, so the magnitude is reasonable.

20. On Earth, a scale shows that you weigh 585 N.
   a. What is your mass?
   b. What would the scale read on the Moon ($g = 1.60$ N/kg)?

21. **CHALLENGE** Use the results from Example Problem 3 to answer questions about a scale in an elevator on Earth. What force would be exerted by the scale on a person in the following situations?
   a. The elevator moves upward at constant speed.
   b. It slows at 2.0 m/s² while moving downward.
   c. It speeds up at 2.0 m/s² while moving downward.
   d. It moves downward at constant speed.
   e. In what direction is the net force as the elevator slows to a stop as it is moving down?

PhysicsLAB

TERMINAL VELOCITY

PROBEWARE LAB How does air resistance affect objects in free fall?

MiniLAB

UPSIDE-DOWN PARACHUTE

How does terminal velocity depend on mass?

Drag Force

It is true that the particles in the air around an object exert forces on that object. Air actually exerts huge forces, but in most cases, it exerts balanced forces on all sides, and therefore it has no net effect. Can you think of any experiences that help to prove that air can exert a force? One such example would be holding a piece of paper at arm's length and blowing hard at the paper. The paper accelerates in response to the air striking it, meaning the air exerts a net force on the paper.

So far, you have neglected the force of air on an object moving through the air. In actuality, when an object moves through any fluid, such as air or water, the fluid exerts a force on the moving object in the direction opposite the object's motion. A **drag force** is the force exerted by a fluid on an object opposing motion through the fluid. This force is dependent on the motion of the object, the properties of the object, and the properties of the fluid that the object is moving through. For example, as the speed of the object increases, so does the magnitude of the drag force. The size and shape of the object also affect the drag force. The fluid's properties, such as its density and viscosity, also affect the drag force.

PHYSICS CHALLENGE

A 415-kg container of food and water is dropped from an airplane at an altitude of 300 m. First, consider the situation ignoring air resistance. Then calculate the more realistic situation involving a drag force provided by a parachute.

1. If you ignore air resistance, how long will it take the container to fall 300 m to the ground?

2. Again, ignoring air resistance, what is the speed of the container just before it hits the ground?

3. The container is attached to a parachute designed to produce a drag force that allows the container to reach a constant downward velocity of 6 m/s. What is the magnitude of the drag force when the container is falling at a constant 6 m/s down?

Terminal velocity If you drop a tennis ball, as in **Figure 12**, it has very little velocity at the start and thus only a small drag force. The downward force of gravity is much stronger than the upward drag force, so there is a downward acceleration. As the ball's velocity increases, so does the drag force. Soon the drag force equals the force of gravity. When this happens, there is no net force, and so there is no acceleration. The constant velocity that is reached when the drag force equals the force of gravity is called the **terminal velocity.**

When light objects with large surface areas are falling, the drag force has a substantial effect on their motion, and they quickly reach terminal velocity. Heavier, more-compact objects are not affected as much by the drag force. For example, the terminal velocity of a table-tennis ball in air is 9 m/s, that of a basketball is 20 m/s, and that of a baseball is 42 m/s. Competitive skiers increase their terminal velocities by decreasing the drag force on them. They hold their bodies in an egg shape and wear smooth clothing and streamlined helmets.

Skydivers can increase or decrease their terminal velocity by changing their body orientation and shape. A horizontal, spread-eagle shape produces the slowest terminal velocity, about 60 m/s. After opening up the parachute, the skydiver becomes part of a very large object with a correspondingly large drag force and a terminal velocity of about 5 m/s.

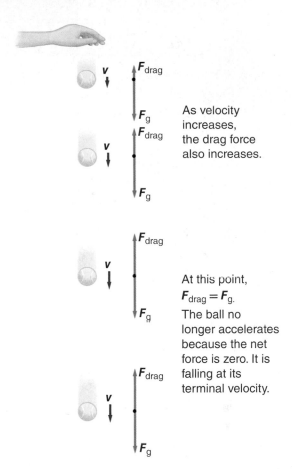

As velocity increases, the drag force also increases.

At this point, $F_{drag} = F_g$. The ball no longer accelerates because the net force is zero. It is falling at its terminal velocity.

Figure 12 The drag force on an object increases as its velocity increases. When the drag force equals the gravitational force, the object is in equilibrium so it no longer accelerates.

View an **animation of terminal velocity.**

Concepts In Motion

SECTION 2 **REVIEW**

Section Self-Check Check your understanding.

22. **MAIN**IDEA The skydiver shown in **Figure 13** falls at a constant speed in the spread-eagle position. Immediately after opening the parachute, is the skydiver accelerating? If so, in which direction? Explain your answer using Newton's laws.

Figure 13

23. **Lunar Gravity** Compare the force holding a 10.0-kg rock on Earth and on the Moon. The gravitational field on the Moon is 1.6 N/kg.

24. **Motion of an Elevator** You are riding in an elevator holding a spring scale with a 1-kg mass suspended from it. You look at the scale and see that it reads 9.3 N. What, if anything, can you conclude about the elevator's motion at that time?

25. **Apparent Weight** You take a ride in a fast elevator to the top of a tall building and ride back down. During which parts of the ride will your apparent and real weights be the same? During which parts will your apparent weight be less than your real weight? More than your real weight? Sketch free-body diagrams to support your answers.

26. **Acceleration** Tecle, with a mass of 65.0 kg, is standing on an ice-skating rink. His friend applies a force of 9.0 N to him. What is Tecle's resulting acceleration?

27. **Critical Thinking** You have a job at a meat warehouse loading inventory onto trucks for shipment to grocery stores. Each truck has a weight limit of 10,000 N of cargo. You push each crate of meat along a low-resistance roller belt to a scale and weigh it before moving it onto the truck. One night, right after you weigh a 1000-N crate, the scale breaks. Describe a way in which you could apply Newton's laws to approximate the masses of the remaining crates.

Newton's Third Law

PHYSICS 4 YOU

If you push against a wall while sitting in a chair that has wheels, you will accelerate across the floor. What applies the unbalanced force that causes your acceleration? Newton's third law helps answer this question.

MAINIDEA

All forces occur in interaction pairs.

Essential Questions

• What is Newton's third law?

• What is the normal force?

Review Vocabulary

symmetry correspondence of parts on opposite sides of a dividing line

New Vocabulary

interaction pair
Newton's third law
tension
normal force

Interaction Pairs

Figure 14 illustrates the idea of forces as interaction pairs. There is a force from the boy on the dog's toy, and there is a force from the dog's toy on the boy. Forces always come in pairs similar to this example. Consider the boy (A) as one system and the toy (B) as another. What forces act on each of the two systems? Looking at the force diagrams in **Figure 14,** you can see that each system exerts a force on the other. The two forces, $F_{A \text{ on } B}$ and $F_{B \text{ on } A}$, are the forces of interaction between the two. Notice the symmetry in the subscripts: A on B and B on A.

The forces $F_{A \text{ on } B}$ and $F_{B \text{ on } A}$ are an **interaction pair,** which is a set of two forces that are in opposite directions, have equal magnitudes, and act on different objects. Sometimes, an interaction pair is called an action-reaction pair. This might suggest that one causes the other; however, this is not true. For example, the force of the boy pulling on the toy doesn't cause the toy to pull on the boy. The two forces either exist together or not at all.

☑ **READING CHECK** **Predict** the magnitude and direction of the force applied on you if you push against a tree with a force of 15 N directed to the left.

Definition of Newton's third law In **Figure 14,** the force exerted by the boy on the toy is equal in magnitude and opposite in direction to the force exerted by the toy on the boy. Such an interaction pair is an example of **Newton's third law,** which states that all forces come in pairs. The two forces in a pair act on different objects and are equal in strength and opposite in direction.

Figure 14 The force that the toy exerts on the boy and the force that the boy exerts on the toy are an interaction pair.

NEWTON'S THIRD LAW

The force of A on B is equal in magnitude and opposite in direction of the force of B on A.

$$\boldsymbol{F}_{\text{A on B}} = -\boldsymbol{F}_{\text{B on A}}$$

Using Newton's third law Consider the situation of holding a book in your hand. You can draw one free-body diagram for you and one for the book. Are there any interaction pairs? When identifying interaction pairs, keep in mind that they always occur in two different free-body diagrams, and they always will have the symmetry of subscripts noted on the previous page. In this case, the interaction pair is $\boldsymbol{F}_{\text{book on hand}}$ and $\boldsymbol{F}_{\text{hand on book}}$.

The ball in **Figure 15** interacts with the table and with Earth. First, analyze the forces acting on one system, the ball. The table exerts an upward force on the ball, and the mass of Earth exerts a downward gravitational force on the ball. Even though these forces are in opposite directions, they are not an interaction pair because they act on the same object. Now consider the ball and the table together. In addition to the upward force exerted by the table on the ball, the ball exerts a downward force on the table. This is an interaction pair.

Notice also that the ball has a weight. If the ball experiences a force due to Earth's mass, then there must be a force on Earth's mass due to the ball. In other words, they are an interaction pair.

$$\boldsymbol{F}_{\text{Earth's mass on ball}} = -\boldsymbol{F}_{\text{ball on Earth's mass}}$$

An unbalanced force on Earth would cause Earth to accelerate. But acceleration is inversely proportional to mass. Because Earth's mass is so huge in comparison to the masses of other objects that we normally consider, Earth's acceleration is so small that it can be neglected. In other words, Earth can be often treated as part of the external world rather than as a second system. The problem-solving strategies below summarize how to deal with interaction pairs.

☑ **READING CHECK Explain** why Earth's acceleration is usually very small compared to the acceleration of the object that Earth interacts with.

PROBLEM-SOLVING STRATEGIES

Use these strategies to solve problems in which there is an interaction between objects in two different systems.

1. Separate the system or systems from the external world.

2. Draw a pictorial model with coordinate systems for each system.

3. Draw a physical model that includes free-body diagrams for each system.

4. Connect interaction pairs by dashed lines.

5. To calculate your answer, use Newton's second law to relate the net force and acceleration for each system.

6. Use Newton's third law to equate the magnitudes of the interaction pairs and give the relative direction of each force.

7. Solve the problem and check the reasonableness of the answers' units, signs, and magnitudes.

Newton's Third Law

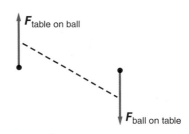

The two forces acting on the ball are $F_{\text{table on ball}}$ and $F_{\text{Earth's mass on ball}}$. These forces are not an interaction pair.

Force interaction pair between ball and table.

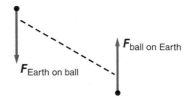

Force interaction pair between ball and Earth.

Figure 15 A ball resting on a table is part of two interaction pairs.

EXAMPLE PROBLEM 4

Find help with **operations with scientific notation.** **Math Handbook**

EARTH'S ACCELERATION A softball has a mass of 0.18 kg. What is the gravitational force on Earth due to the ball, and what is Earth's resulting acceleration? Earth's mass is 6.0×10^{24} kg.

1 ANALYZE AND SKETCH THE PROBLEM

- Draw free-body diagrams for the two systems: the ball and Earth.
- Connect the interaction pair by a dashed line.

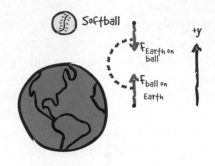

KNOWN

$m_{ball} = 0.18$ kg
$m_{Earth} = 6.0 \times 10^{24}$ kg
$g = 9.8$ N/kg

UNKNOWN

$F_{Earth\ on\ ball} = ?$
$a_{Earth} = ?$

2 SOLVE FOR THE UNKNOWN

Use Newton's second law to find the weight of the ball.

$F_{Earth\ on\ ball} = m_{ball}g$

$\qquad = (0.18\ kg)(-9.8\ N/kg)$ ◀ Substitute $m_{ball} = 0.18$ kg, $g = -9.8$ N/kg.

$\qquad = -1.8\ N$

Use Newton's third law to find $F_{ball\ on\ Earth}$.

$F_{ball\ on\ Earth} = -F_{Earth\ on\ ball}$

$\qquad = -(-1.8\ N)$ ◀ Substitute $F_{Earth\ on\ ball} = -1.8$ N.

$\qquad = +1.8\ N$

Use Newton's second law to find a_{Earth}.

$a_{Earth} = \dfrac{F_{net}}{m_{Earth}}$

$\qquad = \dfrac{1.8\ N}{6.0 \times 10^{24}\ kg}$ ◀ Substitute $F_{net} = 1.8$ N, $m_{Earth} = 6.0 \times 10^{24}$ kg.

$\qquad = 2.9 \times 10^{-25}$ m/s² toward the softball

3 EVALUATE THE ANSWER

- **Are the units correct?** Force is in N and acceleration is in m/s².
- **Do the signs make sense?** Force and acceleration should be positive.
- **Is the magnitude realistic?** It makes sense that Earth's acceleration should be so small because Earth is so massive.

PRACTICE PROBLEMS

Do additional problems. **Online Practice**

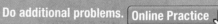

28. You lift a relatively light bowling ball with your hand, accelerating it upward. What are the forces on the ball? What forces does the ball exert? What objects are these forces exerted on?

29. A brick falls from a construction scaffold. Identify any forces acting on the brick. Also identify any forces the brick exerts and the objects on which these forces are exerted. (Air resistance may be ignored.)

30. A suitcase sits on a stationary airport luggage cart, as in **Figure 16**. Draw a free-body diagram for each object and specifically indicate any interaction pairs between the two.

31. CHALLENGE You toss a ball up in the air. Draw a free-body diagram for the ball after it has lost contact with your hand but while it is still moving upward. Identify any forces acting on the ball. Also identify any forces that the ball exerts and the objects on which these forces are exerted. Assume that air resistance is negligible.

Figure 16

Tension

Tension is simply a specific name for the force that a string or rope exerts. A simplification within this textbook is the assumption that all strings and ropes are massless. In **Figure 17,** the rope is about to break in the middle. If the rope breaks, the bucket will fall; before it breaks, there must be forces holding the rope together. The force that the top part of the rope exerts on the bottom part is $F_{\text{top on bottom}}$. Newton's third law states that this force must be part of an interaction pair. The other member of the pair is the force that the bottom part of the rope exerts on the top, $F_{\text{bottom on top}}$. These forces, equal in magnitude but opposite in direction, also are shown in **Figure 17.**

Think about this situation in another way. Before the rope breaks, the bucket is in equilibrium. This means that the force of its weight downward must be equal in magnitude but opposite in direction to the tension in the rope upward. Similarly, if you look at the point in the rope just above the bucket, it also is in equilibrium. Therefore, the tension of the rope below it pulling down must be equal to the tension of the rope above it pulling up. You can move up the rope, considering any point in the rope, and see that the tension forces at any point in the rope are pulling equally in both directions. Thus, the tension in the rope equals the weight of all objects below it.

Examine the tension forces shown in **Figure 18.** If team A is exerting a 500-N force and the rope does not accelerate, then team B also must be pulling with a force of 500 N. What is the tension in the rope? If each team pulls with 500 N of force, is the tension 1000 N? To decide, think of the rope as divided into two halves. The left side is not accelerating, so the net force on it is zero. Thus, $F_{\text{A on left side}} = F_{\text{right side on left side}} = 500$ N. Similarly, $F_{\text{B on right side}} = F_{\text{left side on right side}} = 500$ N. But the two tensions, $F_{\text{right side on left side}}$ and $F_{\text{left side on right side}}$, are an interaction pair, so they are equal and opposite. Thus, the tension in the rope equals the force with which each team pulls, or 500 N. To verify this, you could cut the rope in half and tie the ends to a spring scale. The scale would read 500 N.

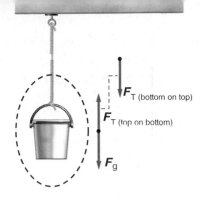

Figure 17 The tension in the rope is equal to the weight of all the objects hanging from it.

Figure 18 The rope is not accelerating, so the tension in the rope equals the force with which each team pulls.

$F_{\text{A on left side}}$ $F_{\text{right side on left side}}$ $F_{\text{left side on right side}}$ $F_{\text{B on right side}}$

Interaction Pair

Team A   Team B

LIFTING A BUCKET A 50.0-kg bucket is being lifted by a rope. The rope will not break if the tension is 525 N or less. The bucket started at rest, and after being lifted 3.0 m, it moves at 3.0 m/s. If the acceleration is constant, is the rope in danger of breaking?

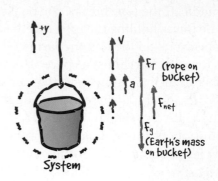

1 ANALYZE AND SKETCH THE PROBLEM

- Draw the situation, and identify the forces on the system.
- Establish a coordinate system with the positive axis upward.
- Draw a motion diagram; include **v** and **a**.
- Draw the free-body diagram, and label the forces.

KNOWN	UNKNOWN
$m = 50.0$ kg $v_f = 3.0$ m/s	$F_T = ?$
$v_i = 0.0$ m/s $d = 3.0$ m	

2 SOLVE FOR THE UNKNOWN

F_{net} is the sum of the positive force of the rope pulling up (F_T) and the negative weight force ($-F_g$) pulling down as defined by the coordinate system.

$$F_{net} = F_T + (-F_g)$$

$$F_T = F_{net} + F_g$$

$$= ma + mg \qquad \blacktriangleleft \text{ Substitute } F_{net} = ma,\ F_g = mg$$

v_i, v_f, and d are known.

$$v_f^2 = v_i^2 + 2ad$$

$$a = \frac{v_f^2 - v_i^2}{2d}$$

$$= \frac{v_f^2}{2d} \qquad \blacktriangleleft \text{ Substitute } v_i = 0.0 \text{ m/s}^2.$$

$$F_T = ma + mg$$

$$= m\left(\frac{v_f^2}{2d}\right) + mg \qquad \blacktriangleleft \text{ Substitute } a = v_f^2 / (2d).$$

$$= (50.0 \text{ kg})\left(\frac{(3.0 \text{ m/s})^2}{2(3.0 \text{ m})}\right) + (50.0 \text{ kg})(9.8 \text{ N/kg}) \qquad \blacktriangleleft \text{ Substitute } m = 50.0 \text{ kg},\ v_f = 3.0 \text{ m/s},\ d = 3.0 \text{ m},\ g = 9.8 \text{ N/kg.}$$

$$= 560 \text{ N}$$

The rope is in danger of breaking because the tension exceeds 525 N.

3 EVALUATE THE ANSWER

- **Are the units correct?** Dimensional analysis verifies kg·m/s^2, which is N.
- **Does the sign make sense?** The upward force should be positive.
- **Is the magnitude realistic?** The magnitude is a little larger than 490 N, which is the weight of the bucket. $F_g = mg = (50.0 \text{ kg})(9.8 \text{ N/kg}) = 490$ N

PRACTICE PROBLEMS   Do additional problems. Online Practice

32. Diego and Mika are trying to fix a tire on Diego's car, but they are having trouble getting the tire loose. When they pull together in the same direction, Mika with a force of 23 N and Diego with a force of 31 N, they just barely get the tire to move off the wheel. What is the magnitude of the force between the tire and the wheel?

33. **CHALLENGE** You are helping to repair a roof by loading equipment into a bucket that workers hoist to the rooftop. If the rope is guaranteed not to break as long as the tension does not exceed 450 N and you fill the bucket until it has a mass of 42 kg, what is the greatest acceleration that the workers can give the bucket as they pull it to the roof?

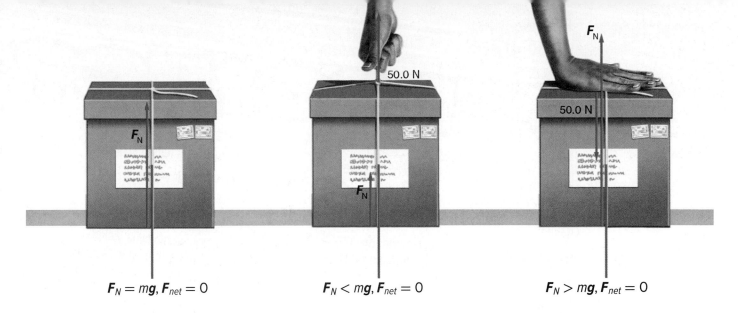

$$F_N = mg, F_{net} = 0$$ $$F_N < mg, F_{net} = 0$$ $$F_N > mg, F_{net} = 0$$

Figure 19 The normal force is not always equal to the object's weight.

The Normal Force

Any time two objects are in contact, they exert a force on each other. Think about a box sitting on a table. There is a downward force on the box due to Earth's gravitational attraction. There also is an upward force that the table exerts on the box. This force must exist because the box is in equilibrium. The **normal force** is the perpendicular contact force that a surface exerts on another surface.

The normal force always is perpendicular to the plane of contact between two objects, but is it always equal to the weight of an object? **Figure 19** shows three situations involving a box with the same weight. What if you tied a string to the box and pulled up on it a little bit, but not enough to accelerate the box, as shown in the middle panel in **Figure 19?** When you apply Newton's second law to the box and the forces acting on the box, you see $F_N + F_{string\ on\ box} - F_g = ma = 0$ N, which can be rearranged to show $F_N = F_g - F_{string\ on\ box}$.

You can see that in this case the normal force that the table exerts on the box is less than the box's weight (F_g). Similarly, if you pushed down on the box on the table as shown in the final panel in **Figure 19,** the normal force would be more than the box's weight. Finding the normal force will be important when you study friction in detail.

PhysicsLAB

NEWTON'S THIRD LAW
What are the interaction pairs between train cars?

SECTION 3 REVIEW

Section Self-Check   Check your understanding.

34. **MAIN**IDEA Hold a ball motionless in your hand in the air as in **Figure 20.** Identify each force acting on the ball and its interaction pair.

Figure 20

35. **Force** Imagine lowering the ball in **Figure 20** at increasing speed. Do any of the forces or their interaction-pair partners change? Draw separate free-body diagrams for the forces acting on the ball and for each set of interaction pairs.

36. **Tension** A block hangs from the ceiling by a massless rope. A second block is attached to the first block and hangs below it on another piece of massless rope. If each of the two blocks has a mass of 5.0 kg, what is the tension in each rope?

37. **Tension** A block hangs from the ceiling by a massless rope. A 3.0-kg block is attached to the first block and hangs below it on another piece of massless rope. The tension in the top rope is 63.0 N. Find the tension in the bottom rope and the mass of the top block.

38. **Critical Thinking** A curtain prevents two tug-of-war teams from seeing each other. One team ties its end of the rope to a tree. If the other team pulls with a 500-N force, what is the tension in the rope? Explain.

SUPERSONIC?

On August 16, 1960, Joe Kittinger ascended 31.3 km into the stratosphere inside a capsule suspended from a helium balloon… and stepped out. He fell for 4 min 36 s, reaching 274 m/s before opening his parachute, landing safely 11 min later.

One small step Kittinger described the surreal experience of first stepping out of the capsule. He had the sensation that he was floating in space, but that was only because his eyes lacked a reference point to gauge his actual speed; in reality, he was dropping faster than anyone ever had before. The jump set records for fastest free fall, longest free fall, and highest parachute jump. Kittinger's jump set the stage for crewed space programs by proving that humans, with proper equipment, can survive the extreme conditions of such high altitudes.

Terminal velocity Terminal velocity is achieved when the upward drag force equals the downward gravitational force. Terminal velocity changes depending on the temperature and density of air at given altitudes. From a normal jump altitude of about 3500 m, the fastest speed a skydiver can achieve is around 90 m/s (nearly 200 mph). Due to the low air density a jumper will experience at extreme altitudes, they could reach speeds of 300 m/s (680 mph) or more.

Going supersonic Kittinger's maximum free-fall speed of 274 m/s (614 mph) was nearly fast enough to break the sound barrier. The speed of sound in air is not a constant but depends on the temperature of the air at a given altitude. At sea level, where the temperature is about 15°C, sound travels at 340 m/s. **Figure 1** shows that at much higher altitudes, where air is less dense and much colder, sound travels more slowly. Breaking the speed of sound during free fall is one of the goals of those attempting to beat Kittinger's records.

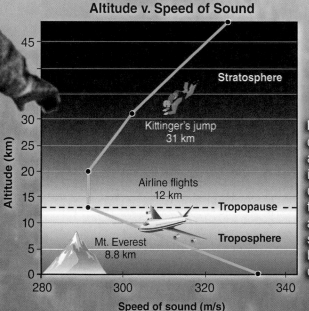

Altitude v. Speed of Sound

FIGURE 1 The speed of sound changes with altitude due to changes in air temperature and density. Suits similar to those worn by astronauts protect stunt skydivers attempting to break the sound barrier during free fall.

GOING**FURTHER** >>>

Research Colonel Kittinger's involvement in Project Excelsior and Project Manhigh. Project Excelsior is especially interesting due to the technical problems during the Excelsior I and Excelsior III jumps.

U.S. Air Force

STUDY GUIDE

BIGIDEA Net forces cause changes in motion.

VOCABULARY
- **force** *(p. 90)*
- **system** *(p. 91)*
- **free-body diagram** *(p. 92)*
- **net force** *(p. 93)*
- **Newton's second law** *(p. 96)*
- **Newton's first law** *(p. 98)*
- **inertia** *(p. 98)*
- **equilibrium** *(p. 99)*

SECTION 1 Force and Motion

MAINIDEA A force is a push or a pull.

- A force is a push or a pull. Forces have both direction and magnitude. A force might be either a contact force or a field force.

- Newton's second law states that the acceleration of a system equals the net force acting on it divided by its mass.

$$a = \frac{F_{net}}{m}$$

- Newton's first law states that an object that is at rest will remain at rest and an object that is moving will continue to move in a straight line with constant speed, if and only if the net force acting on that object is zero. An object with zero net force acting on it is in equilibrium.

VOCABULARY
- **weight** *(p. 100)*
- **gravitational field** *(p. 100)*
- **apparent weight** *(p. 102)*
- **weightlessness** *(p. 102)*
- **drag force** *(p. 104)*
- **terminal velocity** *(p. 105)*

SECTION 2 Weight and Drag Force

MAINIDEA Newton's second law can be used to explain the motion of falling objects.

- The object's weight (F_g) depends on the object's mass and the gravitational field at the object's location.

$$F_g = mg$$

- An object's apparent weight is the magnitude of the support force exerted on it. An object with no apparent weight experiences weightlessness.

- A falling object reaches a constant velocity when the drag force is equal to the object's weight. The constant velocity is called the terminal velocity. The drag force on an object is determined by the object's weight, size, and shape as well as the fluid through which it moves.

VOCABULARY
- **interaction pair** *(p. 106)*
- **Newton's third law** *(p. 106)*
- **tension** *(p. 109)*
- **normal force** *(p. 111)*

SECTION 3 Newton's Third Law

MAINIDEA All forces occur in interaction pairs.

- Newton's third law states that the two forces that make up an interaction pair of forces are equal in magnitude, but opposite in direction and act on different objects. In an interaction pair, $F_{A\ on\ B}$ does not cause $F_{B\ on\ A}$. The two forces either exist together or not at all.

$$F_{A\ on\ B} = -F_{B\ on\ A}$$

- The normal force is a support force resulting from the contact between two objects. It is always perpendicular to the plane of contact between the two objects.

Games and Multilingual eGlossary

Vocabulary Practice

SECTION 1 Force and Motion

Mastering Concepts

39. BIGIDEA You kick a soccer ball across a field. It slows down and comes to a stop. You ask your younger brother to explain what happened to the ball. He says, "The force of your foot was transferred to the ball, which made it move. When that force ran out, the ball stopped." Would Newton agree with that explanation? If not, explain how Newton's laws would describe it.

40. Cycling Imagine riding a single-speed bicycle. Why do you have to push harder on the pedals to start the bicycle moving than to keep it moving at a constant velocity?

Mastering Problems

41. What is the net force acting on a 1.0-kg ball moving at a constant velocity?

42. Skating Joyce and Efua are skating. Joyce pushes Efua, whose mass is 40.0 kg, with a force of 5.0 N. What is Efua's resulting acceleration?

43. A 2300-kg car slows down at a rate of 3.0 m/s² when approaching a stop sign. What is the magnitude of the net force causing it to slow down?

44. Breaking the Wishbone After Thanksgiving, Kevin and Gamal use the turkey's wishbone to make a wish. If Kevin pulls on it with a force 0.17 N larger than the force Gamal pulls with in the opposite direction and the wishbone has a mass of 13 g, what is the wishbone's initial acceleration?

SECTION 2 Weight and Drag Force

Mastering Concepts

45. Suppose that the acceleration of an object is zero. Does this mean that there are no forces acting on the object? Give an example using an everyday situation to support your answer.

46. Basketball When a basketball player dribbles a ball, it falls to the floor and bounces up. Is a force required to make it bounce? Why? If a force is needed, what is the agent involved?

47. A cart has a net horizontal force acting on it to the right. Jon says that it must be moving to the right. Joanne says no, it could be moving in either direction. Is either of these two correct? If so, explain and describe the velocity and acceleration (if any) of the cart.

48. Before a skydiver opens her parachute, she might be falling at a velocity higher than the terminal velocity that she will have after the parachute deploys.

   a. Describe what happens to her velocity as she opens the parachute.

   b. Describe the sky diver's velocity from when her parachute has been open for a time until she is about to land.

49. Three objects are dropped simultaneously from the top of a tall building: a shot put, an air-filled balloon, and a basketball.

   a. Rank the objects in the order in which they will reach terminal velocity, from first to last.

   b. Rank the objects according to the order in which they will reach the ground, from first to last.

   c. What is the relationship between your answers to parts a and b?

Mastering Problems

50. What is your weight in newtons? Show your work.

51. A rescue helicopter lifts two people using a winch and a rescue ring as shown in **Figure 21.**

   a. The winch is capable of exerting a 2000-N force. What is the maximum mass it can lift?

   b. If the winch applies a force of 1200 N, what is the rescuer's and victim's acceleration? Draw a free-body diagram for the people being lifted.

   c. Using the acceleration from part b, how long does it take to pull the people up to the helicopter? Assume the people are initially at rest.

12 m

$m = 120$ kg

Figure 21

52. What force would a scale in an elevator on Earth exert on a 53-kg person standing on it during the following situations?

a. The elevator moves up at a constant speed.

b. It slows at 2.0 m/s² while moving upward.

c. It speeds up at 2.0 m/s² while moving downward.

d. The elevator moves down at a constant speed.

e. It slows to a stop while moving downward with a constant acceleration of 2.5 m/s².

53. Astronomy On the surface of Mercury, the gravitational field is 0.38 times its value on Earth.

a. What would a 6.0-kg mass weigh on Mercury?

b. If the gravitational field on the surface of Pluto is 0.08 times that of Mercury, what would a 7.0-kg mass weigh on Pluto?

54. A 65-kg diver jumps off of a 10.0-m tower. Assume that air resistance is negligible.

a. Find the diver's velocity when the diver hits the water.

b. The diver comes to a stop 2.0 m below the surface. Find the net force exerted by the water.

SECTION 3 Newton's Third Law

Mastering Concepts

55. A rock is dropped from a bridge. Earth pulls on the rock and accelerates it downward. According to Newton's third law, the rock also pulls on Earth, but Earth does not seem to accelerate. Explain.

56. Explain why the tension in a massless rope is constant throughout the rope.

57. Ramon pushes on a bed as shown in **Figure 22**. Draw a free-body diagram for the bed and identify all the forces acting on it. Make a separate list of all the forces that the bed applies to other objects.

Figure 22

58. Baseball A batter swings a bat and hits a baseball. Draw free-body diagrams for the baseball and the bat at the moment of contact. Specifically indicate any interaction pairs between the two diagrams.

59. Ranking Task **Figure 23** shows a block in three different situations. Rank them according to the magnitude of the normal force between the block (or spring) and the floor, greatest to least. Specifically indicate any ties.

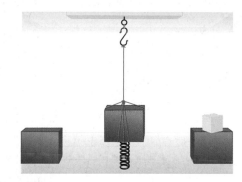

Figure 23

Mastering Problems

60. A 6.0-kg block rests on top of a 7.0-kg block, which rests on a horizontal table.

a. What is the force (magnitude and direction) exerted by the 7.0-kg block on the 6.0-kg block?

b. What is the force (magnitude and direction) exerted by the 6.0-kg block on the 7.0-kg block?

61. Rain A 2.45-mg raindrop falls to the ground. As it is falling, what magnitude of force does it exert on Earth?

62. Male lions and human sprinters can both accelerate at about 10.0 m/s². If a typical lion weighs 170 kg and a typical sprinter weighs 75 kg, what is the difference in the force exerted by the ground during a race between these two species? (Both the forward and normal forces should be calculated.)

63. A 4500-kg helicopter accelerates upward at 2.0 m/s². What lift force is exerted by the air on the propellers?

64. Three blocks are stacked on top of one another. The top block has a mass of 4.6 kg, the middle one has a mass of 1.2 kg, and the bottom one has a mass of 3.7 kg. Identify and calculate any normal forces between the objects.

Applying Concepts

65. Whiplash If you are in a car that is struck from behind, you can receive a serious neck injury called whiplash.

a. Using Newton's laws, explain what happens to cause such an injury.

b. How does a headrest reduce whiplash?

Chapter Self-Check

66. When you look at the label of the product in **Figure 24** to get an idea of how much the box contains, does it tell you its mass, weight, or both? Would you need to make any changes to this label to make it correct for consumption on the Moon?

Figure 24

67. From the top of a tall building, you drop two table-tennis balls, one filled with air and the other with water. Both experience air resistance as they fall. Which ball reaches terminal velocity first? Do both hit the ground at the same time?

68. It can be said that 1 kg is equivalent to 2.21 lb. What does this statement mean? What would be the proper way of making the comparison?

69. You toss a ball straight up into the air. Assume that air resistance is negligible.

    a. Draw a free-body diagram for the ball at three points: on the way up, at the very top, and on the way down. Specifically identify the forces and agents acting on the ball.

    b. What is the ball's velocity at the very top of the motion?

    c. What is the ball's acceleration at this point?

70. When receiving a basketball pass, a player doesn't hold his or her hands still but moves them in the direction of the moving ball. Explain in terms of acceleration and Newton's second law why the player moves his or hands in this manner.

Mixed Review

71. A dragster completed a 402.3-m (0.2500-mi) run in 5.023 s. If the car had a constant acceleration, what was its acceleration and final velocity?

72. Space Station Pratish weighs 588 N on Earth but is currently weightless in a space station. If she pushes off the wall with a vertical acceleration of 3.00 m/s^2, determine the force exerted by the wall during her push off.

73. Jet A 2.75×10^6-N jet plane is ready for takeoff. If the jet's engines supply a constant forward force of 6.35×10^6 N, how much runway will it need to reach its minimum takeoff speed of 285 km/h?

74. Drag Racing A 873-kg dragster, starting from rest, attains a speed of 26.3 m/s in 0.59 s.

    a. Find the average acceleration of the dragster.

    b. What is the magnitude of the average net force on the dragster during this time?

    c. What horizontal force does the seat exert on the driver if the driver has a mass of 68 kg?

75. The dragster in the previous problem completed a 402.3-m track in 4.936 s. It crossed the finish line going 126.6 m/s. Does the assumption of constant acceleration hold true? What information is needed to determine whether the acceleration was constant?

76. Suppose a 65-kg boy and a 45-kg girl use a massless rope in a tug-of-war on an icy, resistance-free surface as in **Figure 25.** If the acceleration of the girl toward the boy is 3.0 m/s^2, find the magnitude of the acceleration of the boy toward the girl.

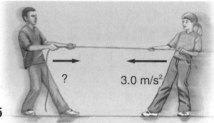

    ?   3.0 m/s^2

Figure 25

77. Baseball As a baseball is being caught, its speed goes from 30.0 m/s to 0.0 m/s in about 0.0050 s. The mass of the baseball is 0.145 kg.

    a. What is the baseball's acceleration?

    b. What are the magnitude and the direction of the force acting on it?

    c. What are the magnitude and the direction of the force acting on the player who caught it?

78. An automobile accelerates uniformly from 0 to 24 m/s in 6.0 s. If the car has a mass of 2.0×10^3 kg, what is the force accelerating it?

79. Air Hockey An air-hockey table works by pumping air through thousands of tiny holes in a table to support light pucks. This allows the pucks to move around on cushions of air with very little resistance. One of these pucks has a mass of 0.25 kg and is pushed along by a 12.0-N force for 0.90 s.

    a. What is the puck's acceleration?

    b. What is the puck's final velocity?

80. Weather Balloon The instruments attached to a weather balloon in **Figure 26** have a mass of 8.0 kg. The balloon is released and exerts an upward force of 98 N on the instruments.

   a. What is the acceleration of the balloon and the instruments?

   b. After the balloon has accelerated for 10.0 s, the instruments are released. What is the velocity of the instruments at the moment of their release?

   c. What net force acts on the instruments after their release?

   d. When does the direction of the instruments' velocity first become downward?

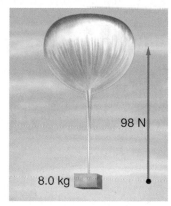

98 N

8.0 kg

Figure 26

81. When a horizontal force of 4.5 N acts on a block on a resistance-free surface, it produces an acceleration of 2.5 m/s². Suppose a second 4.0-kg block is dropped onto the first. What is the magnitude of the acceleration of the combination if the same force continues to act? Assume that the second block does not slide on the first block.

82. Figure 27 shows two blocks, masses 4.3 kg and 5.4 kg, being pushed across a frictionless surface by a 22.5-N horizontal force applied to the 4.3-kg block.

   a. What is the acceleration of the blocks?

   b. What is the force of the 4.3-kg block on the 5.4-kg block?

   c. What is the force of the 5.4-kg block on the 4.3-kg block?

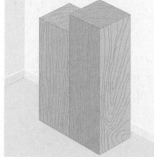

Figure 27

83. A student stands on a bathroom scale in an elevator at rest on the 64th floor of a building. The scale reads 836 N.

   a. As the elevator moves up, the scale reading increases to 936 N. Find the acceleration of the elevator.

   b. As the elevator approaches the 74th floor, the scale reading drops to 782 N. What is the acceleration of the elevator?

   c. Using your results from parts a and b, explain which change in velocity, starting or stopping, takes the longer time.

84. Two blocks, one of mass 5.0 kg and the other of mass 3.0 kg, are tied together with a massless rope as in **Figure 28.** This rope is strung over a massless, resistance-free pulley. The blocks are released from rest. Find the following:

   a. the tension in the rope

   b. the acceleration of the blocks

Hint: You will need to solve two simultaneous equations.

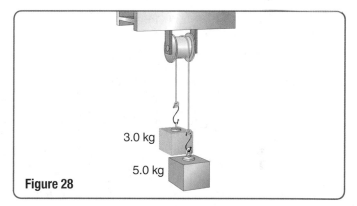

3.0 kg

5.0 kg

Figure 28

Thinking Critically

85. Reverse Problem Write a physics problem with real-life objects for which the following equation would be part of the solution:

$$F = (23 \text{ kg})(1.8 \text{ m/s}^2)$$

86. Formulate Models A 2.0-kg mass (m_A) and a 3.0-kg mass (m_B) are connected to a lightweight cord that passes over a frictionless pulley. The pulley only changes the direction of the force exerted by the rope. The hanging masses are free to move. Choose coordinate systems for the two masses with the positive direction being up for m_A and down for m_B.

   a. Create a pictorial model.

   b. Create a physical model with motion and free-body diagrams.

   c. What is the acceleration of the smaller mass?

87. Use Models Suppose that the masses in the previous problem are now 1.00 kg and 4.00 kg. Find the acceleration of the larger mass.

88. Observe and Infer Three blocks that are connected by massless strings are pulled along a frictionless surface by a horizontal force, as shown in **Figure 29**.

a. What is the acceleration of each block?

b. What are the tension forces in each of the strings? *Hint: Draw a separate free-body diagram for each block.*

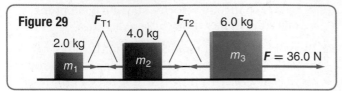

Figure 29 F_{T1} F_{T2} 6.0 kg
2.0 kg 4.0 kg
m_1 m_2 m_3 $F = 36.0$ N

89. Critique Using the Example Problems in this chapter as models, write a solution to the following problem. A 3.46-kg block is suspended from two vertical ropes attached to the ceiling. What is the tension in each rope?

90. Think Critically You are serving as a scientific consultant for a new science-fiction TV series about space exploration. In episode 3, the heroine, Misty Moonglow, has been asked to be the first person to ride in a new interplanetary transport ship. She wants to be sure that the transport actually takes her to the planet she wants to get to, so she needs a device to measure the force of gravity when she arrives. To measure the force of gravity, the script writers would like Misty to perform an experiment involving a scale. It is your job to design a quick experiment Misty can conduct involving a scale to determine which planet she is on. Describe the experiment and include what the results would be for Venus ($g = 8.9$ N/kg), which is where she is supposed to go, and Mercury ($g = 3.7$ N/kg), which is where the transport takes her.

91. Apply Concepts Develop a lab that uses a motion detector and either a calculator or a computer program that graphs the distance a free-falling object moves over equal intervals of time. Also graph velocity versus time. Compare and contrast your graphs. Using your velocity graph, determine the gravitational field. Does it equal g?

92. Problem Posing Complete this problem so that it must be solved using the concept listed below: "A worker unloading a truck gives a 10-kg crate of oranges a push across the floor …"

a. Newton's second law

b. Newton's third law

Writing in Physics

93. Research Newton's contributions to physics and write a one-page summary. Do you think his three laws of motion were his greatest accomplishments? Explain why or why not.

94. Review, analyze, and critique Newton's first law. Can we prove this law? Explain. Be sure to consider the role of resistance.

95. Physicists classify all forces into four fundamental categories: gravitational, electromagnetic, weak nuclear, and strong nuclear. Investigate these forces and describe the situations in which they are found.

Cumulative Review

96. Cross-Country Skiing Your friend is training for a cross-country skiing race, and you and some other friends have agreed to provide him with food and water along his training route. It is a bitterly cold day, so none of you wants to wait outside longer than you have to. Taro, whose house is the stop before yours, calls you at 8:25 A.M. to tell you that the skier just passed his house and is planning to move at an average speed of 8.0 km/h. If it is 5.2 km from Taro's house to yours, when should you expect the skier to pass your house?

97. Figure 30 is a position-time graph of the motion of two cars on a road.

a. At what time(s) does one car pass the other?

b. Which car is moving faster at 7.0 s?

c. At what time(s) do the cars have the same velocity?

d. Over what time interval is car B speeding up all the time?

e. Over what time interval is car B slowing down all the time?

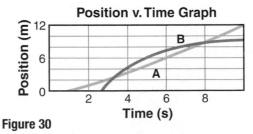

Position v. Time Graph

Figure 30

98. Refer to **Figure 30** to find the instantaneous speed for the following:

a. car B at 2.0 s

b. car B at 9.0 s

c. car A at 2.0 s

MULTIPLE CHOICE

1. What is the acceleration of the car described by the graph below?

   A. 0.20 m/s²   **C.** 1.0 m/s²

   B. 0.40 m/s²   **D.** 2.5 m/s²

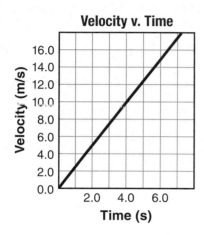

2. What distance will a sprinter travel in 4.0 s if his or her acceleration is 2.5 m/s²? Assume the sprinter starts from rest.

   A. 13 m   **C.** 80 m

   B. 20 m   **D.** 90 m

3. If a motorcycle starts from rest and maintains a constant acceleration of 3 m/s², what will its velocity be after 10 s?

   A. 10 m/s   **C.** 90 m/s

   B. 30 m/s   **D.** 100 m/s

4. How does an object's acceleration change if the net force on the object is doubled?

   A. The acceleration is cut in half.

   B. The acceleration does not change.

   C. The acceleration is doubled.

   D. The acceleration is multiplied by four.

5. What is the weight of a 225-kg space probe on the Moon? The gravitational field on the Moon is 1.62 N/kg.

   A. 139 N   **C.** 1.35×10³ N

   B. 364 N   **D.** 2.21×10³ N

6. A 73-kg woman stands on a scale in an elevator. The scale reads 810 N. What is the magnitude and direction of the elevator's acceleration?

   A. 0.23 m/s² up   **C.** 6.5 m/s² down

   B. 1.3 m/s² up   **D.** 11 m/s² down

7. A 45-kg child sits on a 3.2-kg tire swing. What is the tension in the rope that hangs from a tree branch?

   A. 310 N   **C.** 4.5×10² N

   B. 4.4×10² N   **D.** 4.7×10² N

8. The tree branch in the previous problem sags, and the child's feet rest on the ground. If the tension in the rope is reduced to 220 N, what is the value of the normal force being exerted on the child's feet?

   A. 2.2×10² N   **C.** 4.3×10² N

   B. 2.5×10² N   **D.** 6.9×10² N

9. In the graph below, what is the force being exerted on the 16-kg cart?

   A. 4 N   **C.** 16 N

   B. 8 N   **D.** 32 N

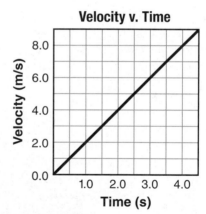

FREE RESPONSE

10. Draw a free-body diagram of a dog sitting on a scale in an elevator. Using words and mathematical formulas, describe what happens to the apparent weight of the dog when the elevator accelerates upward, when the elevator travels at a constant speed downward, and when the elevator falls freely downward.

NEED EXTRA HELP?

If you Missed Question	1	2	3	4	5	6	7	8	9	10
Review Section	1	1	1	1	2	2	3	3	1	2

Displacement and Force in Two Dimensions

BIGIDEA   Forces in two dimensions can be described using vector addition and vector resolution.

SECTIONS

1 Vectors

2 Friction

3 Forces in Two Dimensions

LaunchLAB

ADDING VECTORS

How can you add two force vectors with magnitudes of 2 N and find a sum of 2 N?

WATCH THIS!

FORCES AND MOTION

A hurried commuter leaves a full coffee cup on top of his car before leaving for work. As the car starts moving, what will happen to the coffee cup? How long will it stay on the car before falling off?

PHYSICS T.V.

(l)McGraw-Hill Education; (r)©Pixtal/age fotostock

Vectors

Hanging by your fingertips hundreds of meters above the ground may not be your idea of fun, but every year millions of people enjoy the sport of rock climbing. During their ascents, climbers apply forces in many different directions to overcome the force of gravity pulling them down.

MAIN IDEA

All vectors can be broken into *x*- and *y*-components.

Essential Questions

- How are vectors added graphically?
- What are the components of a vector?
- How are vectors added algebraically?

Review Vocabulary

vector a quantity that has magnitude and direction

New Vocabulary

components
vector resolution

Vectors in Two Dimensions

How do climbers cling to a rock wall? Often the climber has more than one support point. This means there are multiple forces acting on him. Because he grips crevices in the rock, the rock pulls back on him. Also the rope secures him to the rock, so there are two contact forces acting on him. Gravity pulls on him as well, so there are a total of three forces acting on the climber.

One aspect of this situation that is different from the ones that you might have studied earlier is that the forces exerted by the rock face on the climber do not push or pull only in the horizontal or vertical direction. You know from previous study that you can pick your coordinate system and orient it in the way that is most useful to analyze a situation. But how can you set up a coordinate system for a net force when you are dealing with more than one dimension? And what happens when the forces are not at right angles to each other?

Vectors revisited Let's begin by reviewing force vectors in one dimension. Consider the case in **Figure 1** in which you and a friend both push on a table together. Suppose that you each exert a 40-N force to the right. The sum of the forces is 80 N to the right, which is what you probably expected. But how do you find the sum of these force vectors? In earlier discussions of adding vectors along a straight line, you learned the resultant vector always points from the tail of the first vector to the tip of the final vector. Does this rule still apply when the vectors do not lie along a straight line?

Figure 1 The sum of the two applied forces is 80 N to the right.

$$F_{\text{A on table}} + F_{\text{B on table}} = F_{\text{A and B on table}}$$

| 40 N | 40 N | = | 80 N |

$F_{\text{B on table}}$

$F_{\text{A on table}}$

Adding vectors in two dimensions Even when vectors do not lie on a straight line, the resultant vector always points from the tail of the first vector to the tip of the final vector. You can use a protractor and a ruler both to draw the vectors at the correct angles and also to measure the magnitude and the direction of the resultant vector. **Figure 2** illustrates how to add vectors graphically in two dimensions. Notice that when vector **B** is moved, its magnitude and direction are unchanged. This is always the case—you do not change a vector's length or direction when you move that vector.

☑ **READING CHECK** **Describe** the process of graphically adding vectors using a protractor and a ruler.

Perpendicular vectors You can also use trigonometry to determine the length and the direction of resultant vectors. Remember that you can use the Pythagorean theorem to find the lengths of a right triangle's sides. If you were adding together two vectors at right angles, such as vector **A** pointing east and vector **B** pointing north in **Figure 2,** you could use the Pythagorean theorem to find the resultant's magnitude (R).

$$R^2 = A^2 + B^2$$

Angles other than 90° If you are adding two vectors that are at an angle other than 90°, as in **Figure 3,** then you can use the law of sines or the law of cosines. It is best to use the law of sines when you are given two angle measurements and only one vector magnitude. The law of cosines is particularly useful when given two vectors and the angle between the two vectors.

Law of Sines

$$\frac{R}{\sin \theta} = \frac{A}{\sin a} = \frac{B}{\sin b}$$

Law of Cosines

$$R^2 = A^2 + B^2 - 2AB \cos \theta$$

What happens if you apply the law of cosines to a triangle in which $\theta = 90°$? Notice that the first three terms in the law of cosines are the same three terms found in the Pythagorean theorem. The final term in the law of cosines, $-2AB \cos \theta$, equals zero if $\theta = 90°$ because $\cos (90°) = 0$.

$$R^2 = A^2 + B^2 - 2AB \cos 90°$$
$$R^2 = A^2 + B^2 - 2AB (0)$$
$$R^2 = A^2 + B^2$$

If $\theta = 90°$, the triangle is a right triangle and the law of cosines reduces to the Pythagorean theorem.

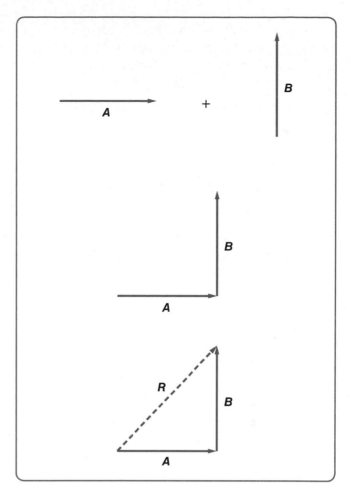

Figure 2 When adding vectors in two dimensions, follow the same process as adding vectors in one dimension: place vectors tip to tail and then connect the tail of the first vector to the tip of the final vector to find the resultant.

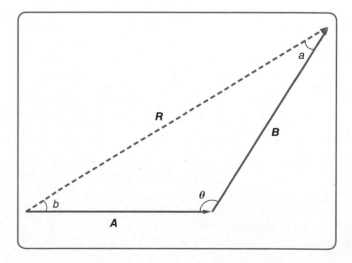

Figure 3 If the vectors are not at right angles, the Pythagorean theorem does not apply. Instead, use the law of cosines or the law of sines with the variables as shown below.

EXAMPLE PROBLEM

FINDING THE MAGNITUDE OF THE SUM OF TWO VECTORS Find the magnitude of the sum of a 15-km displacement and a 25-km displacement when the angle θ between them is 90° and when the angle θ between them is 135°.

1 **ANALYZE AND SKETCH THE PROBLEM**

- Sketch the two displacement vectors, A and B, and the angle between them.

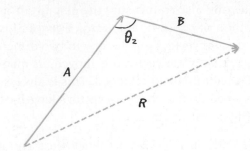

KNOWN	UNKNOWN
$A = 25$ km $\theta = 90°$ or $\theta_2 = 135°$	$R = ?$
$B = 15$ km	

2 **SOLVE FOR THE UNKNOWN**

When the angle θ is 90°, use the Pythagorean theorem to find the magnitude of the resultant vector.

$$R^2 = A^2 + B^2$$

$$R = \sqrt{A^2 + B^2}$$

$$= \sqrt{(25 \text{ km})^2 + (15 \text{ km})^2}$$ ◀ Substitute $A = 25$ km, $B = 15$ km

$$= 29 \text{ km}$$

When the angle θ does not equal 90°, use the law of cosines to find the magnitude of the resultant vector.

$$R^2 = A^2 + B^2 - 2AB(\cos \theta_2)$$

$$R = \sqrt{A^2 + B^2 - 2AB(\cos \theta_2)}$$

$$= \sqrt{(25 \text{ km})^2 + (15 \text{ km})^2 - 2(25 \text{ km})(15 \text{ km})(\cos 135°)}$$ ◀ Substitute $A = 25$ km, $B = 15$ km, $\theta_2 = 135°$

$$= 37 \text{ km}$$

3 **EVALUATE THE ANSWER**

- **Are the units correct?** Each answer is a length measured in kilometers.
- **Do the signs make sense?** The sums are positive.
- **Are the magnitudes realistic?** From the sketch, you can see the resultant should be longer than either vector.

PRACTICE PROBLEMS   Do additional problems.   [Online Practice]

PRACTICE PROBLEMS

1. A car is driven 125.0 km due west then 65.0 km due south. What is the magnitude of its displacement? Solve this problem both graphically and mathematically, and check your answers against each other.

2. Two shoppers walk from the door of the mall to their car. They walk 250.0 m down a lane of cars, and then turn 90° to the right and walk an additional 60.0 m. How far is the shoppers' car from the mall door? Solve this problem both graphically and mathematically, and check your answers against each other.

3. A hiker walks 4.5 km in one direction then makes a 45° turn to the right and walks another 6.4 km. What is the magnitude of the hiker's displacement?

4. **CHALLENGE** An ant crawls on the sidewalk. It first moves south a distance of 5.0 cm. It then turns southwest and crawls 4.0 cm. What is the magnitude of the ant's displacement?

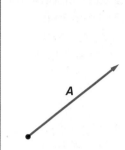

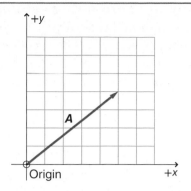

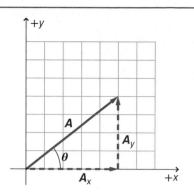

You may place the vector on any coordinate system as long as the vector's direction and magnitude remain unchanged.

The coordinate system can be oriented to make the problem easier to solve.

If the components of a vector are added together, they result in the original vector.

Vector Components

What is the direction of the vector shown in the left panel in **Figure 4?** The vector isn't pointing directly up or to the right, but somewhere between the two. To determine the exact direction, we need to choose a coordinate system. This coordinate system, such as the one in the center panel of **Figure 4,** is similar to laying a grid drawn on a sheet of transparent plastic on top of a vector problem. You choose where to put the center of the grid (the origin) and establish the directions in which the axes point. Notice that the x-axis is drawn through the origin with an arrow pointing in the positive direction. The positive y-axis is located 90° from the positive x-axis and crosses the x-axis at the origin.

How do you choose the direction of the x-axis? There is never a single correct answer, but some choices make the problem easier to solve than others. When the motion you are describing is confined to the surface of Earth, it is often convenient to have the x-axis point east and the y-axis point north. If the motion is on an incline, it's convenient to place the positive x-axis in the direction of the motion parallel to the surface of the incline.

Component vectors Defining a coordinate system allows you to describe a vector in a very useful way. Vector **A,** shown in the right panel of **Figure 4,** for example, could now be described as going 5 units in the positive x-direction and 4 units in the positive y-direction. You can represent this information in the form of two vectors like the ones labeled A_x and A_y in the diagram. Notice that A_x is parallel to the x-axis and A_y is parallel to the y-axis. Further, you can see that if you add A_x and A_y, the resultant is the original vector, **A.** A vector can be broken into its **components,** which are a vector parallel to the x-axis and another parallel to the y-axis. This can always be done, and the following vector equation is always true.

$$A = A_x + A_y$$

This process of breaking a vector into its components is sometimes called **vector resolution.** Notice that the original vector is the hypotenuse of a right triangle. This means that the magnitude of the original vector will always be greater than or equal to the magnitude of either component vector.

Figure 4 Vector **A** is placed on a coordinate system. Notice that **A**'s direction is measured counterclockwise from the positive x-axis.

Describe a vector whose y-component is zero.

View an **animation of component vectors.**

Direction in coordinate systems In a two-dimensional force problem, you should carefully select your coordinate system because the direction of any vector will be specified relative to those coordinates. We define the direction of a vector as the angle that the vector makes with the *x*-axis, measured counterclockwise from the positive *x*-axis. In **Figure 5,** the angle (θ) represents the direction of the vector (**A**). All algebraic calculations involve only the components of vectors, not the vectors themselves.

In addition to measuring the lengths of the component vectors graphically, you can find the components by using trigonometry. The components are calculated using the equations below.

$$\cos \theta = \frac{\text{adjacent side}}{\text{hypotenuse}} = \frac{A_x}{A}; \text{ therefore, } A_x = A \cos \theta$$

$$\sin \theta = \frac{\text{opposite side}}{\text{hypotenuse}} = \frac{A_y}{A}; \text{ therefore, } A_y = A \sin \theta$$

When the angle that a vector makes with the *x*-axis is larger than 90°, the sign of one or more components is negative, as shown in the top coordinate system of **Figure 5.**

✅ **READING CHECK** **Explain** how you should measure the direction of a vector.

Figure 5 The coordinate plane is divided into four quadrants. A vector's components will be positive or negative depending on the vector's quadrant.

Classify as positive or negative the components of a vector whose angle is 280°. In which quadrant does the vector lie?

■ **Coordinate System**

Second quadrant
$90° < \theta < 180°$

A_x is negative.

A_y is positive.

$\tan \theta$ is negative.

II

First quadrant
$0° < \theta < 90°$

A_x is positive.

A_y is positive.

$\tan \theta$ is positive.

I

III

IV

A_x is negative.

A_y is negative.

$\tan \theta$ is positive.

Third quadrant
$180° < \theta < 270°$

A_x is positive.

A_y is negative.

$\tan \theta$ is negative.

Fourth quadrant
$270° < \theta < 360°$

Example:

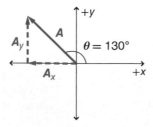

- **A** is in 2nd quadrant.

- Expect A_x to be negative:

$$A_x = A \cos\theta = (5.0 \text{ N}) \cos 130° = -3.2 \text{ N}$$

- Expect A_y to be positive.

$$A_y = A \sin\theta = (5.0 \text{ N}) \sin 130° = 3.8 \text{ N}$$

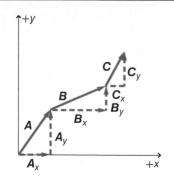

Add the vectors graphically by placing them tip to tail.

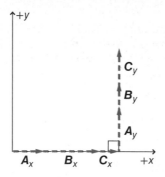

Add the x-components together and the y-components together.

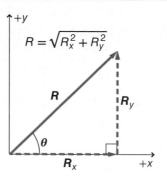

The magnitude of **R** can be calculated using the Pythagorean theorem.

Algebraic Addition of Vectors

You might be wondering why you should resolve vectors into their components. **Figure 6** shows how resolving vectors into components makes adding vectors together much easier. Two or more vectors (**A, B, C,** etc.) may be added by first resolving each vector into its x- and y-components. You add the x-components to form the x-component of the resultant:

$$R_x = A_x + B_x + C_x.$$

Similarly, you add the y-components to form the y-component of the resultant:

$$R_y = A_y + B_y + C_y.$$

Because **R_x** and **R_y** are at a right angle (90°), you can calculate the magnitude of the resultant vector using the Pythagorean theorem,

$$R^2 = R_x{}^2 + R_y{}^2$$

To find the resultant vector's direction, recall that the angle the vector makes with the x-axis is given by the following equation.

$$\theta = \tan^{-1}\left(\frac{R_y}{R_x}\right)$$

You can find the angle by using the $\tan^{-1}$ key on your calculator. Note that when $\tan \theta > 0$, most calculators give the angle between 0° and 90°. When $\tan \theta < 0$, the reported angle will be between 0° and −90°.

Figure 6 The vector sum of **A, B,** and **C** is the same as the vector sum of **R_x** and **R_y**.

PROBLEM-SOLVING STRATEGIES

VECTOR ADDITION
Use the following technique to solve problems for which you need to add or subtract vectors.

1. Choose a coordinate system.

2. Resolve the vectors into their x-components using $A_x = A \cos \theta$ and their y-components using $A_y = A \sin \theta$, where θ is the angle measured counterclockwise from the positive x-axis.

3. Add or subtract the component vectors in the x-direction.

4. Add or subtract the component vectors in the y-direction.

5. Use the Pythagorean theorem, $R = \sqrt{R_x{}^2 + R_y{}^2}$, to find the magnitude of the resultant vector.

6. Use $\theta = \tan^{-1}\left(\frac{R_y}{R_x}\right)$ to find the angle of the resultant vector.

FINDING YOUR WAY HOME You are on a hike. Your camp is 15.0 km away, in the direction 40.0° north of west. The only path through the woods leads directly north. If you follow the path 5.0 km before it opens into a field, how far, and in what direction, would you have to walk to reach your camp?

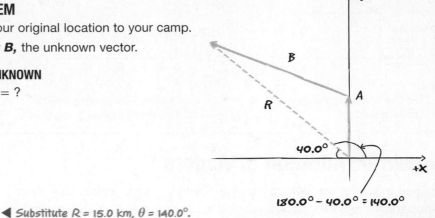

1 ANALYZE AND SKETCH THE PROBLEM

- Draw the resultant vector, **R,** from your original location to your camp.
- Draw **A,** the known vector, and draw **B,** the unknown vector.

KNOWN

A = 5.0 km, due north
R = 15.0 km, 40.0° north of west
$\theta = 140.0°$

UNKNOWN

B = ?

2 SOLVE FOR THE UNKNOWN

Find the components of **R.**

$R_x = R \cos \theta$
$\quad = (15.0 \text{ km}) \cos 140.0°$ ◀ Substitute R = 15.0 km, θ = 140.0°.
$\quad = -11.5 \text{ km}$
$R_y = R \sin \theta$
$\quad = (15.0 \text{ km}) \sin 140.0°$ ◀ Substitute R = 15.0 km, θ = 140.0°.
$\quad = 9.64 \text{ km}$

Because **A** is due north, A_x = 0.0 km and A_y = 5.0 km.
Use the components of **R** and **A** to find the components of **B.**

$B_x = R_x - A_x$
$\quad = -11.5 \text{ km} - 0.0 \text{ km}$ ◀ Substitute R_x = −11.5 km, A_x = 0.0 km.
$\quad = -11.5 \text{ km}$ ◀ The negative sign means that this component points west.
$B_y = R_y - A_y$
$\quad = 9.64 \text{ km} - 5.0 \text{ km}$ ◀ Substitute R_y = 9.64 km, A_y = 5.0 km.
$\quad = 4.6 \text{ km}$ ◀ This component points north.

Use the components of vector **B** to find the magnitude of vector **B.**

$B = \sqrt{B_x^2 + B_y^2}$
$\quad = \sqrt{(-11.5 \text{ km})^2 + (4.6 \text{ km})^2}$ ◀ Substitute B_x = −11.5 km, B_y = 4.6 km.
$\quad = 12.4 \text{ km}$

Locate the tail of vector **B** at the origin of a coordinate system, and draw the components B_x and B_y. The vector **B** is in the second quadrant. Use the tangent to find the direction of vector **B.**

$\theta = \tan^{-1}\left(\dfrac{B_y}{B_x}\right)$
$\quad = \tan^{-1}\left(\dfrac{4.6 \text{ km}}{-11.5 \text{ km}}\right)$ ◀ Substitute B_y = 4.6 km, B_x = −11.5 km.
$\quad = -22° \text{ or } 158°$ ◀ Tangent of an angle is negative in quadrants II and IV, so two answers are possible.

Since **B** is in the second quadrant, θ, measured from the positive x-axis, must be 158°. This direction can also be given as 22° north of west. Thus, **B** = 12.4 km at 22° north of west.

3 EVALUATE THE ANSWER

- **Are the units correct?** Kilometers and degrees are correct.
- **Do the signs make sense?** They agree with the diagram.
- **Is the magnitude realistic?** The length of **B** should be longer than R_x because the angle between **A** and **B** is greater than 90°.

Do additional problems. **Online Practice** 👆

Solve problems 5–10 algebraically. You may also solve some of them graphically to check your answers.

5. Sudhir walks 0.40 km in a direction 60.0° west of north then goes 0.50 km due west. What is his displacement?

6. You first walk 8.0 km north from home then walk east until your displacement from home is 10.0 km. How far east did you walk?

7. In a coordinate system in which the positive x-axis is east, for what range of angles is the x-component positive? For what range is it negative?

8. Could a vector ever be shorter than one of its components? Could a vector ever be equal in length to one of its components? Explain.

9. Two ropes tied to a tree branch hold up a child's swing as shown in **Figure 7**. The tension in each rope is 2.28 N. What is the combined force (magnitude and direction) of the two ropes on the swing?

10. CHALLENGE Afua and Chrissy are going to sleep overnight in their tree house and are using some ropes to pull up a 3.20-kg box containing their pillows and blankets. The girls stand on different branches, as shown in **Figure 8,** and pull at the angles with the forces indicated. Find the x- and y-components of the initial net force on the box. *Hint: Draw a free-body diagram so you do not leave out a force.*

Figure 7

Figure 8

SECTION 1 REVIEW

Section Self-Check 👆 Check your understanding.

11. MAINIDEA Find the components of vector **M,** shown in **Figure 9.**

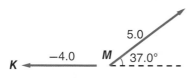

Figure 9

12. Components of Vectors Find the components of vectors **K** and **L** in **Figure 9.**

13. Vector Sum Find the sum of the three vectors shown in **Figure 9.**

14. Vector Difference Subtract vector **K** from vector **L,** shown in **Figure 9.**

15. Commutative Operations Mathematicians say that vector addition is commutative because the order in which vectors are added does not matter.

   a. Use the vectors from **Figure 9** to show graphically that **M** + **L** = **L** + **M.**

   b. Which ordinary arithmetic operations (addition, subtraction, multiplication, and division) are commutative? Which are not? Give an example of each operation to support your conclusion.

16. Distance v. Displacement Is the distance you walk equal to the magnitude of your displacement? Give an example that supports your conclusion.

17. Critical Thinking You move a box through one displacement and then through a second displacement. The magnitudes of the two displacements are unequal. Could the displacements have directions such that the resultant displacement is zero? Suppose you move the box through three displacements of unequal magnitude. Could the resultant displacement be zero? Support your conclusion with a diagram.

Friction

Imagine trying to play basketball while wearing socks instead of athletic shoes. You would slip and slide all over the basketball court. Shoes help provide the forces necessary to quickly change directions while running up and down the court.

Kinetic and Static Friction

Shove your book across a desk. When you stop pushing, the book will quickly come to rest. The friction force of the desk acting on the book accelerated it in the direction opposite to the one in which the book was moving. So far, you have not considered friction in solving problems, but friction is all around you.

Types of friction There are two types of friction. When you moved your book across the desk, the book experienced a type of friction that acts on moving objects. This force is known as **kinetic friction,** and it is exerted on one surface by another when the two surfaces rub against each other because one or both surfaces are moving.

To understand the other type of friction, imagine trying to push a couch across the floor as shown on the left in **Figure 10.** You push on it with a small force, but it does not move. Because it is not accelerating, Newton's laws tell you that the net force on the couch must be zero. There must be a second horizontal force acting on the couch, one that opposes your force and is equal in size. This force is **static friction,** which is the force exerted on one surface by another when there is no motion between the two surfaces. You need to push harder.

If the couch still does not move, the force of static friction must be increasing in response to your applied force. Finally when you push hard enough, the couch will begin to move as in the right side of **Figure 10.** Evidently there is a limit to how large the static friction force can be. Once your force is greater than this maximum static friction, the couch begins moving and kinetic friction begins to act on it.

MAIN IDEA

Friction is a type of force between two touching surfaces.

Essential Questions

• What is the friction force?

• How do static and kinetic friction differ?

Review Vocabulary

force push or pull exerted on an object

New Vocabulary

kinetic friction
static friction
coefficient of kinetic friction
coefficient of static friction

Figure 10 An applied force is balanced by static friction up to a maximum limit. When this limit is exceeded, the object begins to move.

Identify the type of friction force acting on the couch when it begins to move.

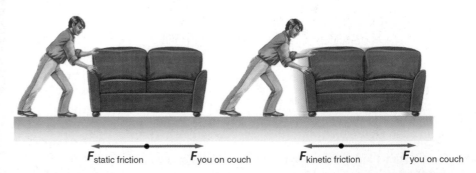

$F_{\text{static friction}}$ $F_{\text{you on couch}}$ $F_{\text{kinetic friction}}$ $F_{\text{you on couch}}$

Static friction increases up to a maximum to balance the applied force.

The couch accelerates when the applied force exceeds the maximum static friction force.

Mathematical models for friction forces On what does a friction force depend? The materials that the surfaces are made of play a role. For example, there is more friction between your shoes and concrete than there is between your socks and a polished wood floor. It might seem reasonable that the force of friction depends on either the surface area in contact or the speed at which the surfaces move past each other. Experiments have shown that this is not true. The normal force between the two objects does matter, however. The harder one object is pushed against the other, the greater the force of friction that results.

✓ **READING CHECK** **Identify** the two factors that affect friction forces.

Kinetic friction Imagine pulling a block along a surface at a constant velocity. Since the block is not accelerating, according to Newton's laws, the friction force must be equal and opposite to the force with which you pull. **Figure 11** shows one way in which you can measure the force you exert as you pull a block of known mass along a table at a constant velocity. The spring scale will indicate the force you exert on the block. You can then stack additional blocks on top of the first block to increase the normal force and repeat the measurement. **Table 1** shows the results of such an experiment.

Plotting the data will yield a graph like the one in **Figure 12.** There is a linear relationship between the kinetic friction force and the normal force. The different lines on the graph correspond to dragging the block along different surfaces. Note that the line for the sandpaper surface has a steeper slope than the line for the highly polished table. You would expect it to be much harder to pull the block along sandpaper than along a polished table, so the slope must be related to the magnitude of the resulting friction force. The slope of the line on a kinetic friction force v. normal force graph, designated μ_k, is called the **coefficient of kinetic friction** and relates the friction force to the normal force.

KINETIC FRICTION FORCE
The kinetic friction force equals the product of the coefficient of kinetic friction and the normal force.

$$F_{f,\,kinetic} = \mu_k F_N$$

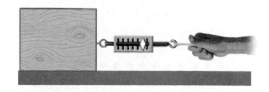

Figure 11 The spring scale applies a constant force on the block.

PhysicsLAB

HIT-AND-RUN DRIVER
FORENSICS LAB How can you determine a car's speed by examining the tire marks left at the scene of an accident?

Figure 12 A plot of kinetic friction v. normal force for a block pulled along different surfaces shows a linear relationship between the two forces for each surface. The slope of the line is μ_k.

Compare the coefficient of kinetic friction for the three surfaces shown on the graph.

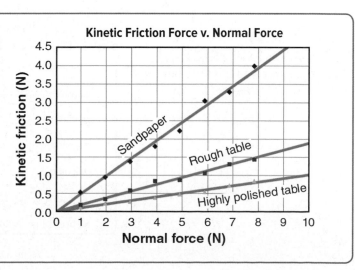

Table 1 Kinetic Friction v. Normal Force (sandpaper)		
Number of blocks	**Normal force (N)**	**Kinetic friction (N)**
1	0.98	0.53
2	1.96	0.95
3	2.94	1.4
4	3.92	1.8
5	4.90	2.3
6	5.88	3.1
7	6.86	3.3
8	7.84	4.0

Kinetic Friction Force v. Normal Force

Sandpaper

Rough table

Highly polished table

Kinetic friction (N)

Normal force (N)

Static friction The maximum static friction force relates to the normal force in a similar way as the kinetic friction force. Remember that the static friction force acts in response to a force trying to cause a stationary object to start moving. If there is no such force acting on an object, the static friction force is zero. If there is a force trying to cause motion, the static friction force will increase up to a maximum value before it is overcome and motion starts.

STATIC FRICTION FORCE
The static friction force is less than or equal to the product of the coefficient of static friction and the normal force.

$$F_{f, \text{static}} \leq \mu_s F_N$$

In the equation for the maximum static friction force, μ_s is the **coefficient of static friction** between the two surfaces. The maximum static friction force that must be overcome before motion can begin is $\mu_s F_N$. In the example of pushing the couch shown in **Figure 10**, the maximum static friction force balances the force of the person pushing on the couch the instant before the couch begins to move.

Note that the equations for the kinetic and maximum static friction forces involve only the magnitudes of the forces. The friction forces (F_f) are always perpendicular to the normal force (F_N).

Measuring coefficients of friction **Table 2** shows coefficients of friction between various surfaces. These coefficients are estimates for each combination of surfaces. Measurements of coefficients of friction are quite sensitive to the conditions of the surfaces. Surface impurities, such as dust or small amounts of oil from experimenters' hands, can significantly affect the results of the measurements. Another important fact regarding **Table 2** is that all the measurements were made on dry surfaces (with the exception of the oiled steel). Wet surfaces behave quite differently than dry surfaces. The linear relationship between the kinetic friction force and the normal force does not always apply to wet surfaces or surfaces treated with oil. All problems in this text assume dry surfaces and a linear relationship between the kinetic friction force and the normal force.

PhysicsLAB

COEFFICIENT OF FRICTION
How are coefficients of static and kinetic friction measured?

Table 2 Typical Coefficients of Friction*		
Surfaces	**Coefficient of static friction (μ_s)**	**Coefficient of kinetic friction (μ_k)**
Cast iron on cast iron	1.1	0.15
Glass on glass	0.94	0.4
Leather on oak	0.61	0.52
Nonstick coating on steel	0.04	0.04
Oak on oak	0.62	0.48
Steel on steel	0.78	0.42
Steel on steel (with castor oil)	0.15	0.08

*All measurements are for dry surfaces unless otherwise stated.

EXAMPLE PROBLEM 3

Find help with **significant figures.** Math Handbook

BALANCED FRICTION FORCES You push a 25.0-kg wooden box across a wooden floor at a constant speed of 1.0 m/s. The coefficient of kinetic friction is 0.20: How large is the force that you exert on the box?

1 ANALYZE AND SKETCH THE PROBLEM

- Identify the forces, and establish a coordinate system.
- Draw a motion diagram indicating constant v and $a = 0$.
- Draw the free-body diagram.

KNOWN		UNKNOWN
$m = 25.0$ kg	$v = 1.0$ m/s	$F_{\text{Person on box}} = ?$
$a = 0.0$ m/s^2	$\mu_k = 0.20$	

2 SOLVE FOR THE UNKNOWN

The normal force is in the y-direction, and the box does not accelerate in that direction.

$$F_N = -F_g$$

$$= -mg \qquad \blacktriangleleft \text{Substitute } F_g = mg$$

$$= -(25.0 \text{ kg})(-9.8 \text{ N/kg}) \qquad \blacktriangleleft \text{Substitute } m = 25.0 \text{ kg, } g = -9.8 \text{ N/kg}$$

$$= +245 \text{ N}$$

The pushing force is in the x-direction; v is constant, thus the box does not accelerate.

$$F_{\text{Person on box}} = \mu_k F_N$$

$$= (0.20)(245 \text{ N}) \qquad \blacktriangleleft \text{Substitute } \mu_k = 0.20, F_N = 245 \text{ N}$$

$$= 49 \text{ N}$$

$$F_{\text{Person on box}} = 49 \text{ N, to the right}$$

3 EVALUATE THE ANSWER

- **Are the units correct?** Force is measured in newtons.
- **Does the sign make sense?** The positive sign agrees with the sketch.
- **Is the magnitude realistic?** The pushing force is $\frac{1}{5}$ the weight of the box. This corresponds with $\mu_k = 0.20 = \frac{1}{5}$.

PRACTICE PROBLEMS

Do additional problems. Online Practice

18. Gwen exerts a 36-N horizontal force as she pulls a 52-N sled across a cement sidewalk at constant speed. What is the coefficient of kinetic friction between the sidewalk and the metal sled runners? Ignore air resistance.

19. Mr. Ames is dragging a box full of books from his office to his car. The box and books together have a combined weight of 134 N. If the coefficient of static friction between the pavement and the box is 0.55, how hard must Mr. Ames push horizontally on the box in order to start it moving?

20. Thomas sits on a small rug on a polished wooden floor. The coefficient of kinetic friction between the rug and the slippery wooden floor is only 0.12. If Thomas weighs 650 N, what horizontal force is needed to pull the rug and Thomas across the floor at a constant speed?

21. **CHALLENGE** You need to move a 105-kg sofa to a different location in the room. It takes a 403-N force to start the sofa moving. What is the coefficient of static friction between the sofa and the carpet?

Find help with **isolating a variable.** Math Handbook

UNBALANCED FRICTION FORCES Imagine the force you exert on the 25.0-kg box in Example Problem 3 is doubled.

a. What is the resulting acceleration of the box?

b. How far will you push the box if you push it for 3 s?

1 ANALYZE AND SKETCH THE PROBLEM
- Draw a motion diagram showing v and a.
- Draw the free-body diagram with a doubled $F_{\text{person on box}}$

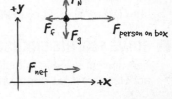

KNOWN		UNKNOWN
$m = 25.0$ kg	$\mu_k = 0.20$	$a = ?$
$v = 1.0$ m/s	$F_{\text{person on box}} = 2(49\text{ N}) = 98$ N	
$t = 3.0$ s		

2 SOLVE FOR THE UNKNOWN

a. The normal force is in the y-direction, and the box does not accelerate in that direction.

$$\mathbf{F}_N = -\mathbf{F}_g$$

$$= -m\mathbf{g}$$

The box does accelerate in the x-direction. So the forces must be unequal.

$$F_{\text{net}} = F_{\text{person on box}} - F_f$$

$$ma = F_{\text{person on box}} - F_f \qquad \blacktriangleleft \text{ Substitute } F_{\text{net}} = ma.$$

$$a = \frac{F_{\text{person on box}} - F_f}{m}$$

Find F_f and substitute it into the expression for a.

$$F_f = \mu_k F_N$$

$$= \mu_k mg \qquad \blacktriangleleft \text{ Substitute } F_N = mg.$$

$$a = \frac{F_{\text{person on box}} - \mu_k mg}{m} \qquad \blacktriangleleft \text{ Substitute } F_f = \mu_k mg.$$

$$= \frac{98\text{ N} - (0.20)(25.0\text{ kg})(9.8\text{ N/kg})}{25.0\text{ kg}} \qquad \blacktriangleleft \text{ Substitute } F_{\text{person on box}} = 98\text{ N}, \mu_k = 0.20, m = 25.0\text{ kg}, g = 9.8\text{ N/kg}.$$

$$= 2.0\text{ m/s}^2$$

b. Use the relationship between distance, time, and constant acceleration.

$$x_f = x_i + v_i t_f + \frac{1}{2}a t_f^2$$

$$x_f = v_i t + \left(\frac{1}{2}\right)a t^2 \qquad \blacktriangleleft \text{ Substitute } x_i = 0\text{ m.}$$

$$= (1.0\text{ m/s})(3.0\text{ s}) + \left(\frac{1}{2}\right)(2.0\text{ m/s}^2)(3.0\text{ s})^2 \qquad \blacktriangleleft \text{ Substitute } v_i = 1\text{ m/s}, t = 3.0\text{ s}, a = 2.0\text{ m/s}^2.$$

$$= 12\text{ m}$$

3 EVALUATE THE ANSWER
- **Are the units correct?** a is measured in m/s², and x_f is measured in meters.
- **Does the sign make sense?** In this coordinate system, the sign should be positive.
- **Is the magnitude realistic?** 12 m is about the length of two full-sized cars. It is realistic to push a box of this distance in 3 s.

22. A 1.4-kg block slides freely across a rough surface such that the block slows down with an acceleration of -1.25 m/s^2. What is the coefficient of kinetic friction between the block and the surface?

23. You want to move a 41-kg bookcase to a different place in the living room. If you push with a force of 65 N and the bookcase accelerates at 0.12 m/s^2, what is the coefficient of kinetic friction between the bookcase and the carpet?

24. Consider the force pushing the box in Example Problem 4. How long would it take for the velocity of the box to double to 2.0 m/s?

25. Ke Min is driving at 23 m/s. He sees a tree branch lying across the road. He slams on the brakes when the branch is 60.0 m in front of him. If the coefficient of kinetic friction between the car's locked tires and the road is 0.41, will the car stop before hitting the branch? The car has a mass of 1200 kg.

26. **CHALLENGE** Isabel pushes a shuffleboard disk, accelerating it to a speed of 6.5 m/s before releasing it as indicated in **Figure 13.** If the coefficient of kinetic friction between the disk and the concrete court is 0.31, how far does the disk travel before it comes to a stop? Will Isabel's shot stop in the 10-point section of the board?

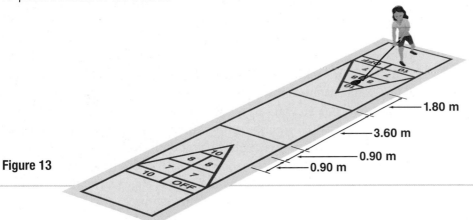

Figure 13

1.80 m
3.60 m
0.90 m
0.90 m

SECTION 2 REVIEW

Section Self-Check Check your understanding.

27. **MAIN**IDEA Compare static friction and kinetic friction. How are the frictional forces similar, and how do the forces differ?

28. **Friction** At a wedding reception, you notice a boy who looks like his mass is about 25 kg running across the dance floor then sliding on his knees until he stops. If the coefficient of kinetic friction between the boy's pants and the floor is 0.15, what is the friction force acting on him as he slides?

29. **Velocity** Dinah is playing cards with her friends, and it is her turn to deal. A card has a mass of 2.3 g, and it slides 0.35 m along the table before it stops. If the coefficient of kinetic friction between the card and the table is 0.24, what was the initial speed of the card as it left Dinah's hand?

30. **Force** The coefficient of static friction between a 40.0-kg picnic table and the ground below that table is 0.43. How large is the greatest horizontal force that could be exerted on the table without moving the table?

31. **Acceleration** You push a 13-kg table in the cafeteria with a horizontal force of 20 N, but the table does not move. You then push the table with a horizontal force of 25 N, and it accelerates at 0.26 m/s^2. What, if anything, can you conclude about the coefficients of static and kinetic friction?

32. **Critical Thinking** Rachel is moving to a new apartment and puts a dresser in the back of her pickup truck. When the truck accelerates forward, what force accelerates the dresser? Under what circumstances could the dresser slide? In which direction?

Forces in Two Dimensions

The person to the left is riding on a zip line. The tension in the rope provides the upward force necessary to balance the person's weight. If the tension in the rope increases, how would the angle the rope makes with the horizontal change?

MAINIDEA

An object is in equilibrium if the net forces in the *x*-direction and in the *y*-direction are zero.

Essential Questions

- How can you find the force required for equilibrium?
- How do you resolve force vector components for motion along an inclined plane?

Review Vocabulary

equilibrium the condition in which the net force on an object is zero

New Vocabulary

equilibrant

Equilibrium Revisited

You have already studied several situations dealing with forces in two dimensions. For example, when friction acts between two surfaces, you must take into account both the friction force that is parallel to the surface and the normal force that is perpendicular to that surface. So far, you have considered only motion along a horizontal surface. Now you will analyze situations in which the forces acting on an object are at angles other than 90°.

Recall that when the net force on an object is zero, the object is in equilibrium. According to Newton's laws, the object will not accelerate because there is no net force acting on it; an object in equilibrium moves with constant velocity. (Remember that staying at rest is a state of constant velocity.) You have already analyzed several equilibrium situations in which two forces act on an object. It is important to realize that equilibrium can also occur if more than two forces act on an object. As long as the net force on the object is zero, the object is in equilibrium.

What is the net force acting on the ring in **Figure 14**? The free-body diagram in **Figure 14** shows the three forces acting on the ring. The ring is not accelerating, so you know the net force must be zero. The free-body diagram, however, does not immediately indicate that the net force is zero. To find the net force, you must add all the vectors together. Remember that vectors may be moved if you do not change their direction (angle) or length. **Figure 15** on the next page shows the process of adding the force vectors to discover the net force.

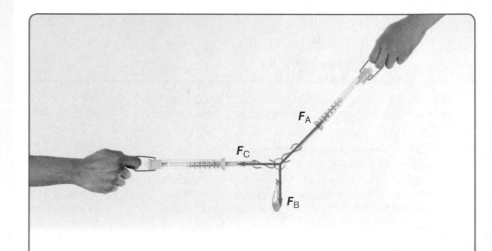

Figure 14 The ring does not accelerate, so the net force acting on it must be zero.

Compare the vertical component of the force pulling up and to the right to the weight of the mass hanging from the ring.

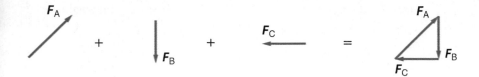

Figure 15 The forces acting on the ring sum to a zero net force.

Equilibrants Figure 15 shows the addition of the three forces, F_A, F_B, and F_C, acting on the ring. Note that the three vectors form a closed triangle. The sum $F_A + F_B + F_C$ is zero. Now suppose you have a situation where two forces are exerted on an object and the sum is not zero, as shown in the free-body diagram in **Figure 16.** How could you find a third force that, when added to the other two, would add up to zero and put the object in equilibrium? To find this force, first add the forces already being exerted on the object. This single force that produces the same effect as the two or more individual forces together is called the resultant force. In order to put the object in equilibrium, you must add another force, called the **equilibrant,** that has the same magnitude as the resultant force but is in the opposite direction. Notice that in **Figure 15** the force F_C is the equilibrant of $F_A + F_B$. **Figure 16** illustrates the procedure for finding the equilibrant for two vectors. Note that this general procedure for finding the equilibrant works for any number of vectors.

☑ **READING CHECK Identify** the relationship between the equilibrant and the resultant vector.

MiniLAB

EQUILIBRIUM
How can you find the equilibrant of two forces?

Figure 16 The equilibrant is the force required to put an object in equilibrium.

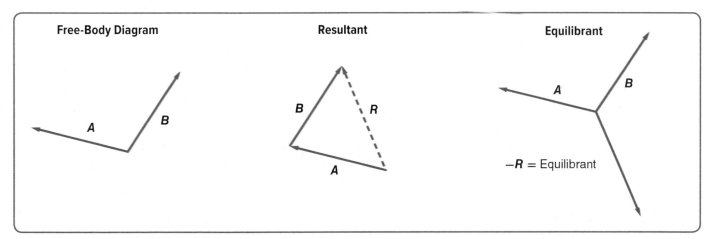

Free-Body Diagram | Resultant | Equilibrant

$-R$ = Equilibrant

PHYSICS CHALLENGE

Find the equilibrant for the entire set of following forces shown in the figure below.

F_1 = 61.0 N at 17.0° north of east
F_2 = 38.0 N at 64.0° north of east
F_3 = 54.0 N at 8.0° west of north
F_4 = 93.0 N at 53.0° west of north
F_5 = 65.0 N at 21.0° south of west

F_6 = 102.0 N at 15.0° west of south
F_7 = 26.0 N south
F_8 = 77.0 N at 22.0° east of south
F_9 = 51.0 N at 33.0° east of south
F_{10} = 82.0 N at 5.0° south of east

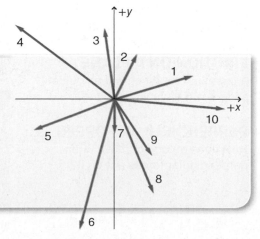

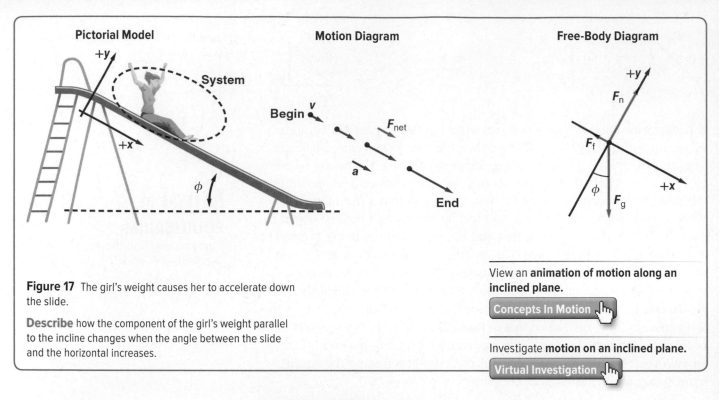

Pictorial Model

+y
System
+x
φ

Motion Diagram

Begin v
F_{net}
a
End

Free-Body Diagram

+y
F_n
F_f
φ
F_g
+x

Figure 17 The girl's weight causes her to accelerate down the slide.

Describe how the component of the girl's weight parallel to the incline changes when the angle between the slide and the horizontal increases.

View an **animation of motion along an inclined plane.**

Concepts In Motion

Investigate **motion on an inclined plane.**

Virtual Investigation

Inclined Planes

You have applied Newton's laws to a variety of situations but only to motions that were either horizontal or vertical. How would you apply Newton's laws in a situation such as the one in **Figure 17,** in which you want to find the net force on the girl?

First, identify the forces acting on the system. In this case, the girl is the system. The gravitational force on the girl is downward, toward the center of Earth. The normal force acts perpendicular to the slide, and the kinetic friction force acts parallel to the slide. You can see the resulting free-body diagram in the final panel of **Figure 17.** You know from experience that the girl's acceleration will be along the slide.

Choosing a coordinate system You should next choose your coordinate system carefully. Because the girl's acceleration is parallel to the incline, one axis, the x-axis, should be in that direction. The y-axis is perpendicular to the x-axis and perpendicular to the incline's surface. With this coordinate system, you now have two forces, the normal force and friction force, in the directions of the coordinate axes. The weight, however, has components in both the x- and y-directions. This means that when you place an object on an inclined plane, the magnitude of the normal force between the object and the plane usually will not equal the object's weight.

☑ **READING CHECK** **Explain** why you would choose the x-direction to be parallel to the slope of an inclined plane.

You will need to apply Newton's laws once in the x-direction and once in the y-direction. Because the weight does not point in either of these directions, you will need to break this vector into its x- and y-components before you can sum the forces in these two directions. Example Problem 5 and Example Problem 6 both show this procedure.

Get help with **components of weight.** Personal Tutor

COMPONENTS OF WEIGHT FOR AN OBJECT ON AN INCLINE A 562-N crate is resting on a plane inclined 30.0° above the horizontal. Find the components of the crate's weight that are parallel and perpendicular to the plane.

1 **ANALYZE AND SKETCH THE PROBLEM**
- Include a coordinate system with the positive x-axis pointing uphill.
- Draw the free-body diagram showing $\boldsymbol{F}_g$, the components $\boldsymbol{F}_{gx}$ and $\boldsymbol{F}_{gy}$, and the angles θ and ϕ.

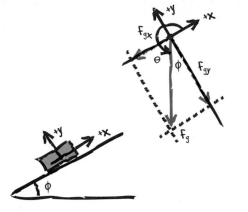

KNOWN

$\boldsymbol{F}_g$ = 562 N down

ϕ = 30.0°

UNKNOWN

F_{gx} = ?

F_{gy} = ?

θ = ?

2 **SOLVE FOR THE UNKNOWN**

$\theta + \phi = 270°$ ◀ The angle from the positive x-axis to the negative y-axis is 270°.

$\theta = 270° - 30°$ ◀ Substitute ϕ = 30° and rearrange.

$= 240°$

$F_{gx} = F_g(\cos \theta)$

$= (562\ \text{N})(\cos 240.0°)$ ◀ Substitute F_g = 562 N, θ = 240.0°.

$= -281\ \text{N}$

$F_{gy} = F_g(\sin \theta)$

$= (562\ \text{N})(\sin 240.0°)$ ◀ Substitute F_g = 562 N, θ = 240.0°.

$= -487\ \text{N}$

3 **EVALUATE THE ANSWER**
- **Are the units correct?** Force is measured in newtons.
- **Do the signs make sense?** The components point in directions opposite to the positive axes.
- **Are the magnitudes realistic?** The values are less than F_g, as expected.

Do additional problems. Online Practice

33. An ant climbs at a steady speed up the side of its anthill, which is inclined 30.0° from the vertical. Sketch a free-body diagram for the ant.

34. Ryan and Becca are moving a folding table out of the sunlight. A cup of lemonade, with a mass of 0.44 kg, is on the table. Becca lifts her end of the table before Ryan does, and as a result, the table makes an angle of 15.0° with the horizontal. Find the components of the cup's weight that are parallel and perpendicular to the plane of the table.

35. Fernando, who has a mass of 43.0 kg, slides down the banister at his grandparents' house. If the banister makes an angle of 35.0° with the horizontal, what is the normal force between Fernando and the banister?

36. **CHALLENGE** A suitcase is on an inclined plane as shown in **Figure 18**. At what angle θ will the component of the suitcase's weight parallel to the plane be equal to half the component of its weight perpendicular to the plane?

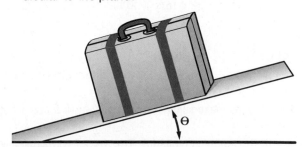

Figure 18

Find help with **isolating a variable.** [Math Handbook]

SLIDE Ichiko, who has a mass of 45 kg, is going down a slide sloped at 27°. The coefficient of kinetic friction is 0.23. How fast does she slide 1.0 s after starting from rest?

1 ANALYZE AND SKETCH THE PROBLEM

- Establish a coordinate system.
- Draw a free-body diagram that shows the forces acting on the girl as she travels down the slide.
- The three forces acting on the girl are the normal force, kinetic friction, and the girl's weight.

KNOWN		UNKNOWN
$m = 45$ kg	$\phi = 27°$	$a = ?$
$\mu_k = 0.23$	$v_i = 0.0$ m/s	$v_f = ?$
$t = 1.0$ s		$\theta = ?$

2 SOLVE FOR THE UNKNOWN

$\theta = 270° + \phi$ ◀ The angle from the positive x-axis to the negative y-axis is 270°.

$= 270° + 27°$ ◀ Substitute $\phi = 27°$.

$= 297°$

y-direction:

$F_{net, y} = ma_y$ ◀ There is no acceleration in the y-direction, so $a_y = 0.0$ m/s².

$= 0.0$ N

Add forces in the *y*-direction to find F_N.

$F_N + F_{gy} = F_{net, y}$

$F_N = -F_{gy}$ ◀ Substitute $F_{net, y} = 0.0$ N and rearrange.

$= -mg(\sin \theta)$ ◀ Substitute $F_{gy} = mg \sin \theta$.

x-direction:

Use the net force in the *x*-direction and Newton's second law to solve for *a*.

$F_{net, x} = F_{gx} - F_f$ ◀ F_f is negative because it is in the negative x-direction.

$ma_x = mg(\cos \theta) - \mu_k F_N$ ◀ Substitute $F_{net, x} = ma_x$, $F_{gx} = mg \cos \theta$, $F_f = \mu_k F_N$.

$ma = mg(\cos \theta) + \mu_k mg(\sin \theta)$ ◀ Substitute $a_x = a$ (all acceleration is in the x-direction), $F_N = -mg \sin \theta$.

$a = g(\cos \theta + \mu_k \sin \theta)$

$= (9.8 \text{ m/s}^2)(\cos 297° + (0.23)\sin 297°)$ ◀ Substitute $g = 9.8$ m/s², $\theta = 297°$, $\mu_k = 0.23$.

$= 2.4 \text{ m/s}^2$

Because v_i, *a*, and *t* are all known, use the relationship between velocity, acceleration, and time.

$v_f = v_i + at$

$= 0.0 \text{ m/s} + (2.4 \text{ m/s}^2)(1.0 \text{ s})$ ◀ Substitute $v_i = 0.0$ m/s, $a = 2.4$ m/s², $t = 1.0$ s.

$= 2.4 \text{ m/s}$

3 EVALUATE THE ANSWER

- **Are the units correct?** Performing dimensional analysis on the units verifies that v_f is in m/s and *a* is in m/s².

- **Do the signs make sense?** Because v_f and *a* are both in the +*x* direction, the signs do make sense.

- **Are the magnitudes realistic?** The velocity is similar to a person's running speed, which is realistic for a steep slide and a low coefficient of kinetic friction.

Do additional problems. | Online Practice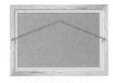

37. Consider the crate on the incline in Example Problem 5. Calculate the magnitude of the acceleration. After 4.00 s, how fast will the crate be moving?

38. Jorge decides to try the slide discussed in Example Problem 6. Jorge's trip down the slide is quite different from Ichiko's. After giving himself a push to get started, Jorge slides at a constant speed. What is the coefficient of kinetic friction between Jorge's pants and the slide?

39. Stacie, who has a mass of 45 kg, starts down a slide that is inclined at an angle of 45° with the horizontal. If the coefficient of kinetic friction between Stacie's shorts and the slide is 0.25, what is her acceleration?

40. **CHALLENGE** You stack two physics books on top of each other as shown in **Figure 19**. You tilt the bottom book until the top book just begins to slide. You perform five trials and measure the angles given in **Table 3**.

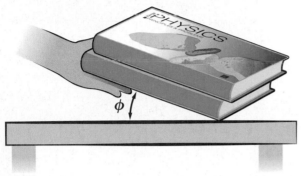

Figure 19

Table 3 Trial Number and Angle of Tilt

Trial	ϕ
1	21°
2	17°
3	21°
4	18°
5	19°

a. What is the average ϕ measured during the five trials?

b. What is the coefficient of static friction between the covers of the two books? Use the average ϕ found in part **a**.

c. You measure the top book's acceleration down the incline to be 1.3 m/s². What is the coefficient of kinetic friction? Assume ϕ is the average value found in part **a**.

SECTION 3 REVIEW

Section Self-Check  Check your understanding.

41. **MAIN**IDEA A rope pulls a 63-kg water skier up a 14.0° incline with a tension of 512 N. The coefficient of kinetic friction between the skier and the ramp is 0.27. What are the magnitude and direction of the skier's acceleration?

42. **Forces** One way to get a car unstuck is to tie one end of a strong rope to the car and the other end to a tree, then pull the rope at its midpoint at right angles to the rope. Draw a free-body diagram and explain how even a small force on the rope can exert a large force on the car.

43. **Mass** A large scoreboard is suspended from the ceiling of a sports arena by ten strong cables. Six of the cables make an angle of 8.0° with the verticals while the other four make an angle of 10.0°. If the tension in each cable is 1300 N, what is the scoreboard's mass?

44. **Vector Addition** What is the sum of three vectors that, when placed tip to tail, form a triangle? If these vectors represent forces on an object, what does this imply about the object? Describe the motion resulting from these three forces acting on the object.

45. **Equilibrium** You are hanging a painting using two lengths of wire. The wires will break if the force is too great. Should you hang the painting as shown in the top or the bottom image of **Figure 20**? Explain.

Figure 20

46. **Critical Thinking** Can the coefficient of friction ever have a value such that a child would be able to slide *up* a slide at a constant velocity? Explain why or why not. Assume that no one pushes or pulls on the child.

Out on a Limb

Have you ever heard the saying, "What goes up must come down"? Architects rely on a thorough understanding of forces to make sure that buildings and other structures stay up.

Balanced forces You may recall that an object accelerates because of a net force acting on it. But just because an object isn't moving doesn't mean forces aren't acting on it. When an object is at rest, the forces acting on the object are balanced.

For example, at this very moment the gravitational force is pulling you down. At the same time, this gravitational force is balanced by the normal force from your chair pushing up against you. Because these forces are balanced, there is no net force and you do not accelerate. When designing a building, bridge, or other structure, an architect must make sure that the gravitational force is properly balanced by other forces.

Cantilevers Have you ever stood on a balcony such as the one shown in **Figure 1?** With nothing pushing up from below to balance the gravitational force, why doesn't it fall? A cantilever is a beam that is supported on only one end. The weight of the cantilever must be balanced by forces exerted by the rest of the structure.

FIGURE 1 These people are looking out from one of the cantilevers on the Willis Tower in Chicago.

GOING**FURTHER** >>>

Research how forces are distributed in a catenary arch, such as the Gateway Arch in St. Louis, Missouri. Create a poster that shows the forces throughout the arch.

STUDY GUIDE

BIGIDEA   Forces in two dimensions can be described using vector addition and vector resolution.

SECTION 1   **Vectors**

MAINIDEA   All vectors can be broken into *x*- and *y*-components.

- Vectors are added graphically by placing the tail of the second vector on the tip of the first vector. The resultant is the vector pointing from the tail of the first vector to the tip of the final vector.

- The components of a vector are projections of the component vectors onto axes. Vectors can be summed by separately adding the *x*- and *y*-components.

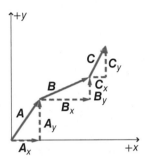 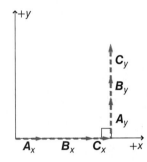 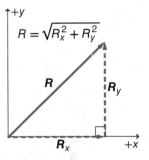

- When two vectors are at right angles, you can use the Pythagorean theorem to determine the magnitude of the resultant vector. The law of cosines and the law of sines can be used to find the resultant of any two vectors.

SECTION 2   **Friction**

MAINIDEA   Friction is a type of force between two touching surfaces.

- Friction is a force that acts parallel to the surfaces when two surfaces touch.

- The kinetic friction force is equal to the coefficient of kinetic friction times the normal force. The static friction force is less than or equal to the coefficient of static friction times the normal force.

$$F_{f,\text{kinetic}} = \mu_k F_N$$

$$F_{f,\text{static}} \le \mu_s F_N$$

SECTION 3   **Forces in Two Dimensions**

MAINIDEA   An object is in equilibrium if the net forces in the *x*-direction and in the *y*-direction are zero.

- The equilibrant is a force of the same magnitude but opposite direction as the sum of all the other forces acting on an object.

- Friction forces are parallel to an inclined plane but point up the plane if the motion of the object is down the plane. An object on an inclined plane has a component of the force of gravity parallel to the plane; this component can accelerate the object down the plane.

Games and Multilingual eGlossary

Vocabulary Practice

ASSESSMENT

SECTION 1 Vectors

Mastering Concepts

47. BIGIDEA How would you add two vectors graphically?

48. Which of the following actions is permissible when you graphically add one vector to another: moving the vector, rotating the vector, or changing the vector's length?

49. In your own words, write a clear definition of the resultant of two or more vectors. Do not explain how to find it; explain what it represents.

50. How is the resultant displacement affected when two displacement vectors are added in a different order?

51. Explain the method you would use to subtract two vectors graphically.

52. Explain the difference between **A** and A.

53. The Pythagorean theorem usually is written $c^2 = a^2 + b^2$. If this relationship is used in vector addition, what do a, b, and c represent?

54. When using a coordinate system, how is the angle or direction of a vector determined with respect to the axes of the coordinate system?

Mastering Problems

55. Cars A car moves 65 km due east then 45 km due west. What is its total displacement?

56. Find the horizontal and vertical components of the following vectors shown in **Figure 21.** In all cases, assume that up and right are positive directions.

  a. E

  b. F

  c. A

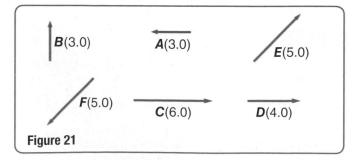

Figure 21

57. Graphically find the sum of the following pairs of vectors, shown in **Figure 21.**

  a. D and A   **c.** C and A

  b. C and D   **d.** E and F

58. Graphically add the following sets of vectors shown in **Figure 21.**

  a. A, C, and D

  b. A, B, and E

  c. B, D, and F

59. Ranking Task Rank the following according to the magnitude of the net force, from least to greatest. Specifically indicate any ties.

  A. 20 N up + 10 N down

  B. 20 N up + 10 N left

  C. 20 N up + 10 N up

  D. 20 N up + 10 N 20° below the horizontal

  E. 20 N up

60. You walk 30 m south and 30 m east. Find the magnitude and direction of the resultant displacement both graphically and algebraically.

61. A hiker's trip consists of three segments. Path A is 8.0 km long heading 60.0° north of east. Path B is 7.0 km long in a direction due east. Path C is 4.0 km long heading 315° counterclockwise from east.

  a. Graphically add the hiker's displacements in the order A, B, C.

  b. Graphically add the hiker's displacements in the order C, B, A.

  c. What can you conclude about the resulting displacements?

62. Two forces are acting on the ring in **Figure 22.** What is the net force acting on the ring?

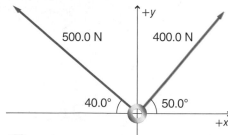

Figure 22

63. Space Exploration A descent vehicle landing on Mars has a vertical velocity toward the surface of Mars of 5.5 m/s. At the same time, it has a horizontal velocity of 3.5 m/s.

  a. At what speed does the vehicle move along its descent path?

  b. At what angle with the vertical is this path?

  c. If the vehicle is 230 m above the surface, how long until it reaches the surface?

64. Three forces are acting on the ring in **Figure 23**. What is the net force acting on the ring?

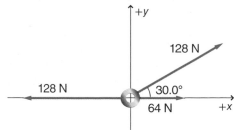

Figure 23

65. A Ship at Sea A ship at sea is due into a port 500.0 km due south in two days. However, a severe storm comes in and blows it 100.0 km due east from its original position. How far is the ship from its destination? In what direction must it travel to reach its destination?

66. Navigation Alfredo leaves camp and, using a compass, walks 4 km E, then 6 km S, 3 km E, 5 km N, 10 km W, 8 km N, and, finally, 3 km S. At the end of two days, he is planning his trip back. By drawing a diagram, compute how far Alfredo is from camp and which direction he should take to get back to camp.

SECTION 2 **Friction**
Mastering Concepts

67. What is the meaning of a coefficient of friction that is greater than 1.0? How would you measure it?

68. Cars Using the model of friction described in this textbook, would the friction between a tire and the road be increased by a wide rather than a narrow tire? Explain. Assume the tires have the same mass .

Mastering Problems

69. If you use a horizontal force of 30.0 N to slide a 12.0-kg wooden crate across a floor at a constant velocity, what is the coefficient of kinetic friction between the crate and the floor?

70. A 225-kg crate is pushed horizontally with a force of 710 N. If the coefficient of friction is 0.20, calculate the acceleration of the crate.

71. A force of 40.0 N accelerates a 5.0-kg block at 6.0 m/s^2 along a horizontal surface.

   a. What would the block's acceleration be if the surface were frictionless?

   b. How large is the kinetic friction force?

   c. What is the coefficient of kinetic friction?

72. Moving Appliances Your family just had a new refrigerator delivered. The delivery man has left and you realize that the refrigerator is not quite in the right position, so you plan to move it several centimeters. If the refrigerator has a mass of 88 kg, the coefficient of kinetic friction between the bottom of the refrigerator and the floor is 0.13, and the static coefficient of friction between these same surfaces is 0.21, how hard do you have to push horizontally to get the refrigerator to start moving?

73. Stopping at a Red Light You are driving a 1200.0-kg car at a constant speed of 14.0 m/s along a straight, level road. As you approach an intersection, the traffic light turns red. You slam on the brakes. The car's wheels lock, the tires begin skidding, and the car slides to a halt in a distance of 25.0 m. What is the coefficient of kinetic friction between your tires and the road?

SECTION 3 **Forces in Two Dimensions**
Mastering Concepts

74. Describe a coordinate system that would be suitable for dealing with a problem in which a ball is thrown up into the air.

75. If a coordinate system is set up such that the positive *x*-axis points in a direction 30° above the horizontal, what should be the angle between the *x*-axis and the *y*-axis? What should be the direction of the positive *y*-axis?

76. Explain how you would set up a coordinate system for motion on a hill.

77. If your textbook is in equilibrium, what can you say about the forces acting on it?

78. Can an object that is in equilibrium be moving? Explain.

79. You are asked to analyze the motion of a book placed on a sloping table.

   a. Describe the best coordinate system for analyzing the motion.

   b. How are the components of the weight of the book related to the angle of the table?

80. For a book on a sloping table, describe what happens to the component of the weight force parallel to the table and the force of friction on the book as you increase the angle the table makes with the horizontal.

   a. Which components of force(s) increase when the angle increases?

   b. Which components of force(s) decrease?

Mastering Problems

81. An object in equilibrium has three forces exerted on it. A 33.0-N force acts at 90.0° from the x-axis, and a 44.0-N force acts at 60.0° from the x-axis. Both angles are measured counterclockwise from the positive x-axis. What are the magnitude and the direction of the third force?

82. Five forces act on the object in **Figure 24:** (1) 60.0 N at 90.0°, (2) 40.0 N at 0.0°, (3) 80.0 N at 270.0°, (4) 40.0 N at 180.0°, and (5) 50.0 N at 60.0°. What are the magnitude and the direction of a sixth force that would produce equilibrium?

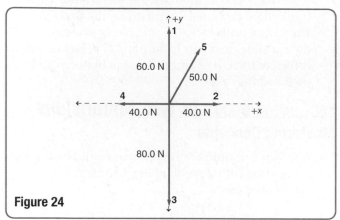

Figure 24

83. Advertising Joe wishes to hang a sign weighing 7.50×10^2 N so that cable A, attached to the store, makes a 30.0° angle, as shown in **Figure 25.** Cable B is horizontal and attached to an adjoining building. What is the tension in cable B?

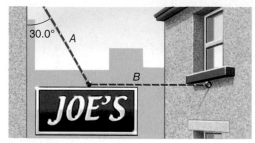

Figure 25

84. A street lamp weighs 150 N. It is supported by two wires that form an angle of 120.0° with each other. The tensions in the wires are equal.

  a. What is the tension in each wire?

  b. If the angle between the wires supporting the street lamp is reduced to 90.0°, what is the tension in each wire?

85. A 215-N box is placed on an incline that makes a 35.0° angle with the horizontal. Find the component of the weight parallel to the incline.

86. Emergency Room You are job-shadowing a nurse in the emergency room of a local hospital. An orderly wheels in a patient who has been in a very serious accident and has had severe bleeding. The nurse quickly explains to you that in a case like this, the patient's bed will be tilted with the head downward to make sure the brain gets enough blood. She tells you that, for most patients, the largest angle that the bed can be tilted without the patient beginning to slide off is 45.0° from the horizontal.

  a. On what factor or factors does this angle of tilting depend?

  b. Find the coefficient of static friction between a typical patient and the bed's sheets.

87. Two blocks are connected by a string over a friction-less, massless pulley such that one is resting on an inclined plane and the other is hanging over the top edge of the plane, as shown in **Figure 26.** The hanging block has a mass of 16.0 kg, and the one on the plane has a mass of 8.0 kg. The coefficient of kinetic friction between the block and the inclined plane is 0.23. The blocks are released from rest.

  a. What is the acceleration of the blocks?

  b. What is the tension in the string connecting the blocks?

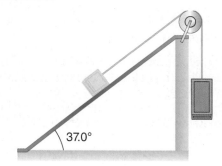

Figure 26

88. In **Figure 27,** a block of mass M is pushed with a force (F) such that the smaller block of mass m does not slide down the front of it. There is no friction between the larger block and the surface below it, but the coefficient of static friction between the two blocks is μ_s. Find an expression for F in terms of M, m, μ_s, and g.

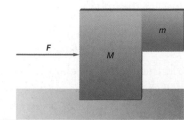

Figure 27

Applying Concepts

89. A vector that is 1 cm long represents a displacement of 5 km. How many kilometers are represented by a 3-cm vector drawn to the same scale?

90. A vector drawn 15 mm long represents a velocity of 30 m/s. How long should you draw a vector to represent a velocity of 20 m/s?

91. What is the largest possible net displacement resulting from two displacements with magnitudes 3 m and 4 m? What is the smallest possible resultant? Draw sketches to demonstrate your answers.

92. How does the resultant displacement change as the angle between two vectors increases from 0° to 180°?

93. A and B are two sides of a right triangle, where $\tan \theta = \frac{A}{B}$.

a. Which side of the triangle is longer if $\tan \theta$ is greater than 1.0?

b. Which side is longer if $\tan \theta$ is less than 1.0?

c. What does it mean if $\tan \theta$ is equal to 1.0?

94. Traveling by Car A car has a velocity of 50 km/h in a direction 60° north of east. A coordinate system with the positive x-axis pointing east and a positive y-axis pointing north is chosen. Which component of the velocity vector is larger, x or y?

95. Under what conditions can the Pythagorean theorem, rather than the law of cosines, be used to find the magnitude of a resultant vector?

96. A problem involves a car moving up a hill, so a coordinate system is chosen with the positive x-axis parallel to the surface of the hill. The problem also involves a stone that is dropped onto the car. Sketch the problem and show the components of the velocity vector of the stone.

97. Pulling a Cart According to legend, a horse learned Newton's laws. When the horse was told to pull a cart, it refused, saying that if it pulled the cart forward, according to Newton's third law, there would be an equal force backwards. Thus, there would be balanced forces, and, according to Newton's second law, the cart would not accelerate. How would you reason with this horse?

98. Tennis When stretching a tennis net between two posts, it is relatively easy to pull one end of the net hard enough to remove most of the slack, but you need a winch to take the last bit of slack out of the net to make the top almost completely horizontal. Why is this true?

99. The weight of a book on an inclined plane can be resolved into two vector components, one along the plane and the other perpendicular to it.

a. At what angle are the components equal?

b. At what angle is the parallel component equal to zero?

100. TV Towers A TV station's transmitting tower is held upright by guy wires that extend from the ground to the top of the tower at an angle of 67° above the horizontal. The force along the guy wires can be resolved into perpendicular and parallel components with respect to the ground. Which one is larger?

Mixed Review

101. The scale in **Figure 28** is being pulled on by three ropes. What net force does the scale read?

27.0°   27.0°

75.0 N   75.0 N

150.0 N

Figure 28

102. Mythology Sisyphus was a character in Greek mythology who was doomed in Hades to push a boulder to the top of a steep mountain. When he reached the top, the boulder would slide back down the mountain and he would have to start all over again. Assume that Sisyphus slides the boulder up the mountain without being able to roll it, even though in most versions of the myth, he rolled it.

a. If the coefficient of kinetic friction between the boulder and the mountainside is 0.40, the mass of the boulder is 20.0 kg, and the slope of the mountain is a constant 30.0°, what is the force that Sisyphus must exert on the boulder to move it up the mountain at a constant velocity?

b. Sisyphus pushes the boulder at a velocity of 0.25 m/s and it takes him 8.0 h to reach the top of the mountain, what is the mythical mountain's vertical height?

103. Landscaping A tree is being transported on a flat-bed trailer by a landscaper. If the base of the tree slides on the trailer, the tree will fall over and be damaged. The coefficient of static friction between the tree and the trailer is 0.50. The truck's initial speed is 55 km/h.

a. The truck must come to a stop at a traffic light without the tree sliding forward and falling on the trailer. What is the maximum possible acceleration the truck can experience?

b. What is the truck's minimum stopping distance if the truck accelerates uniformly at the maximum acceleration calculated in part **a?**

Thinking Critically

104. Use Models Using the Example Problems in this chapter as models, write an example problem to solve the following problem. Include the following sections: Analyze and Sketch the Problem, Solve for the Unknown (with a complete strategy), and Evaluate the Answer. A driver of a 975-kg car traveling 25 m/s puts on the brakes. What is the shortest distance it will take for the car to stop if the coefficient of kinetic friction is 0.65? Assume that the force of friction of the road on the tires is constant and the tires do not slip.

105. Analyze and Conclude Margaret Mary, Doug, and Kako are at a local amusement park and see an attraction called the Giant Slide, which is simply a very long and high inclined plane. Visitors at the amusement park climb a long flight of steps to the top of the 27° inclined plane and are given canvas sacks. They sit on the sacks and slide down the 70-m-long plane. At the time when the three friends walk past the slide, a 135-kg man and a 20-kg boy are each at the top preparing to slide down. "I wonder how much less time it will take the man to slide down than it will take the boy," says Margaret Mary. "I think the boy will take less time," says Doug. "You're both wrong," says Kako. "They will reach the bottom at the same time."

a. Perform the appropriate analysis to determine who is correct.

b. If the man and the boy do not take the same amount of time to reach the bottom of the slide, calculate how many seconds of difference there will be between the two times.

106. Problem Posing Complete this problem so it can be solved using two-dimensional vector addition: "Jeff is cleaning his room when he finds his favorite basketball card under his bed ..."

107. Reverse Problem Write a physics problem with real-life objects for which the **Figure 29** would be part of the solution.

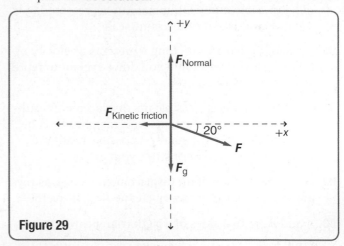

Figure 29

Writing in Physics

108. Investigate some of the techniques used in industry to reduce the friction between various parts of machines. Describe two or three of these techniques and explain the physics of how they work.

109. Olympics In recent years, many Olympic athletes, such as sprinters, swimmers, skiers, and speed skaters, have used modified equipment to reduce the effects of friction and air or water drag. Research a piece of equipment used by one of these types of athletes and the way it has changed over the years. Explain how physics has impacted these changes.

Cumulative Review

110. Perform the indicated addition, subtraction, multiplication, or division and state the answer with the correct number of significant digits.

a. 85.26 g + 4.7 g

b. 1.07 km + 0.608 km

c. 186.4 kg − 57.83 kg

d. (1.26 kg)(9.8 N/kg)

e. $\frac{10 \text{ m}}{4.5 \text{ s}}$

111. You ride your bike for 1.5 h at an average velocity of 10 km/h, then for 30 min at 15 km/h. What is your average velocity?

112. A 45-N force is exerted in the upward direction on a 2.0-kg briefcase. What is the acceleration of the briefcase?

MULTIPLE CHOICE

1. What is the term used to describe a force that equals the sum of two or more forces?

A. equilibrant

B. normal force

C. resultant force

D. tension force

2. For a winter fair, some students decide to build 30.0-kg wooden pull carts on sled skids. If two 90.0-kg passengers get in, how much force will the puller have to exert to move a pull cart? The coefficient of maximum static friction between the sled skids and the road is 0.15.

A. 1.8×10^2 N

B. 3.1×10^2 N

C. 2.1×10^3 N

D. 1.4×10^4 N

3. It takes a minimum force of 280 N to move a 50.0-kg crate. What is the coefficient of maximum static friction between the crate and the floor?

A. 0.18

B. 0.57

C. 1.8

D. 5.6

4. Two tractors in the figure below pull against a 1.00×10^3-kg log. If the angle of the tractors' chains in relation to each other is 18.0° and each tractor pulls with a force of 8×10^2 N, how large is the force exerted by the tractors on the log?

A. 250 N

B. 7.90×10^2 N

C. 1.58×10^3 N

D. 9.80×10^3 N

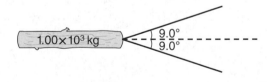

5. An airplane pilot tries to fly directly east with a velocity of 800.0 km/h. If a wind comes from the southwest at 80.0 km/h, what is the relative velocity of the airplane to the surface of Earth?

A. 804 km/h, 5.7° N of E

B. 858 km/h, 3.8° N of E

C. 859 km/h, 4.0° N of E

D. 880 km/h, 45° N of E

6. What is the *y*-component of a 95.3-N force that is exerted at 57.1° to the horizontal?

A. 51.8 N

B. 80.0 N

C. 114 N

D. 175 N

7. As shown in the figure below, a string exerts a force of 18 N on a box at an angle of 34° from the horizontal. What is the horizontal component of the force on the box?

A. 10 N

B. 15 N

C. 21.7 N

D. 32 N

8. Sukey is riding her bicycle on a path when she comes around a corner and sees that a fallen tree is blocking the way 42 m ahead. If the coefficient of friction between her bicycle's tires and the gravel path is 0.36 and she is traveling at 25.0 km/h, how much stopping distance will she require? Sukey and her bicycle, together, have a mass of 95 kg.

A. 3.0 m

B. 4.5 m

C. 6.8 m

D. 9.8 m

FREE RESPONSE

9. A man starts from a position 310 m north of his car and walks for 2.7 min in a westward direction at a constant velocity of 10 km/h. How far is he from his car when he stops?

10. Jeeves is tired of his 41.2-kg son sliding down the banister, so he decides to apply an extremely sticky paste to the top of the banister. The paste increases the coefficient of static friction to 0.72. What will be the magnitude of the static friction force on the boy if the banister is at an angle of 52.4° from the horizontal?

NEED EXTRA HELP?

If you Missed Question	1	2	3	4	5	6	7	8	9	10
Review Section	3	2	2	1	1	1	1	2	1	3

Online Test Practice

CHAPTER 6

Motion in Two Dimensions

BIGIDEA You can use vectors and Newton's laws to describe projectile motion and circular motion.

SECTIONS

LaunchLAB

PROJECTILE MOTION

What does the path of a projectile, such as a ball that is thrown, look like?

WATCH THIS!

Video

PROJECTILE PHYSICS

Have you ever seen a catapult or trebuchet in action? Discover the physics of launching projectiles!

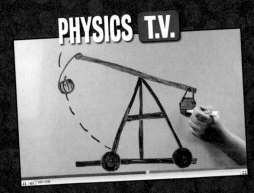

PHYSICS T.V.

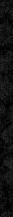

(l)McGraw-Hill Education; (r)GustoImages/Photo Researchers

Projectile Motion

When you throw a softball or a football, it travels in an arc. These tossed balls are projectiles. The word *projectile* comes from the Latin prefix *pro–*, meaning "forward", and the Latin root *ject* meaning "to throw."

MAINIDEA

A projectile's horizontal motion is independent of its vertical motion.

Essential Questions

- How are the vertical and horizontal motions of a projectile related?

- What are the relationships between a projectile's height, time in the air, initial velocity, and horizontal distance traveled?

Review Vocabulary

motion diagram a series of images showing the positions of a moving object taken at regular time intervals

New Vocabulary

projectile
trajectory

Path of a Projectile

A hopping frog, a tossed snowball, and an arrow shot from a bow all move along similar paths. Each path rises and then falls, always curving downward along a parabolic path. An object shot through the air is called a **projectile.** You can draw a free-body diagram of a launched projectile and identify the forces acting on it. If you ignore air resistance, after an initial force launches a projectile, the only force on it as it moves through the air is gravity. Gravity causes the object to curve downward. Its path through space is called its **trajectory.** You can determine a projectile's trajectory if you know its initial velocity. In this chapter, you will study two types of projectile motion. The top of **Figure 1** shows water that is launched as a projectile horizontally. The bottom of the figure shows water launched as a projectile at an angle. In both cases, gravity curves the path downward along a parabolic path.

Figure 1 A projectile launched horizontally immediately curves downward, but if it is launched upward at an angle, it rises and then falls, always curving downward.

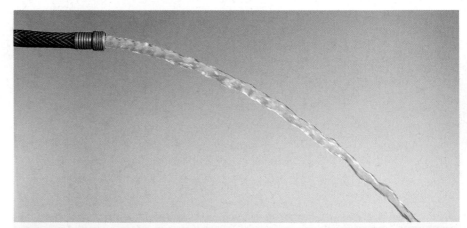

Independence of Motion in Two Dimensions

Think about two softball players warming up for a game, tossing high fly balls back and forth. What does the path of the ball through the air look like? Because the ball is a projectile, it has a parabolic path. Imagine you are standing directly behind one of the players and you are watching the softball as it is being tossed. What would the motion of the ball look like? You would see it go up and back down, just like any object that is tossed straight up in the air. If you were watching the softball from a hot-air balloon high above the field, what motion would you see then? You would see the ball move from one player to the other at a constant speed, just like any object that is given an initial horizontal velocity, such as a hockey puck sliding across ice. The motion of projectiles is a combination of these two motions.

Why do projectiles behave in this way? After a softball leaves a player's hand, what forces are exerted on the ball? If you ignore air resistance, there are no contact forces on the ball. There is only the field force of gravity in the downward direction. How does this affect the ball's motion? Gravity causes the ball to have a downward acceleration.

Comparing motion diagrams The trajectories of two balls are shown in **Figure 2.** The red ball was dropped, and the blue ball was given an initial horizontal velocity. What is similar about the two paths? Look at their vertical positions. The horizontal lines indicate the equal vertical distances. At each moment that a picture was taken, the heights of the two balls were the same. Because the change in vertical position was the same for both, their average vertical velocities during each interval were also the same. The increasingly large distance traveled vertically by the balls, from one time interval to the next, shows that they were accelerating downward due to the force of gravity.

Notice that the horizontal motion of the launched ball does not affect its vertical motion. A projectile launched horizontally has initial horizontal velocity, but it has no initial vertical velocity. Therefore, its vertical motion is like that of an object dropped from rest. Just like the red ball, the blue ball has a downward velocity that increases regularly because of the acceleration due to gravity.

☑ **READING CHECK Explain** why a dropped object has the same vertical velocity as an object launched horizontally.

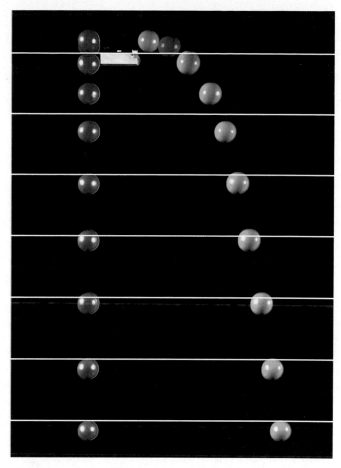

Figure 2 The ball on the left was dropped with no initial velocity. The ball on the right was given an initial horizontal velocity. The balls have the same vertical motion as they fall.

Identify What is the vertical velocity of the balls after falling for 1 s?

Investigate a **soccer kick.**

Virtual Investigation 🖱

MiniLABs

OVER THE EDGE
Does mass affect the motion of a projectile?

PROJECTILE PATH
Are the horizontal and vertical motions of a projectile related?

McGraw-Hill Education

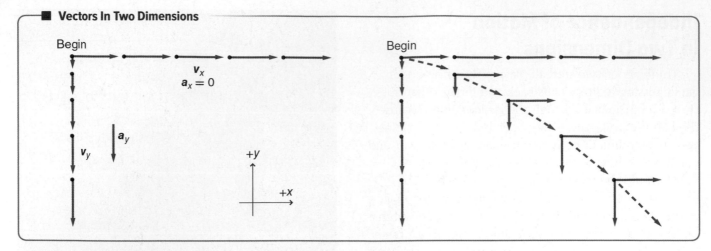

Figure 3 To describe the motion of a horizontally launched projectile, the *x*- and *y*-components can be treated independently. The resultant vectors of the projectile are tangent to a parabola.

Decide What is the value of a_y?

PhysicsLABs

LAUNCH AN INVESTIGATION

FORENSICS LAB How can physics reconstruct a projectile's launch?

ON TARGET

DESIGN YOUR OWN LAB What factors affect projectile motion?

Horizontally Launched Projectiles

Imagine a person standing near the edge of a cliff and kicking a pebble horizontally. Like all horizontally launched projectiles, the pebble will have an initial horizontal velocity, but it will not have an initial vertical velocity. What will happen to the pebble as it falls from the cliff?

Separate motion diagrams Recall that the horizontal motion of a projectile does not affect its vertical motion. It is therefore easier to analyze the horizontal motion and the vertical motion separately. Separate motion diagrams for the *x*-components and *y*-components of a horizontally launched projectile, such as a pebble kicked off a cliff, are shown on the left in **Figure 3.**

Horizontal motion Notice the horizontal vectors in the diagram on the left. Each of the velocity vectors is the same length, which indicates that the object's velocity is not changing. The pebble is not accelerating horizontally. This constant velocity in the horizontal direction is exactly what should be expected, because after the initial kick, there is no horizontal force acting on the pebble. (In reality, the pebble's speed would decrease slightly because of air resistance, but remember that we are ignoring air resistance in this chapter.)

☑ **READING CHECK** Explain why the horizontal motion of a projectile is constant.

Vertical motion Now look at the vertical velocity vectors in the diagram on the left. Each velocity vector has a slightly longer length than the one above it. The changing length shows that the object's velocity is increasing and accelerating downward. Again, this is what should be expected, because in this case the force of gravity is acting on the pebble.

Parabolic path When the *x*- and *y*-components of the object's motion are treated independently, each path is a straight line. The diagram on the right in **Figure 3** shows the actual parabolic path. The horizontal and vertical components at each moment are added to form the total velocity vector at that moment. You can see how the combination of constant horizontal velocity and uniform vertical acceleration produces a trajectory that has a parabolic shape.

PROBLEM-SOLVING STRATEGIES

MOTION IN TWO DIMENSIONS
When solving projectile problems, use the following strategies.

1. Draw a motion diagram with vectors for the projectile at its initial position and its final position. If the projectile is launched at an angle, also show its maximum height and the initial angle.

2. Consider vertical and horizontal motion independently. List known and unknown variables.

3. For horizontal motion, the acceleration is $a_x = 0.0$ m/s^2. If the projectile is launched at an angle, its initial vertical velocity and its vertical velocity when it falls back to that same height have the same magnitude but different direction: $v_{yi} = -v_{yf}$.

4. For vertical motion, $a_y = -9.8$ m/s^2 (if you choose up as positive). If the projectile is launched at an angle, its vertical velocity at its highest point is zero: $v_{y,\,max} = 0$.

5. Choose the motion equations that will enable you to find the unknown variables. Apply them to vertical and horizontal motion separately. Remember that time is the same for horizontal and vertical motion. Solving for time in one of the dimensions identifies the time for the other dimension.

6. Sometimes it is useful to apply the motion equations to part of the projectile's path. You can choose any initial and final points to use in the equations.

> **Motion Equations**
> **Horizontal (constant speed)**
> $$x_f = vt_f + x_i$$
> **Vertical (constant acceleration)**
> $$v_f = v_i + at_f$$
> $$x_f = x_i + v_it_f + \frac{1}{2}at_f^2$$
> $$v_f^2 = v_i^2 + 2a(x_f - x_i)$$

EXAMPLE PROBLEM 1

Find help with **square roots**. Math Handbook

A SLIDING PLATE You are preparing breakfast and slide a plate on the countertop. Unfortunately, you slide it too fast, and it flies off the end of the countertop. If the countertop is 1.05 m above the floor and it leaves the top at 0.74 m/s, how long does it take to fall, and how far from the end of the counter does it land?

1 ANALYZE AND SKETCH THE PROBLEM

Draw horizontal and vertical motion diagrams. Choose the coordinate system so that the origin is at the top of the countertop. Choose the positive x direction in the direction of horizontal velocity and the positive y direction up.

KNOWN		UNKNOWN
$x_i = y_i = 0$ m	$a_x = 0$ m/s^2	$t = ?$
$v_{xi} = 0.75$ m/s	$a_y = -9.8$ m/s^2	$x_f = ?$
$v_{yi} = 0$ m/s	$y_f = -1.05$ m	

2 SOLVE FOR THE UNKNOWN

Use the equation of motion in the y direction to find the time of fall.

$$y_f = y_i + \frac{1}{2}a_yt^2$$

$$t = \sqrt{\frac{2(y_f - y_i)}{a_y}}$$ ◄ Rearrange the equation to solve for time.

$$= \sqrt{\frac{2(-1.05 \text{ m} - 0 \text{ m})}{-9.8 \text{ m/s}^2}} = 0.46 \text{ s}$$ ◄ Substitute y_f = -1.05 m, y_i = 0 m, a_y = -9.8 m/s^2.

Use the equation of motion in the x direction to find where the plate hits the floor.

$$x_f = v_xt = (0.74 \text{ m/s to the right})(0.46 \text{ s}) = 0.34 \text{ m to the right of the counter}$$

3 EVALUATE THE ANSWER

- **Are the units correct?** Time is measured in seconds. Position in measured in meters.

- **Do the signs make sense?** Both are positive. The position sign agrees with the coordinate choice.

- **Are the magnitudes realistic?** A fall of about a meter takes about 0.5 s. During this time, the horizontal displacement of the plate would be about 0.5 s × 0.74 m/s.

PRACTICE PROBLEMS

1. You throw a stone horizontally at a speed of 5.0 m/s from the top of a cliff that is 78.4 m high.

   a. How long does it take the stone to reach the bottom of the cliff?

   b. How far from the base of the cliff does the stone hit the ground?

   c. What are the horizontal and vertical components of the stone's velocity just before it hits the ground?

2. Lucy and her friend are working at an assembly plant making wooden toy giraffes. At the end of the line, the giraffes go horizontally off the edge of a conveyor belt and fall into a box below. If the box is 0.60 m below the level of the conveyor belt and 0.40 m away from it, what must be the horizontal velocity of giraffes as they leave the conveyor belt?

3. **CHALLENGE** You are visiting a friend from elementary school who now lives in a small town. One local amusement is the ice-cream parlor, where Stan, the short-order cook, slides his completed ice-cream sundaes down the counter at a constant speed of 2.0 m/s to the servers. (The counter is kept very well polished for this purpose.) If the servers catch the sundaes 7.0 cm from the edge of the counter, how far do they fall from the edge of the counter to the point at which the servers catch them?

Figure 4 When a projectile is launched at an upward angle, its parabolic path is upward and then downward. The up-and-down motion is clearly represented in the vertical component of the vector diagram.

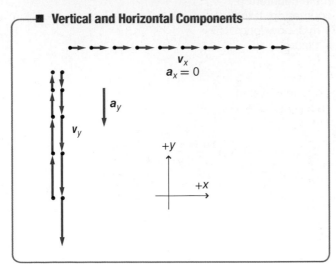

■ **Vertical and Horizontal Components**

v_x
$a_x = 0$

a_y

v_y

+y

+x

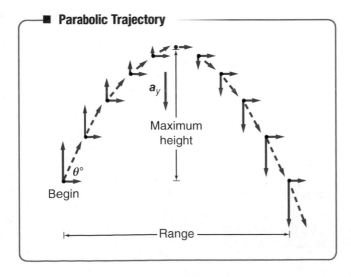

■ **Parabolic Trajectory**

a_y

Maximum height

$\theta°$

Begin

Range

Angled Launches

When a projectile is launched at an angle, the initial velocity has a vertical component as well as a horizontal component. If the object is launched upward, like a ball tossed straight up in the air, it rises with slowing speed, reaches the top of its path where its speed is momentarily zero, and descends with increasing speed.

Separate motion diagrams The upper diagram of **Figure 4** shows the separate vertical- and horizontal-motion diagrams for the trajectory. In the coordinate system, the *x*-axis is horizontal and the *y*-axis is vertical. Note the symmetry. At each point in the vertical direction, the velocity of the object as it is moving upward has the same magnitude as when it is moving downward. The only difference is that the directions of the two velocities are opposite. When solving problems, it is sometimes useful to consider symmetry to determine unknown quantities.

Parabolic path The lower diagram of **Figure 4** defines two quantities associated with the trajectory. One is the maximum height, which is the height of the projectile when the vertical velocity is zero and the projectile has only its horizontal-velocity component. The other quantity depicted is the range (*R*), which is the horizontal distance the projectile travels when the initial and final heights are the same. Not shown is the flight time, which is how much time the projectile is in the air. For football punts, flight time often is called hang time.

☑ **READING CHECK** At what point of a projectile's trajectory is its vertical velocity zero?

EXAMPLE PROBLEM 2

Get help with **the flight of a ball**. Personal Tutor

THE FLIGHT OF A BALL A ball is launched at 4.5 m/s at 66° above the horizontal. It starts and lands at the same distance from the ground. What are the maximum height above its launch level and the flight time of the ball?

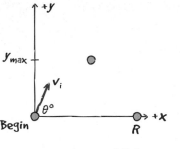

1 ANALYZE AND SKETCH THE PROBLEM

- Establish a coordinate system with the initial position of the ball at the origin.
- Show the positions of the ball at the beginning, at the maximum height, and at the end of the flight. Show the direction of F_{net}.
- Draw a motion diagram showing v and a.

KNOWN	UNKNOWN
$y_i = 0.0$ m $\quad \theta_i = 66°$ $\quad v_{y, max} = 0.0$ m	$y_{max} = ?$
$v_i = 4.5$ m/s $\quad a_y = -9.8$ m/s^2	$t = ?$

2 SOLVE FOR THE UNKNOWN

Find the y-component of v_i.

$$v_{yi} = v_i(\sin \theta_i)$$
$$= (4.5 \text{ m/s})(\sin 66°) = 4.1 \text{ m/s}$$ ◀ Substitute $v_i = 4.5$ m/s, $\theta_i = 66°$.

Use symmetry to find the y-component of v_f.

$$v_{yf} = -v_{yi} = -4.1 \text{ m/s}$$

Solve for the maximum height.

$$v_{y, max}^2 = v_{yi}^2 + 2a_y(y_{max} - y_i)$$
$$(0.0 \text{ m/s})^2 = v_{yi}^2 + 2a_y(y_{max} - 0.0 \text{ m})$$
$$y_{max} = -\frac{v_{yi}^2}{2a_y}$$
$$= -\frac{(4.1 \text{ m/s})^2}{2(-9.8 \text{ m/s}^2)} = 0.86 \text{ m}$$ ◀ Substitute $v_{yi} = 4.1$ m/s; $a_y = -9.8$ m/s^2.

Solve for the time to return to the launching height.

$$v_{yf} = v_{yi} + a_y t$$
$$t = \frac{v_{yf} - v_{yi}}{a_y}$$
$$= \frac{-4.1 \text{ m/s} - 4.1 \text{ m/s}}{-9.8 \text{ m/s}^2} = 0.84 \text{ s}$$ ◀ Substitute $v_{yf} = -4.1$ m/s; $v_{yi} = 4.1$ m/s; $a_y = -9.8$ m/s^2.

3 EVALUATE THE ANSWER

Are the magnitudes realistic? For an object that rises less than 1 m, a time of less than 1 s is reasonable.

PRACTICE PROBLEMS

Do additional problems. Online Practice

4. A player kicks a football from ground level with an initial velocity of 27.0 m/s, 30.0° above the horizontal, as shown in **Figure 5**. Find each of the following. Assume that forces from the air on the ball are negligible.

   a. the ball's hang time
   b. the ball's maximum height
   c. the horizontal distance the ball travels before hitting the ground

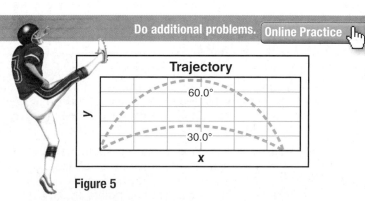

Figure 5

5. The player in the previous problem then kicks the ball with the same speed but at 60.0° from the horizontal. What is the ball's hang time, horizontal distance traveled, and maximum height?

6. **CHALLENGE** A rock is thrown from a 50.0-m-high cliff with an initial velocity of 7.0 m/s at an angle of 53.0° above the horizontal. Find its velocity when it hits the ground below.

Figure 6 Forces from air can increase or decrease the velocity of a moving object.

No Effect from Air

Force of Air that Increases Velocity

Force of Air that Decreases Velocity

Forces from Air

The effect of forces due to air has been ignored so far in this chapter, but think about why a kite stays in the air or why a parachute helps a skydiver fall safely to the ground. Forces from the air can significantly change the motion of an object.

What happens if there is wind? Moving air can change the motion of a projectile. Consider the three cases shown in **Figure 6.** In the top photo, water is flowing from the hose pipe with almost no effect from air. In the middle photo, wind is blowing in the same direction as the water's initial movement. The path of the water changes because the air exerts a force on the water in the same direction as its motion. The horizontal distance the water travels increases because the force increases the water's horizontal speed. The direction of the wind changes in the bottom photo. The horizontal distance the water travels decreases because the air exerts a force in the direction opposite the water's motion.

What if the direction of wind is at an angle relative to a moving object? The horizontal component of the wind affects only the horizontal motion of an object. The vertical component of the wind affects only the vertical motion of the object. In the case of the water, for example, a strong updraft could decrease the downward speed of the water.

The effects shown in **Figure 6** occur because the air is moving enough to significantly change the motion of the water. Even air that is not moving, however, can have a significant effect on some moving objects. A piece of paper held horizontally and dropped, for example, falls slowly because of air resistance. The air resistance increases as the surface area of the object that faces the moving air increases.

SECTION 1 REVIEW

Section Self-Check 🖱 Check your understanding.

7. **MAIN IDEA** Two baseballs are pitched horizontally from the same height but at different speeds. The faster ball crosses home plate within the strike zone, but the slower ball is below the batter's knees. Why do the balls pass the batter at different heights?

8. **Free-Body Diagram** An ice cube slides without friction across a table at a constant velocity. It slides off the table and lands on the floor. Draw free-body and motion diagrams of the ice cube at two points on the table and at two points in the air.

9. **Projectile Motion** A tennis ball is thrown out a window 28 m above the ground at an initial velocity of 15.0 m/s and 20.0° below the horizontal. How far does the ball move horizontally before it hits the ground?

10. **Projectile Motion** A softball player tosses a ball into the air with an initial velocity of 11.0 m/s, as shown in **Figure 7.** What will be the ball's maximum height?

Figure 7

11. **Critical Thinking** Suppose an object is thrown with the same initial velocity and direction on Earth and on the Moon, where the acceleration due to gravity is one-sixth its value on Earth. How will vertical velocity, time of flight, maximum height, and horizontal distance change?

Circular Motion

PHYSICS 4 YOU

Many amusement park and carnival rides spin. When the ride is spinning, forces from the walls or sides of the ride keep the riders moving in a circular path.

MAIN IDEA

An object in circular motion has an acceleration toward the circle's center due to an unbalanced force toward the circle's center.

Essential Questions

• Why is an object moving in a circle at a constant speed accelerating?

• How does centripetal acceleration depend upon the object's speed and the radius of the circle?

• What causes centripetal acceleration?

Review Vocabulary

average velocity the change in position divided by the time during which the change occurred; the slope of an object's position-time graph

New Vocabulary

uniform circular motion
centripetal acceleration
centripetal force

Describing Circular Motion

Consider an object moving in a circle at a constant speed, such as a stone being whirled on the end of a string or a fixed horse on a carousel. Are these objects accelerating? At first, you might think they are not because their speeds do not change. But remember that acceleration is related to the change in velocity, not just the change in speed. Because their directions are changing, the objects must be accelerating.

Uniform circular motion is the movement of an object at a constant speed around a circle with a fixed radius. The position of an object in uniform circular motion, relative to the center of the circle, is given by the position vector **r.** Remember that a position vector is a displacement vector with its tail at the origin. Two position vectors, r_1 and r_2, at the beginning and end of a time interval are shown on the left in **Figure 8.** As the object moves around the circle, the length of the position vector does not change, but its direction does. The diagram also shows two instantaneous velocity vectors. Notice that each velocity vector is tangent to the circular path, at a right angle to the corresponding position vector.

To determine the object's velocity, you first need to find its displacement vector over a time interval. You know that a moving object's average velocity is defined as $\frac{\Delta x}{\Delta t}$, so for an object in circular motion, $\bar{v} = \frac{\Delta r}{\Delta t}$. The right side of **Figure 8** shows Δr drawn as the displacement from r_1 to r_2 during a time interval. The velocity for this time interval has the same direction as the displacement, but its length would be different because it is divided by Δt.

Position and Velocity Vectors　　　　**Displacement Vector**

Figure 8 For an object in uniform circular motion, the velocity is tangent to the circle. It is in the same direction as the displacement.

Analyze How can you tell from the diagram that the motion is uniform?

View an **animation of circular motion and centripetal acceleration.**

Concepts In Motion

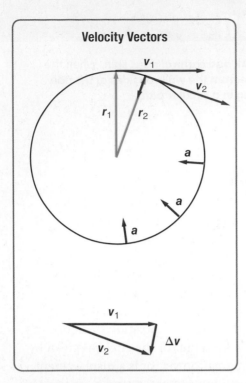

Velocity Vectors

Figure 9 The acceleration of an object in uniform circular motion is the change in velocity divided by the time interval. The direction of centripetal acceleration is always toward the center of the circle.

Centripetal Acceleration

You have read that a velocity vector of an object in uniform circular motion is tangent to the circle. What is the direction of the acceleration? **Figure 9** shows the velocity vectors v_1 and v_2 at the beginning and end of a time interval. The difference in the two vectors (Δv) is found by subtracting the vectors, as shown at the bottom of the figure. The average acceleration $\left(\bar{a} = \frac{\Delta v}{\Delta t}\right)$ for this time interval is in the same direction as Δv. For a very small time interval, Δv is so small that a points toward the center of the circle.

Repeat this process for several other time intervals when the object is in different locations on the circle. As the object moves around the circle, the direction of the acceleration vector changes, but it always points toward the center of the circle. For this reason, the acceleration of an object in uniform circular motion is called center-seeking or **centripetal acceleration.**

Magnitude of acceleration What is the magnitude of an object's centripetal acceleration? Look at the starting points of the velocity vectors in the top of **Figure 9.** Notice the triangle the position vectors at those points make with the center of the circle. An identical triangle is formed by the velocity vectors in the bottom of **Figure 9.** The angle between r_1 and r_2 is the same as that between v_1 and v_2. Therefore, similar triangles are formed by subtracting the two sets of vectors, and the ratios of the lengths of two corresponding sides are equal. Thus, $\frac{\Delta r}{r} = \frac{\Delta v}{v}$. The equation is not changed if both sides are divided by Δt.

$$\frac{\Delta r}{r \Delta t} = \frac{\Delta v}{v \Delta t}$$

But, $v = \frac{\Delta r}{\Delta t}$ and $a = \frac{\Delta v}{\Delta t}$.

$$\left(\frac{1}{r}\right)\left(\frac{\Delta r}{\Delta t}\right) = \left(\frac{1}{v}\right)\left(\frac{\Delta v}{\Delta t}\right)$$

Substituting $v = \frac{\Delta r}{\Delta t}$ in the left-hand side and $a = \frac{\Delta v}{\Delta t}$ in the right-hand side gives the following equation:

$$\frac{v}{r} = \frac{a}{v}.$$

Solve for acceleration, and use the symbol a_c for centripetal acceleration.

CENTRIPETAL ACCELERATION
Centripetal acceleration always points to the center of the circle. Its magnitude is equal to the square of the speed divided by the radius of motion.

$$a_c = \frac{v^2}{r}$$

Period of revolution One way to describe the speed of an object moving in a circle is to measure its period (T), the time needed for the object to make one complete revolution. During this time, the object travels a distance equal to the circumference of the circle ($2\pi r$). The speed, then, is represented by $v = \frac{2\pi r}{T}$. If you substitute for v in the equation for centripetal acceleration, you obtain the following equation:

$$a_c = \frac{\left(2\pi r/T\right)^2}{r} = \frac{4\pi^2 r}{T^2}.$$

Figure 10 As the hammer thrower swings the ball around, tension in the chain is the force that causes the ball to have an inward acceleration.

Predict Neglecting air resistance, how would the horizontal acceleration and velocity of the hammer change if the thrower released the chain?

Centripetal force Because the acceleration of an object moving in a circle is always in the direction of the net force acting on it, there must be a net force toward the center of the circle. This force can be provided by any number of agents. For Earth circling the Sun, the force is the Sun's gravitational force on Earth. When a hammer thrower swings the hammer, as in **Figure 10,** the force is the tension in the chain attached to the massive ball. When an object moves in a circle, the net force toward the center of the circle is called the **centripetal force.** To accurately analyze centripetal acceleration situations, you must identify the agent of the force that causes the acceleration. Then you can apply Newton's second law for the component in the direction of the acceleration in the following way.

NEWTON'S SECOND LAW FOR CIRCULAR MOTION
The net centripetal force on an object moving in a circle is equal to the object's mass times the centripetal acceleration.

$$F_{net} = ma_c$$

Direction of acceleration When solving problems, you have found it useful to choose a coordinate system with one axis in the direction of the acceleration. For circular motion, the direction of the acceleration is always toward the center of the circle. Rather than labeling this axis x or y, call it c, for centripetal acceleration. The other axis is in the direction of the velocity, tangent to the circle. It is labeled *tang* for tangential. You will apply Newton's second law in these directions, just as you did in the two-dimensional problems you have solved before. Remember that centripetal force is just another name for the net force in the centripetal direction. It is the sum of all the real forces, those for which you can identify agents that act along the centripetal axis.

In the case of the hammer thrower in **Figure 10,** in what direction does the hammer fly when the chain is released? Once the contact force of the chain is gone, there is no force accelerating the hammer toward the center of the circle, so the hammer flies off in the direction of its velocity, which is tangent to the circle. Remember, if you cannot identify the agent of a force, then it does not exist.

PhysicsLAB

CENTRIPETAL FORCE
What keeps an object moving when you swing it in a circle?

EXAMPLE PROBLEM 3

Find help with **significant figures**. Math Handbook

UNIFORM CIRCULAR MOTION A 13-g rubber stopper is attached to a 0.93-m string. The stopper is swung in a horizontal circle, making one revolution in 1.18 s. Find the magnitude of the tension force exerted by the string on the stopper.

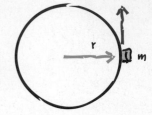

1 ANALYZE AND SKETCH THE PROBLEM

- Draw a free-body diagram for the swinging stopper.
- Include the radius and the direction of motion.
- Establish a coordinate system labeled *tang* and *c*. The directions of a_c and F_T are parallel to *c*.

KNOWN	UNKNOWN
$m = 13$ g	$F_T = ?$
$r = 0.93$ m	
$T = 1.18$ s	

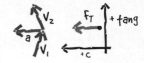

2 SOLVE FOR THE UNKNOWN

Find the magnitude of the centripetal acceleration.

$$a_c = \frac{4\pi^2 r}{T^2}$$

$$= \frac{4\pi^2(0.93 \text{ m})}{(1.18 \text{ s})^2} \qquad \blacktriangleleft \text{ Substitute } r = 0.93 \text{ m}, T = 1.18 \text{ s.}$$

$$= 26 \text{ m/s}^2$$

Use Newton's second law to find the magnitude of the tension in the string.

$$F_T = ma_c$$

$$= (0.013 \text{ kg})(26 \text{ m/s}^2) \qquad \blacktriangleleft \text{ Substitute } m = 0.013 \text{ kg}, a_c = 26 \text{ m/s}^2.$$

$$= 0.34 \text{ N}$$

3 EVALUATE THE ANSWER

- **Are the units correct?** Dimensional analysis verifies that a_c is in meters per second squared and F_T is in newtons.
- **Do the signs make sense?** The signs should all be positive.
- **Are the magnitudes realistic?** The force is almost three times the weight of the stopper, and the acceleration is almost three times that of gravity, which is reasonable for such a light object.

PRACTICE PROBLEMS

Do additional problems. Online Practice

12. A runner moving at a speed of 8.8 m/s rounds a bend with a radius of 25 m. What is the centripetal acceleration of the runner, and what agent exerts the centripetal force on the runner?

13. An airplane traveling at 201 m/s makes a turn. What is the smallest radius of the circular path (in kilometers) the pilot can make and keep the centripetal acceleration under 5.0 m/s²?

14. A 45-kg merry-go-round worker stands on the ride's platform 6.3 m from the center, as shown in **Figure 11**. If her speed (v_{worker}) as she goes around the circle is 4.1 m/s, what is the force of friction (F_f) necessary to keep her from falling off the platform?

15. A 16-g ball at the end of a 1.4-m string is swung in a horizontal circle. It revolves once every 1.09 s. What is the magnitude of the string's tension?

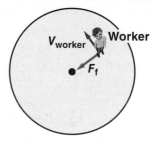

Figure 11

16. CHALLENGE A car racing on a flat track travels at 22 m/s around a curve with a 56-m radius. Find the car's centripetal acceleration. What minimum coefficient of static friction between the tires and the road is necessary for the car to round the curve without slipping?

Centrifugal "Force"

If a car makes a sharp left turn, a passenger on the right side might be thrown against the right door. Is there an outward force on the passenger? Consider a similar situation. If a car in which you are riding stops suddenly, you will be thrown forward into your safety belt. Is there a forward force on you? No, because according to Newton's first law, you will continue moving with the same velocity unless there is a net force acting on you. The safety belt applies the force that accelerates you to a stop.

Figure 12 shows a car turning left as viewed from above. A passenger in the car would continue to move straight ahead if it were not for the force of the door acting in the direction of the acceleration. As the car goes around the curve, the car and the passenger are in circular motion, and the passenger experiences centripetal acceleration. Recall that centripetal acceleration is always directed toward the center of the circle. There is no outward force on the passenger.

If, however, you think about similar situations that you have experienced, you know that it feels as if a force is pushing you outward. The so-called centrifugal, or outward, force is a fictitious, nonexistent force. You feel as if you are being pushed only because you are accelerating relative to your surroundings. There is no real force because there is no agent exerting a force.

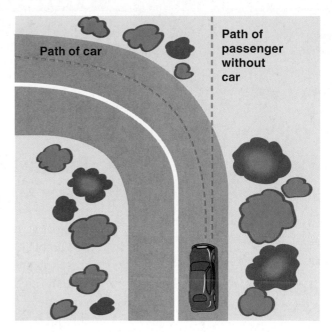

Figure 12 When a car moves around a curve, a passenger feels a fictitious centrifugal force directed outward. In fact, the force on the passenger is the centripetal force, which is directed toward the center of the circle and is exerted by the seat on which the person Is sitting.

SECTION 2 REVIEW

Section Self-Check Check your understanding.

17. **MAINIDEA** If you attach a ball to a rope and swing it at a constant speed in a circle above your head, the ball is in uniform circular motion. In which direction does it accelerate? What force causes the acceleration?

18. **Uniform Circular Motion** What is the direction of the force that acts on the clothes in the spin cycle of a top-load washing machine? What exerts the force?

19. **Centripetal Acceleration** A newspaper article states that when turning a corner, a driver must be careful to balance the centripetal and centrifugal forces to keep from skidding. Write a letter to the editor that describes physics errors in this article.

20. **Free-Body Diagram** You are sitting in the back seat of a car going around a curve to the right. Sketch motion and free-body diagrams to answer these questions:

    a. What is the direction of your acceleration?

    b. What is the direction of the net force on you?

    c. What exerts this force?

21. **Centripetal Acceleration** An object swings in a horizontal circle, supported by a 1.8-m string. It completes a revolution in 2.2 s. What is the object's centripetal acceleration?

22. **Centripetal Force** The 40.0-g stone in **Figure 13** is whirled horizontally at a speed of 2.2 m/s. What is the tension in the string?

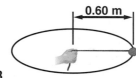

Figure 13

23. **Amusement-Park Ride** A ride at an amusement park has people stand around a 4.0-m radius circle with their backs to a wall. The ride then spins them with a 1.7-s period of revolution. What are the centripetal acceleration and velocity of the riders?

24. **Centripetal Force** A bowling ball has a mass of 7.3 kg. What force must you exert to move it at a speed of 2.5 m/s around a circle with a radius of 0.75 m?

25. **Critical Thinking** Because of Earth's daily rotation, you always move with uniform circular motion. What is the agent that supplies the force that causes your centripetal acceleration? If you are standing on a scale, how does the circular motion affect the scale's measure of your weight?

Relative Velocity

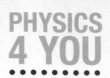

PHYSICS 4 YOU Have you ever noticed that people riding along with you on an escalator don't seem to be moving, while people going in the opposite direction seem to be moving very fast? These people may in fact have the same speed relative to the ground, but their velocities relative to you are very different.

MAINIDEA

An object's velocity depends on the reference frame chosen.

Essential Questions

• What is relative velocity?

• How do you find the velocities of an object in different reference frames?

Review Vocabulary

resultant a vector that results from the sum of two other vectors

New Vocabulary

reference frame

Relative Motion in One Dimension

Suppose you are in a school bus that is traveling at a velocity of 8 m/s in a positive direction. You walk with a velocity of 1 m/s toward the front of the bus. If a friend is standing on the side of the road watching the bus go by, how fast would your friend say you are moving? If the bus is traveling at 8 m/s, its speed as measured by your friend in a coordinate system fixed to the road is 8 m/s. When you are standing still on the bus, your speed relative to the road is also 8 m/s, but your speed relative to the bus is zero. How can your speed be different?

Different reference frames In this example, your motion is viewed from different coordinate systems. A coordinate system from which motion is viewed is a **reference frame.** Walking at 1 m/s toward the front of the bus means your velocity is measured in the reference frame of the bus. Your velocity in the road's reference frame is different. You can rephrase the problem as follows: given the velocity of the bus relative to the road and your velocity relative to the bus, what is your velocity relative to the road?

A vector representation of this problem is shown in **Figure 14.** If right is positive, your speed relative to the road is 9 m/s, the sum of 8 m/s and 1 m/s. Suppose that you now walk at the same speed toward the rear of the bus. What would be your velocity relative to the road? **Figure 14** shows that because the two velocities are in opposite directions, the resultant speed is 7 m/s, the difference between 8 m/s and 1 m/s. You can see that when the velocities are along the same line, simple addition or subtraction can be used to determine the relative velocity.

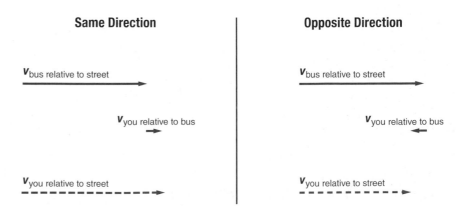

Figure 14 When an object moves in a moving reference frame, you add the velocities if they are in the same direction. You subtract one velocity from the other if they are in opposite directions.

Recall What do the lengths of the velocity vectors indicate?

Combining velocity vectors Take a closer look at how the relative velocities in **Figure 14** were obtained. Can you find a mathematical rule to describe how velocities are combined when the motion is in a moving reference frame? For the situation in which you are walking in a bus, you can designate the velocity of the bus relative to the road as $\boldsymbol{v}_{b/r}$. You can designate your velocity relative to the bus as $\boldsymbol{v}_{y/b}$ and the velocity of you relative to the road as $\boldsymbol{v}_{y/r}$. To find the velocity of you relative to the road in both cases, you added the velocity vectors of you relative to the bus and the bus relative to the road. Mathematically, this is represented as $\boldsymbol{v}_{y/b} + \boldsymbol{v}_{b/r} = \boldsymbol{v}_{y/r}$. The more general form of this equation is as follows.

RELATIVE VELOCITY
The relative velocity of object a to object c is the vector sum of object a's velocity relative to object b and object b's velocity relative to object c.

$$\boldsymbol{v}_{a/b} + \boldsymbol{v}_{b/c} = \boldsymbol{v}_{a/c}$$

Relative Motion in Two Dimensions

Adding relative velocities also applies to motion in two dimensions. As with one-dimensional motion, you first draw a vector diagram to describe the motion, and then you solve the problem mathematically.

Vector diagrams The method of drawing vector diagrams for relative motion in two dimensions is shown in **Figure 15.** The velocity vectors are drawn tip-to-tail. The reference frame from which you are viewing the motion, often called the ground reference frame, is considered to be at rest. One vector describes the velocity of the second reference frame relative to ground. The second vector describes the motion in that moving reference frame. The resultant shows the relative velocity, which is the velocity relative to the ground reference frame.

☑ **READING CHECK Decide** Can a moving car be a reference frame?

An example is the relative motion of an airplane. Airline pilots cannot expect to reach their destinations by simply aiming their planes along a compass direction. They must take into account the plane's speed relative to the air, which is given by their airspeed indicators, and their direction of flight relative to the air. They also must consider the velocity of the wind at the altitude they are flying relative to the ground. These two vectors must be combined to obtain the velocity of the airplane relative to the ground. The resultant vector tells the pilot how fast and in what direction the plane must travel relative to the ground to reach its destination. A similar situation occurs for boats that are traveling on water that is flowing.

PhysicsLAB

MOVING REFERENCE FRAME
How can you describe motion in a moving reference frame?

Figure 15 Vectors are placed tip-to-tail to find the relative velocity vector for two-dimensional motion. The subscript o/g refers to an object relative to ground, o/m refers to an object relative to a moving reference frame, and m/g refers to the moving frame relative to ground.

Analyze How would the resultant vector change if the ground reference frame were considered to be the moving reference frame?

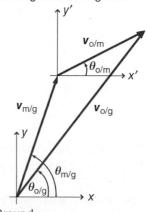

Reference frame moving relative to ground

Ground reference frame

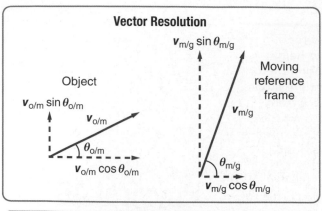

Vector Resolution

$v_{o/m} \sin \theta_{o/m}$

Object

$v_{o/m}$

$\theta_{o/m}$

$v_{o/m} \cos \theta_{o/m}$

$v_{m/g} \sin \theta_{m/g}$

Moving reference frame

$v_{m/g}$

$\theta_{m/g}$

$v_{m/g} \cos \theta_{m/g}$

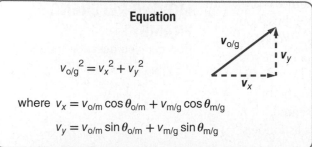

Equation

$v_{o/g}$

v_y

v_x

$$v_{o/g}^2 = v_x^2 + v_y^2$$

where $v_x = v_{o/m} \cos \theta_{o/m} + v_{m/g} \cos \theta_{m/g}$

$v_y = v_{o/m} \sin \theta_{o/m} + v_{m/g} \sin \theta_{m/g}$

Figure 16 To find the velocity of an object in a moving reference frame, resolve the vectors into x-and y-components.

Combining velocities You can use the equations in **Figure 16** to solve problems for relative motion in two dimensions. The velocity of a reference frame moving relative to the ground is labeled $v_{m/g}$. The velocity of an object in the moving frame is labeled $v_{o/m}$. The relative velocity equation gives the object's velocity relative to the ground: $v_{m/g} = v_{o/m} + v_{m/g}$.

To determine the magnitude of the object's velocity relative to the ground ($v_{m/g}$), first resolve the velocity vectors of the object and the moving reference frame into x- and y-components. Then apply the Pythagorean theorem. The general equation is shown in **Figure 16,** but for many problems the equation is simpler because the vectors are along an axis. As shown in the example problem below, you can find the angle of the object's velocity relative to the ground by observing the vector diagram and applying a trigonometric relationship.

☑ **READING CHECK Explain** How are vectors used to describe relative motion in two dimensions?

EXAMPLE PROBLEM 4

Find help with **inverse tangent.** **Math Handbook**

RELATIVE VELOCITY OF A MARBLE Ana and Sandra are riding on a ferry boat traveling east at 4.0 m/s. Sandra rolls a marble with a velocity of 0.75 m/s north, straight across the deck of the boat to Ana. What is the velocity of the marble relative to the water?

1 ANALYZE AND SKETCH THE PROBLEM

Establish a coordinate system. Draw vectors for the velocities.

KNOWN		UNKNOWN
$v_{b/w} = 4.0$ m/s	$v_{m/b} = 0.75$ m/s	$v_{m/w} = ?$

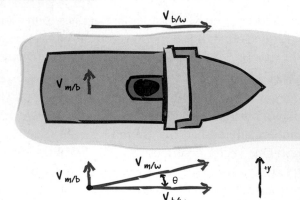

2 SOLVE FOR THE UNKNOWN

The velocities are perpendicular, so we can use the Pythagorean theorem.

$v_{m/w}^2 = v_{b/w}^2 + v_{m/b}^2$

$v_{m/w} = \sqrt{v_{b/w}^2 + v_{m/b}^2}$

$= \sqrt{(4.0 \text{ m/s})^2 + (0.75 \text{ m/s})^2} = 4.1$ m/s

◀ Substitute $v_{b/w} = 4.0$ m/s, $v_{m/b} = 0.75$ m/s.

Find the angle of the marble's velocity.

$\theta = \tan^{-1}\left(\dfrac{v_{m/b}}{v_{b/w}}\right)$

$= \tan^{-1}\left(\dfrac{0.75 \text{ m/s}}{4.0 \text{ m/s}}\right) = 11°$ north of east

The marble travels 4.1 m/s at 11° north of east.

3 EVALUATE THE ANSWER

- **Are the units correct?** Dimensional analysis verifies units of meters per second for velocity.

- **Do the signs make sense?** The signs should all be positive.

- **Are the magnitudes realistic?** The resulting velocity is of the same order of magnitude as the velocities given in the problem and slightly larger than the larger of the two.

26. You are riding in a bus moving slowly through heavy traffic at 2.0 m/s. You hurry to the front of the bus at 4.0 m/s relative to the bus. What is your speed relative to the street?

27. Rafi is pulling a toy wagon through a neighborhood at a speed of 0.75 m/s. A caterpillar in the wagon is crawling toward the rear of the wagon at a rate of 2.0 cm/s. What is the caterpillar's velocity relative to the ground?

28. A boat is rowed directly upriver at a speed of 2.5 m/s relative to the water. Viewers on the shore see that the boat is moving at only 0.5 m/s relative to the shore. What is the speed of the river? Is it moving with or against the boat?

29. A boat is traveling east at a speed of 3.8 m/s. A person walks across the boat with a velocity of 1.3 m/s south.

   a. What is the person's speed relative to the water?

   b. In what direction, relative to the ground, does the person walk?

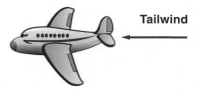
Tailwind

30. An airplane flies due north at 150 km/h relative to the air. There is a wind blowing at 75 km/h to the east relative to the ground. What is the plane's speed relative to the ground?

31. **CHALLENGE** The airplane in **Figure 17** flies at 200.0 km/h relative to the air. What is the velocity of the plane relative to the ground if it flies during the following wind conditions?

   a. a 50.0-km/h tailwind

   b. a 50.0-km/h headwind

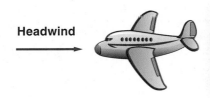
Headwind

Figure 17

SECTION 3 REVIEW

32. **MAIN**IDEA A plane has a speed of 285 km/h west relative to the air. A wind blows 25 km/h east relative to the ground. What are the plane's speed and direction relative to the ground?

33. **Relative Velocity** A fishing boat with a maximum speed of 3 m/s relative to the water is in a river that is flowing at 2 m/s. What is the maximum speed the boat can obtain relative to the shore? The minimum speed? Give the direction of the boat, relative to the river's current, for the maximum speed and the minimum speed relative to the shore.

34. **Relative Velocity of a Boat** A motorboat heads due west at 13 m/s relative to a river that flows due north at 5.0 m/s. What is the velocity (both magnitude and direction) of the motorboat relative to the shore?

35. **Boating** You are boating on a river that flows toward the east. Because of your knowledge of physics, you head your boat 53° west of north and have a velocity of 6.0 m/s due north relative to the shore.

   a. What is the velocity of the current?

   b. What is the speed of your boat relative to the water?

36. **Boating** Martin is riding on a ferry boat that is traveling east at 3.8 m/s. He walks north across the deck of the boat at 0.62 m/s. What is Martin's velocity relative to the water?

37. **Relative Velocity** An airplane flies due south at 175 km/h relative to the air. There is a wind blowing at 85 km/h to the east relative to the ground. What are the plane's speed and direction relative to the ground?

38. **A Plane's Relative Velocity** An airplane flies due north at 235 km/h relative to the air. There is a wind blowing at 65 km/h to the northeast relative to the ground. What are the plane's speed and direction relative to the ground?

39. **Critical Thinking** You are piloting the boat in **Figure 18** across a fast-moving river. You want to reach a pier directly opposite your starting point. Describe how you would navigate the boat in terms of the components of your velocity relative to the water.

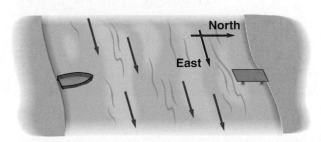
North

East

Figure 18

Need for SPEED

Race-Car Driver

The job of a race-car driver is more than just pushing down the gas pedal and following the curve of the track. Managing the extreme forces at work while driving a car at speeds of nearly 320 kilometers per hour takes endurance, strength, and fast reflexes—especially during the turns.

1 Heat A race-car driver wears a helmet to protect the head from impact, a full-body suit that protects against fire, and gloves to improve steering-wheel grip. As if all this gear isn't hot enough, the cockpit of a race car can become as hot as the Sahara.

2 Force It can take more than 40,000 N of force to turn a race car moving 290 km/h on a banked—or angled—race track.

3 Drag Race cars achieve their greatest speeds along the straight parts of the race track. The force from the road on the tires pushes the car forward, while the drag force from air resistance pushes the car backward.

5 Turns When the wheels change orientation, the road exerts a force on the tires that turns the car. Friction between the tires and the track allows the car to grip the road and turn. The greater the friction, the faster the driver can take the turn.

4 Grip The gravitational force pulls the car downward, producing friction between the track and the small area of the tire that touches it, called the contact patch. Air flowing around the car body also produces a downward force on the car, which results in an increased normal force and increased friction at the contact patch.

GOING FURTHER >>>

Research Compare at least two different kinds of auto racing events, such as drag racing and stock car, in terms of the different forces at work due to the different car styles, track structures, and racing rules.

STUDY GUIDE

BIGIDEA   You can use vectors and Newton's laws to describe projectile motion and circular motion.

VOCABULARY
- **projectile** *(p. 152)*
- **trajectory** *(p. 152)*

SECTION 1 Projectile Motion

MAINIDEA   A projectile's horizontal motion is independent of its vertical motion.

- The vertical and horizontal motions of a projectile are independent. When there is no air resistance, the horizontal motion component does not experience an acceleration and has constant velocity; the vertical motion component of a projectile experiences a constant acceleration under these same conditions.

- The curved flight path a projectile follows is called a trajectory and is a parabola. The height, time of flight, initial velocity, and horizontal distance of this path are related by the equations of motion. The horizontal distance a projectile travels before returning to its initial height depends on the acceleration due to gravity and on both components of the initial velocity.

VOCABULARY
- **uniform circular motion** *(p. 159)*
- **centripetal acceleration** *(p. 160)*
- **centripetal force** *(p. 161)*

SECTION 2 Circular Motion

MAINIDEA   An object in circular motion has an acceleration toward the circle's center due to an unbalanced force toward the circle's center.

- An object moving in a circle at a constant speed has an acceleration toward the center of the circle because the direction of its velocity is constantly changing.

- Acceleration toward the center of the circle is called centripetal acceleration. It depends directly on the square of the object's speed and inversely on the radius of the circle.

$$a_c = \frac{v^2}{r}$$

- A net force must be exerted by external agents toward the circle's center to cause centripetal acceleration.

$$F_{net} = ma_c$$

VOCABULARY
- **reference frame** *(p. 164)*

SECTION 3 Relative Velocity

MAINIDEA   An object's velocity depends on the reference frame chosen.

- A coordinate system from which you view motion is called a reference frame. Relative velocity is the velocity of an object observed in a different, moving reference frame.

- You can use vector addition to solve motion problems of an object in a moving reference frame.

Games and Multilingual eGlossary

Vocabulary Practice

SECTION 1 **Projectile Motion**

Mastering Concepts

40. Some students believe the force that starts the motion of a projectile, such as the kick given a soccer ball, remains with the ball. Is this a correct viewpoint? Present arguments for or against.

41. Consider the trajectory of the cannonball shown in **Figure 19.**

   a. Where is the magnitude of the vertical-velocity component largest?

   b. Where is the magnitude of the horizontal-velocity component largest?

   c. Where is the vertical velocity smallest?

   d. Where is the magnitude of the acceleration smallest?

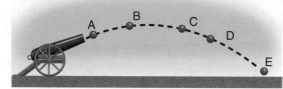

Figure 19

42. Trajectory Describe how forces cause the trajectory of an object launched horizontally to be different from the trajectory of an object launched upward at an angle.

43. Reverse Problem Write a physics problem with real-life objects for which the following equations would be part of the solution. *Hint: The two equations describe the same object.*

$$x = (1.5 \text{ m/s})t \qquad 8.0 \text{ m} = \frac{1}{2}(9.8 \text{ m/s}^2)t^2$$

44. An airplane pilot flying at constant velocity and altitude drops a heavy crate. Ignoring air resistance, where will the plane be relative to the crate when the crate hits the ground? Draw the path of the crate as seen by an observer on the ground.

Mastering Problems

45. You accidentally throw your car keys horizontally at 8.0 m/s from a cliff 64 m high. How far from the base of the cliff should you look for the keys?

46. A dart player throws a dart horizontally at 12.4 m/s. The dart hits the board 0.32 m below the height from which it was thrown. How far away is the player from the board?

47. The toy car in **Figure 20** runs off the edge of a table that is 1.225 m high. The car lands 0.400 m from the base of the table.

   a. How long did it take the car to fall?

   b. How fast was the car going on the table?

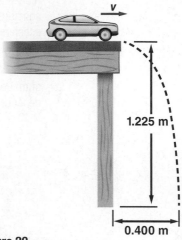

Figure 20

48. Swimming You took a running leap off a high-diving platform. You were running at 2.8 m/s and hit the water 2.6 s later. How high was the platform, and how far from the edge of the platform did you hit the water? Assume your initial velocity is horizontal. Ignore air resistance.

49. Archery An arrow is shot at 30.0° above the horizontal. Its velocity is 49 m/s, and it hits the target.

   a. What is the maximum height the arrow will attain?

   b. The target is at the height from which the arrow was shot. How far away is it?

50. BIGIDEA A pitched ball is hit by a batter at a 45° angle and just clears the outfield fence, 98 m away. If the top of the fence is at the same height as the pitch, find the velocity of the ball when it left the bat. Ignore air resistance.

51. At-Sea Rescue An airplane traveling 1001 m above the ocean at 125 km/h is going to drop a box of supplies to shipwrecked victims below.

   a. How many seconds before the plane is directly overhead should the box be dropped?

   b. What is the horizontal distance between the plane and the victims when the box is dropped?

52. Diving Divers in Acapulco dive from a cliff that is 61 m high. What is the minimum horizontal velocity a diver must have to enter the water at least 23 m from the cliff?

53. Jump Shot A basketball player is trying to make a half-court jump shot and releases the ball at the height of the basket. Assume that the ball is launched at an angle of 51.0° above the horizontal and a horizontal distance of 14.0 m from the basket. What speed must the player give the ball in order to make the shot?

54. The two baseballs in **Figure 21** were hit with the same speed, 25 m/s. Draw separate graphs of y versus t and x versus t for each ball.

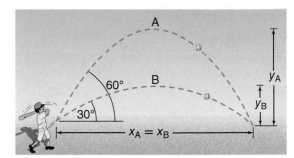

Figure 21

SECTION 2 Circular Motion

Mastering Concepts

55. Can you go around a curve with the following accelerations? Explain.

a. zero acceleration vector

b. constant acceleration vector

56. To obtain uniform circular motion, how must the net force that acts on a moving object depend on the speed of the object?

57. Suppose you whirl a yo-yo about your head in a horizontal circle.

a. In what direction must a force act on the yo-yo?

b. What exerts the force?

c. If you let go of the string on the yo-yo, in which direction would the toy travel? Use Newton's laws in your answer.

Mastering Problems

58. Car Racing A 615-kg racing car completes one lap in a time of 14.3 s around a circular track that has a radius of 50.0 m. Assume the race car moves at a constant speed.

a. What is the acceleration of the car?

b. What force must the track exert on the tires to produce this acceleration?

59. Ranking Task Rank the following objects according to their centripetal accelerations, from least to greatest. Specifically indicate any ties.

A: a 0.50-kg stone moving in a circle of radius 0.6 m at a speed of 2.0 m/s

B: a 0.50-kg stone moving in a circle of radius 1.2 m at a speed of 3.0 m/s

C: a 0.60-kg stone moving in a circle of radius 0.8 m at a speed of 2.4 m/s

D: a 0.75-kg stone moving in a circle of radius 1.2 m at a speed of 3.0 m/s

E: a 0.75-kg stone moving in a circle of radius 0.6 m at a speed of 2.4 m/s

60. Hammer Throw An athlete whirls a 7.00 kg hammer 1.8 m from the axis of rotation in a horizontal circle, as shown in **Figure 22.** If the hammer makes one revolution in 1.0 s, what is the centripetal acceleration of the hammer? What is the tension in the chain?

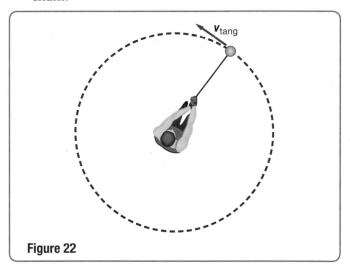

Figure 22

61. A rotating rod that is 15.3 cm long is spun with its axis through one end of the rod. The other end of the rod has a constant speed of 2010 m/s (4500 mph).

a. What is the centripetal acceleration of the end of the rod?

b. If you were to attach a 1.0-g object to the end of the rod, what force would be needed to hold it on the rod?

62. A carnival clown rides a motorcycle down a ramp and then up and around a large, vertical loop. If the loop has a radius of 18 m, what is the slowest speed the rider can have at the top of the loop so that the motorcycle stays in contact with the track and avoids falling? *Hint: At this slowest speed, the track exerts no force on the motorcycle at the top of the loop.*

63. A 75-kg pilot flies a plane in a loop as shown in **Figure 23.** At the top of the loop, when the plane is completely upside-down for an instant, the pilot hangs freely in the seat and does not push against the seat belt. The airspeed indicator reads 120 m/s. What is the radius of the plane's loop?

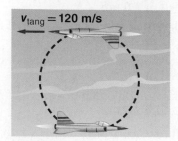

$v_{tang} = 120$ m/s

Figure 23

SECTION 3 Relative Velocity

Mastering Concepts

64. Why is it that a car traveling in the opposite direction as the car in which you are riding on the freeway often looks like it is moving faster than the speed limit?

Mastering Problems

65. Odina and LaToya are sitting by a river and decide to have a race. Odina will run down the shore to a dock, 1.5 km away, then turn around and run back. LaToya will also race to the dock and back, but she will row a boat in the river, which has a current of 2.0 m/s. If Odina's running speed is equal to LaToya's rowing speed in still water, which is 4.0 m/s, what will be the outcome of the race? Assume they both turn instantaneously.

66. Crossing a River You row a boat, such as the one in **Figure 24,** perpendicular to the shore of a river that flows at 3.0 m/s. The velocity of your boat is 4.0 m/s relative to the water.

   a. What is the velocity of your boat relative to the shore?

   b. What is the component of your velocity parallel to the shore? Perpendicular to it?

v_b v_w

Figure 24

67. Air Travel You are piloting a small plane, and you want to reach an airport 450 km due south in 3.0 h. A wind is blowing from the west at 50.0 km/h. What heading and airspeed should you choose to reach your destination in time?

68. Problem Posing Complete this problem so that it can be solved using the concept of relative velocity: "Hannah is on the west bank of a 55-m-wide river with a current of 0.7 m/s ...."

Applying Concepts

69. Projectile Motion Explain how horizontal motion can be uniform while vertical motion is accelerated. How will projectile motion be affected when drag due to air resistance is taken into consideration?

70. Baseball A batter hits a pop-up straight up over home plate at an initial speed of 20 m/s. The ball is caught by the catcher at the same height at which it was hit. At what velocity does the ball land in the catcher's mitt? Neglect air resistance.

71. Fastball In baseball, a fastball takes about $\frac{1}{2}$ s to reach the plate. Assuming that such a pitch is thrown horizontally, compare the distance the ball falls in the first $\frac{1}{4}$ s with the distance it falls in the second $\frac{1}{4}$ s.

72. You throw a rock horizontally. In a second horizontal throw, you throw the rock harder and give it even more speed.

   a. How will the time it takes the rock to hit the ground be affected? Ignore air resistance.

   b. How will the increased speed affect the distance from where the rock left your hand to where the rock hits the ground?

73. Field Biology A zoologist standing on a cliff aims a tranquilizer gun at a monkey hanging from a tree branch that is in the gun's range. The barrel of the gun is horizontal. Just as the zoologist pulls the trigger, the monkey lets go and begins to fall. Will the dart hit the monkey? Ignore air resistance.

74. Football A quarterback throws a football at 24 m/s at a 45° angle. If it takes the ball 3.0 s to reach the top of its path and the ball is caught at the same height at which it is thrown, how long is it in the air? Ignore air resistance.

75. Track and Field You are working on improving your performance in the long jump and believe that the information in this chapter can help. Does the height that you reach make any difference to your jump? What influences the length of your jump?

76. Driving on a Freeway Explain why it is that when you pass a car going in the same direction as you on the freeway, it takes a longer time than when you pass a car going in the opposite direction.

77. Imagine you are sitting in a car tossing a ball straight up into the air.

  a. If the car is moving at a constant velocity, will the ball land in front of, behind, or in your hand?

  b. If the car rounds a curve at a constant speed, where will the ball land?

78. You swing one yo-yo around your head in a horizontal circle. Then you swing another yo-yo with twice the mass of the first one, but you don't change the length of the string or the period. How do the tensions in the strings differ?

79. Car Racing The curves on a race track are banked to make it easier for cars to go around the curves at high speeds. Draw a free-body diagram of a car on a banked curve. From the motion diagram, find the direction of the acceleration.

  a. What exerts the force in the direction of the acceleration?

  b. Can you have such a force without friction?

Mixed Review

80. Early skeptics of the idea of a rotating Earth said that the fast spin of Earth would throw people at the equator into space. The radius of Earth is about 6.38×10^3 km. Show why this idea is wrong by calculating the following.

  a. the speed of a 97-kg person at the equator

  b. the force needed to accelerate the person in the circle

  c. the weight of the person

  d. the normal force of Earth on the person, that is, the person's apparent weight

81. Firing a Missile An airplane moving at 375 m/s relative to the ground fires a missile forward at a speed of 782 m/s relative to the plane. What is the missile's speed relative to the ground?

82. Rocketry A rocket in outer space that is moving at a speed of 1.25 km/s relative to an observer fires its motor. Hot gases are expelled out the back at 2.75 km/s relative to the rocket. What is the speed of the gases relative to the observer?

83. A 1.13-kg ball is swung vertically from a 0.50-m cord in uniform circular motion at a speed of 2.4 m/s. What is the tension in the cord at the bottom of the ball's motion?

84. Two dogs, initially separated by 500.0 m, are running toward each other, each moving with a constant speed of 2.5 m/s. A dragonfly, moving with a constant speed of 3.0 m/s, flies from the nose of one dog to the other, then turns around instantaneously and flies back to the other dog. It continues to fly back and forth until the dogs run into each other. What distance does the dragonfly fly during this time?

85. Banked Roads Curves on roads often are banked to help prevent cars from slipping off the road. If the speed limit for a particular curve of radius 36.0 m is 15.7 m/s (35 mph), at what angle should the road be banked so that cars will stay on a circular path even if there were no friction between the road and the tires? If the speed limit was increased to 20.1 m/s (45 mph), at what angle should the road be banked?

86. The 1.45-kg ball in **Figure 25** is suspended from a 0.80-m string and swung in a horizontal circle at a constant speed.

  a. What is the tension in the string?

  b. What is the speed of the ball?

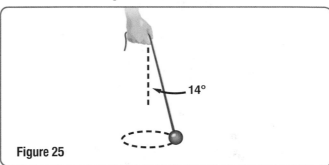

Figure 25

87. A baseball is hit directly in line with an outfielder at an angle of 35.0° above the horizontal with an initial speed of 22.0 m/s. The outfielder starts running as soon as the ball is hit at a constant speed of 2.5 m/s and barely catches the ball. Assuming that the ball is caught at the same height at which it was hit, what was the initial separation between the hitter and the outfielder? *Hint: There are two possible answers.*

88. A Jewel Heist You are a technical consultant for a locally produced cartoon. In one episode, two criminals, Shifty and Crafty, have stolen some jewels. Crafty has the jewels when the police start to chase him. He runs to the top of a 60.0-m tall building in his attempt to escape. Meanwhile, Shifty runs to the convenient hot-air balloon 20.0 m from the base of the building and untethers it, so it begins to rise at a constant speed. Crafty tosses the bag of jewels horizontally with a speed of 7.3 m/s just as the balloon begins its ascent. What must the velocity of the balloon be for Shifty to easily catch the bag?

Chapter Self-Check

Thinking Critically

89. Apply Concepts Consider a roller-coaster loop like the one in **Figure 26.** Are the cars traveling through the loop in uniform circular motion? Explain.

Figure 26

90. Apply Computers and Calculators A baseball player hits a belt-high (1.0 m) fastball down the left-field line. The player hits the ball with an initial velocity of 42.0 m/s at an angle 26° above the horizontal. The left-field wall is 96.0 m from home plate at the foul pole and is 14 m high. Write the equation for the height of the ball (y) as a function of its distance from home plate (x). Use a computer or graphing calculator to plot the path of the ball. Trace along the path to find how high above the ground the ball is when it is at the wall.

 a. Is the hit a home run?

 b. What is the minimum speed at which the ball could be hit and clear the wall?

 c. If the initial velocity of the ball is 42.0 m/s, for what range of angles will the ball go over the wall?

91. Analyze Albert Einstein showed that the rule you learned for the addition of velocities does not work for objects moving near the speed of light. For example, if a rocket moving at speed v_A releases a missile that has speed v_B relative to the rocket, then the speed of the missile relative to an observer that is at rest is given by $v = \dfrac{v_A + v_B}{1 + \dfrac{v_A v_B}{c^2}}$, where c is the speed of light, 3.00×10^8 m/s. This formula gives the correct values for objects moving at slow speeds as well. Suppose a rocket moving at 11 km/s shoots a laser beam out in front of it. What speed would an unmoving observer find for the laser light? Suppose that a rocket moves at a speed $\frac{c}{2}$, half the speed of light, and shoots a missile forward at a speed of $\frac{c}{2}$ relative to the rocket. How fast would the missile be moving relative to a fixed observer?

92. Analyze and Conclude A ball on a light string moves in a vertical circle. Analyze and describe the motion of this system. Be sure to consider the effects of gravity and tension. Is this system in uniform circular motion? Explain your answer.

Writing In Physics

93. Roller Coasters The vertical loops on most roller coasters are not circular in shape. Research and explain the physics behind this design choice.

94. Many amusement-park rides utilize centripetal acceleration to create thrills for the park's customers. Choose two rides other than roller coasters that involve circular motion, and explain how the physics of circular motion creates the sensations for the riders.

Cumulative Review

95. Multiply or divide, as indicated, using significant figures correctly.

 a. $(5 \times 10^8 \text{ m})(4.2 \times 10^7 \text{ m})$

 b. $(1.67 \times 10^{-2} \text{ km})(8.5 \times 10^{-6} \text{ km})$

 c. $\dfrac{2.6 \times 10^4 \text{ kg}}{9.4 \times 10^3 \text{ m}^3}$

 d. $\dfrac{6.3 \times 10^{-1} \text{ m}}{3.8 \times 10^2 \text{ s}}$

96. Plot the data in **Table 1** on a position-time graph. Find the average speed in the time interval between 0.0 s and 5.0 s.

Table 1 Position v. Time	
Clock Reading t (s)	**Position x (m)**
0.0	30
1.0	30
2.0	35
3.0	45
4.0	60
5.0	70

97. Carlos and his older brother Ricardo are at the grocery store. Carlos, with mass 17.0 kg, likes to hang on the front of the cart while Ricardo pushes it, even though both boys know this is not safe. Ricardo pushes the 12.4-kg cart with his brother on it such that they accelerate at a rate of 0.20 m/s².

 a. With what force is Ricardo pushing?

 b. What is the force the cart exerts on Carlos?

MULTIPLE CHOICE

1. A 1.60-m-tall girl throws a football at an angle of 41.0° from the horizontal and at an initial speed of 9.40 m/s. What is the horizontal distance between the girl and the spot when the ball is again at the height above the ground from which the girl threw it?

A. 4.55 m	**C.** 8.90 m
B. 5.90 m	**D.** 10.5 m

2. A dragonfly is sitting on a merry-go-round 2.8 m from the center. If the tangential velocity of the ride is 0.89 m/s, what is the centripetal acceleration of the dragonfly?

A. 0.11 m/s^2	**C.** 0.32 m/s^2
B. 0.28 m/s^2	**D.** 2.2 m/s^2

3. The force exerted by a 2.0-m massless string on a 0.82-kg object being swung in a horizontal circle is 4.0 N. What is the tangential velocity of the object?

A. 2.8 m/s	**C.** 4.9 m/s
B. 3.1 m/s	**D.** 9.8 m/s

4. A 1000-kg car enters an 80.0-m-radius curve at 20.0 m/s. What centripetal force must be supplied by friction so the car does not skid?

A. 5.0 N	**C.** 5.0×10^3 N
B. 2.5×10^2 N	**D.** 1.0×10^3 N

5. A jogger on a riverside path sees a rowing team coming toward him. Relative to the ground, the jogger is running at 10 km/h west and the boat is sailing at 20 km/h east. How quickly does the jogger approach the boat?

A. 10 km/h	**C.** 20 km/h
B. 30 km/h	**D.** 40 km/h

6. What is the maximum height obtained by a 125-g apple that is slung from a slingshot at an angle of 78° from the horizontal with an initial velocity of 18 m/s?

A. 0.70 m	**C.** 32 m
B. 16 m	**D.** 33 m

7. An orange is dropped at the same time and from the same height that a bullet is shot from a gun. Which of the following is true?

A. The acceleration due to gravity is greater for the orange because the orange is heavier.

B. Gravity acts less on the bullet than on the orange because the bullet is moving so quickly.

C. The velocities will be the same.

D. The two objects will hit the ground at the same time.

FREE RESPONSE

8. A lead cannonball is shot horizontally at a speed of 25 m/s out of the circus cannon, shown in the figure, on the high-wire platform on one side of a circus ring. If the platform is 52 m above the 80-m diameter ring, will the performers need to adjust their cannon so that the ball will land inside the ring instead of past it? Explain.

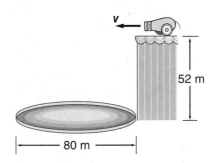

9. A mythical warrior swings a 5.6-kg mace on the end of a magically massless 86-cm chain in a horizontal circle above his head. The mace makes one full revolution in 1.8 s. Find the tension in the magical chain.

NEED EXTRA HELP?

If You Missed Question	1	2	3	4	5	6	7	8	9
Review Section	1	2	2	2	3	1	1	1	2

Online Test Practice 👆

Gravitation

BIGIDEA   Gravity is an attractive field force that acts between objects with mass.

SECTIONS

1 Planetary Motion and Gravitation

2 Using the Law of Universal Gravitation

LaunchLAB

MODEL MERCURY'S MOTION

How can measurements of angles and distances be used to draw a model of an orbit?

WATCH THIS!

PLANETARY MOTION

What can looking through a telescope tell you about gravitation? Explore the physics of amateur astronomy and find out!

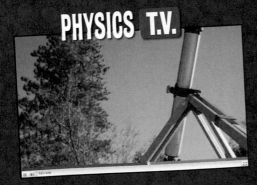

PHYSICS T.V.

(l)PhotoLink/Getty Images, (r)NASA

Planetary Motion and Gravitation

Our solar system includes the Sun, Earth and seven other major planets, dwarf planets, and interplanetary dust and gas. Various moons orbit the planets. What holds all this together?

MAINIDEA

The gravitational force between two objects is proportional to the product of their masses divided by the square of the distance between them.

Essential Questions

• What is the relationship between a planet's orbital radius and period?

• What is Newton's law of universal gravitation, and how does it relate to Kepler's laws?

• Why was Cavendish's investigation important?

Review Vocabulary

Newton's third law states all forces come in pairs and that the two forces in a pair act on different objects, are equal in strength, and are opposite in direction

New Vocabulary

Kepler's first law
Kepler's second law
Kepler's third law
gravitational force
law of universal gravitation

Early Observations

In ancient times, the Sun, the Moon, the planets, and the stars were assumed to revolve around Earth. Nicholas Copernicus, a Polish astronomer, noticed that the best available observations of the movements of planets did not fully agree with the Earth-centered model.

The results of his many years of work were published in 1543, when Copernicus was on his deathbed. His book showed that the motion of planets is much more easily understood by assuming that Earth and other planets revolve around the Sun. His model helped explain phenomena such as the inner planets Mercury and Venus always appearing near the Sun. Copernicus's view advanced our understanding of planetary motion. He incorrectly assumed, however, that planetary orbits are circular. This assumption did not fit well with observations, and modification of Copernicus's model was necessary to make it accurate.

Tycho Brahe was born a few years after Copernicus died. As a boy of 14 in Denmark, Tycho observed an eclipse of the Sun on August 21, 1560. The fact that it had been predicted inspired him toward a career in astronomy.

As Tycho studied astronomy, he realized that the charts of the time did not accurately predict astronomical events. Tycho recognized that measurements were required from one location over a long period of time. He was granted an estate on the Danish island of Hven and the funding to build an early research institute. Telescopes had not been invented, so to make measurements, Tycho used huge instruments that he designed and built in his own shop, such as those shown in **Figure 1.** Tycho is credited with the most accurate measurements of the time.

Figure 1 Instruments such as these were used by Tycho to measure the positions of planets.

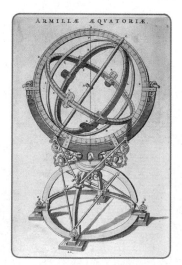

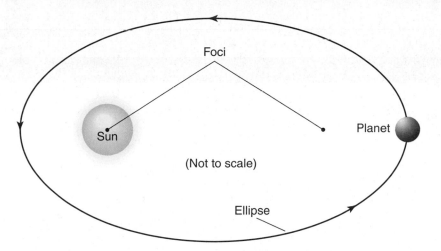

Figure 2 The orbit of each planet is an ellipse, with the Sun at one focus.

Kepler's Laws

View an **animation of Kepler's first law.**

Concepts In Motion 🖱

In 1600 Tycho moved to Prague where Johannes Kepler, a 29-year-old German, became one of his assistants. Kepler analyzed Tycho's observations. After Tycho's death in 1601, Kepler continued to study Tycho's data and used geometry and mathematics to explain the motion of the planets. After seven years of careful analysis of Tycho's data on Mars, Kepler discovered the laws that describe the motion of every planet and satellite, natural or artificial. Here, the laws are presented in terms of planets.

Kepler's first law states that the paths of the planets are ellipses, with the Sun at one focus. An ellipse has two foci, as shown in **Figure 2.** Although exaggerated ellipses are used in the diagrams, Earth's actual orbit is very nearly circular. You would not be able to distinguish it from a circle visually.

Kepler found that the planets move faster when they are closer to the Sun and slower when they are farther away from the Sun. **Kepler's second law** states that an imaginary line from the Sun to a planet sweeps out equal areas in equal time intervals, as illustrated in **Figure 3.**

☑ **READING CHECK Compare** the distances traveled from point 1 to point 2 and from point 6 to point 7 in **Figure 3.** Through which distance would Earth be traveling fastest?

A period is the time it takes for one revolution of an orbiting body. Kepler also discovered a mathematical relationship between periods of planets and their mean distances away from the Sun.

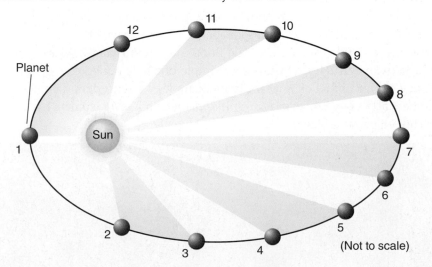

Figure 3 Kepler found that elliptical orbits sweep out equal areas in equal time periods.

Explain why the equal time areas are shaped differently.

View an **animation and a simulation of Kepler's second law.**

Concepts In Motion 🖱

Table 1 Solar System Data

Name	Average Radius (m)	Mass (kg)	Average Distance from the Sun (m)
Sun	6.96×10^8	1.99×10^{30}	—
Mercury	2.44×10^6	3.30×10^{23}	5.79×10^{10}
Venus	6.05×10^6	4.87×10^{24}	1.08×10^{11}
Earth	6.38×10^6	5.97×10^{24}	1.50×10^{11}
Mars	3.40×10^6	6.42×10^{23}	2.28×10^{11}
Jupiter	7.15×10^7	1.90×10^{27}	7.78×10^{11}
Saturn	6.03×10^7	5.69×10^{26}	1.43×10^{12}
Uranus	2.56×10^7	8.68×10^{25}	2.87×10^{12}
Neptune	2.48×10^7	1.02×10^{26}	4.50×10^{12}

Kepler's third law states that the square of the ratio of the periods of any two planets revolving about the Sun is equal to the cube of the ratio of their average distances from the Sun. Thus, if the periods of the planets are T_A and T_B and their average distances from the Sun are r_A and r_B, Kepler's third law can be expressed as follows.

View an **animation and a simulation of Kepler's third law.**

Concepts In Motion

KEPLER'S THIRD LAW
The square of the ratio of the period of planet A to the period of planet B is equal to the cube of the ratio of the distance between the centers of planet A and the Sun to the distance between the centers of planet B and the Sun.

$$\left(\frac{T_A}{T_B} \right)^2 = \left(\frac{r_A}{r_B} \right)^3$$

Note that Kepler's first two laws apply to each planet, moon, and satellite individually. The third law, however, relates the motion of two objects around a single body. For example, it can be used to compare the planets' distances from the Sun, shown in **Table 1,** to their periods around the Sun. It also can be used to compare distances and periods of the Moon and artificial satellites orbiting Earth.

Comet periods Comets are classified as long-period comets or short-period comets based on orbital periods. Long-period comets have orbital periods longer than 200 years and short-period comets have orbital periods shorter than 200 years. Comet Hale-Bopp, shown in **Figure 4,** with a period of approximately 2400 years, is an example of a long-period comet. Comet Halley, with a period of 76 years, is an example of a short-period comet. Comets also obey Kepler's laws. Unlike planets, however, comets have highly elliptical orbits.

PhysicsLAB

MODELING ORBITS
What is the shape of the orbits of planets and satellites in the solar system?

Figure 4 Hale-Bopp is a long-period comet, with a period of 2400 years. This photo was taken in 1997, when Hale-Bopp was highly visible.

EXAMPLE PROBLEM 1

Find help with **isolating a variable.** Math Handbook

CALLISTO'S DISTANCE FROM JUPITER Galileo measured the orbital radii of Jupiter's moons using the diameter of Jupiter as a unit of measure. He found that Io, the closest moon to Jupiter, has a period of 1.8 days and is 4.2 units from the center of Jupiter. Callisto, the fourth moon from Jupiter, has a period of 16.7 days. Using the same units that Galileo used, predict Callisto's distance from Jupiter.

1 ANALYZE AND SKETCH THE PROBLEM

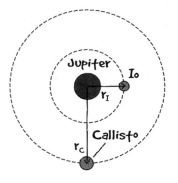

- Sketch the orbits of Io and Callisto.
- Label the radii.

KNOWN

$T_C = 16.7$ days
$T_I = 1.8$ days
$r_I = 4.2$ units

UNKNOWN

$r_C = ?$

2 SOLVE FOR CALLISTO'S DISTANCE FROM JUPITER

Solve Kepler's third law for r_C.

$$\left(\frac{T_C}{T_I}\right)^2 = \left(\frac{r_C}{r_I}\right)^3$$

$$r_C{}^3 = r_I{}^3\left(\frac{T_C}{T_I}\right)^2$$

$$r_C = \sqrt[3]{r_I{}^3\left(\frac{T_C}{T_I}\right)^2}$$ ◄ Substitute $r_I = 4.2$ units, $T_C = 16.7$ days, $T_I = 1.8$ days

$$= \sqrt[3]{(4.2\text{ units})^3\left(\frac{16.7\text{ days}}{1.8\text{ days}}\right)^2}$$

$$= \sqrt[3]{6.4\times10^3\text{ units}^3}$$

$$= 19\text{ units}$$

3 EVALUATE THE ANSWER

- **Are the units correct?** r_C should be in Galileo's units, like r_I.
- **Is the magnitude realistic?** The period is larger, so the radius should be larger.

PRACTICE PROBLEMS

Do additional problems. Online Practice

1. If Ganymede, one of Jupiter's moons, has a period of 32 days, how many units is its orbital radius? Use the information given in Example Problem 1.

2. An asteroid revolves around the Sun with a mean orbital radius twice that of Earth's. Predict the period of the asteroid in Earth years.

3. Venus has a period of revolution of 225 Earth days. Find the distance between the Sun and Venus as a multiple of Earth's average distance from the Sun.

4. Uranus requires 84 years to circle the Sun. Find Uranus's average distance from the Sun as a multiple of Earth's average distance from the Sun.

5. From **Table 1** you can find that, on average, Mars is 1.52 times as far from the Sun as Earth is. Predict the time required for Mars to orbit the Sun in Earth days.

6. The Moon has a period of 27.3 days and a mean distance of 3.9×10^5 km from its center to the center of Earth.

   a. Use Kepler's laws to find the period of a satellite in orbit 6.70×10^3 km from the center of Earth.

   b. How far above Earth's surface is this satellite?

7. **CHALLENGE** Using the data in the previous problem for the period and radius of revolution of the Moon, predict what the mean distance from Earth's center would be for an artificial satellite that has a period of exactly 1.00 day.

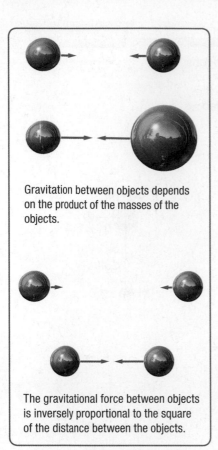

Gravitation between objects depends on the product of the masses of the objects.

The gravitational force between objects is inversely proportional to the square of the distance between the objects.

Figure 5 Mass and distance affect the magnitude of the gravitational force between objects.

View an **animation of the law of universal gravitation.**

Figure 6 This is a graphical representation of the inverse square relationship.

Inverse Square Law

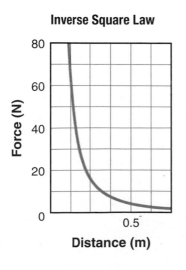

Newton's Law of Universal Gravitation

In 1666, Isaac Newton began his studies of planetary motion. It has been said that seeing an apple fall made Newton wonder if the force that caused the apple to fall might extend to the Moon, or even beyond. He found that the magnitude of the force (F_g) on a planet due to the Sun varies inversely with the square of the distance (r) between the centers of the planet and the Sun. That is, F_g is proportional to $\frac{1}{r^2}$. The force ($\mathbf{F_g}$) acts in the direction of the line connecting the centers of the two objects, as shown in **Figure 5.**

Newton found that both the apple's and the Moon's accelerations agree with the $\frac{1}{r^2}$ relationship. According to his own third law, the force Earth exerts on the apple is exactly the same as the force the apple exerts on Earth. Even though these forces are exactly the same, you can easily observe the effect of the force on the apple because it has much lower mass than Earth. The force of attraction between two objects must be proportional to the objects' masses and is known as the **gravitational force.**

Newton was confident that the same force of attraction would act between any two objects anywhere in the universe. He proposed the **law of universal gravitation,** which states that objects attract other objects with a force that is proportional to the product of their masses and inversely proportional to the square of the distance between them as shown below.

LAW OF UNIVERSAL GRAVITATION
The gravitational force is equal to the universal gravitational constant, times the mass of object 1, times the mass of object 2, divided by the distance between the centers of the objects, squared.

$$F_g = \frac{Gm_1m_2}{r^2}$$

According to Newton's equation, F is directly proportional to m_1 and m_2. If the mass of a planet near the Sun doubles, the force of attraction doubles. Use the Connecting Math to Physics feature below to examine how changing one variable affects another. **Figure 6** illustrates the inverse square relationship graphically. The term G is the universal gravitational constant and will be discussed in the next sections.

CONNECTING MATH TO PHYSICS

Direct and Inverse Relationships Newton's law of universal gravitation has both direct and inverse relationships.

$F_g \propto m_1m_2$		$F_g \propto \frac{1}{r^2}$	
Change	**Result**	**Change**	**Result**
$(2m_1)m_2$	$2F_g$	$2r$	$\frac{1}{4}F_g$
$(3m_1)m_2$	$3F_g$	$3r$	$\frac{1}{9}F_g$
$(2m_1)(3m_2)$	$6F_g$	$\frac{1}{2}r$	$4F_g$
$\left(\frac{1}{2}\right)m_1m_2$	$\frac{1}{2}F_g$	$\frac{1}{3}r$	$9F_g$

Universal Gravitation and Kepler's Third Law

Newton stated the law of universal gravitation in terms that applied to the motion of planets about the Sun. This agreed with Kepler's third law and confirmed that Newton's law fit the best observations of the day.

Consider a planet orbiting the Sun, as shown in **Figure 7**. Newton's second law of motion, $F_{net} = ma$, can be written as $F_{net} = m_p a_c$, where F_{net} is the magnitude of the gravitational force, m_p is the mass of the planet, and a_c is the centripetal acceleration of the planet. For simplicity, assume circular orbits. Recall from your study of uniform circular motion that for a circular orbit $a_c = \frac{4\pi^2 r}{T^2}$. This means that $F_{net} = m_p a_c$ may now be written $F_{net} = \frac{m_p 4\pi^2 r}{T^2}$. In this equation, T is the time in seconds required for the planet to make one complete revolution about the Sun. If you set the right side of this equation equal to the right side of the law of universal gravitation, you arrive at the following result:

$$\frac{Gm_S m_p}{r^2} = \frac{m_p 4\pi^2 r}{T^2}$$

$$T^2 = \left(\frac{4\pi^2}{Gm_S}\right) r^3$$

$$T = \sqrt{\left(\frac{4\pi^2}{Gm_S}\right) r^3}$$

The period of a planet orbiting the Sun can be expressed as follows.

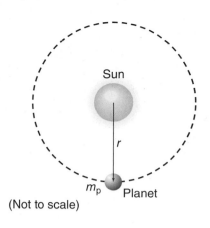

Figure 7 A planet with mass m_p and average distance from the Sun r orbits the Sun. The mass of the Sun is m_S.

PERIOD OF A PLANET ORBITING THE SUN
The period of a planet orbiting the Sun is equal to 2π times the square root of the average distance from the Sun cubed, divided by the product of the universal gravitational constant and the mass of the Sun.

$$T = 2\pi\sqrt{\frac{r^3}{Gm_S}}$$

Squaring both sides makes it apparent that this equation is Kepler's third law of planetary motion: the square of the period is proportional to the cube of the distance that separates the masses. The factor $\frac{4\pi^2}{Gm_S}$ depends on the mass of the Sun and the universal gravitational constant. Newton found that this factor applied to elliptical orbits as well.

PHYSICS CHALLENGE

Astronomers have detected three planets that orbit the star Upsilon Andromedae. Planet B has an average orbital radius of 0.0595 AU and a period of 4.6171 days. Planet C has an average orbital radius of 0.832 AU and a period of 241.33 days. Planet D has an average orbital radius of 2.53 AU and a period of 1278.1 days. (Distances are given in astronomical units (AU)—Earth's average distance from the Sun. The distance from Earth to the Sun is 1.00 AU.)

1. Do these planets obey Kepler's third law?

2. Find the mass of the star Upsilon Andromedae in units of the Sun's mass. *Hint: compare* $\frac{r^3}{T^2}$ *for these planets with that of Earth in the same units (AU and days).*

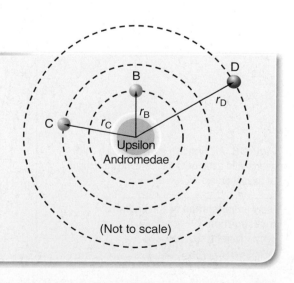

(Not to scale)

Measuring the Universal Gravitational Constant

How large is the constant G? As you know, the force of gravitational attraction between two objects on Earth is relatively small. The slightest attraction, even between two massive bowling balls, is almost impossible to detect. In fact, it took 100 years from the time of Newton's work for scientists to develop an apparatus that was sensitive enough to measure the force of gravitational attraction.

Cavendish's apparatus In 1798 English scientist Henry Cavendish used equipment similar to the apparatus shown in **Figure 8** to measure the gravitational force between two objects. The apparatus has a horizontal rod with small lead spheres attached to each end. The rod is suspended at its midpoint so that it can rotate. Because the rod is suspended by a thin wire, the rod and spheres are very sensitive to horizontal forces.

To measure G, two large spheres are placed in a fixed position close to each of the two small spheres, as shown in **Figure 8.** The force of attraction between the large and small spheres causes the rod to rotate. When the force required to twist the wire equals the gravitational force between the spheres, the rod stops rotating. By measuring the angle through which the rod turns, the attractive force between the objects can be calculated.

✔ **READING CHECK Explain** why the rod and sphere in Cavendish's apparatus must be sensitive to horizontal forces.

The angle through which the rod turns is measured by using a beam of light that is reflected from the mirror. The distances between the sphere's centers and the force can both be measured. The masses of the spheres are known. By substituting the values for force, mass, and distance into Newton's law of universal gravitation, an experimental value for G is found: when m_1 and m_2 are measured in kilograms, r in meters, and F in newtons, $G = 6.67 \times 10^{-11}$ N·m²/kg².

Investigate **universal gravitation.**

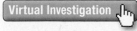
Virtual Investigation

■ **Cavendish Balance**

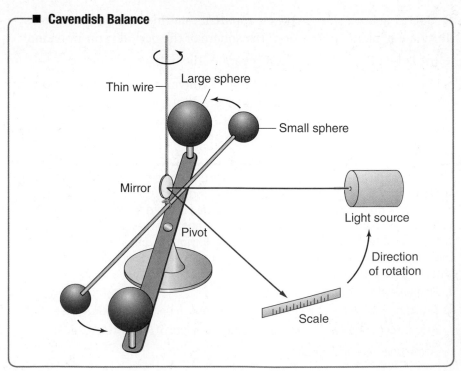

Figure 8 A Cavendish balance uses a light source and a mirror to measure the movement of the spheres.

View an **animation of Cavendish's investigation.**

Concepts In Motion

The importance of G Cavendish's investigation often is called "weighing Earth" because it helped determine Earth's mass. Once the value of G is known, not only the mass of Earth, but also the mass of the Sun can be determined. In addition, the gravitational force between any two objects can be calculated by using Newton's law of universal gravitation. For example, the attractive gravitational force (F_g) between two bowling balls of mass 7.26 kg, with their centers separated by 0.30 m, can be calculated as follows:

$$F_g = \frac{(6.67{\times}10^{-11}\ \text{N}{\cdot}\text{m}^2/\text{kg}^2)(7.26\ \text{kg})(7.26\ \text{kg})}{(0.30\ \text{m})^2} = 3.9{\times}10^{-8}\ \text{N}$$

You know that on Earth's surface, the weight of an object of mass m is a measure of Earth's gravitational attraction: $F_g = mg$. If Earth's mass is represented by m_E and Earth's radius is represented by r_E, the following is true:

$$F_g = \frac{Gm_E m}{r_E^2} = mg, \text{ and so } g = \frac{Gm_E}{r_E^2}$$

This equation can be rearranged to solve for m_E.

$$m_E = \frac{g r_E^2}{G}$$

Using $g = 9.8$ N/kg, $r_E = 6.38{\times}10^6$ m, and $G = 6.67{\times}10^{-11}$ N·m²/kg², the following result is obtained for Earth's mass:

$$m_E = \frac{(9.8\ \text{N/kg})(6.38{\times}10^6\ \text{m})^2}{6.67{\times}10^{-11}\ \text{N}{\cdot}\text{m}^2/\text{kg}^2} = 5.98{\times}10^{24}\ \text{kg}$$

When you compare the mass of Earth to that of a bowling ball, you can see why the gravitational attraction between everyday objects is not easily observed. Cavendish's investigation determined the value of G, confirmed Newton's prediction that a gravitational force exists between any two objects, and helped calculate the mass of Earth **(Figure 9)**.

Figure 9 Cavendish's investigations helped calculate the mass of Earth.

SECTION 1 REVIEW

Section Self-Check Check your understanding.

8. **MAIN IDEA** What is the gravitational force between two 15-kg balls whose centers are 35 m apart? What fraction is this of the weight of one ball?

9. **Neptune's Orbital Period** Neptune orbits the Sun at an average distance given in **Figure 10,** which allows gases, such as methane, to condense and form an atmosphere. If the mass of the Sun is $1.99{\times}10^{30}$ kg, calculate the period of Neptune's orbit.

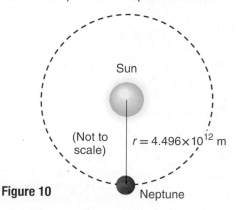

Sun

(Not to scale) $r = 4.496{\times}10^{12}$ m

Figure 10   Neptune

10. **Gravity** If Earth began to shrink, but its mass remained the same, what would happen to the value of g on Earth's surface?

11. **Universal Gravitational Constant** Cavendish did his investigation using lead spheres. Suppose he had replaced the lead spheres with copper spheres of equal mass. Would his value of G be the same or different? Explain.

12. Kepler's three statements and Newton's equation for gravitational attraction are called laws. Were they ever theories? Will they ever become theories?

13. **Critical Thinking** Picking up a rock requires less effort on the Moon than on Earth.

   a. How will the Moon's gravitational force affect the path of the rock if it is thrown horizontally?

   b. If the thrower accidentally drops the rock on her toe, will it hurt more or less than it would on Earth? Explain.

Section 1 • Planetary Motion and Gravitation **185**

Using the Law of Universal Gravitation

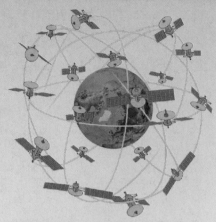

MAINIDEA

All objects are surrounded by a gravitational field that affects the motions of other objects.

Essential Questions

- How can you describe orbital motion?

- How are gravitational mass and inertial mass alike, and how are they different?

- How is gravitational force explained, and what did Einstein propose about gravitational force?

Review Vocabulary

centripetal acceleration the center-seeking acceleration of an object moving in a circle at a constant speed

New Vocabulary

inertial mass

gravitational mass

APPLYING PRACTICES

Use Mathematical or Computational Representations Go to the resources tab in ConnectED to find the Applying Practices worksheet *Planetary Orbits*.

PHYSICS 4 YOU

Have you ever used a device to locate your position or to map where you want to go? Where does that device get information? The Global Positioning System (GPS) consists of many satellites circling Earth. GPS satellites give accurate position data anywhere on or near Earth.

Orbits of Planets and Satellites

The planet Uranus was discovered in 1781. By 1830 it was clear that the law of gravitation didn't correctly predict its orbit. Two astronomers proposed that Uranus was being attracted by the Sun and by an undiscovered planet. They calculated the orbit of such a planet in 1845, and, one year later, astronomers at the Berlin Observatory found the planet now called Neptune. How is it possible for planets, such as Neptune and Uranus, to remain in orbit around the Sun?

Newton used a drawing similar to the one shown in **Figure 11** to illustrate a thought experiment on the motion of satellites. Imagine a cannon, perched high atop a mountain, firing a cannonball horizontally with a given horizontal speed. The cannonball is a projectile, and its motion has both vertical and horizontal components. Like all projectiles on Earth, it would follow a parabolic trajectory and fall back to the ground.

If the cannonball's horizontal speed were increased, it would travel farther across the surface of Earth and still fall back to the ground. If an extremely powerful cannon were used, however, the cannonball would travel all the way around Earth and keep going. It would fall toward Earth at the same rate that Earth's surface curves away. In other words, the curvature of the projectile would continue to just match the curvature of Earth so that the cannonball would never get any closer to or farther away from Earth's curved surface. The cannonball would, therefore, be in orbit.

Figure 11 Newton imagined a projectile launched parallel to Earth. If it has enough speed it will fall toward Earth with a curvature that matches the curvature of Earth's surface.

Identify the factor that is not considered in this example.

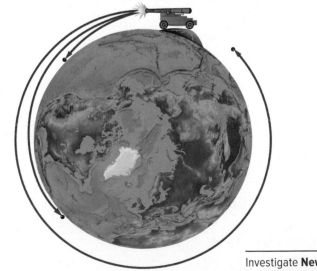

Investigate **Newton's cannon.**

Virtual Investigation

Newton's thought experiment ignored air resistance. For the cannonball to be free of air resistance, the mountain on which the cannon is perched would have to be more than 150 km above Earth's surface. By way of comparison, the mountain would have to be much taller than the peak of Mount Everest, the world's tallest mountain, which is only 8.85 km in height. A cannonball launched from a mountain that is 150 km above Earth's surface would encounter little or no air resistance at an altitude of 150 km because the mountain would be above most of the atmosphere. Thus, a cannonball or any object or satellite at or above this altitude could orbit Earth.

A satellite's speed A satellite in an orbit that is always the same height above Earth moves in uniform circular motion. Recall that its centripetal acceleration is given by $a_c = \frac{v^2}{r}$. Newton's second law, $F_{net} = ma_c$, can thus be rewritten $F_{net} = \frac{mv^2}{r}$. If Earth's mass is m_E, then this expression combined with Newton's law of universal gravitation produces the following equation:

$$\frac{Gm_Em}{r^2} = \frac{mv^2}{r}$$

Solving for the speed of a satellite in circular orbit about Earth (v) yields the following.

SPEED OF A SATELLITE ORBITING EARTH
The speed of a satellite orbiting Earth is equal to the square root of the universal gravitational constant times the mass of Earth, divided by the radius of the orbit.

$$v = \sqrt{\frac{Gm_E}{r}}$$

A satellite's orbital period A satellite's orbit around Earth is similar to a planet's orbit about the Sun. Recall that the period of a planet orbiting the Sun is expressed by the following equation:

$$T = 2\pi\sqrt{\frac{r^3}{Gm_S}}$$

Thus, the period for a satellite orbiting Earth is given by the following equation.

PERIOD OF A SATELLITE ORBITING EARTH
The period for a satellite orbiting Earth is equal to 2π times the square root of the radius of the orbit cubed, divided by the product of the universal gravitational constant and the mass of Earth.

$$T = 2\pi\sqrt{\frac{r^3}{Gm_E}}$$

The equations for the speed and period of a satellite can be used for any object in orbit about another. The mass of the central body will replace m_E in the equations, and r will be the distance between the centers of the orbiting body and the central body. Orbital speed (v) and period (T) are independent of the mass of the satellite.

☑ **READING CHECK Describe** how the mass of a satellite affects that satellite's orbital speed and period.

REAL-WORLD
PHYSICS
●●●●●●●●●●●●●●

GEOSYNCHRONOUS ORBIT The GOES weather satellites orbit Earth once a day at an altitude of 35,785 km. The orbital speed of the satellite matches Earth's rate of rotation. Thus, to an observer on Earth, the satellite appears to remain above one spot. Dish antennas on Earth can be directed to one point in the sky and remain in a fixed position as the satellite orbits.

Get help with **orbital speed.** Personal Tutor

EXAMPLE PROBLEM

ORBITAL SPEED AND PERIOD Assume that a satellite orbits Earth 225 km above its surface. Given that the mass of Earth is 5.97×10^{24} kg and the radius of Earth is 6.38×10^6 m, what are the satellite's orbital speed and period?

1 ANALYZE AND SKETCH THE PROBLEM

Sketch the situation showing the height of the satellite's orbit.

KNOWN

$h = 2.25 \times 10^5$ m
$r_E = 6.38 \times 10^6$ m
$m_E = 5.97 \times 10^{24}$ kg
$G = 6.67 \times 10^{-11}$ N·m²/kg²

UNKNOWN

$v = ?$
$T = ?$

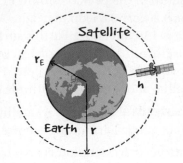

2 SOLVE FOR ORBITAL SPEED AND PERIOD

Determine the orbital radius by adding the height of the satellite's orbit to Earth's radius.

$r = h + r_E$
$= 2.25 \times 10^5$ m $+ 6.38 \times 10^6$ m $= 6.60 \times 10^6$ m

◀ Substitute $h = 2.25 \times 10^5$ m and $r_E = 6.38 \times 10^6$ m.

Solve for the speed.

$v = \sqrt{\dfrac{Gm_E}{r}}$

$= \sqrt{\dfrac{(6.67 \times 10^{-11} \text{ N·m}^2/\text{kg}^2)(5.97 \times 10^{24} \text{ kg})}{6.60 \times 10^6 \text{ m}}}$

$= 7.77 \times 10^3$ m/s

◀ Substitute $G = 6.67 \times 10^{-11}$ N·m²/kg², $m_E = 5.97 \times 10^{24}$ kg, and $r = 6.60 \times 10^6$ m.

Solve for the period.

$T = 2\pi\sqrt{\dfrac{r^3}{Gm_E}}$

$= 2\pi\sqrt{\dfrac{(6.60 \times 10^6 \text{ m})^3}{(6.67 \times 10^{-11} \text{ N·m}^2/\text{kg}^2)(5.97 \times 10^{24} \text{ kg})}}$

$= 5.34 \times 10^3$ s

◀ Substitute $r = 6.60 \times 10^6$ m, $G = 6.67 \times 10^{-11}$ N·m²/kg², and $m_E = 5.97 \times 10^{24}$ kg.

This is approximately 89 min, or 1.5 h.

3 EVALUATE THE ANSWER

Are the units correct? The unit for speed is meters per second, and the unit for period is seconds.

PRACTICE PROBLEMS

Do additional problems. Online Practice

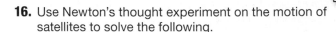

For the following problems, assume a circular orbit for all calculations.

14. Suppose that the satellite in Example Problem 2 is moved to an orbit that is 24 km larger in radius than its previous orbit.

  a. What is its speed?

  b. Is this faster or slower than its previous speed?

  c. Why do you think this is so?

15. Uranus has 27 known moons. One of these moons is Miranda, which orbits at a radius of 1.29×10^8 m. Uranus has a mass of 8.68×10^{25} kg. Find the orbital speed of Miranda. How many Earth days does it take Miranda to complete one orbit?

16. Use Newton's thought experiment on the motion of satellites to solve the following.

  a. Calculate the speed that a satellite shot from a cannon must have to orbit Earth 150 km above its surface.

  b. How long, in seconds and minutes, would it take for the satellite to complete one orbit and return to the cannon?

17. CHALLENGE Use the data for Mercury in **Table 1** to find the following.

  a. the speed of a satellite that is in orbit 260 km above Mercury's surface

  b. the period of the satellite

▶ **CONNECTION TO EARTH SCIENCE** *Landsat 7,* shown in **Figure 12,** is an artificial satellite that provides images of Earth's continental surfaces. *Landsat 7* images have been used to create maps, study land use, and monitor resources and global changes. The *Landsat 7* system enables researchers to monitor small-scale processes, such as deforestation, on a global scale. Satellites, such as *Landsat 7,* are accelerated to the speeds necessary for them to achieve orbit by large rockets, such as shuttle-booster rockets. Because the acceleration of any mass must follow Newton's second law of motion, $F_{net} = ma$, more force is required to launch a more massive satellite into orbit. Thus, the mass of a satellite is limited by the capability of the rocket used to launch it.

Figure 12 *Landsat 7* is capable of providing up to 532 images of Earth per day.

Free-Fall Acceleration

The acceleration of objects due to Earth's gravity can be found by using Newton's law of universal gravitation and his second law of motion. For a free-falling object of mass m, the following is true:

$$F = \frac{Gm_Em}{r^2} = ma, \text{ so } a = \frac{Gm_E}{r^2}$$

If you set $a = g_E$ and $r = r_E$ on Earth's surface, the following equation can be written:

$$g = \frac{Gm_E}{r_E^2}, \text{ thus, } m_E = \frac{gr_E^2}{G}$$

You saw above that $a = \frac{Gm_E}{r^2}$ for a free-falling object. Substitution of the above expression for m_E yields the following:

$$a = \frac{G\left(\frac{gr_E^2}{G}\right)}{r^2}$$

$$a = g\left(\frac{r_E}{r}\right)^2$$

On the surface of Earth, $r = r_E$ and so $a = g$. But, as you move farther from Earth's center, r becomes larger than r_E, and the free-fall acceleration is reduced according to this inverse square relationship. What happens to your mass as you move farther and farther from Earth's center?

Figure 13 Astronauts in orbit around Earth are in free fall because their spacecraft and everything in it is accelerating toward Earth along with the astronauts. That is, the floor is constantly falling from beneath their feet.

Weight and weightlessness You may have seen photos similar to **Figure 13** in which astronauts are on a spacecraft in an environment often called zero-g or weightlessness. The spacecraft orbits about 400 km above Earth's surface. At that distance, $g = 8.7$ N/kg, only slightly less than that on Earth's surface. Earth's gravitational force is certainly not zero in the shuttle. In fact, gravity causes the shuttle to orbit Earth. Why, then, do the astronauts appear to have no weight?

Remember that you sense weight when something, such as the floor or your chair, exerts a contact force on you. But if you, your chair, and the floor all are accelerating toward Earth together, then no contact forces are exerted on you. Thus, your apparent weight is zero and you experience apparent weightlessness. Similarly, the astronauts experience apparent weightlessness as the shuttle and everything in it falls freely toward Earth.

WEIGHTLESS WATER
What are the effects of weightlessness in free fall?

WEIGHT IN FREE FALL
What is the effect of free fall on mass?

The Gravitational Field

Recall from studying motion that many common forces are contact forces. Friction is exerted where two objects touch; for example, the floor and your chair or desk push on you when you are in contact with them. Gravity, however, is different. It acts on an apple falling from a tree and on the Moon in orbit. In other words, gravity acts over a distance. It acts between objects that are not touching or that are not close together. Newton was puzzled by this concept. He wondered how the Sun could exert a force on planet Earth, which is hundreds of millions of kilometers away.

Field concept The answer to the puzzle arose from a study of magnetism. In the nineteenth century, Michael Faraday developed the concept of a field to explain how a magnet attracts objects. Later, the field concept was applied to gravity.

Any object with mass is surrounded by a gravitational field, which exerts a force that is directly proportional to the mass of the object and inversely proportional to the square of the distance from the object's center. Another object experiences a force due to the interaction between its mass and the gravitational field (g) at its location. The direction of g and the gravitational force is toward the center of the object producing the field. Gravitational field strength is expressed by the following equation.

Figure 14 Earth's gravitational field can be represented by vectors pointing toward Earth's center. The decrease in g's magnitude follows an inverse-square relationship as the distance from Earth's center increases.

Explain why the value of g never reaches zero.

GRAVITATIONAL FIELD
The gravitational field strength produced by an object is equal to the universal gravitational constant times the object's mass, divided by the square of the distance from the object's center.

$$g = \frac{Gm}{r^2}$$

Suppose the gravitational field is created by the Sun. Then a planet of mass m in the Sun's gravitational field has a force exerted on it that depends on its mass and the magnitude of the gravitational field at its location. That is, $F_g = mg$, toward the Sun. The force is caused by the interaction of the planet's mass with the gravitational field at its location, not with the Sun millions of kilometers away. To find the gravitational field caused by more than one object, calculate all gravitational fields and add them as vectors.

The gravitational field is measured by placing an object with a small mass (m) in the gravitational field and measuring the force (F_g) on it. The gravitational field is calculated using $g = \frac{F_g}{m}$. The gravitational field is measured in units of newtons per kilogram (N/kg).

On Earth's surface, the strength of the gravitational field is 9.8 N/kg, and its direction is toward Earth's center. The field can be represented by a vector of length g pointing toward the center of the object producing the field. You can picture the gravitational field produced by Earth as a collection of vectors surrounding Earth and pointing toward it, as shown in **Figure 14.** The strength of Earth's gravitational field varies inversely with the square of the distance from Earth's center. Earth's gravitational field depends on Earth's mass but not on the mass of the object experiencing it.

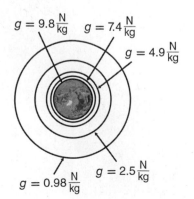

$g = 9.8 \frac{N}{kg}$ $g = 7.4 \frac{N}{kg}$

$g = 4.9 \frac{N}{kg}$

$g = 0.98 \frac{N}{kg}$ $g = 2.5 \frac{N}{kg}$

Two Kinds of Mass

You read that mass can be defined as the slope of a graph of force versus acceleration. That is, mass is equal to the net force exerted on an object divided by its acceleration. This kind of mass, related to the inertia of an object, is called inertial mass and is represented by the following equation.

INERTIAL MASS
Inertial mass is equal to the net force exerted on the object divided by the acceleration of the object.

$$m_{\text{inertial}} = \frac{F_{\text{net}}}{a}$$

Inertial mass You know that it is much easier to push an empty cardboard box across the floor than it is to push one that is full of books. The full box has greater inertial mass than the empty one. The **inertial mass** of an object is a measure of the object's resistance to any type of force. Inertial mass of an object is measured by exerting a force on the object and measuring the object's acceleration. The more inertial mass an object has, the less acceleration it undergoes as a result of a net force exerted on it.

Gravitational mass Newton's law of universal gravitation, $F_g = \frac{Gm_1m_2}{r^2}$, also involves mass–but a different kind of mass. Mass as used in the law of universal gravitation is a quantity that measures an object's response to gravitational force and is called **gravitational mass.** It can be measured by using a simple balance, such as the one shown in **Figure 15.** If you measure the magnitude of the attractive force exerted on an object by another object of mass m, at a distance r, then you can define the gravitational mass in the following way.

GRAVITATIONAL MASS
The gravitational mass of an object is equal to the distance between the centers of the objects squared, times the gravitational force, divided by the product of the universal gravitational constant, times the mass of the other object.

$$m_g = \frac{r^2F_g}{Gm}$$

How different are these two kinds of mass? Suppose you have a watermelon in the trunk of your car. If you accelerate the car forward, the watermelon will roll backward relative to the trunk. This is a result of its inertial mass–its resistance to acceleration. Now, suppose your car climbs a steep hill at a constant speed. The watermelon will again roll backward. But this time, it moves as a result of its gravitational mass. The watermelon is pulled downward toward Earth.

Newton made the claim that inertial mass and gravitational mass are equal in magnitude. This hypothesis is called the principle of equivalence. All investigations conducted so far have yielded data that support this principle. Most of the time we refer simply to the mass of an object. Albert Einstein also was intrigued by the principle of equivalence and made it a central point in his theory of gravity.

PhysicsLABs

HOW CAN YOU MEASURE MASS?
How is an inertial balance used to measure mass?

INERTIAL MASS AND GRAVITATIONAL MASS
How can you determine the relationship between inertial mass and gravitational mass?

Figure 15 A simple balance is used to determine the gravitational mass of an object.

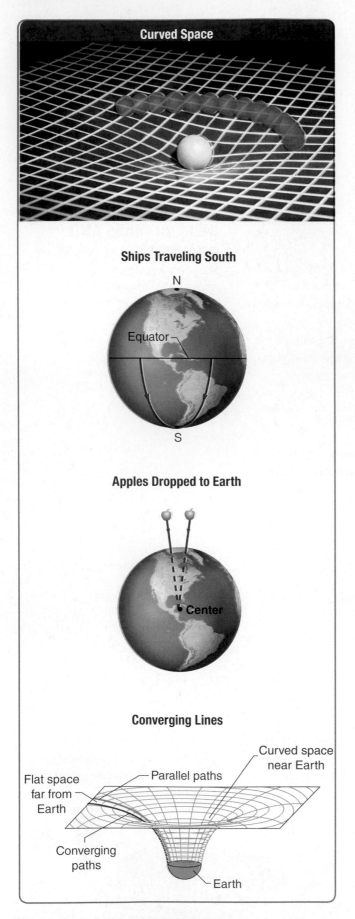

Figure 16 Visualizing how space is curved is difficult. Analogies can help you understand difficult concepts.

Einstein's Theory of Gravity

Newton's law of universal gravitation allows us to calculate the gravitational force that exists between two objects because of their masses. Newton was puzzled, however, as to how two objects could exert forces on each other if those two objects were millions of kilometers away from each other. Albert Einstein proposed that gravity is not a force but rather an effect of space itself. According to Einstein's explanation of gravity, mass changes the space around it. Mass causes space to be curved, and other bodies are accelerated because of the way they follow this curved space.

Curved space One way to picture how mass affects space is to model three-dimensional space as a large, two-dimensional sheet, as shown in the top part of **Figure 16.** The yellow ball on the sheet represents a massive object. The ball forms an indentation on the sheet. A red ball rolling across the sheet simulates the motion of an object in space. If the red ball moves near the sagging region of the sheet, it will be accelerated. In a similar way, Earth and the Sun are attracted to each other because of the way space is distorted by the two objects.

Ships traveling south The following is another analogy that might help you understand the curvature of space. Suppose you watch from space as two ships travel due south from the equator. At the equator, the ships are separated by 4000 km. As they approach the South Pole, the distance decreases to 1 km. To the sailors, their paths are straight lines, but because of Earth's curvature, they travel in a curve, as viewed far from Earth's surface, as in **Figure 16.**

Apples dropped to Earth Consider a similar motion. Two apples are dropped to Earth, initially traveling in parallel paths, as in **Figure 16.** As they approach Earth, they are pulled toward Earth's center. Their paths converge.

Converging lines This convergence can be attributed to the curvature of space near Earth. Far from any massive object, such as a planet or star, space is flat, and parallel lines remain parallel. Then they begin to converge. In flat space, the parallel lines would remain parallel. In curved space, the lines converge.

Einstein's theory or explanation, called the general theory of relativity, makes many predictions about how massive objects affect one another. In every test conducted to date, Einstein's theory has been shown to give the correct results.

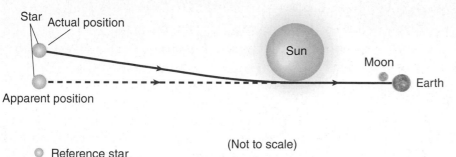

Star Actual position

Sun

Moon

Earth

Apparent position

Reference star

(Not to scale)

Deflection of light Einstein's theory predicts that massive objects deflect and bend light. Light follows the curvature of space around the massive object and is deflected, as shown in **Figure 17.** In 1919, during an eclipse of the Sun, astronomers found that light from distant stars that passed near the Sun was deflected an amount that agreed with Einstein's predictions.

Another result of general relativity is the effect of gravity on light from extremely massive objects. If an object is massive and dense enough, the light leaving it is totally bent back to the object. No light ever escapes the object. Objects such as these, called black holes, have been identified as a result of their effects on nearby stars. Black holes have been detected through the radiation produced when matter is pulled into them.

While Einstein's theory provides very accurate predictions of gravity's effects, it is still incomplete. It does not explain the origin of mass or how mass curves space. Physicists are working to understand the deeper meaning of gravity and the origin of mass itself.

SECTION 2 REVIEW

Section Self-Check   Check your understanding.

18. **MAIN**IDEA The Moon is 3.9×10^5 km from Earth's center and Earth is 14.96×10^7 km from the Sun's center. The masses of Earth and the Sun are 5.97×10^{24} kg and 1.99×10^{30} kg, respectively. During a full moon, the Sun, Earth, and the Moon are in line with each other, as shown in **Figure 18.**

   a. Find the ratio of the gravitational fields due to Earth and the Sun at the center of the Moon.

   b. What is the net gravitational field due to the Sun and Earth at the center of the Moon?

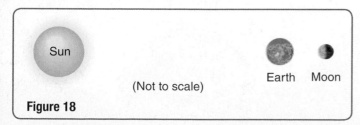

Sun

Earth   Moon

(Not to scale)

Figure 18

19. **Apparent Weightlessness** Chairs in an orbiting spacecraft are weightless. If you were on board such a spacecraft and you were barefoot, would you stub your toe if you kicked a chair? Explain.

20. **Gravitational Field** The mass of the Moon is 7.3×10^{22} kg and its radius is 1785 km. What is the strength of the gravitational field on the surface of the Moon?

21. **Orbital Period and Speed** Two satellites are in circular orbits about Earth. One is 150 km above the surface, the other is 160 km.

   a. Which satellite has the larger orbital period?

   b. Which has the greater speed?

22. **Theories and Laws** Why is Einstein's description of gravity called a theory, while Newton's is a law?

23. **Astronaut** What would be the strength of Earth's gravitational field at a point where an 80.0-kg astronaut would experience a 25.0 percent reduction in weight?

24. **A Satellite's Mass** When the first artificial satellite was launched into orbit by the former Soviet Union in 1957, U.S. president Dwight D. Eisenhower asked scientists to calculate the mass of the satellite. Would they have been able to make this calculation? Explain.

25. **Critical Thinking** It is easier to launch a satellite from Earth into an orbit that circles eastward than it is to launch one that circles westward. Explain.

NO ESCAPE

Is a BLACK HOLE really a hole?

You may have wondered about black holes.
What are they and where do they come from?

A star explodes What happens when a giant star runs low on fuel? The star cannot maintain its temperature, causing it to collapse under its own weight. The resulting explosion, called a supernova, is usually brighter than the entire galaxy it is in.

If what's left of the star is massive enough (more than three times the mass of the Sun), the remnant becomes one of the strangest objects in the universe: a black hole.

Escape from Earth Imagine standing on the surface of Earth and throwing a ball straight up. As the ball moves up, gravitational force robs the ball of its upward velocity. Finally, when the upward velocity reaches zero, the ball begins to fall back down. **Figure 1** illustrates this.

Of course, this happens due to the limitations of your throwing arm. If you could throw the ball fast enough (about 11,000 m/s), it would not fall back to Earth but instead would escape Earth entirely. This speed is the escape velocity from the surface of our planet.

Escape from a black hole? A black hole is a very compact, massive object with an escape velocity so high that nothing, not even light, can escape. The image in **Figure 2** shows a star being swallowed by a black hole. Anything that passes through the boundary of the influence of a black hole is truly lost forever.

Are black holes really holes? NO! They are extremely dense objects in space.

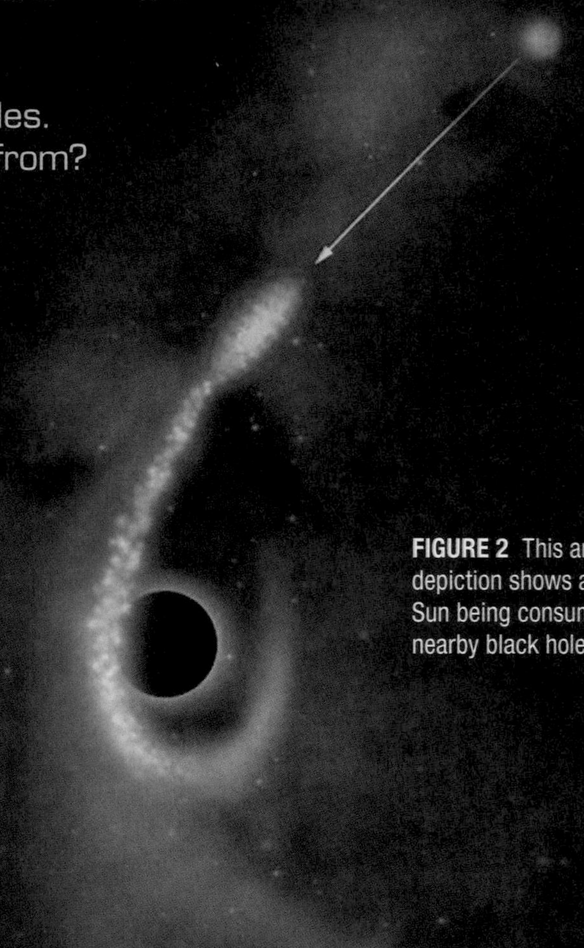

FIGURE 2 This artist's depiction shows a star like our Sun being consumed by a nearby black hole.

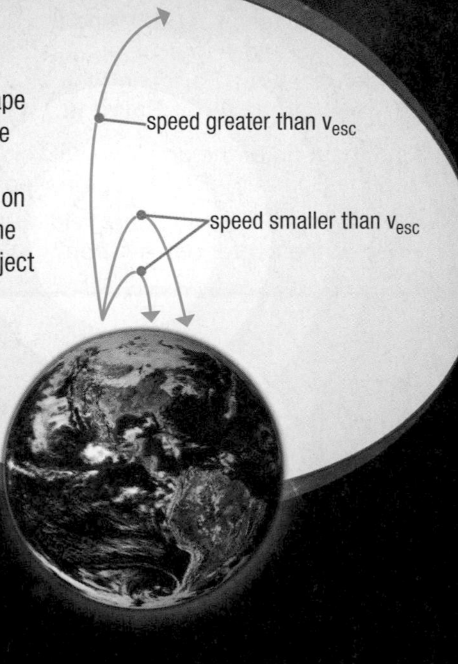

FIGURE 1 Escape velocity from the surface of any object depends on the mass and the radius of the object in question.

speed greater than v_{esc}

speed smaller than v_{esc}

GOING**FURTHER** >>>

Research Light traveling through a vacuum cannot slow down. So how, then, does a black hole prevent a light beam from escaping? To find out, investigate and write about an effect known as the gravitational red shift.

STUDY GUIDE

BIGIDEA   Gravity is an attractive field force that acts between objects with mass.

SECTION 1   Planetary Motion and Gravitation

MAINIDEA   The gravitational force between two objects is proportional to the product of their masses divided by the square of the distance between them.

- Kepler's first law states that planets move in elliptical orbits, with the Sun at one focus, and Kepler's second law states that an imaginary line from the Sun to a planet sweeps out equal areas in equal times. Kepler's third law states that the square of the ratio of the periods of any two planets is equal to the cube of the ratio of the distances between the centers of the planets and the center of the Sun.

$$\left(\frac{T_A}{T_B}\right)^2 = \left(\frac{r_A}{r_B}\right)^3$$

- Newton's law of universal gravitation can be used to rewrite Kepler's third law to relate the radius and period of a planet to the mass of the Sun. Newton's law of universal gravitation states that the gravitational force between any two objects is directly proportional to the product of their masses and inversely proportional to the square of the distance between their centers. The force is attractive and along a line connecting the centers of the masses.

$$F = \frac{Gm_1m_2}{r^2}$$

- Cavendish's investigation determined the value of G, confirmed Newton's prediction that a gravitational force exists between two objects, and helped calculate the mass of Earth.

SECTION 2   Using the Law of Universal Gravitation

MAINIDEA   All objects are surrounded by a gravitational field that affects the motions of other objects.

- The speed and period of a satellite in circular orbit describe orbital motion. Orbital speed and period for any object in orbit around another are calculated with Newton's second law.

- Gravitational mass and inertial mass are two essentially different concepts. The gravitational and inertial masses of an object, however, are numerically equal.

- All objects have gravitational fields surrounding them. Any object within a gravitational field experiences a gravitational force exerted on it by the gravitational field. Einstein's general theory of relativity explains gravitational force as a property of space itself.

Games and Multilingual eGlossary

Vocabulary Practice

SECTION 1
Planetary Motion and Gravitation
Mastering Concepts

26. Problem Posing Complete this problem so that it can be solved using Kepler's third law: "Suppose a new planet was found orbiting the Sun in the region between Jupiter and Saturn . . ."

27. In 1609 Galileo looked through his telescope at Jupiter and saw four moons. The name of one of the moons that he saw is Io. Restate Kepler's first law for Io and Jupiter.

28. Earth moves more slowly in its orbit during summer in the northern hemisphere than it does during winter. Is it closer to the Sun in summer or in winter?

29. Is the area swept out per unit of time by Earth moving around the Sun equal to the area swept out per unit of time by Mars moving around the Sun? Explain your answer.

30. Why did Newton think that a force must act on the Moon?

31. How did Cavendish demonstrate that a gravitational force of attraction exists between two small objects?

32. What happens to the gravitational force between two masses when the distance between the masses is doubled?

33. According to Newton's version of Kepler's third law, how would the ratio $\frac{T^2}{r^3}$ change if the mass of the Sun were doubled?

Mastering Problems

34. Jupiter is 5.2 times farther from the Sun than Earth is. Find the length of Jupiter's orbital period in Earth years.

35. The dwarf planet Pluto has a mean distance from the Sun of 5.87×10^{12} m. What is its orbital period of Pluto around the Sun in years?

36. Use **Table 1** to compute the gravitational force that the Sun exerts on Jupiter.

37. The gravitational force between two electrons that are 1.00 m apart is 5.54×10^{-71} N. Find the mass of an electron.

38. Two bowling balls each have a mass of 6.8 kg. The centers of the bowling balls are located 21.8 cm apart. What gravitational force do the two bowling balls exert on each other?

39. Figure 19 shows a Cavendish apparatus like the one used to find G. It has a large lead sphere that is 5.9 kg in mass and a small one with a mass of 0.047 kg. Their centers are separated by 0.055 m. Find the force of attraction between them.

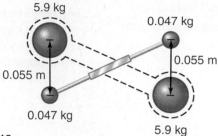

5.9 kg

0.047 kg

0.055 m

0.055 m

0.047 kg

5.9 kg

Figure 19

40. Assume that your mass is 50.0 kg. Earth's mass is 5.97×10^{24} kg, and its radius is 6.38×10^6 m.

   a. What is the force of gravitational attraction between you and Earth?

   b. What is your weight?

41. A 1.0-kg mass weighs 9.8 N on Earth's surface, and the radius of Earth is roughly 6.4×10^6 m.

   a. Calculate the mass of Earth.

   b. Calculate the average density of Earth.

42. Tom's mass is 70.0 kg, and Sally's mass is 50.0 kg. Tom and Sally are standing 20.0 m apart on the dance floor. Sally looks up and sees Tom. She feels an attraction. Supposing that the attraction is gravitational, find its size. Assume that both Tom and Sally can be replaced by spherical masses.

43. BIGIDEA The centers of two balls are 2.0 m apart, as shown in **Figure 20**. One ball has a mass of 8.0 kg. The other has a mass of 6.0 kg. What is the gravitational force between them?

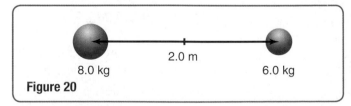

2.0 m

8.0 kg

6.0 kg

Figure 20

44. The star HD102272 has two planets. Planet A has a period of 127.5 days and a mean orbital radius of 0.615 AU. Planet B has a period of 520 days and a mean orbital radius of 1.57 AU. What is the mass of the star in units of the Sun's mass?

45. If a small planet, D, were located 8.0 times as far from the Sun as Earth is, how many years would it take the planet to orbit the Sun?

46. The Moon's center is 3.9×10^8 m from Earth's center. The Moon is 1.5×10^8 km from the Sun's center. If the mass of the Moon is 7.3×10^{22} kg, find the ratio of the gravitational forces exerted by Earth and the Sun on the Moon.

47. Ranking Task Using the solar system data in the reference tables at the end of the book, rank the following pairs of planets according to the gravitational force they exert on each other, from least to greatest. Specifically indicate any ties.

A. Mercury and Venus, when 5.0×10^7 km apart

B. Jupiter and Saturn, when 6.6×10^8 km apart

C. Jupiter and Earth, when 6.3×10^8 km apart

D. Mercury and Earth, when 9.2×10^7 km apart

E. Jupiter and Mercury, when 7.2×10^8 km apart

48. Two spheres are placed so that their centers are 2.6 m apart. The gravitational force between the two spheres is 2.75×10^{-12} N. What is the mass of each sphere if one of the spheres is twice the mass of the other sphere?

49. Toy Boat A force of 40.0 N is required to pull a 10.0-kg wooden toy boat at a constant velocity across a smooth glass surface on Earth. What is the force that would be required to pull the same wooden toy boat across the same glass surface on the planet Jupiter?

50. Mimas, one of Saturn's moons, has an orbital radius of 1.87×10^8 m and an orbital period of 23.0 h. Use Newton's version of Kepler's third law to find Saturn's mass.

51. Halley's Comet Every 76 years, comet Halley is visible from Earth. Find the average distance of the comet from the Sun in astronomical units. (AU is equal to the Earth's average distance from the Sun. The distance from Earth to the Sun is defined as 1.00 AU.)

52. Area is measured in m^2, so the rate at which area is swept out by a planet or satellite is measured in m^2/s.

a. How quickly is an area swept out by Earth in its orbit about the Sun?

b. How quickly is an area swept out by the Moon in its orbit about Earth? Use 3.9×10^8 m as the average distance between Earth and the Moon and 27.33 days as the period of the Moon.

53. The orbital radius of Earth's Moon is 3.9×10^8 m. Use Newton's version of Kepler's third law to calculate the period of Earth's Moon if the orbital radius were doubled.

SECTION 2
Using the Law of Universal Gravitation
Mastering Concepts

54. How do you answer the question, "What keeps a satellite up?"

55. A satellite is orbiting Earth. On which of the following does its speed depend?

a. mass of the satellite

b. distance from Earth

c. mass of Earth

56. What provides the force that causes the centripetal acceleration of a satellite in orbit?

57. During space flight, astronauts often refer to forces as multiples of the force of gravity on Earth's surface. What does a force of 5g mean to an astronaut?

58. Newton assumed that a gravitational force acts directly between Earth and the Moon. How does Einstein's view of the attractive force between the two bodies differ from Newton's view?

59. Show that the units of g, previously given as N/kg, are also m/s^2.

60. If Earth were twice as massive but remained the same size, what would happen to the value of g?

Mastering Problems

61. Satellite A geosynchronous satellite is one that appears to remain over one spot on Earth, as shown in **Figure 21**. Assume that a geosynchronous satellite has an orbital radius of 4.23×10^7 m.

a. Calculate its speed in orbit.

b. Calculate its period.

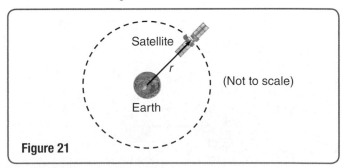

Satellite

r

(Not to scale)

Earth

Figure 21

62. Asteroid The dwarf planet Ceres has a mass of 7×10^{20} kg and a radius of 500 km.

a. What is g on the surface of Ceres?

b. How much would a 90-kg astronaut weigh on Ceres?

63. The Moon's mass is 7.34×10^{22} kg, and it has an orbital radius of 3.9×10^8 m from Earth. Earth's mass is 5.97×10^{24} kg.

  a. Calculate the gravitational force of attraction between Earth and the Moon.

  b. Find the magnitudes of Earth's gravitational field at the Moon.

64. Reverse Problem Write a physics problem with real-life objects for which the following equation would be part of the solution:

$$8.3 \times 10^3 \text{ m/s} = \sqrt{\frac{(6.67 \times 10^{-11} \text{ N} \cdot \text{m}^2/\text{kg}^2)(5.97 \times 10^{24} \text{ kg})}{r}}$$

65. The radius of Earth is about 6.38×10^3 km. A spacecraft with a weight of 7.20×10^3 N travels away from Earth. What is the weight of the spacecraft at each of the following distances from the surface of Earth?

  a. 6.38×10^3 km

  b. 1.28×10^4 km

  c. 2.55×10^4 km

66. Rocket How high does a rocket have to go above Earth's surface before its weight is half of what it is on Earth?

67. Two satellites of equal mass are put into orbit 30.0 m apart. The gravitational force between them is 2.0×10^{-7} N.

  a. What is the mass of each satellite?

  b. What is the initial acceleration given to each satellite by gravitational force?

68. Two large spheres are suspended close to each other. Their centers are 4.0 m apart, as shown in **Figure 22.** One sphere weighs 9.8×10^2 N. The other sphere weighs 1.96×10^2 N. What is the gravitational force between them?

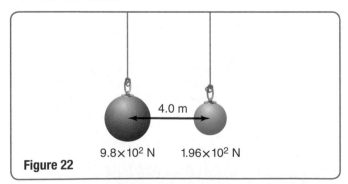

4.0 m

9.8×10^2 N 1.96×10^2 N

Figure 22

69. Suppose the centers of Earth and the Moon are 3.9×10^8 m apart, and the gravitational force between them is about 1.9×10^{20} N. What is the approximate mass of the Moon?

70. On the surface of the Moon, a 91.0-kg physics teacher weighs only 145.6 N. What is the value of the Moon's gravitational field at its surface?

71. The mass of an electron is 9.1×10^{-31} kg. The mass of a proton is 1.7×10^{-27} kg. An electron and a proton are about 0.59×10^{-10} m apart in a hydrogen atom. What gravitational force exists between the proton and the electron of a hydrogen atom?

72. Consider two spherical 8.0-kg objects that are 5.0 m apart.

  a. What is the gravitational force between the two objects?

  b. What is the gravitational force between them when they are 5.0×10^1 m apart?

73. If you weigh 637 N on Earth's surface, how much would you weigh on the planet Mars? Mars has a mass of 6.42×10^{23} kg and a radius of 3.40×10^6 m.

74. Find the value of g, the gravitational field at Earth's surface, in the following situations.

  a. Earth's mass is triple its actual value, but its radius remains the same.

  b. Earth's radius is tripled, but its mass remains the same.

  c. Both the mass and the radius of Earth are doubled.

Applying Concepts

75. Acceleration The force of gravity acting on an object near Earth's surface is proportional to the mass of the object. **Figure 23** shows a table-tennis ball and a golf ball in free fall. Why does a golf ball not fall faster than a table-tennis ball?

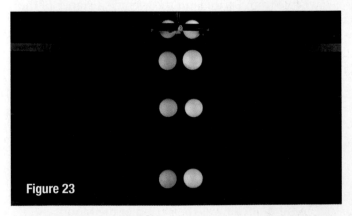

Figure 23

76. What information do you need to find the mass of Jupiter using Newton's version of Kepler's third law?

77. Why was the mass of the dwarf planet Pluto not known until a satellite of Pluto was discovered?

78. A satellite is one Earth radius above Earth's surface. How does the acceleration due to gravity at that location compare to acceleration at the surface of Earth?

79. What would happen to the value of G if Earth were twice as massive, but remained the same size?

80. An object in Earth's gravitational field doubles in mass. How does the force exerted by the field on the object change?

81. Decide whether each of the orbits shown in **Figure 24** is a possible orbit for a planet.

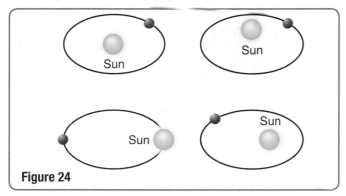

Figure 24

82. The Moon and Earth are attracted to each other by gravitational force. Does the more massive Earth attract the Moon with a greater force than the Moon attracts Earth? Explain.

83. **Figure 25** shows a satellite orbiting Earth. Examine the equation $v = \sqrt{\frac{Gm_E}{r}}$, relating the speed of an orbiting satellite and its distance from the center of Earth. Does a satellite with a large or small orbital radius have the greater velocity?

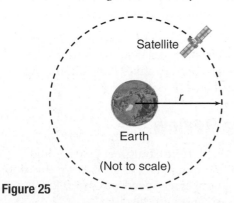

Figure 25

84. **Space Shuttle** If a space shuttle goes into a higher orbit, what happens to the shuttle's period?

85. Jupiter has about 300 times the mass of Earth and about ten times Earth's radius. Estimate the size of g on the surface of Jupiter.

86. Mars has about one-ninth the mass of Earth. **Figure 26** shows satellite M, which orbits Mars with the same orbital radius as satellite E, which orbits Earth. Which satellite has a smaller period?

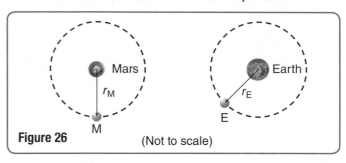

Figure 26 (Not to scale)

87. **Weight** Suppose that yesterday your body had a mass of 50.0 kg. This morning you stepped on a scale and found that you had gained weight. Assume your location has not changed.

   a. What happened, if anything, to your mass?

   b. What happened, if anything, to the ratio of your weight to your mass?

88. As an astronaut in an orbiting space shuttle, how would you go about "dropping" an object down to Earth?

89. **Weather Satellites** The weather pictures you see on television come from a spacecraft that is in a stationary position relative to Earth, 35,700 km above the equator. Explain how the satellite can stay in exactly the same position. What would happen if it were closer? Farther out? *Hint: Draw a pictorial model.*

Mixed Review

90. Use the information for Earth to calculate the mass of the Sun, using Newton's version of Kepler's third law.

91. The Moon's mass is 7.3×10^{22} kg and its radius is 1738 km. Suppose you perform Newton's thought experiment in which a cannonball is fired horizontally from a very high mountain on the Moon.

   a. How fast would the cannonball have to be fired to remain in orbit?

   b. How long would it take the cannonball to return to the cannon?

92. **Car Races** Suppose that a Martian base has been established. The inhabitants want to hold car races for entertainment. They want to construct a flat, circular race track. If a car can reach speeds of 12 m/s, what is the smallest radius of a track for which the coefficient of friction is 0.50?

Chapter Self-Check

93. **Apollo 11** On July 19, 1969, *Apollo 11* was adjusted to orbit the Moon at a height of 111 km. The Moon's radius is 1738 km, and its mass is 7.3×10^{22} kg.

   a. What was the period of *Apollo 11* in minutes?

   b. At what velocity did *Apollo 11* orbit the Moon?

94. The Moon's period is one month. Answer the following assuming the mass of Earth is doubled.

   a. What would the Moon's period be in months?

   b. Where would a satellite with an orbital period of one month be located?

   c. How would the length of a year on Earth change?

95. **Satellite** A satellite is in orbit, as in **Figure 27,** with an orbital radius that is half that of the Moon's. Find the satellite's period in units of the Moon's period.

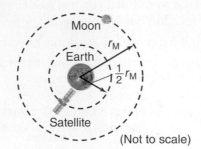

Figure 27   (Not to scale)

96. How fast would a planet of Earth's mass and size have to spin so that an object at the equator would be weightless? Give the period in minutes.

Thinking Critically

97. **Make and Use Graphs** Use Newton's law of universal gravitation to find an equation where x is equal to an object's distance from Earth's center and y is its acceleration due to gravity. Use a graphing calculator to graph this equation, using 6400–6600 km as the range for x and 9–10 m/s² as the range for y.

   The equation should be of the form $y = c\left(\dfrac{1}{x^2}\right)$. Use this graph and find y for these locations: sea level, 6400 km; the top of Mt. Everest, 6410 km; a satellite in typical orbit, 6500 km; a satellite in higher orbit, 6600 km.

98. Suppose the Sun were to disappear—its mass destroyed. If the gravitational force were action at a distance, Earth would experience the loss of the gravitational force of the Sun immediately. But, if the force were caused by a field or Einstein's curvature of space, the information that the Sun was gone would travel at the speed of light. How long would it take this information to reach Earth?

99. **Analyze and Conclude** The tides on Earth are caused by the pull of the Moon. Is this statement true?

   a. Determine the forces (in newtons) that the Moon and the Sun exert on a mass (m) of water on Earth. Answer in terms of m.

   b. Which celestial body, the Sun or the Moon, has a greater pull on the waters of Earth?

   c. What is the difference in force exerted by the Moon on water at the near surface and water at the far surface (on the opposite side) of Earth, as illustrated in **Figure 28.** Answer in terms of m.

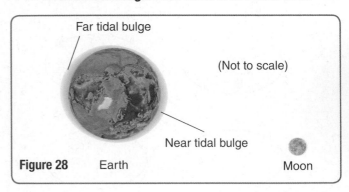

Figure 28   Earth   Moon

   d. Determine the difference in force exerted by the Sun on water at the near surface and on water at the far surface (on the opposite side) of Earth.

   e. Which celestial body has a greater difference in pull from one side of Earth to the other?

   f. Why is it misleading to say the tides result from the pull of the Moon? Make a correct statement to explain how the Moon causes tides on Earth.

Writing in Physics

100. Research and report the history of how the distance between the Sun and Earth was determined.

101. Explore the discovery of planets around other stars. What methods did the astronomers use? What measurements did they take? How did they use Kepler's third law?

Cumulative Review

102. **Airplanes** A jet left Pittsburgh at 2:20 P.M. and landed in Washington, DC, at 3:15 P.M. on the same day. If the jet's average speed was 441.0 km/h, what is the distance between the cities?

103. Carolyn wants to weigh her brother Jared. He agrees to stand on a scale, but only if they ride in an elevator. He steps on the scale while the elevator accelerates upward at 1.75 m/s². The scale reads 716 N. What is Jared's usual weight?

MULTIPLE CHOICE

1. Two satellites are in orbit around a planet. One satellite has an orbital radius of 8.0×10^6 m. The period of revolution for this satellite is 1.0×10^6 s. The other satellite has an orbital radius of 2.0×10^7 m. What is this satellite's period of revolution?

   A. 5.0×10^5 s   **C.** 4.0×10^6 s

   B. 2.5×10^6 s   **D.** 1.3×10^7 s

2. The illustration below shows a satellite in orbit around a small planet. The satellite's orbital radius is 6.7×10^4 km and its speed is 2.0×10^5 m/s. What is the mass of the planet around which the satellite orbits? ($G = 6.67 \times 10^{-11}$ N·m²/kg²)

   A. 2.5×10^{18} kg   **C.** 2.5×10^{23} kg

   B. 4.0×10^{20} kg   **D.** 4.0×10^{28} kg

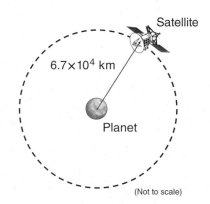

Satellite

6.7×10^4 km

Planet

(Not to scale)

3. Two satellites are in orbit around the same planet. Satellite A has a mass of 1.5×10^2 kg, and satellite B has a mass of 4.5×10^3 kg. The mass of the planet is 6.6×10^{24} kg. Both satellites have the same orbital radius of 6.8×10^6 m. What is the difference in the orbital periods of the satellites?

   A. no difference   **C.** 2.2×10^2 s

   B. 1.5×10^2 s   **D.** 3.0×10^2 s

4. A moon revolves around a planet with a speed of 9.0×10^3 m/s. The distance from the moon to the center of the planet is 5.43×10^6 m. What is the orbital period of the moon?

   A. $1.2\pi \times 10^2$ s   **C.** $1.2\pi \times 10^3$ s

   B. $6.0\pi \times 10^2$ s   **D.** $1.2\pi \times 10^9$ s

5. A moon in orbit around a planet experiences a gravitational force not only from the planet, but also from the Sun. The illustration below shows a moon during a solar eclipse, when the planet, the moon, and the Sun are aligned. The moon has a mass of about 3.9×10^{21} kg. The mass of the planet is 2.4×10^{26} kg, and the mass of the Sun is 1.99×10^{30} kg. The distance from the moon to the center of the planet is 6.0×10^8 m, and the distance from the moon to the Sun is 1.5×10^{11} m. What is the ratio of the gravitational force on the moon due to the planet, compared to its gravitational force due to the Sun during the solar eclipse?

   A. 0.5   **C.** 5.0

   B. 2.5   **D.** 7.5

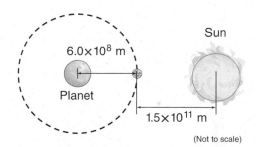

Sun

6.0×10^8 m

Planet

1.5×10^{11} m

(Not to scale)

FREE RESPONSE

6. Two satellites are in orbit around a planet. Satellite S_1 takes 20 days to orbit the planet at a distance of 2×10^5 km from the center of the planet. Satellite S_2 takes 160 days to orbit the planet. What is the distance of satellite S_2 from the center of the planet?

NEED EXTRA HELP?

If you Missed Question	1	2	3	4	5	6
Review Section	1	2	2	2	2	1

Rotational Motion

BIGIDEA   Applying a torque to an object causes a change in that object's angular velocity.

SECTIONS

LaunchLAB

ROLLING OBJECTS

How do different objects rotate as they roll?

WATCH THIS!

Video

DANGEROUS CURVE

Have you ever seen a road sign depicting a truck tipping over? You will usually find them near curves. They warn truck and bus drivers to slow down while rounding the curve. What can happen if these drivers do not slow down on the curve?

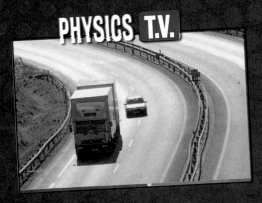

PHYSICS T.V.

CRYSTAL·LIL'S

(l)Pixtal/age fotostock; (r)LMR Group/Alamy

Describing Rotational Motion

Many cars have a tachometer that displays the rate at which the motor's shaft rotates. This speed is measured in thousands of revolutions per minute. Why might a driver need to know this information?

MAINIDEA

Angular displacement, angular velocity, and angular acceleration all help describe angular motion.

Essential Questions

- What is angular displacement?
- What is average angular velocity?
- What is average angular acceleration, and how is it related to angular velocity?

Review Vocabulary

displacement change in position having both magnitude and direction; it is equal to the final position minus the initial position

New Vocabulary

radian
angular displacement
angular velocity
angular acceleration

Angular Displacement

You probably have observed a spinning object many times. How would you measure such an object's rotation? Find a circular object, such as a DVD. Mark one point on the edge of the DVD so that you can keep track of its position. Rotate the DVD to the left (counterclockwise), and as you do so, watch the location of the mark. When the mark returns to its original position, the DVD has made one complete revolution.

Measuring revolution How can you measure a fraction of one revolution? It can be measured in several different ways, but the two most used are degrees and radians. A degree is $\frac{1}{360}$ of a revolution and is the usual scale marking on a protractor. In mathematics and physics, the radian is related to the ratio of the circumference of a circle to its radius. In one revolution, a point on the edge of a wheel travels a distance equal to 2π times the radius of the wheel. For this reason, the **radian** is defined as $\frac{1}{(2\pi)}$ of a revolution. One complete revolution is an angle of 2π radians. The radian is abbreviated *rad*.

The Greek letter theta (θ) is used to represent the angle of revolution. **Figure 1** shows the angles in radians for several common fractions of a revolution. Note that counterclockwise rotation is designated as positive, while clockwise is negative. As an object rotates, the change in the angle is called the object's **angular displacement.**

Figure 1 A fraction of a revolution can be measured in degrees or radians. Some common angles are shown below measured in radians. Each angle is measured in the counterclockwise direction from $\theta = 0$.

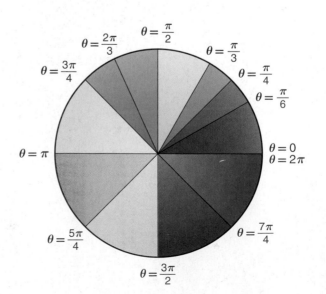

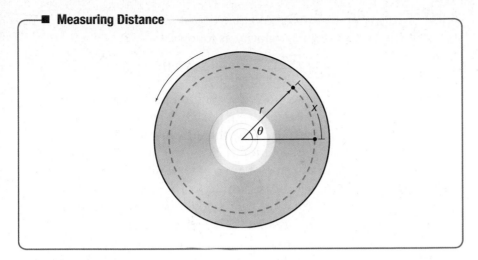

Figure 2 The dashed line shows the path of a point on a DVD as the DVD rotates counter-clockwise about its center. The point is located a distance r from the center of the DVD and moves a distance x as it rotates.

Explain what the variables r, x, and θ represent.

Earth's revolution As you know, Earth turns one complete revolution, or 2π rad, in 24 h. In 12 h, it rotates through π rad. Through what angle does Earth rotate in 6 h? Because 6 h is one-fourth of a day, Earth rotates through an angle of $\frac{\pi}{2}$ rad during that period. Earth's rotation as seen from the North Pole is positive. Is it positive or negative when viewed from the South Pole?

✓ **READING CHECK Identify** the angle that Earth rotates in 48 h.

Measuring distance How far does a point on a rotating object move? You already found that a point on the edge of an object moves 2π times the radius in one revolution. In general, for rotation through an angle (θ), a point at a distance r from the center, as shown in **Figure 2**, moves a distance given by $x = r\theta$. If r is measured in meters, you might think that multiplying it by θ rad would result in x being measured in m·rad. However, this is not the case. Radians indicate the dimensionless ratio between x and r. Thus, x is measured in meters.

Angular Velocity

How fast does a DVD spin? How can you determine its speed of rotation? Recall that velocity is displacement divided by the time taken to make the displacement. Likewise, the **angular velocity** of an object is angular displacement divided by the time taken to make the angular displacement. The angular velocity of an object is given by the following ratio, and is represented by the Greek letter omega (ω).

AVERAGE ANGULAR VELOCITY OF AN OBJECT
The angular velocity equals the angular displacement divided by the time required to make the rotation.

$$\omega = \frac{\Delta\theta}{\Delta t}$$

Recall that if the velocity changes over a time interval, the average velocity is not equal to the instantaneous velocity at any given instant. Similarly, the angular velocity calculated in this way is actually the average angular velocity over a time interval (Δt). Instantaneous angular velocity equals the slope of a graph of angular position versus time.

✓ **READING CHECK Define** angular velocity in your own words.

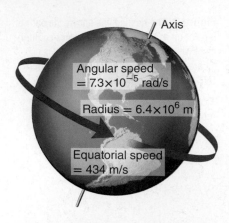

Figure 3 Earth is a rotating, rigid body and all parts rotate at the same rate.

Earth's angular velocity Angular velocity is measured in rad/s. You can calculate Earth's angular velocity as follows:

$$\omega_E = \frac{(2\pi \text{ rad})}{(24.0 \text{ h})(3600 \text{ s/h})} = 7.27\times10^{-5} \text{ rad/s}$$

In the same way that counterclockwise rotation produces positive angular displacement, it also results in positive angular velocity.

If an object's angular velocity is ω, then the linear velocity of a point a distance r from the axis of rotation is given by $v = r\omega$. The speed at which an object on Earth's equator moves as a result of Earth's rotation is given by $v = r\omega = (6.38\times10^6 \text{ m/rad})(7.27\times10^{-5} \text{ rad/s}) = 464 \text{ m/s}$. Earth is an example of a rotating, rigid body, as shown in **Figure 3.** Even though different points at different latitudes on Earth do not move the same distance in each revolution, all points rotate through the same angle. All parts of a rigid body rotate at the same rate. The Sun is not a rigid body. Different parts of the Sun rotate at different angular velocities. Most objects that we will consider in this chapter are rigid bodies.

Angular Acceleration

What if angular velocity is changing? For example, a car could accelerate from 0.0 m/s to 25 m/s in 15 s. In the same 15 s, the angular velocity of the car's 0.64 m diameter wheels would change from 0.0 rad/s to 78 rad/s. The wheels would undergo **angular acceleration,** which is the change in angular velocity divided by the time required to make the change. Angular acceleration (α) is represented by the following equation and is measured in rad/s^2.

AVERAGE ANGULAR ACCELERATION OF AN OBJECT
Angular acceleration is equal to the change in angular velocity divided by the time required to make that change.

$$\alpha = \frac{\Delta\omega}{\Delta t}$$

If the change in angular velocity is positive, then the angular acceleration also is positive. Angular acceleration defined in this way is also the average angular velocity over the time interval Δt. One way to find the instantaneous angular acceleration is to find the slope of a graph of angular velocity as a function of time. The linear acceleration of a point at a distance (r) from the axis of an object with angular acceleration (α) is given by $a = r\alpha$. **Table 1** is a summary of linear and angular relationships discussed previously in this section.

☑ **READING CHECK** **Compare** the angular velocity and angular acceleration of a rotating body.

Table 1 Linear and Angular Measures			
Quantity	**Linear**	**Angular**	**Relationship**
Displacement	x (m)	θ (rad)	$x = r\theta$
Velocity	v (m/s)	ω (rad/s)	$v = r\omega$
Acceleration	a (m/s^2)	α (rad/s^2)	$a = r\alpha$

Do additional problems. Online Practice

1. What is the angular displacement of each of the following hands of a clock in 1.00 h? State your answer in three significant digits.

  a. the second hand

  b. the minute hand

  c. the hour hand

2. A rotating toy above a baby's crib makes one complete counterclockwise rotation in 1 min.

  a. What is its angular displacement in 3 min?

  b. What is the toy's angular velocity in rad/min?

  c. If the toy is turned off, does it have positive or negative angular acceleration? Explain.

3. If a truck has a linear acceleration of 1.85 m/s^2 and the wheels have an angular acceleration of 5.23 rad/s^2, what is the diameter of the truck's wheels?

4. The truck in the previous problem is towing a trailer with wheels that have a diameter of 48 cm.

  a. How does the linear acceleration of the trailer compare with that of the truck?

  b. How do the angular accelerations of the wheels of the trailer and the wheels of the truck compare?

5. **CHALLENGE** You replace the tires on your car with tires of larger diameter. After you change the tires, how will the angular velocity and number of revolutions be different, for trips at the same speed and over the same distance?

Angular frequency An object can revolve many times in a given amount of time. For instance, a spinning wheel may complete several revolutions in 1 min. The number of complete revolutions made by an object in 1 s is called angular frequency. Angular frequency is defined as $f \equiv \frac{\omega}{2\pi}$.

One example of such a rotating object is a computer hard drive. Listen carefully when you start a computer. You often will hear the hard drive spinning. Hard drive frequencies are measured in revolutions per minute (RPM). Inexpensive hard drives rotate at 4800, 5400, and 7200 RPM. More advanced hard drives operate at 10,000 or 15,000 RPM. The faster the hard drive rotates, the quicker the hard drive can access or store information.

SECTION 1 REVIEW

Section Self-Check Check your understanding.

6. **MAIN**IDEA The Moon rotates once on its axis in 27.3 days. Its radius is 1.74×10^6 m.

  a. What is the period of the Moon's rotation in seconds?

  b. What is the frequency of the Moon's rotation in rad/s?

  c. A rock sits on the surface at the Moon's equator. What is its linear speed due to the Moon's rotation?

  d. Compare this speed with the speed of a person at Earth's equator due to Earth's rotation.

7. **Angular Displacement** A movie lasts 2 h. During that time, what is the angular displacement of each of the following?

  a. the hour hand

  b. the minute hand

  c. the second hand

8. **Angular Acceleration** In the spin cycle of a clothes washer, the drum turns at 635 rev/min. If the lid of the washer is opened, the motor is turned off. If the drum requires 8.0 s to slow to a stop, what is the angular acceleration of the drum?

9. **Angular Displacement** Do all parts of the minute hand on a watch, shown in **Figure 4,** have the same angular displacement? Do they move the same linear distance? Explain.

10. **Critical Thinking** A CD-ROM has a spiral track that starts 2.7 cm from the center of the disk and ends 5.5 cm from the center. The disk drive must turn the disk so that the linear velocity of the track is a constant 1.4 m/s. Find the following.

  a. the angular velocity of the disk (in rad/s and rev/min) for the start of the track

  b. the disk's angular velocity at the end of the track

  c. the disk's angular acceleration if the disk is played for 76 min

Figure 4

Rotational Dynamics

PHYSICS 4 YOU

Have you ever watched a washing machine spin? During the rinse cycle, it starts spinning slowly, but its angular velocity soon increases until the clothes are a blur. The rotational motion causes water in the clothes to move to the drum. The holes in the drum allow the water to drain from the washer.

MAIN IDEA

Torques cause changes in angular velocity.

Essential Questions

- What is torque?
- How is the moment of inertia related to rotational motion?
- How are torque, the moment of inertia, and Newton's second law for rotational motion related?

Review Vocabulary

magnitude a measure of size

New Vocabulary

lever arm
torque
moment of inertia
Newton's second law for rotational motion

Force and Angular Velocity

How do you start the rotation of an object? That is, how do you increase its angular velocity? A toy top is a handy round object that is easy to spin. If you wrap a string around it and pull hard, you could make the top spin rapidly. The force of the string is exerted at the outer edge of the top and at right angles to the line from the center of the top to the point where the string leaves the top's surface.

You have learned that a force changes the velocity of a point object. In the case of a toy top, a force that is exerted in a very specific way changes the angular velocity of an extended object. An extended object is an object that has a definite shape and size.

Consider how you open a door: you exert a force. How can you exert the force to open the door most easily? To get the most effect from the least force, you exert the force as far from the axis of rotation as possible, as shown in **Figure 5.** In this case, the axis of rotation is an imaginary vertical line through the hinges. The doorknob is near the outer edge of the door. You exert the force on the doorknob at right angles to the door. Thus, the magnitude of the force, the distance from the axis to the point where the force is exerted, and the direction of the force determine the change in angular velocity.

Figure 5 When opening a door that is free to rotate about its hinges, apply the force farthest from the hinges, at an angle perpendicular to the door.

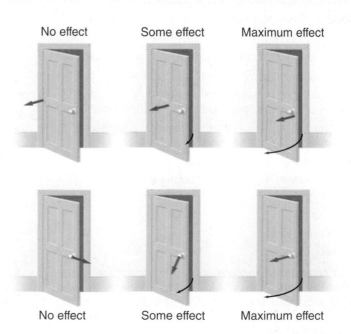

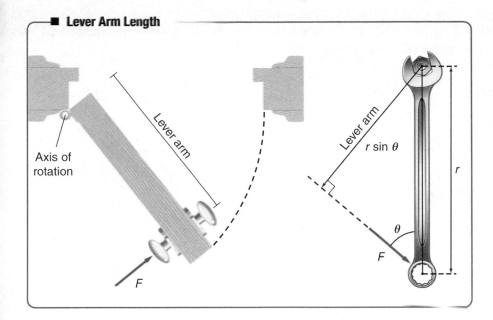

Lever Arm Length

Axis of
rotation

Lever arm

Lever arm

$r \sin \theta$

r

θ

F

F

Figure 6 The lever arm is the perpendicular distance from the axis of rotation to the point where the force is exerted. For the door, the lever arm is along the width of the door, from the hinge to the point where the force is exerted. For the wrench, the lever arm is equal to $r \sin \theta$, when the angle (θ) between the force and the radius of rotation is not equal to 90°.

Explain why the formula $r \sin \theta$ is used to find the length of the lever arm.

Lever arm For a given applied force, the change in angular velocity depends on the **lever arm,** which is the perpendicular distance from the axis of rotation to the point where the force is exerted. If the force is perpendicular to the radius of rotation, as it was with the toy top, then the lever arm is the distance from the axis (r). For the door example, the lever arm is the distance from the hinges to the point where you exert the force, as illustrated on the left in **Figure 6.** If a force is not exerted perpendicular to the radius, the length of the lever arm is reduced. You must use mathematics to find the length of the lever arm.

To find the lever arm, you must extend the line of the force until it forms a right angle with a line from the center of rotation, as shown in **Figure 6.** The distance between this intersection and the axis is the lever arm. Using trigonometry, the lever arm (L) can be calculated by the equation $L = r \sin \theta$. In this equation, r is the distance from the axis of rotation to the point where the force is exerted, and θ is the angle between the force and the radius from the axis of rotation to the point where the force is applied.

✓ **READING CHECK** **Explain** what each of the variables represent in the equation $L = r \sin \theta$.

View a **simulation of a rotating ladder.**

Concepts In Motion

Torque The term **torque** describes the combination of force and lever arm that can cause an object to rotate. The magnitude of a torque is the product of the force and the perpendicular lever arm. Because force is measured in newtons and distance is measured in meters, torque is measured in newton-meters (N•m). Torque is represented by the Greek letter tau (τ) and is represented by the equation shown below.

TORQUE
Torque is equal to the force F times the lever arm ($r \sin \theta$).

$$\tau = Fr \sin \theta$$

✓ **READING CHECK** **Identify** what each of the variables in the torque equation—τ, F, r, and θ—represents.

Find help with **trigonometric ratios.** Math Handbook

LEVER ARM A bolt on a car engine must be tightened with a torque of 35 N·m. You use a 25-cm-long wrench and pull the end of the wrench at an angle of 60.0° to the handle of the wrench. How long is the lever arm, and how much force must you exert?

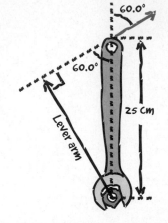

1 ANALYZE AND SKETCH THE PROBLEM

Sketch the situation. Find the lever arm by extending the force vector backward until a line that is perpendicular to it intersects the axis of rotation.

KNOWN	UNKNOWN
$r = 0.25$ m $\quad \tau = 35$ N·m	$L = ?$
$\theta = 60.0°$	$F = ?$

2 SOLVE FOR THE UNKNOWN

Solve for the length of the lever arm.

$L = r \sin \theta$

$\quad = (0.25 \text{ m})(\sin 60.0°)$ ◀ Substitute $r = 0.25$ m and $\theta = 60.0°$ into the equation. Then, solve the equation.

$\quad = 0.22$ m

Solve for the force.

$\tau = Fr \sin \theta$

$F = \dfrac{\tau}{(r \sin \theta)}$

$\quad = \dfrac{(35 \text{ N·m})}{(0.25 \text{ m})(\sin 60.0°)}$ ◀ Substitute $\tau = 35$ N·m, $r = 0.25$ m, and $\theta = 60.0°$ into the equation.

$\quad = 1.6 \times 10^2$ N ◀ Then, solve the equation. Remember to use significant digits.

3 EVALUATE THE ANSWER

- **Are the units correct?** Force is measured in newtons.
- **Does the sign make sense?** Only the magnitude of the force needed to rotate the wrench clockwise is calculated.

PRACTICE PROBLEMS

Do additional problems. Online Practice

11. Consider the wrench in Example Problem 1. What force is needed if it is applied to the wrench pointing perpendicular to the wrench?

12. If a torque of 55.0 N·m is required to turn a bolt and the largest force you can exert is 135 N, how long a lever arm must you use to turn the bolt?

13. You have a 0.234-m-long wrench. A job requires a torque of 32.4 N·m, and you can exert a force of 232 N.

    a. What is the smallest angle, with respect to the handle of the wrench, at which you can pull on the wrench and get the job done?

    b. A friend can exert 275 N. What is the smallest angle she can use to accomplish the job?

14. You stand on a bicycle pedal, as shown in **Figure 7.** Your mass is 65 kg. If the pedal makes an angle of 35° above the horizontal and the pedal is 18 cm from the center of the chain ring, how much torque would you exert?

15. **CHALLENGE** If the pedal in the previous problem is horizontal, how much torque would you exert? How much torque would you exert when the pedal is vertical?

Figure 7

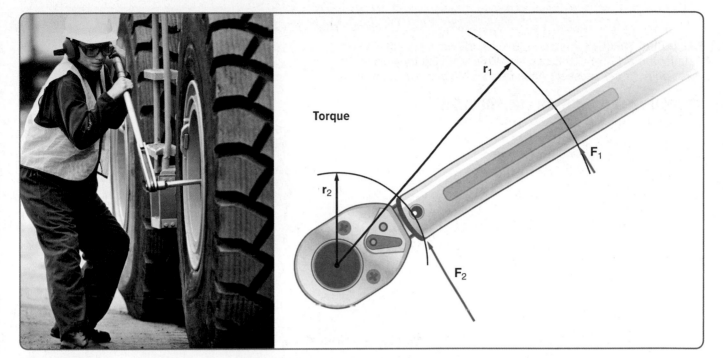

Torque

r_1

F_1

r_2

F_2

Figure 8 This worker uses a long wrench because it requires him to exert less force to tighten and loosen the nut. The wrench has a long lever arm, and less force is required if the force is applied farther from the axis of rotation (the center of the nut).

Finding Net Torque

Figure 8 shows a practical application of increasing torque to make a task easier. For another example of torque, try the following investigation. Collect two pencils, some coins, and some transparent tape. Tape two identical coins to the ends of the pencil and balance it on the second pencil, as shown in **Figure 9.** Each coin exerts a torque that is equal to the distance from the balance point to the center of the coin (r) times its weight (F_g), as follows:

$$\tau = rF_g$$

But the torques are equal and opposite in direction. Thus, the net torque is zero:

$$\tau_1 - \tau_2 = 0$$
$$\text{or}$$
$$r_1F_{g1} - r_2F_{g2} = 0$$

How can you make the pencil rotate? You could add a second coin on top of one of the two coins, thereby making the two forces different. You also could slide the balance point toward one end or the other of the pencil, thereby making the two lever arms of different length.

PhysicsLABs

LEVERAGE
FORENSICS LAB Can a person use a simple lever to open a heavy locked door?

TORQUES
Can you measure forces that produce torque?

View an **animation of net torque.**

Concepts In Motion

r_1 r_2

F_{g1} F_{g2}

Figure 9 The torque exerted by the first coin ($F_{g1}r_1$) is equal and opposite in direction to the torque exerted by the second coin ($F_{g2}r_2$) when the pencil is balanced.

Find help with **isolating a variable.** Math Handbook

BALANCING TORQUES Kariann (56 kg) and Aysha (43 kg) want to balance on a 1.75-m-long seesaw. Where should they place the pivot point of the seesaw? Assume that the seesaw is massless.

1 ANALYZE AND SKETCH THE PROBLEM

- Sketch the situation.
- Draw and label the vectors.

KNOWN

$m_K = 56$ kg
$m_A = 43$ kg
$r_K + r_A = 1.75$ m

UNKNOWN

$r_K = ?$
$r_A = ?$

2 SOLVE FOR THE UNKNOWN

Find the two forces.

Kariann:

$F_{gK} = m_K g$

$= (56$ kg$)(9.8$ N/kg$)$ ◀ Substitute the known values into the equation: $m_K = 56$ kg, $g = 9.8$ N/kg.

$= 5.5 \times 10^2$ N

Aysha:

$F_{gA} = m_A g$

$= (43$ kg$)(9.8$ N/kg$)$ ◀ Substitute the known values into the equation: $m_A = 43$ kg, $g = 9.8$ N/kg.

$= 4.2 \times 10^2$ N

Define Kariann's distance in terms of the length of the seesaw and Aysha's distance.

$r_K = 1.75$ m $- r_A$

When there is no rotation, the sum of the torques is zero.

$$F_{gK} r_K = F_{gA} r_A$$

$$F_{gK} r_K - F_{gA} r_A = 0.0 \text{ N·m}$$

$$F_{gK}(1.75 \text{ m} - r_A) - F_{gA} r_A = 0.0 \text{ N·m}$$

◀ Substitute the relationship between Kariann's distance in terms of Aysha's distance in to the equation: $r_K = 1.75$ m $- r_A$.

Solve for r_A.

$$F_{gK}(1.75 \text{ m}) - F_{gK}(r_A) - F_{gA} r_A = 0.0 \text{ N·m}$$

$$F_{gK} r_A + F_{gA} r_A = F_{gK}(1.75 \text{ m})$$

$$(F_{gK} + F_{gA}) r_A = F_{gK}(1.75 \text{ m})$$

$$r_A = \frac{F_{gK}(1.75 \text{ m})}{(F_{gK} + F_{gA})}$$

$$= \frac{(5.5 \times 10^2 \text{ N})(1.75 \text{ m})}{(5.5 \times 10^2 \text{ N} + 4.2 \times 10^2 \text{ N})}$$

◀ Substitute $F_{gK} = 5.5 \times 10^2$ N and $F_{gA} = 4.2 \times 10^2$ N.

$$= 0.99 \text{ m}$$

3 EVALUATE THE ANSWER

- **Are the units correct?** Distance is measured in meters.
- **Do the signs make sense?** Distances are positive.
- **Is the magnitude realistic?** Aysha is about 1 m from the center, so Kariann is about 0.75 m away from it. Because Kariann's weight is greater than Aysha's weight, the lever arm on Kariann's side should be shorter. Aysha is farther from the pivot, as expected.

16. Ashok, whose mass is 43 kg, sits 1.8 m from a pivot at the center of a seesaw. Steve, whose mass is 52 kg, wants to seesaw with Ashok. How far from the center of the seesaw should Steve sit?

17. A bicycle-chain wheel has a radius of 7.70 cm. If the chain exerts a 35.0-N force on the wheel in the clockwise direction, what torque is needed to keep the wheel from turning?

18. Two stationary baskets of fruit hang from strings going around pulleys of different diameters, as shown in **Figure 10**. What is the mass of basket A?

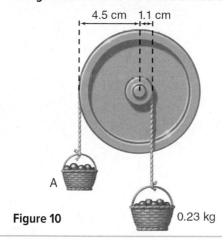

4.5 cm   1.1 cm

A

Figure 10   0.23 kg

19. Suppose the radius of the larger pulley in problem 18 was increased to 6.0 cm. What is the mass of basket A now?

20. CHALLENGE A bicyclist, of mass 65.0 kg, stands on the pedal of a bicycle. The crank, which is 0.170 m long, makes a 45.0° angle with the vertical, as shown in **Figure 11**. The crank is attached to the chain wheel, which has a radius of 9.70 cm. What force must the chain exert to keep the wheel from turning?

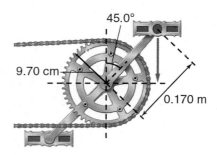

45.0°

9.70 cm

0.170 m

Figure 11

The Moment of Inertia

If you exert a force on a point mass, its acceleration will be inversely proportional to its mass. How does an extended object rotate when a torque is exerted on it? To observe firsthand, recover the pencil, the coins, and the transparent tape that you used earlier in this chapter. First, tape the coins at the ends of the pencil. Hold the pencil between your thumb and forefinger, and wiggle it back and forth. Take note of the forces that your thumb and forefinger exert. These forces create torques that change the angular velocity of the pencil and coins.

Now move the coins so that they are only 1 or 2 cm apart. Wiggle the pencil as before. Did you notice that the pencil is now easier to rotate? The torque that was required was much less this time. The mass of an object is not the only factor that determines how much torque is needed to change its angular velocity; the distribution or location of the mass also is important.

The resistance to rotation is called the **moment of inertia,** which is represented by the symbol I and has units of mass times the square of the distance. For a point object located at a distance (r) from the axis of rotation, the moment of inertia is given by the following equation.

MOMENT OF INERTIA OF A POINT MASS
The moment of inertia of a point mass is equal to the mass of the object times the square of the object's distance from the axis of rotation.

$$I = mr^2$$

MiniLAB

BALANCING TORQUES
Can you find the equilibrium point on a beam?

Table 2 Moments of Inertia for Various Objects

Object	Location of Axis	Diagram	Moment of Inertia
Thin hoop of radius r	through central diameter	Axis	mr^2
Solid, uniform cylinder of radius r	through center	Axis	$\left(\dfrac{1}{2}\right)mr^2$
Uniform sphere of radius r	through center	Axis	$\left(\dfrac{2}{5}\right)mr^2$
Long, uniform rod of length l	through center	Axis	$\left(\dfrac{1}{12}\right)ml^2$
Long, uniform rod of length l	through end	Axis	$\left(\dfrac{1}{3}\right)ml^2$
Thin, rectangular plate of length l and width w	through center	Axis	$\left(\dfrac{1}{12}\right)m(l^2 + w^2)$

Figure 12 The moment of inertia of a book depends on the axis of rotation. The moment of inertia of the book on the top is larger than the moment of inertia of the book on the bottom because the average distance of the book's mass from the rotational axis is larger.

Identify which book requires more torque to rotate it and why.

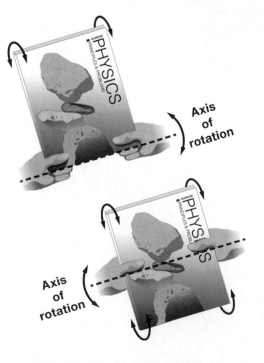

Moment of inertia and mass As you have seen, the moments of inertia for extended objects, such as the pencil and coins, depend on how far the masses are from the axis of rotation. A bicycle wheel, for example, has almost all of its mass in the rim and tire. Its moment of inertia about its axle is almost exactly equal to mr^2, where r is the radius of the wheel. For most objects, however, the mass is distributed closer to the axis so the moment of inertia is less than mr^2. For example, as shown in **Table 2,** for a solid cylinder of radius r, $I = \left(\dfrac{1}{2}\right)mr^2$, while for a solid sphere, $I = \left(\dfrac{2}{5}\right)mr^2$.

☑ **READING CHECK Write** the equation for the moment of inertia of a hoop.

Moment of inertia and rotational axis The moment of inertia also depends on the location and direction of the rotational axis, as illustrated in **Figure 12.** To observe this firsthand, hold a book in the upright position by placing your hands at the bottom of the book. Feel the torque needed to rock the book toward you and then away from you. Now put your hands in the middle of the book and feel the torque needed to rock the book toward you and then away from you. Note that much less torque is needed when your hands are placed in the middle of the book because the average distance of the book's mass from the rotational axis is much less in this case.

MOMENT OF INERTIA A simplified model of a twirling baton is a thin rod with two round objects at each end. The length of the baton is 0.66 m, and the mass of each object is 0.30 kg. Find the moment of inertia of the baton as it is rotated about an axis at the midpoint between the round objects and perpendicular to the rod. What is the moment of inertia of the baton if the axis is moved to one end of the rod? Which is greater? The mass of the rod is negligible compared to the masses of the objects at the ends.

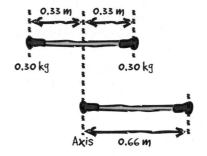

1 ANALYZE AND SKETCH THE PROBLEM

Sketch the situation. Show the baton with the two different axes of rotation and the distances from the axes of rotation to the masses.

KNOWN   **UNKNOWN**

$m = 0.30$ kg $I = ?$

$l = 0.66$ m

2 SOLVE FOR THE UNKNOWN

Calculate the moment of inertia of each mass separately.

Rotating about the center of the rod:

$$r = \left(\frac{1}{2}\right)l$$

$$= \left(\frac{1}{2}\right)(0.66 \text{ m})$$ ◀ Substitute the known value, l = 0.66 m, into the equation.

$$= 0.33 \text{ m}$$

$$I_{\text{single mass}} = mr^2$$

$$= (0.30 \text{ kg})(0.33 \text{ m})^2$$ ◀ Substitute m = 0.30 kg and r = 0.33 m into the equation.

$$= 0.033 \text{ kg·m}^2$$

Find the moment of inertia of the baton.

$$I = 2I_{\text{single mass}}$$

$$= 2(0.033 \text{ kg·m}^2)$$ ◀ Substitute $I_{\text{single mass}}$ = 0.033 kg·m² into the equation.

$$= 0.066 \text{ kg·m}^2$$

Rotating about one end of the rod:

$$I_{\text{single mass}} = mr^2$$

$$= (0.30 \text{ kg})(0.66 \text{ m})^2$$ ◀ Substitute m = 0.30 kg and r = 0.66 m into the equation.

$$= 0.13 \text{ kg·m}^2$$

The $I_{\text{single mass}} = mr^2$

$$= (0.30 \text{ kg})(0.0 \text{ m})^2$$

$$= 0 \text{ for the other mass.}$$

Find the moment of inertia of the baton.

$$I = I_{\text{single mass}} + 0$$

$$= 0.13 \text{ kg·m}^2$$

The moment of inertia is greater when the baton is swung around one end.

3 EVALUATE THE ANSWER

- **Are the units correct?** Moment of inertia is measured in kg·m².

- **Is the magnitude realistic?** Masses and distances are small, and so are the moments of inertia. Doubling the distance increases the moment of inertia by a factor of 4. Thus, doubling the distance increases the moment of inertia more than having only one mass decreases the moment of inertia.

PRACTICE PROBLEMS

21. Two children of equal masses sit 0.3 m from the center of a seesaw. Assuming that their masses are much greater than that of the seesaw, by how much is the moment of inertia increased when they sit 0.6 m from the center? Ignore the moment of inertia for the seesaw.

22. Suppose there are two balls with equal diameters and masses. One is solid, and the other is hollow, with all its mass distributed at its surface. Are the moments of inertia of the balls equal? If not, which is greater?

23. Calculate the moments of inertia for each object below using the formulas in **Table 2**. Each object has a radius of 2.0 m and a mass of 1.0 kg.

   a. a thin hoop

   b. a solid, uniform cylinder

   c. a solid, uniform sphere

24. CHALLENGE Figure 13 shows three equal-mass spheres on a rod of very small mass. Consider the moment of inertia of the system, first when it is rotated about sphere A and then when it is rotated about sphere C.

   a. Are the moments of inertia the same or different? Explain. If the moments of inertia are different, in which case is the moment of inertia greater?

   b. Each sphere has a mass of 0.10 kg. The distance between spheres A and C is 0.20 m. Find the moment of inertia in the following instances: rotation about sphere A, rotation about sphere C.

Figure 13

Newton's Second Law for Rotational Motion

Newton's second law for linear motion is expressed as $a = \frac{F_{net}}{m}$. If you rewrite this equation to represent rotational motion, acceleration is replaced by angular acceleration (α) force is replaced by net torque (τ_{net}) and mass is replaced by moment of inertia (I). Angular acceleration is directly proportional to the net torque and inversely proportional to the moment of inertia as stated in **Newton's second law for rotational motion.** This law is expressed by the following equation.

NEWTON'S SECOND LAW FOR ROTATIONAL MOTION
The angular acceleration of an object about a particular axis equals the net torque on the object divided by the moment of inertia.

$$\alpha = \frac{\tau_{net}}{I}$$

If the torque on an object and the angular velocity of that object are in the same direction, then the angular velocity of the object increases. If the torque and angular velocity are in different directions, then the angular velocity decreases.

PHYSICS CHALLENGE

Moments of Inertia Rank the objects shown in the diagram from least to greatest according to their moments of inertia about the indicated axes. All spheres have equal masses and all separations are the same. Assume that the rod's mass is negligible.

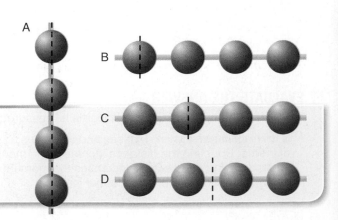

TORQUE A solid steel wheel is free to rotate about a motionless central axis. It has a mass of 15 kg and a diameter of 0.44 m and starts at rest. You want to increase this wheel's rotation about its central axis to 8.0 rev/s in 15 s.

a. What torque must be applied to the wheel?

b. If you apply the torque by wrapping a strap around the outside of the wheel, how much force should you exert on the strap?

1 ANALYZE AND SKETCH THE PROBLEM

Sketch the situation. The torque must be applied in a counterclockwise direction; force must be exerted as shown.

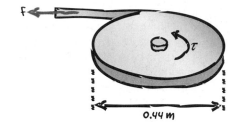

KNOWN	UNKNOWN
$m = 15$ kg	$\alpha = ?$
$r = \left(\frac{1}{2}\right)(0.44 \text{ m}) = 0.22$ m	$I = ?$
$\omega_i = 0.0$ rad/s	$\tau = ?$
$\omega_f = 2\pi(8.0 \text{ rev/s})$	$F = ?$
$t = 15$ s	

2 SOLVE FOR THE UNKNOWN

a. Solve for angular acceleration.

$$\alpha = \frac{\Delta\omega}{\Delta t}$$

$$= \frac{(16\pi \text{ rad/s} - (0.0 \text{ rad/s}))}{15 \text{ s}}$$

◀ *Substitute $\omega_f = 16\pi$ rad/s and $\omega_i = 0.0$ rad/s into the equation.*

$$= 3.4 \text{ rad/s}^2$$

Solve for the moment of inertia.

$$I = \left(\frac{1}{2}\right)mr^2$$

$$= \left(\frac{1}{2}\right)(15 \text{ kg})(0.22 \text{ m})^2$$

◀ *Substitute $m = 15$ kg and $r = 0.22$ m into the equation.*

$$= 0.36 \text{ kg·m}^2$$

Solve for torque.

$$\tau = I\alpha$$

$$= (0.36 \text{ kg·m}^2)(3.4 \text{ rad/s}^2)$$

◀ *Substitute $I = 0.36$ kg·m² and $\alpha = 3.4$ rad/s² into the equation.*

$$= 1.2 \text{ kg·m}^2/\text{s}^2$$

$$= 1.2 \text{ N·m}$$

b. Solve for force.

$$\tau = Fr$$

$$F = \frac{\tau}{r}$$

$$= \frac{(1.2 \text{ N·m})}{(0.22 \text{ m})}$$

◀ *Substitute $\tau = 1.2$ N·m and $r = 0.22$ m into the equation.*

$$= 5.5 \text{ N}$$

3 EVALUATE THE ANSWER

- **Are the units correct?** Torque is measured in N·m and force is measured in N.

- **Is the magnitude realistic?** Despite its large mass, the small size of the wheel makes it relatively easy to spin.

PRACTICE PROBLEMS

25. Consider the wheel in Example Problem 4. If the force on the strap were twice as great, what would be the angular frequency of the wheel after 15 s?

26. A solid wheel accelerates at 3.25 rad/s² when a force of 4.5 N exerts a torque on it. If the wheel is replaced by a wheel which has all of its mass on the rim, the moment of inertia is given by $I = mr^2$. If the same angular velocity were desired, what force should be exerted on the strap?

27. A bicycle wheel on a repair bench can be accelerated either by pulling on the chain that is on the gear or by pulling on a string wrapped around the tire. The tire's radius is 0.38 m, while the radius of the gear is 0.14 m. What force would you need to pull on the string to produce the same acceleration you obtained with a force of 15 N on the chain?

28. The bicycle wheel in the previous problem is used with a smaller gear whose radius is 0.11 m. The wheel can be accelerated either by pulling on the chain that is on the gear or by pulling a string that is wrapped around the tire. If you obtained the needed acceleration with a force of 15 N on the chain, what force would you need to exert on the string?

29. A chain is wrapped around a pulley and pulled with a force of 16.0 N. The pulley has a radius of 0.20 m. The pulley's rotational speed increases from 0.0 to 17.0 rev/min in 5.00 s. What is the moment of inertia of the pulley?

30. **CHALLENGE** A disk with a moment of inertia of 0.26 kg·m² is attached to a smaller disk mounted on the same axle. The smaller disk has a diameter of 0.180 m and a mass of 2.5 kg. A strap is wrapped around the smaller disk, as shown in **Figure 14**. Find the force needed to give this system an angular acceleration of 2.57 rad/s².

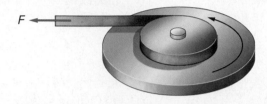

Figure 14

SECTION 2 REVIEW

Section Self-Check Check your understanding.

31. **MAIN**IDEA Vijesh enters a revolving door that is not moving. Explain where and how Vijesh should push to produce a torque with the least amount of force.

32. **Lever Arm** You open a door by pushing at a right angle to the door. Your friend pushes at the same place, but at an angle of 55° from the perpendicular. If both you and your friend exert the same torque on the door, how do the forces you and your friend applied compare?

33. The solid wheel, shown in **Figure 15**, has a mass of 5.2 kg and a diameter of 0.55 m. It is at rest, and you need it to rotate at 12 rev/s in 35 s.

    a. What torque do you need to apply to the wheel?

    b. If a nylon strap is wrapped around the outside of the wheel, how much force do you need to exert on the strap?

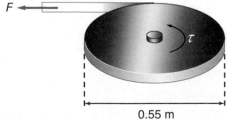

0.55 m

Figure 15

34. **Net Torque** Two people are pulling on ropes wrapped around the edge of a large wheel. The wheel has a mass of 12 kg and a diameter of 2.4 m. One person pulls in a clockwise direction with a 43-N force, while the other pulls in a counterclockwise direction with a 67-N force. What is the net torque on the wheel?

35. **Moment of Inertia** Refer to **Table 2,** and rank the moments of inertia from least to greatest of the following objects: a sphere, a wheel with almost all of its mass at the rim, and a solid disk. All have equal masses and diameters. Explain the advantage of using the object with the least moment of inertia.

36. **Newton's Second Law for Rotational Motion** A rope is wrapped around a pulley and pulled with a force of 13.0 N. The pulley's radius is 0.150 m. The pulley's rotational speed increases from 0.0 to 14.0 rev/min in 4.50 s. What is the moment of inertia of the pulley?

37. **Critical Thinking** A ball on an extremely low-friction, tilted surface will slide downhill without rotating. If the surface is rough, however, the ball will roll. Explain why, using a free-body diagram.

Equilibrium

This gymnast moves and shifts his body during his routine to control his movements. Here, he has increased his moment of inertia to help establish balance. At other times, he decreases his moment of inertia to facilitate rapid changes in position.

MAIN IDEA

An object in static equilibrium experiences a net force of zero and a net torque of zero.

Essential Questions

- What is center of mass?
- How does the location of the center of mass affect the stability of an object?
- What are the conditions for equilibrium?
- How do rotating frames of reference give rise to apparent forces?

Review Vocabulary

torque a measure of how effectively a force causes rotation; the magnitude is equal to the force times the lever arm

New Vocabulary

center of mass
centrifugal "force"
Coriolis "force"

The Center of Mass

Why are some vehicles more likely than others to roll over when involved in an accident? What causes a vehicle to roll over? The answers are important to the engineers who design safe vehicles. In this section, you will learn some of the factors that cause an object to tip over.

How does a freely moving object rotate around its center of mass? A wrench may spin about its handle or end-over-end. Does any single point on the wrench follow a straight path? **Figure 16** shows the path of the wrench. You can see that there is a single point whose path traces a straight line, as if the wrench could be replaced by a point particle at that location. The black *X* in the photo represents this point. The point on the object that moves in the same way that a point particle would move is the **center of mass** of an object.

Locating the center of mass How can you locate the center of mass of an object? First, suspend the object from any point. When the object stops swinging, the center of mass is somewhere along the vertical line drawn from the suspension point. Draw the line. Then, suspend the object from another point. Again, the center of mass must be directly below this point. Draw a second vertical line. The center of mass is at the point where the two lines cross. The wrench and all other objects that are freely moving through space rotate about an axis that goes through their center of mass. Where would you think the center of mass of a person is located?

☑ **READING CHECK** **Paraphrase** the definition of the center of mass.

Figure 16 The path of the center of mass of a wrench is a straight line.

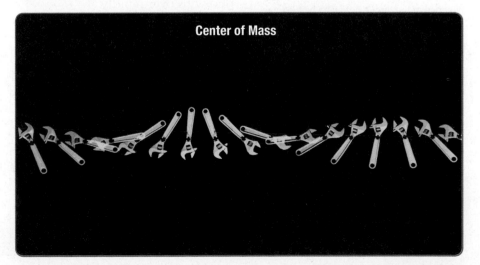

Center of Mass

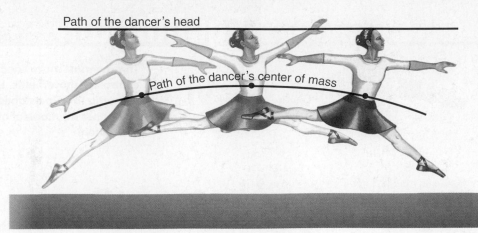

Figure 17 The upward motion of the ballet dancer's head is less than the upward motion of the center of mass. Thus, the head and torso move in a nearly horizontal path. This creates an illusion of floating.

Path of the dancer's head

Path of the dancer's center of mass

The human body's center of mass For a person who is standing with her arms hanging straight down, the center of mass is a few centimeters below the navel, midway between the front and back of the person's body. The center of mass is farther below the navel for women than men, which often results in better balance for women than men. Because the human body is flexible, however, its center of mass is not fixed. If you raise your hands above your head, your center of mass rises 6 to 10 cm. A ballet dancer, for example, can appear to be floating on air by changing her center of mass in a leap. By raising her arms and legs while in the air, as shown in **Figure 17,** the dancer moves her center of mass up. The path of the center of mass is a parabola, but the dancer's head stays at almost the same height for a surprisingly long time.

Center of Mass and Stability

What factors determine whether a vehicle is stable or prone to roll over in an accident? To understand the problem, think about tipping over a box. A tall, narrow box standing on end tips more easily than a low, broad box. Why? To tip a box, as shown in **Figure 18,** you must rotate it about a corner. You pull at the top with a force (F) applying a torque (τ_F). The weight of the box, acting on the center of mass (F_g) applies an opposing torque (τ_w). When the center of mass is directly above the point of support, τ_w is zero. The only torque is the one applied by you. As the box rotates farther, its center of mass is no longer above its base of support, and both torques act in the same direction. At this point, the box tips over rapidly.

Figure 18 The curved arrows show the direction of the torque produced by the force exerted to tip over a box.

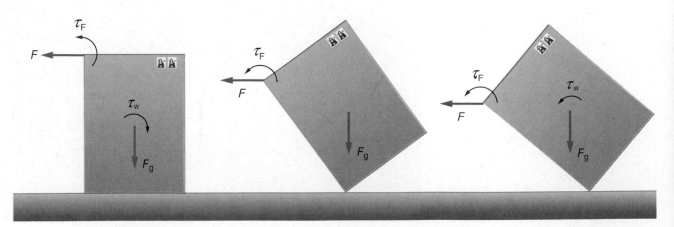

Stability An object is said to be stable if a large external force is required to tip it. The box in **Figure 18** is stable as long as the direction of the torque due to its weight (τ_w) tends to keep it upright. This occurs as long as the box's center of mass lies above its base. To tip the box over, you must rotate its center of mass around the axis of rotation until it is no longer above the base of the box. To rotate the box, you must lift its center of mass. The broader the base, the more stable the object is. Passengers on a city bus, for example, often stand with their feet spread apart to avoid toppling over as the bus starts and stops and weaves through traffic.

Why do vehicles roll over? **Figure 19** shows two vehicles about to roll over. Note that the one with the higher center of mass does not have to be tilted very far for its center of mass to be outside its base—its center of mass does not have to be raised as much as the other vehicle's center of mass. As demonstrated by the vehicles, the lower the location of an object's center of mass, the greater its stability.

You are stable when you stand flat on your feet. When you stand on tiptoe, however, your center of mass moves forward directly above the balls of your feet, and you have very little stability. In judo, aikido, and other martial arts, the fighter uses torque to rotate the opponent into an unstable position, where the opponent's center of mass does not lie above his or her feet. A small person can use torque, rather than force, to defend himself or herself against a stronger person.

In summary, if the center of mass is not located above the base of an object, it is unstable and will roll over without additional torque. If the center of mass is above the base of the object, it is stable. If the base of the object is very narrow and the center of mass is high, then the object might be stable, but the slightest force will cause it to tip over.

☑ **READING CHECK** **Describe** when an object is the most stable.

Conditions for Equilibrium

If your pen is at rest, what is needed to keep it at rest? You could either hold it up or place it on a desk or some other surface. An upward force must be exerted on the pen to balance the downward force of gravity. You must also hold the pen so that it will not rotate. An object is said to be in static equilibrium if both its velocity and angular velocity are zero or constant. Thus, for an object to be in static equilibrium, it must meet two conditions. First, it must be in translational equilibrium; that is, the net force exerted on the object must be zero. Second, it must be in rotational equilibrium; that is, the net torque exerted on the object must be zero.

MiniLAB

SPINNING TOPS
Where is a spinning object's center of mass?

View an **animation about stability**.

Concepts In Motion

PhysicsLAB

EQUILIBRIUM
Can you make the net torque on scaffolding zero so that it will not rotate?

Figure 19 Larger vehicles have a higher center of mass than smaller ones. The higher the center of mass, the smaller the tilt needed to cause the vehicle's center of mass to move outside its base and cause the vehicle to roll over.

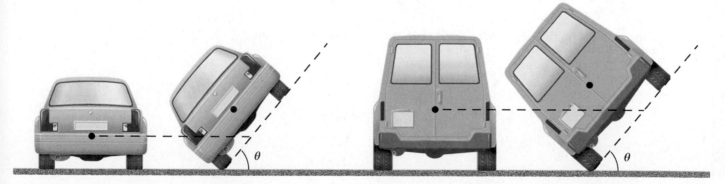

STATIC EQUILIBRIUM A 5.8-kg ladder, 1.80 m long, rests on two saw-horses. Sawhorse A is 0.60 m from one end of the ladder, and saw-horse B is 0.15 m from the other end of the ladder. What force does each sawhorse exert on the ladder?

1 ANALYZE AND SKETCH THE PROBLEM

• Sketch the situation.

• Choose the axis of rotation at the point where F_A acts on the ladder. Thus, the torque due to F_A is zero.

KNOWN   **UNKNOWN**

$m = 5.8$ kg $F_A = ?$

$l = 1.80$ m $F_B = ?$

$l_A = 0.60$ m

$l_B = 0.15$ m

2 SOLVE FOR THE UNKNOWN

For a ladder that has a constant density, the center of mass is at the center rung.

The net force is the sum of all forces on the ladder.

The ladder is in translational equilibrium, so the net force exerted on it is zero.

$F_{net} = F_A + F_B + (-F_g)$

$0.0\ N = F_A + F_B + (-F_g)$ ◀ The ladder is in translational equilibrium, so the net force exerted on it is zero.

Solve for F_A.

$F_A = F_g - F_B$

Find the torques due to F_g and F_B.

$\tau_g = -F_g r_g$ ◀ τ_g is in the clockwise direction.

$\tau_B = +F_B r_B$ ◀ τ_B is in the counterclockwise direction.

The net torque is the sum of all torques on the object.

$\tau_{net} = \tau_B + \tau_g$

$0.0\ N \cdot m = \tau_B + \tau_g$ ◀ The ladder is in rotational equilibrium, so $\tau_{net} = 0.0$ N•m.

$\tau_B = -\tau_g$

$F_B r_B = F_g r_g$ ◀ Substitute $\tau_B = r_B F_B$ and $\tau_g = -r_g F_g$ into the equation.

Solve for F_B.

$F_B = \dfrac{F_g r_g}{r_B}$

$= \dfrac{r_g mg}{r_B}$ ◀ Substitute $F_g = mg$

Using the expression $F_A = F_g - F_B$, substitute in the expressions for F_B and F_g.

$F_A = F_g - F_B$

$= F_g - \dfrac{r_g mg}{r_B}$ ◀ Substitute $F_B = \dfrac{r_g mg}{r_B}$ into the equation.

$= mg - \dfrac{r_g mg}{r_B}$ ◀ Substitute $F_g = mg$ into the equation.

$= mg\left(1 - \dfrac{r_g}{r_B}\right)$

Solve for r_g.

$$r_g = \left(\frac{1}{2}\right)l - l_A$$

◀ For a ladder, which has a constant density, the center of mass is at the center rung.

$$= 0.90 \text{ m} - 0.60 \text{ m}$$

◀ Substitute $\frac{l}{2} = 0.90$ m and $l_A = 0.60$ m into the equation.

$$= 0.30 \text{ m}$$

Solve for r_B.

$$r_B = (0.90 \text{ m} - l_B) + (0.90 \text{ m} - l_A)$$

$$= (0.90 \text{ m} - 0.15 \text{ m}) + (0.90 \text{ m} - 0.60 \text{ m})$$

◀ Substitute $l_B = 0.15$ m and $l_A = 0.60$ m into the equation.

$$= 0.75 \text{ m} + 0.30 \text{ m}$$

$$= 1.05 \text{ m}$$

Calculate F_B.

$$F_B = \frac{r_g mg}{r_B}$$

$$= \frac{(0.30 \text{ m})(5.8 \text{ kg})(9.8 \text{ N/kg})}{(1.05 \text{ m})}$$

◀ Substitute $r_g = 0.30$ m, $m = 5.8$ kg, $g = 9.8$ N/kg, and $r_B = 1.05$ m into the equation.

$$= 16 \text{ N}$$

Calculate F_A.

$$F_A = mg\left(1 - \frac{r_g}{r_B}\right)$$

$$= (5.8 \text{ kg})(9.8 \text{ N/kg})\left(1 - \frac{(0.30 \text{ m})}{(1.05 \text{ m})}\right)$$

◀ Substitute $r_g = 0.30$ m, $m = 5.8$ kg, $g = 9.8$ N/kg, and $r_B = 1.05$ m into the equation.

$$= 41 \text{ N}$$

3 EVALUATE THE ANSWER

- **Are the units correct?** Forces are measured in newtons.
- **Do the signs make sense?** Both forces are upward.
- **Is the magnitude realistic?** The forces add up to the weight of the ladder, and the force exerted by the sawhorse closer to the center of mass is greater, which makes sense.

PRACTICE PROBLEMS Do additional problems. Online Practice

38. What would be the forces exerted by the two sawhorses if the ladder in Example Problem 5 had a mass of 11.4 kg?

39. A 7.3-kg ladder, 1.92 m long, rests on two sawhorses, as shown in **Figure 20**. Sawhorse A, on the left, is located 0.30 m from the end, and sawhorse B, on the right, is located 0.45 m from the other end. Choose the axis of rotation to be the center of mass of the ladder.

 a. What are the torques acting on the ladder?

 b. Write the equation for rotational equilibrium.

 c. Solve the equation for F_A in terms of F_g.

 d. How would the forces exerted by the two sawhorses change if A were moved very close to, but not directly under, the center of mass?

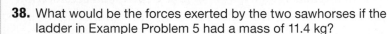
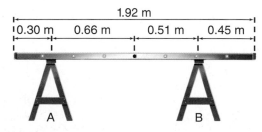

1.92 m
0.30 m 0.66 m 0.51 m 0.45 m

A B

Figure 20

40. A 4.5-m-long wooden plank with a 24-kg mass is supported in two places. One support is directly under the center of the board, and the other is at one end. What are the forces exerted by the two supports?

41. **CHALLENGE** A 85-kg diver walks to the end of a diving board. The board, which is 3.5 m long with a mass of 14 kg, is supported at the center of mass of the board and at one end. What are the forces on the two supports?

Rotating Frames of Reference

When you sit on a spinning amusement-park ride, it feels as if a strong force is pushing you to the outside. A pebble on the floor of the ride would accelerate outward without a horizontal force being exerted on it in the same direction. The pebble would not move in a straight line relative to the floor of the ride. In other words, Newton's laws of motion would not seem to apply. This is because in the ride your point of view, called your frame of reference, is rotating. Newton's laws are valid only in non-rotating or nonaccelerated frames of reference.

Motion in a rotating reference frame is important to us because Earth rotates. The effects of the rotation of Earth are too small to be noticed in the classroom or lab, but they are significant influences on the motion of the atmosphere and, therefore, on climate and weather.

Centrifugal "Force"

Suppose you fasten one end of a spring to the center of a rotating platform. An object lies on the platform and is attached to the other end of the spring. As the platform rotates, an observer on the platform sees the object stretch the spring. The observer might think that some force toward the outside of the platform is pulling on the object. This apparent force seems to pull on a moving object but does not exert a physical outward push on it. This apparent force, which seems to push an object outward, is observed only in rotating frames of reference and is called the **centrifugal "force."** It is not a real force because there is no physical outward push on the object.

As the platform rotates, an observer on the ground would see things differently. This observer sees the object moving in a circle, and it accelerates toward the center because of the force of the spring. As you know, this centripetal acceleration is given by $a_c = \frac{v^2}{r}$. It also can be written in terms of angular velocity, as $a_c = \omega^2 r$. Centripetal acceleration is proportional to the distance from the axis of rotation and depends on the square of the angular velocity. Thus, if you double the rotational frequency, the angular acceleration increases by a factor of four.

☑ **READING CHECK** **Define** centrifugal force in your own words.

The Coriolis "Force"

A second effect of rotation is shown in **Figure 21.** Suppose a person standing at the center of a rotating disk throws a ball toward the edge of the disk. Consider the horizontal motion of the ball as seen by two observers, and ignore the vertical motion of the ball as it falls. An observer standing outside the disk sees the ball travel in a straight line at a constant speed toward the edge of the disk. However, the other observer, who is stationed on the disk and rotating with it, sees the ball follow a curved path at a constant speed. The apparent force that seems to deflect a moving object from its path and is observed only in rotating frames of reference is called the **Coriolis "force."** Like the centrifugal "force," the Coriolis "force" is not a real force. It seems to exist because we observe a deflection in horizontal motion when we are in a rotating frame of reference.

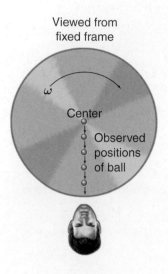

Viewed from fixed frame

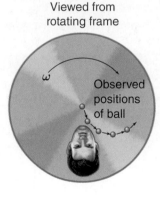

Viewed from rotating frame

Figure 21 The Coriolis "force" is not a real force. It exists only in rotating reference frames.

Coriolis "force" due to Earth's rotation

Suppose a cannon is fired from a point on the equator toward a target due north of it. If the projectile were fired directly northward, it would also have an eastward velocity component because of the rotation of Earth. Recall that Earth is actually rotating beneath the projectile.

This eastward speed is greater at the equator than at any other latitude. Thus, as the projectile moves northward, it also moves eastward faster than points on Earth below it do. The result is that the projectile lands east of the target as shown in **Figure 22.**

☑ **READING CHECK Describe** how you would aim a projectile to compensate for the rotation of Earth.

While an observer in space would see Earth's rotation, an observer on Earth could claim that the projectile missed the target because of the Coriolis "force" on the rocket. Note that for objects moving toward the equator, the direction of the apparent force is westward. A projectile will land west of the target when fired due south.

▶ **CONNECTION TO EARTH SCIENCE** The direction of winds around high- and low-pressure areas results from the Coriolis "force." Winds flow from areas of high to low pressure. Because of the Coriolis "force" in the northern hemisphere, winds from the south go to the east of low-pressure areas. Winds from the north end up west of low-pressure areas. Therefore, winds rotate counter-clockwise around low-pressure areas in the northern hemisphere. In the southern hemisphere, however, winds rotate clockwise around low-pressure areas. This is why tropical cyclones, or hurricanes as they are also called, rotate clockwise in the southern hemisphere and counterclockwise in the northern hemisphere.

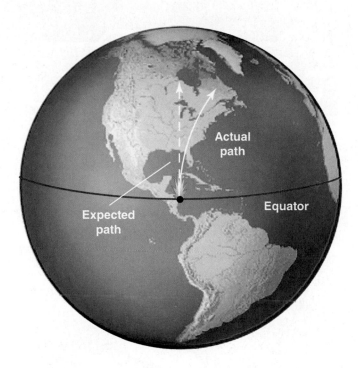

Figure 22 An observer on Earth sees the Coriolis "force" cause a projectile fired due north to deflect to the right of the intended target.

SECTION 3 REVIEW

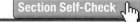 **Section Self-Check** Check your understanding.

42. **MAIN**IDEA Give an example of an object for each of the following conditions.

   a. rotational equilibrium, but not translational equilibrium

   b. translational equilibrium, but not rotational equilibrium

43. **Center of Mass** Can the center of mass of an object be located in an area where the object has no mass? Explain.

44. **Stability of an Object** Why is a modified vehicle with its body raised high on risers less stable than a similar vehicle with its body at normal height?

45. **Center of Mass** Where is the center of mass on a roll of masking tape?

46. **Locating the Center of Mass** Describe how you would find the center of mass of this textbook.

47. **Rotating Frames of Reference** A penny is placed on a rotating, old-fashioned record turntable. At the highest speed, the penny starts sliding outward. What are the forces acting on the penny?

48. **Critical Thinking** You have read about how the spin of Earth on its axis affects the winds. Predict the direction of the flow of surface ocean currents in the northern and southern hemispheres.

Spin-Cycle

CENTRIFUGES

Powerful physics is at work in devices known as laboratory centrifuges.

One of the main uses of laboratory centrifuges is separating blood into its components for analysis. Some labs also use centrifuges to separate DNA and proteins from the samples.

Centrifuges must be balanced with a sample on one side offset by a sample or a blank on the other. Much like a washing machine, an out-of-balance centrifuge can quickly malfunction.

Laboratory centrifuges can spin very quickly, up to thousands of revolutions per minute (rpm), producing accelerations of hundreds or thousands of times the free-fall acceleration.

Centrifuges use a principle called sedimentation. The most dense parts of a mixture move toward the bottom of the sample tube, while the least dense portions remain at the top.

GOING**FURTHER** >>>

Research Interview a technician at a pathology laboratory or a student at a local college or university to learn how centrifuges are used in separating samples to be analyzed.

STUDY GUIDE

BIGIDEA   Applying a torque to an object causes a change in that object's angular velocity.

SECTION 1 Describing Rotational Motion

MAINIDEA   Angular displacement, angular velocity, and angular acceleration all help describe angular motion.

- Angular displacement is the change in the angle (θ) as an object rotates. It is usually measured in degrees or radians.

- Average angular velocity is the object's angular displacement divided by the time taken to make the angular displacement. Average angular velocity is represented by the Greek letter omega (ω) and is determined by the following equation:

$$\omega = \frac{\Delta\theta}{\Delta t}$$

- Average angular acceleration is the change in angular velocity divided by the time required to make the change.

$$\alpha = \frac{\Delta\omega}{\Delta t}$$

SECTION 2 Rotational Dynamics

MAINIDEA   Torques cause changes in angular velocity.

- Torque describes the combination of a force and a lever arm that can cause an object to rotate. Torque is represented by the Greek letter tau (τ) and is determined by the following equation:

$$\tau = Fr \sin\theta$$

- The moment of inertia is a point object's resistance to changes in angular velocity. The moment of inertia is represented by the letter I and for a point mass, it is represented by the following equation:

$$I = mr^2$$

- Newton's second law for rotational motion states that angular acceleration is directly proportional to the net torque and inversely proportional to the moment of inertia.

$$\alpha = \frac{\tau_{net}}{I}$$

SECTION 3 Equilibrium

MAINIDEA   An object in static equilibrium experiences a net force of zero and a net torque of zero.

- The center of mass of an object is the point on the object that moves in the same way that a point particle would move.

- An object is stable against rollover if its center of mass is above its base.

- An object in equilibrium has no net force exerted on it and there is no net torque acting on it.

- Centrifugal "force" and the Coriolis "force" are two apparent, but nonexistent, forces that seem to exist when a object is analyzed from a rotating frame of reference.

Games and Multilingual eGlossary

Vocabulary Practice

SECTION 1 Describing Rotational Motion

Mastering Concepts

49. BIGIDEA A bicycle wheel rotates at a constant 25 rev/min. Is its angular velocity decreasing, increasing, or constant?

50. A toy rotates at a constant 5 rev/min. Is its angular acceleration positive, negative, or zero?

51. Do all parts of Earth move at the same rate? Explain.

52. A unicycle wheel rotates at a constant 14 rev/min. Is the total acceleration of a point on the tire inward, outward, tangential, or zero?

Mastering Problems

53. On a test stand a bicycle wheel is being rotated about its axle so that a point on the edge moves through 0.210 m. The radius of the wheel is 0.350 m, as shown in **Figure 23**. Through what angle (in radians) is the wheel rotated?

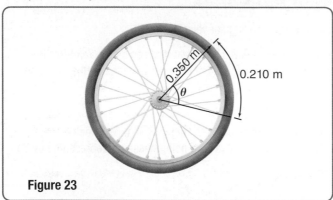

Figure 23

54. The outer edge of a truck tire that has a radius of 45 cm has a velocity of 23 m/s. What is the angular velocity of the tire in rad/s?

55. A steering wheel is rotated through 128°, as shown in **Figure 24**. Its radius is 22 cm. How far would a point on the steering wheel's edge move?

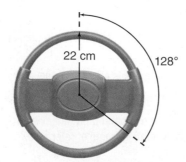

Figure 24

56. Propeller A propeller spins at 1880 rev/min.

 a. What is its angular velocity in rad/s?

 b. What is the angular displacement of the propeller in 2.50 s?

57. The propeller in the previous problem slows from 475 rev/min to 187 rev/min in 4.00 s. What is its angular acceleration?

58. The automobile wheel shown in **Figure 25** rotates at 2.50 rad/s. How fast does a point 7.00 cm from the center travel?

Figure 25

59. Washing Machine A washing machine's two spin cycles are 328 rev/min and 542 rev/min. The diameter of the drum is 0.43 m.

 a. What is the ratio of the centripetal accelerations for the fast and slow spin cycles? Recall that $a_c = \frac{v^2}{r}$ and $v = r\omega$.

 b. What is the ratio of the linear velocity of an object at the surface of the drum for the fast and slow spin cycles?

 c. Find the maximum centripetal acceleration in terms of g for the washing machine.

60. A laboratory ultracentrifuge is designed to produce a centripetal acceleration of $0.35 \times 10^6 \, g$ at a distance of 2.50 cm from the axis. What angular velocity in revolutions per minute is required?

SECTION 2 Rotational Dynamics

Mastering Concepts

61. Think about some possible rotations of your textbook. Are the moments of inertia about these three axes the same or different? Explain.

62. An auto repair manual specifies the torque needed to tighten bolts on an engine. Why does it not mention force?

63. Ranking Task Rank the torques on the five doors shown in **Figure 26** from least to greatest. Note that the magnitude of all the forces is the same.

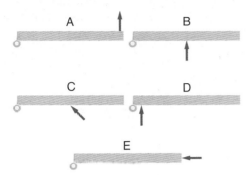

Figure 26

Mastering Problems

64. Wrench A bolt is to be tightened with a torque of 8.0 N·m. If you have a wrench that is 0.35 m long, what is the least amount of force you must exert?

65. What is the torque on a bolt produced by a 15-N force exerted perpendicular to a wrench that is 25 cm long, as shown in **Figure 27?**

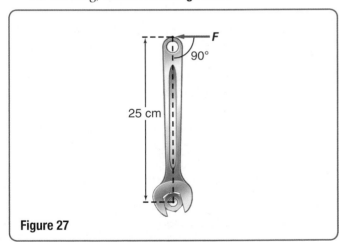

Figure 27

66. A bicycle wheel with a radius of 38 cm is given an angular acceleration of 2.67 rad/s^2 by applying a force of 0.35 N on the edge of the wheel. What is the wheel's moment of inertia?

67. Toy Top A toy top consists of a rod with a diameter of 8.0 mm and a disk of mass 0.0125 kg and a diameter of 3.5 cm. The moment of inertia of the rod can be neglected. The top is spun by wrapping a string around the rod and pulling it with a velocity that increases from zero to 3.0 m/s over 0.50 s.

a. What is the resulting angular velocity of the top?

b. What force was exerted on the string?

68. A toy consisting of two balls, each 0.45 kg, at the ends of a 0.46-m-long, thin, lightweight rod is shown in **Figure 28.** Find the moment of inertia of the toy. The moment of inertia is to be found about the center of the rod.

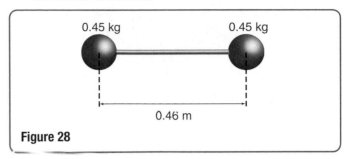

Figure 28

SECTION 3 Equilibrium
Mastering Concepts

69. To balance a car's wheel, it is placed on a vertical shaft and weights are added to make the wheel horizontal. Why is this equivalent to moving the center of mass until it is at the center of the wheel?

70. A stunt driver maneuvers a monster truck so that it is traveling on only two wheels. Where is the center of mass of the truck?

71. Suppose you stand flat-footed, then you rise and balance on tiptoe. If you stand with your toes touching a wall, you cannot balance on tiptoe. Explain.

72. Why does a gymnast appear to be floating on air when she raises her arms above her head in a leap?

73. Why is a vehicle with wheels that have a large diameter more likely to roll over than a vehicle with wheels that have a smaller diameter?

Mastering Problems

74. A 12.5-kg board, 4.00 m long, is being held up on one end by Ahmed. He calls for help in lifting the board, and Judi responds.

a. What is the least force that Judi could exert to lift the board to the horizontal position? What part of the board should she lift to exert this force?

b. What is the greatest force that Judi could exert to lift the board to the horizontal position? What part of the board should she lift to exert this force?

75. A car's specifications state that its weight distribution is 53 percent on the front tires and 47 percent on the rear tires. The wheel base is 2.46 m. Where is the car's center of mass?

76. Two people are holding up the ends of a 4.25-kg wooden board that is 1.75 m long. A 6.00-kg box sits on the board, 0.50 m from one end, as shown in **Figure 29.** What forces do the two people exert?

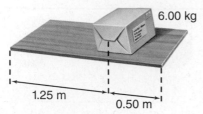

6.00 kg

1.25 m
0.50 m

Figure 29

Applying Concepts

77. Two gears are in contact and rotating. One is larger than the other, as shown in **Figure 30.** Compare their angular velocities. Also compare the linear velocities of two teeth that are in contact.

Figure 30

78. How can you experimentally find the moment of inertia of an object?

79. Bicycle Wheels Three bicycle wheels have masses that are distributed in three different ways: mostly at the rim, uniformly, and mostly at the hub. The wheels all have the same mass. If equal torques are applied to them, which one will have the greatest angular acceleration? Which one will have the least?

80. Bowling Ball When a bowling ball leaves a bowler's hand, it does not spin. After it has gone about half the length of the lane, however, it does spin. Explain how its rotation rate increased.

81. Flat Tire Suppose your car has a flat tire. You find a lug wrench to remove the nuts from the bolt studs. You cannot turn the nuts. Your friend suggests ways you might produce enough torque to turn them. What three ways might your friend suggest?

82. Why can you ignore forces that act on the axis of rotation of an object in static equilibrium when determining the net torque?

83. Tightrope Walkers Tightrope walkers often carry long poles that sag so that the ends are lower than the center as shown in **Figure 31.** How does a pole increase the tightrope walker's stability? *Hint: Consider both center of mass and moment of inertia.*

Figure 31

84. Merry-Go-Round While riding a merry-go-round, you toss a key to a friend standing on the ground. For your friend to be able to catch the key, should you toss it a second or two before you reach the spot where your friend is standing or wait until your friend is directly beside you? Explain.

85. In solving problems about static equilibrium, why is the axis of rotation often placed at a point where one or more forces are acting on the object?

86. Ranking Task Rank the objects in **Figure 32** according to their moments of inertia, from least to greatest. Each has a radius of (R) and total mass (M), which is uniformly distributed throughout the shaded region. Specifically indicate any ties.

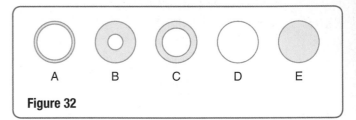

A   B   C   D   E

Figure 32

Mixed Review

87. A wooden door of mass m and length l is held horizontally by Dan and Ajit. Dan suddenly drops his end.

   a. What is the angular acceleration of the door just after Dan lets go?

   b. Is the acceleration constant? Explain.

88. Hard Drive A hard drive on a computer spins at 7200 rpm (revolutions per minute). If the drive is designed to start from rest and reach operating speed in 1.5 s, what is the angular acceleration of the disk?

89. Topsoil Ten bags of topsoil, each weighing 75 N, are placed on a 2.43-m-long sheet of wood. They are stacked 0.50 m from one end of the sheet of wood, as shown in **Figure 33.** Two people lift the sheet of wood, one at each end. Ignoring the weight of the wood, how much force must each person exert?

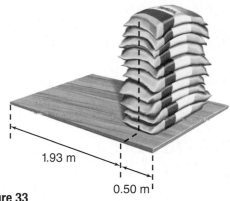

1.93 m

0.50 m

Figure 33

90. A steel beam that is 6.50 m long weighs 325 N. It rests on two supports, 3.00 m apart, with equal amounts of the beam extending from each end. Suki, who weighs 575 N, stands on the beam in the center and then walks toward one end. How close to the end can she walk before the beam begins to tip?

91. The second hand on a watch is 12 mm long. What is the velocity of its tip?

92. A cylinder with a 50-cm diameter, as shown in **Figure 34,** is at rest on a surface. A rope is wrapped around the cylinder and pulled. The cylinder rolls without slipping.

a. After the rope has been pulled a distance of 2.50 m at a constant speed, how far has the center of mass of the cylinder moved?

b. If the rope was pulled a distance of 2.50 m in 1.25 s, how fast was the center of mass of the cylinder moving?

c. What is the angular velocity of the cylinder?

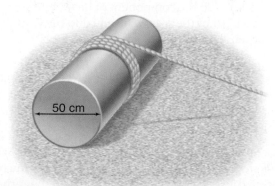

50 cm

Figure 34

93. Basketball A basketball is rolled down the court. A regulation basketball has a diameter of 24.1 cm, a mass of 0.60 kg, and a moment of inertia of 5.8×10^{-3} kg·m². The basketball's initial velocity is 2.5 m/s.

a. What is its initial angular velocity?

b. The ball rolls a total of 12 m. How many revolutions does it make?

c. What is its total angular displacement?

94. The basketball in the previous problem stops rolling after traveling 12 m.

a. If its acceleration was constant, what was its angular acceleration?

b. What torque was acting on it as it was slowing down?

95. Speedometers Most speedometers in automobiles measure the angular velocity of the transmission and convert it to speed. How will increasing the diameter of the tires affect the reading of the speedometer?

96. A box is dragged across the floor using a rope that is a distance h above the floor. The coefficient of friction is 0.35. The box is 0.50 m high and 0.25 m wide. Find the force that just tips the box.

97. Lumber You buy a 2.44-m-long piece of 10 cm × 10 cm lumber. Your friend buys a piece of the same size and cuts it into two lengths, each 1.22 m long, as shown in **Figure 35.** You each carry your lumber on your shoulders.

a. Which load is easier to lift? Why?

b. Both you and your friend apply a torque with your hands to keep the lumber from rotating. Which load is easier to keep from rotating? Why?

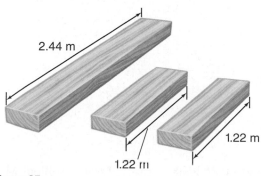

2.44 m

1.22 m

1.22 m

Figure 35

98. Surfboard Harris and Paul carry a surfboard that is 2.43 m long and weighs 143 N. Paul lifts one end with a force of 57 N.

a. What force must Harris exert?

b. What part of the board should Harris lift?

Thinking Critically

99. Analyze and Conclude A banner is suspended from a horizontal, pivoted pole, as shown in **Figure 36.** The pole is 2.10 m long and weighs 175 N. The banner, which weighs 105 N, is suspended 1.80 m from the pivot point or axis of rotation. What is the tension in the cable supporting the pole?

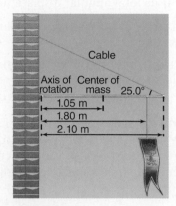

Figure 36

100. Analyze and Conclude A pivoted lamp pole is shown in **Figure 37.** The pole weighs 27 N, and the lamp weighs 64 N.

a. What is the torque caused by each force?

b. Determine the tension in the rope supporting the lamp pole.

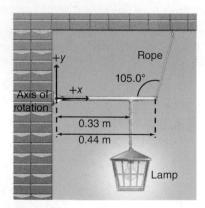

Figure 37

101. Reverse Problem Write a physics problem with real-life objects for which the following equation would be part of the solution:

$$\Delta\theta = \left(20\ \frac{\text{rad}}{\text{s}}\right)(4\ \text{s}) - \frac{1}{2}\left(3.5\ \frac{\text{rad}}{\text{s}^2}\right)(4\ \text{s})^2$$

102. Problem Posing Complete this problem so that it can be solved using the concept of torque: "A painter carries a 3.0-m, 12-kg ladder...."

103. Apply Concepts Consider a point on the edge of a wheel rotating about its axis.

a. Under what conditions can the centripetal acceleration be zero?

b. Under what conditions can the tangential (linear) acceleration be zero?

c. Can the tangential acceleration be nonzero while the centripetal acceleration is zero? Explain.

d. Can the centripetal acceleration be nonzero while the tangential acceleration is zero? Explain.

104. Apply Concepts When you apply the brakes in a car, the front end dips. Why?

Writing in Physics

105. Astronomers know that if a natural satellite is too close to a planet, it will be torn apart by tidal forces. The difference in the gravitational force on the part of the satellite nearest the planet and the part farthest from the planet is stronger than the forces holding the satellite together. Research the Roche limit, and determine how close the Moon would have to orbit Earth to be at the Roche limit.

106. Automobile engines are rated by the torque they produce. Research and explain why torque is an important quantity to measure.

Cumulative Review

107. Two blocks, one of mass 2.0 kg and the other of mass 3.0 kg, are tied together with a massless rope. This rope is strung over a massless, resistance-free pulley. The blocks are released from rest. Find the following:

a. the tension in the rope

b. the acceleration of the blocks

108. Eric sits on a seesaw. At what angle, relative to the vertical, will the component of his weight parallel to the length of the seesaw be equal to one-third the perpendicular component of his weight?

109. The pilot of a plane wants to reach an airport 325 km due north in 2.75 h. A wind is blowing from the west at 30.0 km/h. What heading and airspeed should be chosen to reach the destination on time?

110. A 60.0-kg speed skater with a velocity of 18.0 m/s skates into a curve of 20.0-m radius. How much friction must be exerted between the skates and the ice for her to negotiate the curve?

MULTIPLE CHOICE

1. The illustration below shows two boxes on opposite ends of a massless board that is 3.0 m long. The board is supported in the middle by a fulcrum. The box on the left has a mass (m_1) of 25 kg, and the box on the right has a mass (m_2) of 15 kg. How far should the fulcrum be positioned from the left side of the board in order to balance the masses horizontally?

A. 0.38 m   **C.** 1.1 m

B. 0.60 m   **D.** 1.9 m

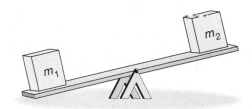

2. A force of 60 N is exerted on one end of a 1.0-m-long lever. The other end of the lever is attached to a rotating rod that is perpendicular to the lever. By pushing down on the end of the lever, you can rotate the rod. If the force on the lever is exerted at an angle of 30° to the perpendicular to the lever, what torque is exerted on the rod? (sin 30° = 0.5; cos 30° = 0.87; tan 30° = 0.58)

A. 30 N   **C.** 60 N

B. 52 N   **D.** 69 N

3. A child attempts to use a wrench to remove a nut on a bicycle. Removing the nut requires a torque of 10 N·m. The maximum force the child is capable of exerting at a 90° angle is 50 N. What is the length of the wrench the child must use to remove the nut?

A. 0.1 m   **C.** 0.2 m

B. 0.15 m   **D.** 0.25 m

4. A car moves a distance of 420 m. Each tire on the car has a diameter of 42 cm. Which shows how many revolutions each tire makes as they move that distance?

A. $\left(\dfrac{(5.0\times10^1)}{\pi}\right)$ rev   **C.** $\left(\dfrac{(1.5\times10^2)}{\pi}\right)$ rev

B. $\left(\dfrac{(1.0\times10^2)}{\pi}\right)$ rev   **D.** $\left(\dfrac{(1.0\times10^3)}{\pi}\right)$ rev

5. A thin hoop with a mass of 5.0 kg rotates about a perpendicular axis through its center. A force of 25 N is exerted tangentially to the hoop. If the hoop's radius is 2.0 m, what is its angular acceleration?

A. 1.3 rad/s   **C.** 5.0 rad/s

B. 2.5 rad/s   **D.** 6.3 rad/s

6. Two of the tires on a farmer's tractor have diameters of 1.5 m. If the farmer drives the tractor at a linear velocity of 3.0 m/s, what is the angular velocity of each tire?

A. 2.0 rad/s   **C.** 4.0 rad/s

B. 2.3 rad/s   **D.** 4.5 rad/s

FREE RESPONSE

7. You use a 25-cm long wrench to remove the lug nuts on a car wheel, as shown in the illustration below. If you pull up on the end of the wrench with a force of 2.0×10^2 N at an angle of 30°, what is the torque on the wrench? (sin 30° = 0.5, cos 30° = 0.87)

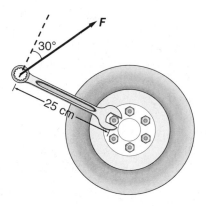

NEED EXTRA HELP?

If You Missed Question	1	2	3	4	5	6	7
Review Section	2	2	2	1	1	1	2

Online Test Practice

Momentum and Its Conservation

BIGIDEA   If the net force on a closed system is zero, the total momentum of that system is conserved.

LaunchLAB

COLLIDING OBJECTS

What factors determine the speed and direction of objects after a collision?

WATCH THIS!

Video

CRASH!

From fender benders to bumper cars, there is a lot of physics involved in collisions. How do forces shape what happens when two objects collide?

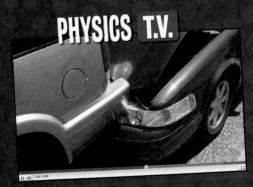

PHYSICS T.V.

(l)Mikael Karlsson/Alamy, (r)MIKE CLARKE/AFP/Getty Images

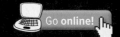

Impulse and Momentum

PHYSICS 4 YOU

Lacrosse players wear helmets and padding to protect themselves from flying balls. Lacrosse balls are not very massive (about 145 g), but players might hurl them at speeds over 40 m/s. Why are lacrosse balls so dangerous?

MAIN IDEA

An object's momentum is equal to its mass multiplied by its velocity.

Essential Questions

- What is impulse?
- What is momentum?
- What is angular momentum?

Review Vocabulary

angular velocity the angular displacement of an object divided by the time needed to make the displacement

New Vocabulary

impulse
momentum
impulse-momentum theorem
angular momentum
angular impulse-angular momentum theorem

APPLYING PRACTICES

PBL Go to the resources tab in ConnectED to find the PBL *Egg Heads*.

Impulse-Momentum Theorem

It can be exciting to watch a baseball player hit a home run. The pitcher hurls the baseball toward the plate. The batter swings, and the ball recoils from the impact. Before the collision, the baseball moves toward the bat. During the collision, the ball is squashed against the bat. After the collision, the ball moves at a high velocity away from the bat, and the bat continues along its path but at a slower velocity.

According to Newton's second law of motion, the force from the bat changed the ball's velocity. This force changes over time, as shown in **Figure 1.** Just after contact, the ball is squeezed, and the force increases to a maximum more than 10,000 times the weight of the ball. The ball then recovers its shape and rebounds from the bat. The force on the ball rapidly returns to zero. This whole event takes place within about 3.0 ms. How can you calculate the change in velocity of the baseball?

Impulse Newton's second law of motion ($F = ma$) can be rewritten in terms of the change in velocity divided by the time for that change.

$$F = ma = m\left(\frac{\Delta v}{\Delta t}\right)$$

Multiplying both sides by the time interval (Δt) results in this equation.

$$F\Delta t = m\Delta v = m(v_f - v_i)$$

The left side of the equation ($F\Delta t$) is defined as the impulse. The **impulse** on an object is the product of the average force on an object and the time interval over which it acts. Impulse is measured in newton-seconds. If the force F varies with time, the magnitude of the impulse equals the area under the curve of a force-time graph, as in **Figure 1.**

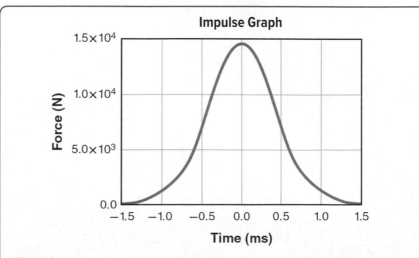

Figure 1 The force on a baseball increases then decreases during a collision.

Momentum The right side of the equation ($m\Delta v$) involves the change in velocity, where $m\Delta v = mv_f - mv_i$. The product of the object's mass (m) and the object's velocity (v) is defined as the **momentum** of the object. Momentum is measured in kg·m/s. An object's momentum, also known as linear momentum, is represented by the following equation.

MOMENTUM
The momentum of an object equals the mass of the object times the object's velocity.

$$p = mv$$

Now, you can rewrite the impulse as $F\Delta t = m\Delta v = mv_f - mv_i = p_f - p_i$. Thus, the impulse on an object is equal to the change in its momentum, which is called the **impulse-momentum theorem.** The following relationship expresses the impulse-momentum theorem.

IMPULSE-MOMENTUM THEOREM
An impulse acting on an object is equal to the object's final momentum minus the object's initial momentum.

$$F\Delta t = p_f - p_i$$

If the force on an object is constant, the impulse is the product of the force multiplied by the time interval over which it acts. Generally, the force is not constant; however, you can find the impulse using an average force multiplied by the time interval over which it acts, or by finding the area under a force-time graph.

Because velocity is a vector, momentum also is a vector. Similarly, impulse is a vector because force is a vector. As with any other vector quantity, make sure you use signs consistently to indicate direction.

Using the impulse-momentum theorem According to the impulse-momentum theorem, the impulse from a bat changes a baseball's momentum. You can calculate the impulse using a force-time graph. In **Figure 1,** the area under the curve is about 15 N·s. The direction of the impulse is in the direction of the force. Therefore, the change in momentum of the ball also is 15 N·s. Because 1 N·s is equal to 1 kg·m/s, the momentum gained by the ball is 15 kg·m/s in the direction of the force acting on it.

Suppose a batter hits a fastball. Assume the positive direction is toward the pitcher. Before the collision of the ball and the bat, the ball, with a mass of 0.145 kg, has a velocity of −47 m/s. Therefore, the baseball's momentum is $p_i = (0.145 \text{ kg})(-47 \text{ m/s}) = -6.8 \text{ kg·m/s}$. The momentum of the ball after the collision is found by solving the impulse-momentum theorem for the final momentum: $p_f = p_i + F\Delta t$.

$$p_f = p_i + F\Delta t = -6.8 \text{ kg·m/s} + 15 \text{ kg·m/s} = +8.2 \text{ kg·m/s}$$

Because $p_f = mv_f$, solving for v_f yields the ball's final velocity:

$$v_f = \frac{p_f}{m} = \frac{+8.2 \text{ kg·m/s}}{+0.145 \text{ kg}} = +57 \text{ m/s}$$

A speed of 52 m/s is fast enough to clear most outfield fences if the batter hits the baseball in the correct direction.

PhysicsLAB

STICKY COLLISIONS
What happens to the momentum of an object during a collision?

AVERAGE FORCE A 2200-kg vehicle traveling at 94 km/h (26 m/s) can be stopped in 21 s by gently applying the brakes. It can be stopped in 3.8 s if the driver slams on the brakes, or in 0.22 s if it hits a concrete wall. What is the impulse exerted on the vehicle? What average force is exerted on the vehicle in each of these stops?

1 ANALYZE AND SKETCH THE PROBLEM

- Sketch the system.
- Include a coordinate axis. Select the car's direction as positive.
- Draw a vector diagram for momentum and impulse.

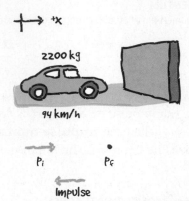

KNOWN		UNKNOWN
$m = 2200$ kg	$\Delta t_{\text{gentle braking}} = 21$ s	$F\Delta t = ?$
$v_i = +26$ m/s	$\Delta t_{\text{hard braking}} = 3.8$ s	$F_{\text{gentle braking}} = ?$
$v_f = 0.0$ m/s	$\Delta t_{\text{hitting a wall}} = 0.22$ s	$F_{\text{hard braking}} = ?$
		$F_{\text{hitting a wall}} = ?$

2 SOLVE FOR THE UNKNOWN

Determine the initial momentum (p_i).

$$p_i = mv_i$$

$$= (2200 \text{ kg})(+26 \text{ m/s})$$ ◄ Substitute $m = 2200$ kg, $v_i = +26$ m/s.

$$= +5.7 \times 10^4 \text{ kg} \cdot \text{m/s}$$

Determine the final momentum (p_f).

$$p_f = mv_f$$

$$= (2200 \text{ kg})(0.0 \text{ m/s})$$ ◄ Substitute $m = 2200$ kg, $v_f = 0.0$ m/s.

$$= 0.0 \text{ kg} \cdot \text{m/s}$$

Apply the impulse-momentum theorem to determine the impulse and the force needed to stop the vehicle.

$$F\Delta t = p_f - p_i$$

$$F\Delta t = (0.0 \text{ kg} \cdot \text{m/s}) - (5.7 \times 10^4 \text{ kg} \cdot \text{m/s})$$ ◄ Substitute $p_f = 0.0$ kg·m/s, $p_i = 5.7 \times 10^4$ kg·m/s.

$$= -5.7 \times 10^4 \text{ kg} \cdot \text{m/s}$$

$$F = \frac{-5.7 \times 10^4 \text{ kg} \cdot \text{m/s}}{\Delta t}$$

$$F_{\text{gentle braking}} = \frac{-5.7 \times 10^4 \text{ kg} \cdot \text{m/s}}{21 \text{ s}}$$ ◄ Substitute $\Delta t_{\text{gentle braking}} = 21$ s.

$$= -2.7 \times 10^3 \text{ N}$$

$$F_{\text{hard braking}} = \frac{-5.7 \times 10^4 \text{ kg} \cdot \text{m/s}}{3.8 \text{ s}}$$ ◄ Δt

$$= -1.5 \times 10^4 \text{ N}$$

$$F_{\text{hitting a wall}} = \frac{-5.7 \times 10^4 \text{ kg} \cdot \text{m/s}}{0.22 \text{ s}}$$ ◄ Substitute $\Delta t_{\text{hitting a wall}} = 0.22$ s.

$$= -2.6 \times 10^5 \text{ N}$$

3 EVALUATE THE ANSWER

- **Are the units correct?** Impulse is in kg·m/s. Force is measured in newtons.
- **Does the direction make sense?** Force is exerted in the direction opposite to the velocity of the car, which means the force is negative.
- **Is the magnitude realistic?** It is reasonable that the force needed to stop a car is thousands or hundreds of thousands of newtons. The impulse is the same for all stops. Thus, as the stopping time decreases by about a factor of 10, the force increases by about a factor of 10.

1. A compact car, with mass 725 kg, is moving at 115 km/h toward the east. Sketch the moving car.

   a. Find the magnitude and direction of its momentum. Draw an arrow on your sketch showing the momentum.

   b. A second car, with a mass of 2175 kg, has the same momentum. What is its velocity?

2. The driver of the compact car in the previous problem suddenly applies the brakes hard for 2.0 s. As a result, an average force of 5.0×10^3 N is exerted on the car to slow it down.

   a. What is the change in momentum, or equivalently, what is the magnitude and direction of the impulse on the car?

   b. Complete the "before" and "after" sketches, and determine the momentum and the velocity of the car now.

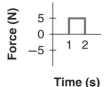

3. A 7.0-kg object, moving at 2.0 m/s, receives two impulses (one after the other) along the direction of its motion. Both of these impulses are illustrated in **Figure 2**. Find the resulting speed and direction of motion of the object after each impulse.

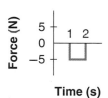

4. The driver accelerates a 240.0-kg snowmobile, which results in a force being exerted that speeds up the snowmobile from 6.00 m/s to 28.0 m/s over a time interval of 60.0 s.

   a. Sketch the event, showing the initial and final situations.

   b. What is the snowmobile's change in momentum? What is the impulse on the snowmobile?

   c. What is the magnitude of the average force that is exerted on the snowmobile?

Figure 2

5. **CHALLENGE** Suppose a 60.0-kg person was in the vehicle that hit the concrete wall in Example Problem 1. The velocity of the person equals that of the car both before and after the crash, and the velocity changes in 0.20 s. Sketch the problem.

   a. What is the average force exerted on the person?

   b. Some people think they can stop their bodies from lurching forward in a vehicle that is suddenly braking by putting their hands on the dashboard. Find the mass of an object that has a weight equal to the force you just calculated. Could you lift such a mass? Are you strong enough to stop your body with your arms?

Using the impulse-momentum theorem to save lives The impulse-momentum theorem shows that a large impulse causes a large change in momentum. This large impulse could result either from a large force acting over a short period of time or from a smaller force acting over a longer period of time. A passenger in a car accident might experience a large impulse, but it is the force on the passenger that causes injuries.

What happens to the driver when a crash suddenly stops a car? An impulse, either from the dashboard or from an air bag, brings the driver's momentum to zero. According to the impulse-momentum equation, $F\Delta t = \boldsymbol{p}_f - \boldsymbol{p}_i$. The final momentum ($\boldsymbol{p}_f$) is zero. The initial momentum ($\boldsymbol{p}_i$) is the same with or without an air bag. Thus, the impulse ($F\Delta t$) also is the same. An air bag, such as the one shown in **Figure 3**, increases the time interval during which the force acts on the passenger. Therefore the required force is less. The air bag also spreads the force over a larger area of the person's body, thereby reducing the likelihood of injuries.

FO4305OZ02

Figure 3 An air bag reduces injuries by making the force on a passenger less and by spreading that force over a larger area.

Angular Momentum

The impulse-momentum theorem is useful if the momentum of an object is linear, but how can you describe momentum that is angular, as it is for the rotating bowling ball in **Figure 4?** Recall from your study of rotational motion that the angular velocity of a rotating object changes if torque is applied to it. This is a statement of Newton's law for rotational motion, $\tau = \frac{I\Delta\omega}{\Delta t}$. You can rearrange this relationship, just as Newton's second law of motion was, to produce $\tau\Delta t = I\Delta\omega$.

The left side of this equation ($\tau\Delta t$) is the rotating object's angular impulse. You can rewrite the right side as $I\Delta\omega = I\omega_f - I\omega_i$. The product of a rotating object's moment of inertia and angular velocity is called **angular momentum,** which is represented by the symbol L. The following relationship describes an object's angular momentum.

ANGULAR MOMENTUM
The angular momentum of an object is defined as the product of the object's moment of inertia and the object's angular velocity.

$$L = I\omega$$

Angular momentum is measured in kg·m²/s. Just as an object's linear momentum changes when an impulse acts on it, the object's angular momentum changes when an angular impulse acts on it. The object's angular impulse is equal to the change in the object's angular momentum, as stated by the **angular impulse-angular momentum theorem.** This theorem can be represented by the following relationship.

ANGULAR IMPULSE-ANGULAR MOMENTUM THEOREM
The angular impulse on an object is equal to the object's final angular momentum minus the object's initial angular momentum.

$$\tau\Delta t = L_f - L_i$$

If the net force on an object is zero, its linear momentum is constant. If the net torque acting on an object is zero, its angular momentum is also constant, but the two situations are slightly different. Because an object's mass cannot change, if its momentum is constant, then its velocity is also constant. In the case of angular momentum, however, the object's angular velocity can change if the shape of the object changes. This is because the moment of inertia depends on the object's mass and the way it is distributed about the axis of rotation or revolution. Thus, the angular velocity of an object can change even if no torques act on it.

▶ **CONNECTION TO ASTRONOMY** Astronomers have discovered many examples of two stars that orbit each other. Together the stars are called a binary star system. The torque on the binary system is zero because the gravitational force acts only directly between the stars. Therefore, the binary system's angular momentum is constant. But sometimes the stars are so close that the strong gravitational force rips some material from one star and deposits it on the other star. This movement of matter changes the moment of inertia of the binary system. As a result, the angular velocity of the system will change, even though the angular momentum is unchanged.

Figure 4 The ball's linear momentum is the product of its mass and its velocity. The ball's angular momentum as it rolls down the lane is the product of its moment of inertia and its angular velocity.

REAL-WORLD PHYSICS
●●●●●●●●●●●●●

PADDLE SPORTS A table-tennis player controls the flight of the ball by putting spin on the ball. The player hits the ball at an angle so that the paddle exerts both an impulse to change the momentum and an angular impulse to make the ball spin. The interaction with the air will cause a forward-spinning ball to drop rapidly, making it more difficult for the opponent to return.

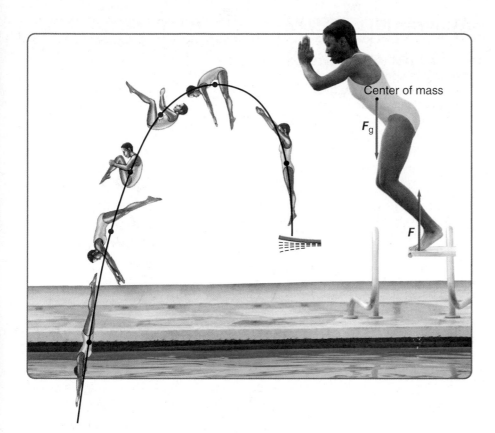

Figure 5 Before she dives, the girl's center of mass is in front of her feet. During her dive, the path of her center of mass forms a parabola. She changes her moment of inertia by moving her arms and legs.

Analyze How does extending her arms affect the diver's motion?

```
                              COLOR CONVENTION
          force  ◄─────────►  blue
```

Diving into a pool

Diving into a pool Consider the diver in **Figure 5.** How does she start rotating her body? She uses the diving board to apply an external force to her body. Then, she moves her center of mass in front of her feet and uses the board to give a final upward push to her feet. This provides a torque that acts over time (Δt) and increases the angular momentum of the diver.

Before the diver reaches the water, she can change her angular velocity by changing her moment of inertia. She may go into a tuck position, grabbing her knees with her hands. By moving her mass closer to the axis of rotation, the diver decreases her moment of inertia and increases her angular velocity. When she nears the water, she stretches her body straight, thereby increasing the moment of inertia and reducing the angular velocity. As a result, she goes straight into the water.

☑ **READING CHECK Explain** why going into a tuck position increases a diver's angular velocity.

Figure 6 When the skater pushes off the ice, the ice exerts a torque on the skater.

Ice-skating

Ice-skating An ice-skater uses a similar method to spin. To begin rotating on one foot, the ice-skater causes the ice to exert a force on her body by pushing a portion of one skate into the ice. If the skater in **Figure 6** pushes on the ice in one direction, the ice will exert a force on her in the opposite direction. The force results in a torque if the force is exerted some distance away from the pivot point and in a direction that is not toward it. The greatest torque for a given force will result if the push is perpendicular to the lever arm.

The ice-skater then can control her angular velocity by changing her moment of inertia. Both arms and one leg can be extended from the body to slow the rotation or can be pulled in close to the axis of rotation to speed it up. To stop spinning, another torque must be exerted by using the free skate to create a way for the ice to exert the needed force.

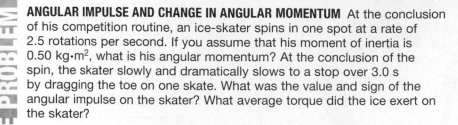

ANGULAR IMPULSE AND CHANGE IN ANGULAR MOMENTUM At the conclusion of his competition routine, an ice-skater spins in one spot at a rate of 2.5 rotations per second. If you assume that his moment of inertia is 0.50 kg·m², what is his angular momentum? At the conclusion of the spin, the skater slowly and dramatically slows to a stop over 3.0 s by dragging the toe on one skate. What was the value and sign of the angular impulse on the skater? What average torque did the ice exert on the skater?

1 ANALYZE AND SKETCH THE PROBLEM

- Sketch the system.
- Draw the axis of rotation and the direction of the angular impulse.

KNOWN

ω_i = 2.5 rotations/s
ω_f = 0 rad/s
I = 0.50 kg·m²
Δt = 3.0 s
L_f = 0.0 kg·m²/s

UNKNOWN

L_i = ?
angular impulse = ?
τ_{avg} = ?

2 SOLVE FOR THE UNKNOWNS

Determine the initial angular momentum.

$L_i = I\omega_i$

$\quad = (0.50 \text{ kg·m}^2)(2.5 \text{ rotations/s})(2\pi \text{ rad/rotation})$ ◀ Substitute I = 0.50 kg·m² and ω_i = 2.5 rotations/s.
Convert from rotations to radians.

$\quad = 7.9 \text{ kg·m}^2\text{/s}$ ◀ Radians is a dimensionless unit and is removed.

Determine the angular impulse that the ice exerted on the skater.

angular impulse = $\tau_{avg} \Delta t = L_f - L_i$ ◀ The angular impulse equals the change in angular momentum.

$\quad = 0.0 \text{ kg·m}^2\text{/s} - 7.9 \text{ kg·m}^2\text{/s}$ ◀ Substitute L_f = 0.0 kg·m²/s, L_i = 7.9 kg·m²/s.

$\quad = -7.9 \text{ kg·m}^2\text{/s}$

Determine the average torque that the ice exerted on the skater.
From the angular impulse, you can rearrange the quantities to find the average torque.

$\tau_{avg} = \dfrac{L_f - L_i}{\Delta t}$ ◀ Solve for the average torque.

$\quad = \dfrac{0.0 \text{ kg·m}^2\text{/s} - 7.9 \text{ kg·m}^2\text{/s}}{3.0 \text{ s}}$ ◀ Substitute L_f = 0.0 kg·m²/s, L_i = 7.9 kg·m²/s, and Δt = 3.0 s.

$\quad = -2.6 \text{ N·m}$ ◀ The unit kg·m²/s² is equivalent to N·m.

3 EVALUATE THE ANSWER

- **Are the units correct?** Yes, a kg·m²/s² is the equivalent of a N·m.
- **Do the signs make sense?** The angular impulse and the torque are each negative, indicating that the angular momentum decreases.
- **Are the magnitudes realistic?** Yes, the value of the torque is small, but the product of the torque and time equals the value of the angular momentum.

6. A 0.25-m-diameter circular saw blade in a workshop rotates at 5.0×10^3 rpm, as shown in **Figure 7.** After the electrical power to the saw is turned off, it takes several seconds for the blade to slow to a complete stop. The moment of inertia of the blade is 8.0×10^{-3} kg·m². Friction in the axle provides an average torque of 2.3×10^{-1} N·m to slow the blade. How many seconds does it take for the blade to stop?

7. A baseball pitcher can throw a 132 km/h (82 mph) curve ball that rotates about 6.0×10^2 rpm. What is the angular velocity of the thrown ball? The pitcher's throwing motion lasts about 0.15 s, and the moment of inertia of the ball is 8.0×10^{-5} kg·m². What average torque did the pitcher exert on the ball?

8. As a bowler releases the ball onto the alley, the ball does not roll but slides. Slowly the friction of the alley surface causes the ball to roll and have a final angular velocity of 7.00×10^1 rad/s. The moment of inertia of the ball is 0.0350 kg·m², and the ball moves down the alley in 2.40 s. What are the angular impulse and the average torque that the alley surface exerts on the bowling ball?

9. A bicycle is clamped upside down on a workbench for the bicycle repair woman to repair a front wheel axle. She gives the front wheel a spin with her hand, and the wheel rotates at 5.0 rev/s. What is the angular velocity of the wheel? If the moment of inertia of the wheel is 0.060 kg·m², what angular impulse did the repair woman give the wheel?

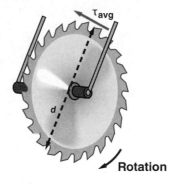

Figure 7

SECTION 1 REVIEW

10. **MAIN**IDEA Which has more momentum, a truck that is parked or a falling raindrop? Explain.

11. **Momentum** Is the momentum of a car traveling south different from that of the same car when it travels north at the same speed? Draw the momentum vectors to support your answer.

12. **Impulse and Momentum** When you jump from a height to the ground, you let your legs bend at the knees as your feet hit the floor. Explain why you do this in terms of the physics concepts introduced in this chapter.

13. **Impulse and Momentum** A 0.174-kg softball is pitched horizontally at 26.0 m/s. The ball moves in the opposite direction at 38.0 m/s after it is hit by the bat.

    a. Draw arrows showing the ball's momentum before and after the bat hits it.

    b. What is the change in momentum of the ball?

    c. What is the impulse delivered by the bat?

    d. If the bat and ball are in contact for 0.80 ms, what is the average force the bat exerts on the ball?

14. **Momentum** The speed of a basketball as it is dribbled is the same just before and just after the ball hits the floor. Is the impulse on and the change in momentum of the basketball equal to zero when the basketball hits the floor? If not, in which direction is the change in momentum? Draw the basketball's momentum vectors before and after it hits the floor.

Spinning slowly   **Spinning quickly**

Figure 8

15. **Angular Momentum** The ice-skater in **Figure 8** spins with his arms outstretched. When he pulls his arms in and raises them above his head, he spins much faster than before. Did a torque act on the ice-skater? What caused his angular velocity to increase?

16. **Critical Thinking** An archer shoots arrows at a target. Some of the arrows stick in the target, while others bounce off. Assuming that the masses of the arrows and the velocities of the arrows are the same, which arrows produce a bigger impulse on the target? *Hint: Draw a diagram to show the momentum of the arrows before and after hitting the target for the two instances.*

Conservation of Momentum

PHYSICS 4 YOU

When a game of billiards is started, the balls are usually arranged in a triangle. A player then shoots the cue ball at them, causing the balls to spread out in all directions. How does the motion of the cue ball before the break affect the motions of the balls after the break?

MAIN IDEA

In a closed, isolated system, linear momentum and angular momentum are conserved.

Essential Questions

- How does Newton's third law relate to conservation of momentum?
- Under which conditions is momentum conserved?
- How can the law of conservation of momentum and the law of conservation of angular momentum help explain the motion of objects?

Review Vocabulary

momentum the product of an object's mass and the object's velocity

New Vocabulary

closed system
isolated system
law of conservation of momentum
law of conservation of angular momentum

APPLYING PRACTICES

Use Mathematical Representations Go to the resources tab in ConnectED to find the Applying Practices worksheet *Conservation of Momentum*.

Two-Particle Collisions

In the first section of this chapter, you learned how a force applied during a time interval changes the momentum of a baseball. In the discussion of Newton's third law of motion, you learned that forces are the result of interactions between two objects. The force of a bat on a ball is accompanied by an equal and opposite force of the ball on the bat. Does the momentum of the bat, therefore, also change?

The bat, the hand and arm of the batter, and the ground on which the batter is standing are all objects that interact when a batter hits the ball. Thus, the bat cannot be considered as a single object. In contrast to this complex system, examine for a moment the much simpler system shown in **Figure 9,** the collision of two balls.

Force and impulse During the collision of the two balls, each one briefly exerts a force on the other. Despite the differences in sizes and velocities of the balls, the forces they exert on each other are equal and opposite, according to Newton's third law of motion. These forces are represented by $F_{\text{red on blue}} = -F_{\text{blue on red}}$.

How do the impulses imparted by both balls compare? Because the time intervals over which the forces are exerted are the same, the impulses the balls exert on each other must also be equal in magnitude but opposite in direction: $(F\Delta t)_{\text{red on blue}} = -(F\Delta t)_{\text{blue on red}}$.

Notice that no mention has been made of the masses of the balls. Even though the balls have different sizes and approach each other with different velocities, and even though they may have different masses, the forces and the impulses they exert on each other have equal strength but opposite directions.

Figure 9 The balls have different sizes, masses, and velocities, but if they interact the forces and impulses they exert on each other are equal and opposite.

Momentum How did the momentums of the two balls the girl and the boy rolled toward each other in **Figure 9** change as a result of the collision? Compare the changes shown in **Figure 10.**

For ball C: $\boldsymbol{p}_{Cf} - \boldsymbol{p}_{Ci} = \boldsymbol{F}_{D \text{ on } C}\Delta t$
For ball D: $\boldsymbol{p}_{Df} - \boldsymbol{p}_{Di} = \boldsymbol{F}_{C \text{ on } D}\Delta t$

According to the impulse-momentum theorem, the change in momentum is equal to the impulse. Because the impulses are equal in magnitude but opposite in direction, we know the following:

$$\boldsymbol{p}_{Cf} - \boldsymbol{p}_{Ci} = -(\boldsymbol{p}_{Df} - \boldsymbol{p}_{Di})$$
$$\boldsymbol{p}_{Cf} + \boldsymbol{p}_{Df} = \boldsymbol{p}_{Ci} + \boldsymbol{p}_{Di}$$

This equation states that the sum of the momentums of the balls in **Figure 10** is the same before and after the collision. That is, the momentum gained by ball D is equal to the momentum lost by ball C. If the system is defined as the two balls, the sum of the gain and loss in momentum is zero, and therefore, momentum is conserved for the system.

Momentum in a Closed, Isolated System

Under what conditions is the momentum of the system of two balls conserved? The first and most obvious condition is that no balls are lost and no balls are gained. Such a system, which does not gain or lose mass, is said to be a **closed system.**

The second condition required to conserve the momentum of a system is that the only forces that are involved are internal forces; that is, there are no unbalanced forces acting on the system from objects outside of it. When the net external force exerted on a closed system is zero, the system is described as an **isolated system.** No system on Earth is absolutely isolated, however, because there will always be some interactions between a system and its surroundings. Often, these interactions are small enough to be ignored when solving physics problems.

Systems can contain any number of objects, and the objects that make up a system can stick together or they can come apart during a collision. But even under these conditions, momentum is conserved. The **law of conservation of momentum** states that the momentum of any closed, isolated system does not change. This law enables you to make a connection between conditions before and after an interaction without knowing any details of the interaction.

☑ **READING CHECK Explain** the difference between a closed system and an isolated system.

Figure 10 When two balls of different mass and velocity collide, the momentum of each ball is changed. If the system is isolated, however, the sum of their momentums before the collision equals the sum of their momentums after the collision.

Analyze How can you tell from the diagram that the impulses exerted by the balls have the same magnitude?

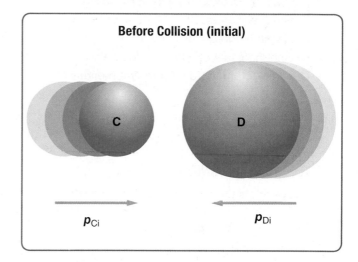

Before Collision (initial)

$\boldsymbol{p}_{Ci}$ $\boldsymbol{p}_{Di}$

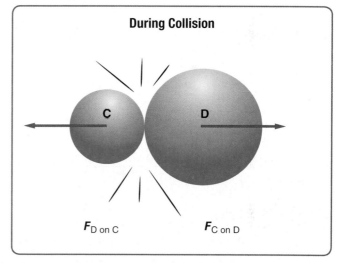

During Collision

$\boldsymbol{F}_{D \text{ on } C}$ $\boldsymbol{F}_{C \text{ on } D}$

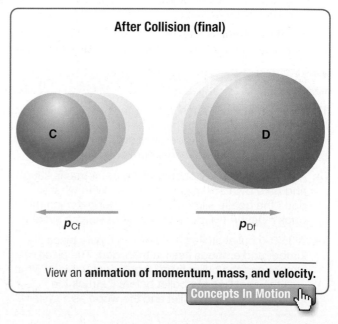

After Collision (final)

$\boldsymbol{p}_{Cf}$ $\boldsymbol{p}_{Df}$

View an **animation of momentum, mass, and velocity.**

Concepts In Motion 🖑

EXAMPLE PROBLEM

SPEED A 1875-kg car going 23 m/s rear-ends a 1025-kg compact car going 17 m/s on ice in the same direction. The two cars stick together. How fast do the two attached cars move immediately after the collision?

1 ANALYZE AND SKETCH THE PROBLEM

- Define the system.
- Establish a coordinate system.
- Sketch the situation showing the before and after states.
- Draw a vector diagram for the momentum.

KNOWN		UNKNOWN
$m_C = 1875$ kg	$m_D = 1025$ kg	$v_f = ?$
$v_{Ci} = +23$ m/s	$v_{Di} = +17$ m/s	

2 SOLVE FOR THE UNKNOWN

The system is the two cars. Momentum is conserved because the ice makes the total external force on the cars nearly zero.

$$p_i = p_f$$

$$p_{Ci} + p_{Di} = p_{Cf} + p_{Df}$$

$$m_C v_{Ci} + m_D v_{Di} = m_C v_{Cf} + m_D v_{Df}$$

Because the two cars stick together, their velocities after the collision, denoted as v_f, are equal.

$$v_{Cf} = v_{Df} = v_f$$

$$m_C v_{Ci} + m_D v_{Di} = (m_C + m_D)v_f$$

Solve for v_f.

$$v_f = \frac{m_C v_{Ci} + m_D v_{Di}}{m_C + m_D}$$

$$= \frac{(1875 \text{ kg})(+23 \text{ m/s}) + (1025 \text{ kg})(+17 \text{ m/s})}{1875 \text{ kg} + 1025 \text{ kg}}$$

◀ Substitute $m_C = 1875$ kg, $v_{Ci} = +23$ m/s, $m_D = 1025$ kg, $v_{Di} = +17$ m/s.

$$= +21 \text{ m/s}$$

3 EVALUATE THE ANSWER

- **Are the units correct?** Velocity is measured in meters per second.
- **Does the direction make sense?** v_{Ci} and v_{Di} are in the positive direction; therefore, v_f should be positive.
- **Is the magnitude realistic?** The magnitude of v_f is between the initial speeds of the two cars, but closer to the speed of the more massive one, so it is reasonable.

PRACTICE PROBLEMS   Do additional problems. Online Practice

17. Two freight cars, each with a mass of 3.0×10^5 kg, collide and stick together. One was initially moving at 2.2 m/s, and the other was at rest. What is their final speed? Define the system as the two cars.

18. A 0.105-kg hockey puck moving at 24 m/s is caught and held by a 75-kg goalie at rest. With what speed does the goalie slide on the ice after catching the puck? Define the puck and the goalie as a system.

19. A 35.0-g bullet strikes a 5.0-kg stationary piece of lumber and embeds itself in the wood. The piece of lumber and the bullet fly off together at 8.6 m/s. What was the speed of the bullet before it struck the lumber? Define the bullet and the wood as a system.

20. A 35.0-g bullet moving at 475 m/s strikes a 2.5-kg bag of flour at rest on ice. The bullet passes through the bag and exits it at 275 m/s. How fast is the bag moving when the bullet exits?

21. The bullet in the previous problem strikes a 2.5-kg steel ball that is at rest. After the collision, the bullet bounces backward at 5.0 m/s. How fast is the ball moving when the bullet bounces backward?

22. CHALLENGE A 0.50-kg ball traveling at 6.0 m/s collides head-on with a 1.00-kg ball moving in the opposite direction at 12.0 m/s. After colliding, the 0.50-kg ball bounces backward at 14 m/s. Find the other ball's speed and direction after the collision.

Before Push

After Push

Figure 11 The boy exerts a force, causing the girl to move to the right. The boy's position relative to the ball shows that the equal but opposite force from the girl caused him to recoil to the left.

Infer How can you tell from the motion that the boy's mass is greater than the girl's mass?

Recoil

It is very important to define a system carefully. The momentum of a baseball changes when the external force of a bat is exerted on it. The baseball, therefore, is not an isolated system. On the other hand, the total momentum of two colliding balls within an isolated system does not change because all forces are between the objects within the system.

Can you find the final velocities of the in-line skaters in **Figure 11?** Assume they are skating on a smooth surface with no external forces. They both start at rest, one behind the other.

Skater C, the boy, gives skater D, the girl, a push. Now, both skaters are moving in opposite directions. Because the push was an internal force, you can use the law of conservation of momentum to find the skaters' relative velocities. The total momentum of the system was zero before the push. Therefore, it must be zero after the push.

Before		After
$p_{Ci} + p_{Di}$	$=$	$p_{Cf} + p_{Df}$
0	$=$	$p_{Cf} + p_{Df}$
p_{Cf}	$=$	$-p_{Df}$
$m_C\boldsymbol{v}_{Cf}$	$=$	$-m_D\boldsymbol{v}_{Df}$

The coordinate system can be chosen so that the positive direction is to the right. The momentums of the skaters after the push are equal in magnitude but opposite in direction. The backward motion of skater C is an example of recoil. Are the skaters' velocities equal and opposite? The last equation can be written to solve for the velocity of skater C.

$$\boldsymbol{v}_{Cf} = \left(\frac{-m_D}{m_C}\right)\boldsymbol{v}_{Df}$$

The velocities depend on the skaters' relative masses. If skater C has a mass of 68.0 kg and skater D's mass is 45.4 kg, then the ratio of their velocities will be 68.0 : 45.4, or 1.50. The less massive skater moves at the greater velocity. Without more information about how hard skater C pushed skater D, however, you cannot find the velocity of each skater.

PhysicsLAB

COLLIDING CARTS

PROBEWARE LAB What happens to the momentums of two carts when they collide?

MiniLAB

REBOUND HEIGHT

How do mass and velocity affect the momentum of a bouncing ball?

Investigate **a hover glider.**

Virtual Investigation

Propulsion in Space

How does a rocket in space change velocity? The rocket carries both fuel and oxidizer. When the fuel and oxidizer combine in the rocket motor, the resulting hot gases leave the exhaust nozzle at high speed. The rocket and the chemicals form a closed system. The forces that expel the gases are internal forces, so the system is also an isolated system. Thus, objects in space can accelerate by using the law of conservation of momentum and Newton's third law of motion.

A newer type of space propulsion uses the recoil that results from the force of ions to move a spaceship. An ion thruster accelerates ions in an electric or magnetic field. As the ions are expelled in one direction, the conservation of momentum results in the movement of the spaceship in the opposite direction. The force exerted by an ion engine might only be a fraction of a newton, but running the engine for long periods of time can significantly change the velocitiy of a spaceship.

View an **animation of thrust and momentum.**

Concepts In Motion

EXAMPLE PROBLEM 4

Find help with **isolating a variable.** **Math Handbook**

EXAMPLE PROBLEM

SPEED An astronaut at rest in space fires a thruster pistol that expels a quick burst of 35 g of hot gas at 875 m/s. The combined mass of the astronaut and pistol is 84 kg. How fast and in what direction is the astronaut moving after firing the pistol?

1 ANALYZE AND SKETCH THE PROBLEM

- Define the system and establish a coordinate axis.
- Sketch before and after conditions. Draw a vector diagram.

KNOWN	UNKNOWN
$m_C = 84$ kg $\mathbf{v}_{Ci} = \mathbf{v}_{Di} = 0.0$ m/s	$\mathbf{v}_{Cf} = ?$
$m_D = 0.035$ kg $\mathbf{v}_{Df} = -875$ m/s	

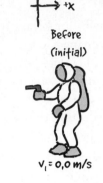

Before (initial)

$v_i = 0.0$ m/s

2 SOLVE FOR THE UNKNOWN

The system is the astronaut, the gun, and the gas.

$$\mathbf{p}_i = \mathbf{p}_{Ci} + \mathbf{p}_{Di} = 0.0 \text{ kg·m/s}$$

◀ Before the pistol is fired, the system is at rest; initial momentum is zero.

Use the law of conservation of momentum to find $\mathbf{p}_{Cf}$.

$$\mathbf{p}_i = \mathbf{p}_f$$

$$0.0 \text{ kg·m/s} = \mathbf{p}_{Cf} + \mathbf{p}_{Df}$$

◀ The momentum of the astronaut is equal in magnitude, but opposite in direction, to the momentum of the gas leaving the pistol.

$$\mathbf{p}_{Cf} = -\mathbf{p}_{Df}$$

Solve for the final velocity of the astronaut ($\mathbf{v}_{Cf}$).

$$m_C\mathbf{v}_{Cf} = -m_D\mathbf{v}_{Df}$$

$$\mathbf{v}_{Cf} = \frac{-m_D\mathbf{v}_{Df}}{m_C}$$

$$= \frac{-(0.035 \text{ kg})(-875 \text{ m/s})}{84 \text{ kg}}$$

◀ Substitute $m_D = 0.035$ kg, $v_{Df} = -875$ m/s, $m_C = 84$ kg.

$$= +0.36 \text{ m/s}$$

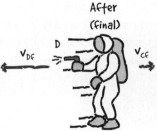

After (final)

v_{Df} D v_{Cf}

Vector diagram

• P_i P_{Df} P_{Cf}

• $P_f = P_{Cf} + P_{Df}$

3 EVALUATE THE ANSWER

- **Are the units correct?** The velocity is measured in meters per second.
- **Does the direction make sense?** The astronaut's direction is opposite that of the gas.
- **Is the magnitude realistic?** The astronaut's mass is much larger than that of the gas, so the velocity of the astronaut is much less than that of the expelled gas.

23. A 4.00-kg model rocket is launched, expelling burned fuel with a mass of 50.0 g at a speed of 625 m/s. What is the velocity of the rocket after the fuel has burned? *Hint: Ignore the external forces of gravity and air resistance.*

24. A thread connects a 1.5-kg cart and a 4.5-kg cart. After the thread is burned, a compressed spring pushes the carts apart, giving the 1.5-kg cart a velocity of 27 cm/s to the left. What is the velocity of the 4.5-kg cart?

25. **CHALLENGE** Carmen and Judi row their canoe alongside a dock. They stop the canoe, but they do not secure it. The canoe can still move freely. Carmen, who has a mass of 80.0 kg, then steps out of the canoe onto the dock. As she leaves the canoe, Carmen moves forward at a speed of 4.0 m/s, causing the canoe, with Judi still in it, to move also. At what speed and in what direction do the canoe and Judi move if their combined mass is 115 kg?

Two-Dimensional Collisions

Until now you have considered momentum in only one dimension. The law of conservation of momentum holds for all closed systems with no external forces. It is valid regardless of the directions of the particles before or after they interact. But what happens in two or three dimensions? **Figure 12** shows a collision of two billiard balls. Consider the billiard balls to be the system. The original momentum of the moving ball is p_{Ci}, and the momentum of the stationary ball is zero. Therefore, the momentum of the system before the collision is p_{Ci}.

After the collision, both balls are moving and have momentums. Ignoring friction with the tabletop, the system is closed and isolated. Thus, the law of conservation of momentum can be used. The initial momentum equals the sum of the final momentums, so $p_{Ci} = p_{Cf} + p_{Df}$.

The sum of the components of the vectors before and after the collision must also be equal. If the x-axis is defined in the direction of the initial momentum, then the y-component of the initial momentum is zero. The sum of the final y-components also must be zero.

$$p_{Cf, y} + p_{Df, y} = 0$$

The y-components are equal in magnitude but are in the opposite direction and, thus, have opposite signs. The sums of the horizontal components before and after the collision also are equal.

$$p_{Ci} = p_{Cf, x} + p_{Df, x}$$

View an **animation of conservation of momentum.**

Concepts In Motion

MiniLAB

MOMENTUM
How can you use the law of conservation of momentum to determine an object's velocity after a collision?

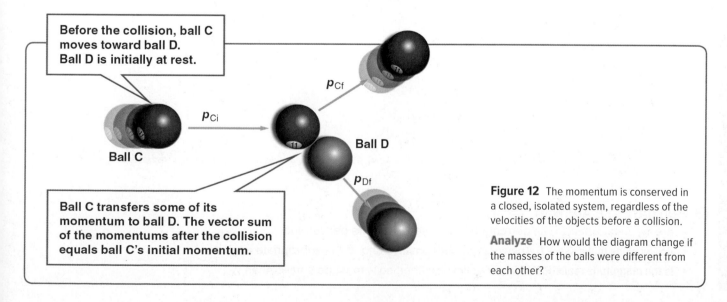

Before the collision, ball C moves toward ball D. Ball D is initially at rest.

p_{Cf}

p_{Ci}

Ball D

Ball C

Ball C transfers some of its momentum to ball D. The vector sum of the momentums after the collision equals ball C's initial momentum.

p_{Df}

Figure 12 The momentum is conserved in a closed, isolated system, regardless of the velocities of the objects before a collision.

Analyze How would the diagram change if the masses of the balls were different from each other?

EXAMPLE PROBLEM

SPEED A 1325-kg car, C, moving north at 27.0 m/s, collides with a 2165-kg car, D, moving east at 11.0 m/s. The collision causes the two cars to stick together. In what direction and with what speed do they move after the collision?

1 ANALYZE AND SKETCH THE PROBLEM

- Define the system.
- Sketch the before and after states.
- Establish the coordinate axis with the y-axis north and the x-axis east.
- Draw a momentum-vector diagram.

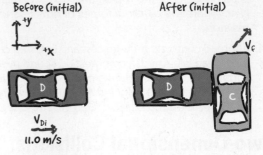

Before (initial)   After (initial)

KNOWN	UNKNOWN
$m_C = 1325$ kg	$v_f = ?$
$m_D = 2165$ kg	$\theta = ?$
$v_{Ci, y} = 27.0$ m/s	
$v_{Di, x} = 11.0$ m/s	

2 SOLVE FOR THE UNKNOWN

Define the system as the two cars. Determine the magnitudes of the initial momenta of the cars and the momentum of the system.

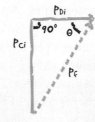

$$p_{Ci} = m_C v_{Ci, y}$$
$$= (1325 \text{ kg})(27.0 \text{ m/s})$$
$$= 3.58 \times 10^4 \text{ kg·m/s (north)}$$

◀ Substitute $m_C = 1325$ kg, $v_{Ci, y} = 27.0$ m/s.

$$p_{Di} = m_D v_{Di, x}$$
$$= (2165 \text{ kg})(11.0 \text{ m/s})$$
$$= 2.38 \times 10^4 \text{ kg·m/s (east)}$$

◀ Substitute $m_D = 2165$ kg, $v_{Di, x} = 11.0$ m/s.

Use the law of conservation of momentum to find p_f.

$$p_{f, x} = p_{i, x} = 2.38 \times 10^4 \text{ kg·m/s}$$

◀ Substitute $p_{i, x} = p_{Di} = 2.38 \times 10^4$ kg·m/s.

$$p_{f, y} = p_{i, y} = 3.58 \times 10^4 \text{ kg·m/s}$$

◀ Substitute $p_{i, y} = p_{Ci} = 3.58 \times 10^4$ kg·m/s.

Use the diagram to set up equations for $p_{f, x}$ and $p_{f, y}$.

$$p_f = \sqrt{(p_{f, x})^2 + (p_{f, y})^2}$$
$$= \sqrt{(2.38 \times 10^4 \text{ kg·m/s})^2 + (3.58 \times 10^4 \text{ kg·m/s})^2}$$
$$= 4.30 \times 10^4 \text{ kg·m/s}$$

◀ Substitute $p_{f, x} = 2.38 \times 10^4$ kg·m/s, $p_{f, y} = 3.58 \times 10^4$ kg·m/s.

Solve for θ.

$$\theta = \tan^{-1}\left(\frac{p_{f, y}}{p_{f, x}}\right)$$
$$= \tan^{-1}\left(\frac{3.58 \times 10^4 \text{ kg·m/s}}{2.38 \times 10^4 \text{ kg·m/s}}\right)$$
$$= 56.4°$$

◀ Substitute $p_{f, y} = 3.58 \times 10^4$ kg·m/s, $p_{f, x} = 2.38 \times 10^4$ kg·m/s.

Determine the final speed.

$$v_f = \frac{p_f}{m_C + m_D}$$
$$= \frac{4.30 \times 10^4 \text{ kg·m/s}}{1325 \text{ kg} + 2165 \text{ kg}}$$
$$= 12.3 \text{ m/s}$$

◀ Substitute $p_f = 4.30 \times 10^4$ kg·m/s, $m_C = 1325$ kg, $m_D = 2165$ kg.

3 EVALUATE THE ANSWER

- **Are the units correct?** The correct unit for speed is meters per second.
- **Do the signs make sense?** Answers are both positive and at the appropriate angles.
- **Is the magnitude realistic?** The cars stick together, so v_f must be smaller than v_{Ci}.

26. A 925-kg car moving north at 20.1 m/s collides with a 1865-kg car moving west at 13.4 m/s. After the collision, the two cars are stuck together. In what direction and at what speed do they move after the collision? Define the system as the two cars.

27. A 1383-kg car moving south at 11.2 m/s is struck by a 1732-kg car moving east at 31.3 m/s. After the collision, the cars are stuck together. How fast and in what direction do they move immediately after the collision? Define the system as the two cars.

28. A 1345-kg car moving east at 15.7 m/s is struck by a 1923-kg car moving north. They stick together and move with a velocity of 14.5 m/s at $\theta = 63.5°$. Was the north-moving car exceeding the 20.1 m/s speed limit?

29. **CHALLENGE** A stationary billiard ball with mass 0.17 kg is struck by an identical ball moving 4.0 m/s. Afterwards, the second ball moves 60.0° to the left of its original direction. The stationary ball moves 30.0° to the right of the moving ball's original direction. What is the velocity of each ball after the collision?

Conservation of Angular Momentum

Like linear momentum, angular momentum can be conserved. The **law of conservation of angular momentum** states that if no net external torque acts on a closed system, then its angular momentum does not change, as represented by the following equation.

LAW OF CONSERVATION OF ANGULAR MOMENTUM
An isolated system's initial angular momentum is equal to its final angular momentum.

$$L_i = L_f$$

The spinning ice-skater in **Figure 13** demonstrates conservation of angular momentum. When he pulls in his arms, he spins faster. Without an external torque, his angular momentum does not change and $L = I\omega$ is constant. The increased angular velocity must be accompanied by a decreased moment of inertia. By pulling in his arms, the skater brings more mass closer to the axis of rotation, decreasing the radius of rotation and decreasing his moment of inertia. You can calculate changes in angular velocity using the law of conservation of angular momentum.

$$L_i = L_f$$
$$\text{thus, } I_i\omega_i = I_f\omega_f$$
$$\frac{\omega_f}{\omega_i} = \frac{I_i}{I_f}$$

Figure 13 When an ice-skater tucks his arms, his moment of inertia decreases. Because angular momentum is conserved, his angular velocity increases.

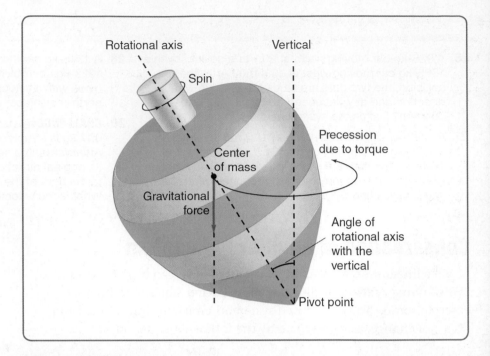

Figure 14 If you spin a top at a tilt, Earth's gravity exerts a torque on it, causing the upper end to precess.

Decide Why does the top precess only when it is tilted?

PhysicsLAB

ROTATION OF A WHEEL
What can set you spinning?

Tops and Gyroscopes

Because of the conservation of angular momentum, the direction of rotation of a spinning object can be changed only by applying a torque. If you played with a top as a child, you may have spun it by twisting its axle between your fingers or by pulling the string wrapped around its axle. When a top is vertical, the force of gravity on the center of mass of the top points through the pivot point. As a result, there is no torque on the top, and the axis of its rotation does not change. If the top is tipped, as shown in **Figure 14,** a torque tends to rotate it downward. Rather than tipping over, however, the axis of the top slowly precesses, which means it revolves slightly away from the vertical line.

Earth's rotation is another example of precession of a spinning object. Earth is not a perfect sphere. Because it has an equatorial bulge, the gravitational pull of the Sun exerts a torque on it, causing it to precess. It takes about 26,000 years for Earth's rotational axis to go through one cycle of precession.

☑ **READING CHECK** **Analyze** Why does Earth precess as it rotates?

PHYSICS CHALLENGE

COLLIDING CARS Your friend was driving her 1265-kg car north on Oak Street when she was hit by a 925-kg compact car going west on Maple Street. The cars stuck together and slid 23.1 m at 42° north of west. The speed limit on both streets is 22 m/s (50 mph). Define the two cars as a system. Assume that momentum was conserved during the collision and that acceleration was constant during the skid. The coefficient of kinetic friction between the tires and the pavement is 0.65.

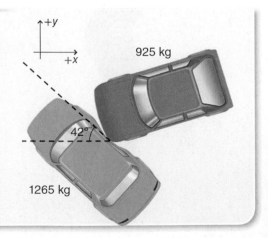

1. Your friend claims she was not speeding but the driver of the other car was. How fast was your friend driving before the crash?

2. How fast was the other car moving before the crash? Can you support your friend's case in court?

A gyroscope, such as the one shown in **Figure 15**, is a wheel or disk that spins rapidly around one axis while being free to rotate around one or two other axes. The direction of its large angular momentum can be changed only by applying an appropriate torque. Without such a torque, the direction of the axis of rotation does not change.

Gyroscopes are used in airplanes, submarines, and spacecraft to keep an unchanging reference direction. Giant gyroscopes are used in cruise ships to detect changes in orientation so that the ship can reduce its motion in rough water. Gyroscopic compasses, unlike magnetic compasses, maintain direction even when they are not on a level surface.

A football quarterback uses the gyroscope effect to make an accurate forward pass. A spiral is a pass in which the football spins around its longer axis. If the quarterback throws the ball in the direction of its spin axis of rotation, the ball keeps its pointed end forward, thereby reducing air resistance. Thus, the ball can be thrown far and accurately. If its spin direction is slightly off, the ball wobbles. If the ball is not spun, it tumbles end over end.

Spin also stabilizes the flight of a flying disk. A well-spun plastic disk can fly many meters through the air without wobbling. Some people are able to perform tricks with a yo-yo because the yo-yo's fast rotational speed keeps it rotating in one plane.

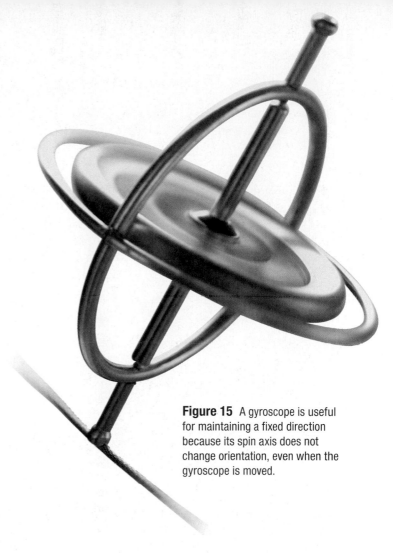

Figure 15 A gyroscope is useful for maintaining a fixed direction because its spin axis does not change orientation, even when the gyroscope is moved.

SECTION 2 **REVIEW**

Section Self-Check Check your understanding.

30. **MAIN**IDEA The outer rim of a plastic disk is thick and heavy. Besides making it easier to catch, how does this affect the rotational properties of the plastic disk?

31. **Speed** A cart, weighing 24.5 N, is released from rest on a 1.00-m ramp, inclined at an angle of 30.0° as shown in **Figure 16.** The cart rolls down the incline and strikes a second cart weighing 36.8 N.

   a. Define the two carts as the system. Calculate the speed of the first cart at the bottom of the incline.

   b. If the two carts stick together, with what initial speed will they move along?

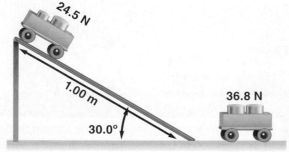

Figure 16

32. **Conservation of Momentum** During a tennis serve, the racket of a tennis player continues forward after it hits the ball. Is momentum conserved in the collision between the tennis racket and the ball? Explain, making sure you define the system.

33. **Momentum** A pole-vaulter runs toward the launch point with horizontal momentum. Where does the vertical momentum come from as the athlete vaults over the crossbar?

34. **Initial Momentum** During a soccer game, two players come from opposite directions and collide when trying to head the ball. The players come to rest in midair and fall to the ground. Describe their initial momentums.

35. **Critical Thinking** You catch a heavy ball while you are standing on a skateboard, and then you roll backward. If you were standing on the ground, however, you would be able to avoid moving while catching the ball.

   a. Identify the system you use in each case.

   b. Explain both situations using the law of conservation of momentum.

FIRE IN THE
SKY

Imagine a giant asteroid the size of a small city crashing into the planet. Its momentum would certainly move matter on Earth.

1 Scientists think such a collision did occur 65 million years ago, and its effects may have caused the extinction of the dinosaurs. This impact left the crater at Chicxulub, Mexico.

Outer Ring of Chicxulub crater

Trough

Mexico

Cenotes (sinkholes)

2 Astronomers are tracking more than 1,000 Near Earth Objects (NEOs) that are greater than 1 km in diameter—big enough to devastate the area near where they land. So far, none are predicted to collide with Earth.

4 Should scientists detect an NEO that is a threat, world leaders hope to find a way to send spacecraft to meet it and alter its course.

5 Some worry that a dangerous NEO might have too much momentum for human technology to significantly change that NEO's path. Scientists hope that early detection will give people plenty of time to prepare for any natural disasters that could be caused by a collision.

3 The impact of a giant asteroid would send up debris that might darken the sky for months. The impulse from its crash might also set in motion giant ocean waves that would flood shores around the globe.

GOING**FURTHER** >>>

Research Find out more about the impact that resulted in the creation of Earth's moon. What evidence is there that such an impact occurred?

STUDY GUIDE

BIGIDEA   If the net force on a closed system is zero, the total momentum of that system is conserved.

SECTION 1 **Impulse and Momentum**

MAINIDEA   An object's momentum is equal to its mass multiplied by its velocity.

- The impulse on an object is the average net force exerted on the object multiplied by the time interval over which the force acts.

$$impulse = F\Delta t$$

- The momentum of an object is the product of its mass and velocity and is a vector quantity.

$$\boldsymbol{p} = m\boldsymbol{v}$$

When solving a momentum problem, first define the objects in the system and examine their momentum before and after the event. The impulse on an object is equal to the change in momentum of the object.

$$F\Delta t = \boldsymbol{p}_f - \boldsymbol{p}_i$$

- The angular momentum of a rotating object is the product of its moment of inertia and its angular velocity.

$$L = I\omega$$

The angular impulse-angular momentum theorem states that the angular impulse on an object is equal to the change in the object's angular momentum.

$$\tau\Delta t = L_f - L_i$$

SECTION 2 **Conservation of Momentum**

MAINIDEA   In a closed, isolated system, linear momentum and angular momentum are conserved.

- According to Newton's third law of motion and the law of conservation of momentum, the forces exerted by colliding objects on each other are equal in magnitude and opposite in direction.

- Momentum is conserved in a closed, isolated system.

$$\boldsymbol{p}_f = \boldsymbol{p}_i$$

- The law of conservation of momentum relates the momentums of objects before and after a collision. Use vector analysis to solve momentum-conservation problems in two dimensions. The law of conservation of angular momentum states that if there are no external torques acting on a system, then the angular momentum is conserved.

$$L_f = L_i$$

Because angular momentum is conserved, the direction of rotation of a spinning object can be changed only by applying a torque.

Games and Multilingual eGlossary

Vocabulary Practice

SECTION 1 Impulse and Momentum

Mastering Concepts

36. Can a bullet have the same momentum as a truck? Explain.

37. During a baseball game, a pitcher throws a curve ball to the catcher. Assume that the speed of the ball does not change in flight.

   a. Which player exerts the larger impulse on the ball?

   b. Which player exerts the larger force on the ball?

38. Newton's second law of motion states that if no net force is exerted on a system, no acceleration is possible. Does it follow that no change in momentum can occur?

39. Why are cars made with bumpers that can be pushed in during a crash?

40. An ice-skater is doing a spin.

   a. How can the skater's angular momentum be changed?

   b. How can the skater's angular velocity be changed without changing the angular momentum?

Mastering Problems

41. Golf Rocío strikes a 0.058-kg golf ball with a force of 272 N and gives it a velocity of 62.0 m/s. How long was Rocío's club in contact with the ball?

42. A 0.145-kg baseball is pitched at 42 m/s. The batter then hits the ball horizontally toward the pitcher at 58 m/s.

   a. Find the change in momentum of the ball.

   b. If the ball and the bat are in contact for 4.6×10^{-4} s, what is the average force during contact?

43. A 0.150-kg ball, moving in the positive direction at 12 m/s, is acted on by the impulse illustrated in the graph in **Figure 17**. What is the ball's speed at 4.0 s?

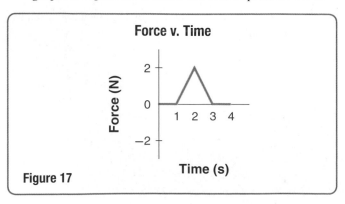

Force v. Time

Figure 17

44. Bowling A force of 186 N acts on a 7.3-kg bowling ball for 0.40 s. What is the ball's change in momentum? What is its change in velocity?

45. Hockey A hockey puck has a mass of 0.115 kg and strikes the pole of the net at 37 m/s. It bounces off in the opposite direction at 25 m/s, as shown in **Figure 18**.

   a. What is the impulse on the puck?

   b. If the collision takes 5.0×10^{-4} s, what is the average force on the puck?

0.115 kg

25 m/s

Figure 18

46. A 5500-kg freight truck accelerates from 4.2 m/s to 7.8 m/s in 15.0 s by the application of a constant force.

   a. What change in momentum occurs?

   b. How large a force is exerted?

47. In a ballistics test at the police department, Officer Ríos fires a 6.0-g bullet at 350 m/s into a container that stops it in 1.8 ms. What is the average force that stops the bullet?

48. Volleyball A 0.24-kg volleyball approaches Tina with a velocity of 3.8 m/s. Tina bumps the ball, giving it a speed of 2.4 m/s but in the opposite direction. What average force did she apply if the interaction time between her hands and the ball was 0.025 s?

49. Before a collision, a 25-kg object was moving at 112 m/s. Find the impulse that acted on the object if, after the collision, it moved at the following velocities.

   a. +8.0 m/s

   b. −8.0 m/s

50. Baseball A 0.145-kg baseball is moving at 35 m/s when it is caught by a player.

   a. Find the change in momentum of the ball.

   b. If the ball is caught with the mitt held in a stationary position so that the ball stops in 0.050 s, what is the average force exerted on the ball?

   c. If, instead, the mitt is moving backward so that the ball takes 0.500 s to stop, what is the average force exerted by the mitt on the ball?

51. Ranking Task Rank the following objects according to the amount of momentum they have, from least to greatest. Specifically indicate any ties.

Object A: mass 2.5 kg, velocity 1.0 m/s east

Object B: mass 3.0 kg, velocity 0.90 m/s west

Object C: mass 3.0 kg, velocity 1.2 m/s west

Object D: mass 4.0 kg, velocity 0.50 m/s north

Object E: mass 4.0 kg, velocity 0.90 m/s east

52. Hockey A hockey player makes a slap shot, exerting a constant force of 30.0 N on the puck for 0.16 s. What is the magnitude of the impulse on the puck?

53. Skateboarding Your brother's mass is 35.6 kg, and he has a 1.3-kg skateboard. What is the combined momentum of your brother and his skateboard if they are moving at 9.50 m/s?

54. A hockey puck has a mass of 0.115 kg and is at rest. A hockey player makes a shot, exerting a constant force of 30.0 N on the puck for 0.16 s. With what speed does it head toward the goal?

55. A nitrogen molecule with a mass of 4.7×10^{-26} kg, moving at 550 m/s, strikes the wall of a container and bounces back at the same speed.

a. What is the molecule's impulse on the wall?

b. If there are 1.5×10^{23} of these collisions each second, what is the average force on the wall?

56. Rockets Small rockets are used to slightly adjust the speeds of spacecraft. A rocket with a thrust of 35 N is fired to change a 72,000-kg spacecraft's speed by 63 cm/s. For how long should it be fired?

57. An animal rescue plane flying due east at 36.0 m/s drops a 175-N bale of hay from an altitude of 60.0 m, as shown in **Figure 19**. What is the momentum of the bale the moment before it strikes the ground? Give both magnitude and direction.

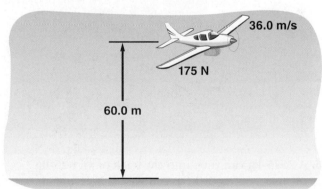

Figure 19

58. Accident A car moving at 10.0 m/s crashes into a barrier and stops in 0.050 s. There is a 20.0-kg child in the car. Assume that the child's velocity is changed by the same amount as that of the car, and in the same time period.

a. What is the impulse needed to stop the child?

b. What is the average force on the child?

c. What is the approximate mass of an object whose weight equals the force in part **b?**

d. Could you lift such a weight with your arm?

e. Why is it advisable to use a proper restraining seat rather than hold a child on your lap?

59. Reverse Problem Write a physics problem with real-life objects for which the following equation would be part of the solution:

$$F = \frac{(1.3 \text{ kg})(20.0 \text{ cm/s} - 0.0 \text{ cm/s})}{0.55 \text{ s}}$$

SECTION 2 Conservation of Momentum

Mastering Concepts

60. What is meant by "an isolated system"?

61. A spacecraft in outer space increases its velocity by firing its rockets. How can hot gases escaping from its rocket engine change the velocity of the craft when there is nothing in space for the gases to push against?

62. A cue ball travels across a pool table and collides with the stationary eight ball. The two balls have equal masses. After the collision, the cue ball is at rest. What must be true regarding the speed of the eight ball?

63. BIGIDEA Consider a ball falling toward Earth.

a. Why is the momentum of the ball not conserved?

b. In what system that includes the falling ball is the momentum conserved?

64. A falling basketball hits the floor. Just before it hits, the momentum is in the downward direction, and after it hits the floor, the momentum is in the upward direction.

a. Why isn't the momentum of the basketball conserved even though the bounce is a collision?

b. In what system is the momentum conserved?

65. Only an external force can change the momentum of a system. Explain how the internal force of a car's brakes brings the car to a stop.

66. Children's playgrounds often have circular-motion rides. How could a child change the angular momentum of such a ride as it is turning?

67. Problem Posing Complete this problem so that it can be solved using conservation of momentum: "Armando, mass 60.0 kg, is at the ice-skating rink . . ."

Mastering Problems

68. A 12.0-g rubber bullet travels at a forward velocity of 150 m/s, hits a stationary 8.5-kg concrete block resting on a frictionless surface, and ricochets in the opposite direction with a velocity of −110 m/s, as shown in **Figure 20**. How fast will the concrete block be moving?

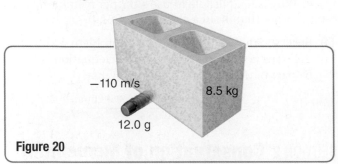

−110 m/s

8.5 kg

12.0 g

Figure 20

69. Football A 95-kg fullback, running at 8.2 m/s, collides in midair with a 128-kg defensive tackle moving in the opposite direction. Both players end up with zero speed.

  a. Identify the before and after situations, and draw a diagram of both.

  b. What was the fullback's momentum before the collision?

  c. What was the change in the fullback's momentum?

  d. What was the change in the defensive tackle's momentum?

  e. What was the defensive tackle's original momentum?

  f. How fast was the defensive tackle moving originally?

70. Marble C, with mass 5.0 g, moves at a speed of 20.0 cm/s. It collides with a second marble, D, with mass 10.0 g, moving at 10.0 cm/s in the same direction. After the collision, marble C continues with a speed of 8.0 cm/s in the same direction.

  a. Sketch the situation, and identify the system. Identify the before and after situations, and set up a coordinate system.

  b. Calculate the marbles' momentums before the collision.

  c. Calculate the momentum of marble C after the collision.

  d. Calculate the momentum of marble D after the collision.

  e. What is the speed of marble D after the collision?

71. Two lab carts are pushed together with a spring mechanism compressed between them. Upon being released, the 5.0-kg cart repels with a velocity of 0.12 m/s in one direction, while the 2.0-kg cart goes in the opposite direction. What is the velocity of the 2.0-kg cart?

72. A 50.0-g projectile is launched with a horizontal velocity of 647 m/s from a 4.65-kg launcher moving in the same direction at 2.00 m/s. What is the launcher's velocity after the launch?

73. Skateboarding Kofi, with mass 42.00 kg, is riding a skateboard with a mass of 2.00 kg and traveling at 1.20 m/s. Kofi jumps off, and the skateboard stops dead in its tracks. In what direction and with what velocity did he jump?

74. In-line Skating Diego and Keshia are on in-line skates. They stand face-to-face and then push each other away with their hands. Diego has a mass of 90.0 kg, and Keshia has a mass of 60.0 kg.

  a. Sketch the event, identifying the before and after situations, and set up a coordinate axis.

  b. Find the ratio of the skaters' velocities just after their hands lose contact.

  c. Which skater has the greater speed?

  d. Which skater pushed harder?

75. Billiards A cue ball, with mass 0.16 kg, rolling at 4.0 m/s, hits a stationary eight ball of similar mass. If the cue ball travels 45° to the left of its original path and the eight ball travels 45° in the opposite direction, as shown in **Figure 21,** what is the velocity of each ball after the collision?

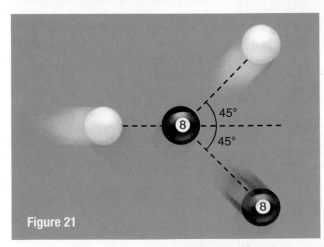

45°

45°

Figure 21

76. A 2575-kg van runs into the back of an 825-kg compact car at rest. They move off together at 8.5 m/s. Assuming that the friction with the road is negligible, calculate the initial speed of the van.

77. A 0.200-kg plastic ball has a forward velocity of 0.30 m/s. It collides with a second plastic ball of mass 0.100 kg, which is moving along the same line at a speed of 0.10 m/s. After the collision, both balls continue moving in the same, original direction. The speed of the 0.100-kg ball is 0.26 m/s. What is the new velocity of the 0.200-kg ball?

Applying Concepts

78. Explain the concept of impulse using physical ideas rather than mathematics.

79. An object initially at rest experiences the impulses described by the graph in **Figure 22.** Describe the object's motion after impulses A, B, and C.

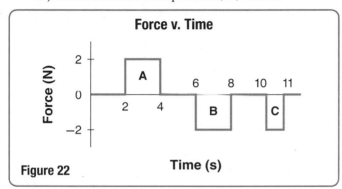

Force v. Time

Figure 22

80. Is it possible for an object to obtain a larger impulse from a smaller force than it does from a larger force? Explain.

81. Foul Ball You are sitting at a baseball game when a foul ball comes in your direction. You prepare to catch it bare-handed. To catch it safely, should you move your hands toward the ball, hold them still, or move them in the same direction as the moving ball? Explain.

82. A 0.11-g bullet leaves a pistol at 323 m/s, while a similar bullet leaves a rifle at 396 m/s. Explain the difference in exit speeds of the two bullets, assuming that the forces exerted on the bullets by the expanding gases have the same magnitude.

83. During a space walk, the tether connecting an astronaut to the spaceship breaks. Using a gas pistol, the astronaut manages to get back to the ship. Use the language of the impulse-momentum theorem and a diagram to explain why this method was effective.

84. Tennis Ball As a tennis ball bounces off a wall, its momentum is reversed. Explain this action in terms of the law of conservation of momentum. Define the system and draw a diagram as a part of your explanation.

85. Two trucks that appear to be identical collide on an icy road. One was originally at rest. The trucks are stuck together and move at more than half the original speed of the moving truck. What can you conclude about the contents of the two trucks?

86. Bullets Two bullets of equal mass are shot at equal speeds at equal blocks of wood on a smooth ice rink. One bullet, made of rubber, bounces off the wood. The other bullet, made of aluminum, burrows into the wood. In which case does the block of wood move faster? Explain.

Mixed Review

87. A constant force of 6.00 N acts on a 3.00-kg object for 10.0 s. What are the changes in the object's momentum and velocity?

88. An external, constant force changes the speed of a 625-kg car from 10.0 m/s to 44.0 m/s in 68.0 s.
  a. What is the car's change in momentum?
  b. What is the magnitude of the force?

89. Gymnastics Figure 23 shows a gymnast performing a routine. First, she does giant swings on the upper bar, holding her body straight and pivoting around her hands. Then, she lets go of the high bar and grabs her knees with her hands in the tuck position. Finally, she straightens up and lands on her feet.
  a. In the second and final parts of the gymnast's routine, around what axis does she spin?
  b. Rank in order, from greatest to least, her moments of inertia for the three positions.
  c. Rank in order, from greatest to least, her angular velocities in the three positions.

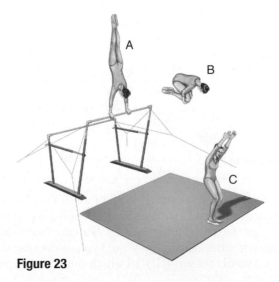

Figure 23

90. Dragster An 845-kg dragster accelerates on a race track from rest to 100.0 km/h in 0.90 s.

   a. What is the change in momentum of the dragster?

   b. What is the average force exerted on the dragster?

   c. What exerts that force?

91. Ice Hockey A 0.115-kg hockey puck, moving at 35.0 m/s, strikes a 0.365-kg jacket that is thrown onto the ice by a fan of a certain hockey team. The puck and jacket slide off together. Find their velocity.

92. A 50.0-kg woman, riding on a 10.0-kg cart, is moving east at 5.0 m/s. The woman jumps off the front of the cart and lands on the ground at 7.0 m/s eastward, relative to the ground.

   a. Sketch the before and after situations, and assign a coordinate axis to them.

   b. Find the cart's velocity after the woman jumps off.

93. A 60.0-kg dancer leaps 0.32 m high.

   a. With what momentum does he reach the ground?

   b. What impulse is needed to stop the dancer?

   c. As the dancer lands, his knees bend, lengthening the stopping time to 0.050 s. Find the average force exerted on the dancer's body.

   d. Compare the stopping force with his weight.

Thinking Critically

94. Analyze and Conclude Two balls during a collision are shown in **Figure 24,** which is drawn to scale. The balls enter from the left of the diagram, collide, and then bounce away. The heavier ball, at the bottom of the diagram, has a mass of 0.600 kg, and the other has a mass of 0.400 kg. Using a vector diagram, determine whether momentum is conserved in this collision. Explain any difference in the momentum of the system before and after the collision.

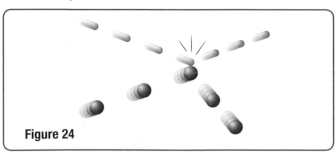

Figure 24

95. Analyze and Conclude A student, holding a bicycle wheel with its axis vertical, sits on a stool that can rotate with negligible friction. She uses her hand to get the wheel spinning. Would you expect the student and stool to turn? If so, in which direction? Explain.

96. Apply Concepts A 92-kg fullback, running at a speed of 5.0 m/s, attempts to dive directly across the goal line for a touchdown. Just as he reaches the line, he is met head-on in midair by two 75-kg linebackers, both moving in the direction opposite the fullback. One is moving at 2.0 m/s, and the other at 4.0 m/s. They all become entangled as one mass.

   a. Sketch the before and after situations.

   b. What is the players' velocity after the collision?

   c. Does the fullback score a touchdown?

Writing in Physics

97. How can highway barriers be designed to be more effective in saving people's lives? Research this issue and describe how impulse and change in momentum can be used to analyze barrier designs.

98. While air bags save many lives, they also have caused injuries and even death. Research the arguments and responses of automobile makers to this statement. Determine whether the problems involve impulse and momentum or other issues.

Cumulative Review

99. The 0.72-kg ball in **Figure 25** is swung vertically from a 0.60-m string in uniform circular motion at a speed of 3.3 m/s. What is the tension in the cord at the top of the ball's motion?

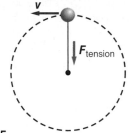

v

F_{tension}

Figure 25

100. You wish to launch a satellite that will remain above the same spot on Earth's surface. This means the satellite must have a period of exactly one day. Calculate the radius of the circular orbit this satellite must have. *Hint: The Moon also circles Earth, and both the Moon and the satellite will obey Kepler's third law. The Moon is 3.9×10^8 m from Earth, and its period is 27.33 days.*

101. A rope is wrapped around a drum that is 0.600 m in diameter. A machine pulls the rope for 2.00 s with a constant 40.0-N force. In that time, 5.00 m of rope unwinds. Find α, ω at 2.00 s, and I.

MULTIPLE CHOICE

1. When a star that is much larger than the Sun nears the end of its lifetime, it begins to collapse but continues to rotate. Which of the following describes the conditions of the collapsing star's moment of inertia (I), angular momentum (L), and angular velocity (ω)?

   A. I increases, L stays constant, ω decreases.

   B. I decreases, L stays constant, ω increases.

   C. I increases, L increases, ω increases.

   D. I increases, L increases, ω stays constant.

2. A 40.0-kg ice-skater glides with a speed of 2.0 m/s toward a 10.0-kg sled at rest on the ice. The ice-skater reaches the sled and holds on to it. The ice-skater and the sled then continue sliding in the same direction in which the ice-skater was originally skating. What is the speed of the ice-skater and the sled after they collide?

   A. 0.4 m/s   **C.** 1.6 m/s

   B. 0.8 m/s   **D.** 3.2 m/s

3. A bicyclist applies the brakes and slows the motion of the wheels. The angular momentum of each wheel then decreases from 7.0 kg·m²/s to 3.5 kg·m²/s over a period of 5.0 s. What is the angular impulse on each wheel?

   A. -0.7 kg·m²/s

   B. -1.4 kg·m²/s

   C. -2.1 kg·m²/s

   D. -3.5 kg·m²/s

4. A 45.0-kg ice-skater stands at rest on the ice. A friend tosses the skater a 5.0-kg ball. The skater and the ball then move backward across the ice with a speed of 0.50 m/s. What was the speed of the ball at the moment just before the skater caught it?

   A. 2.5 m/s   **C.** 4.0 m/s

   B. 3.0 m/s   **D.** 5.0 m/s

5. What is the difference in momentum between a 50.0-kg runner moving at a speed of 3.00 m/s and a 3.00×10^3-kg truck moving at a speed of only 1.00 m/s?

   A. 1275 kg·m/s   **C.** 2850 kg·m/s

   B. 2550 kg·m/s   **D.** 2950 kg·m/s

6. When the large gear in the diagram below rotates, it turns the small gear in the opposite direction at the same linear speed. The larger gear has twice the radius and four times the mass of the smaller gear. What is the angular momentum of the larger gear as a function of the angular momentum of the smaller gear?

   Hint: The moment of inertia for a disk is $\left(\frac{1}{2}\right) mr^2$, where m is mass and r is the radius of the disk.

   A. $-2L_{\text{small}}$ **C.** $-8L_{\text{small}}$

   B. $-4L_{\text{small}}$ **D.** $-16L_{\text{small}}$

7. A force of 16 N exerted against a rock with an impulse of 0.8 kg·m/s causes the rock to fly off the ground with a speed of 4.0 m/s. What is the mass of the rock?

   A. 0.2 kg   **C.** 1.6 kg

   B. 0.8 kg   **D.** 4.0 kg

8. An 82-kg hockey goalie, standing at rest, catches a 0.105-kg hockey puck that is moving at a speed of 46 m/s. With what speed does the goalie slide on the ice?

   A. 0.059 m/s   **C.** 1.2 m/s

   B. 0.56 m/s   **D.** 5.3 m/s

FREE RESPONSE

9. A 12.0-kg rock falls from a cliff to the ground directly below. Assuming the rock does not bounce, what is the impulse on the rock if its velocity at the moment it strikes the ground is 20.0 m/s downward?

NEED EXTRA HELP?

If you Missed Question	1	2	3	4	5	6	7	8	9
Review Section	2	2	1	2	1	1	1	2	1

Online Test Practice

Work, Energy, and Machines

BIGIDEA   Doing work on a system changes the system's energy.

SECTIONS

1 **Work and Energy**

2 **Machines**

LaunchLAB

ENERGY AND FALLING

What factors affect the size of the crater a meteor leaves?

WATCH THIS!

Video

MACHINES

What happens when simple machines are combined to do work? Explore the physics of everyday gadgets and learn more about work, energy, and machines.

PHYSICS T.V.

(l)©Lawrence Manning/Corbis; (r)Digital Vision/PunchStock

Go online!

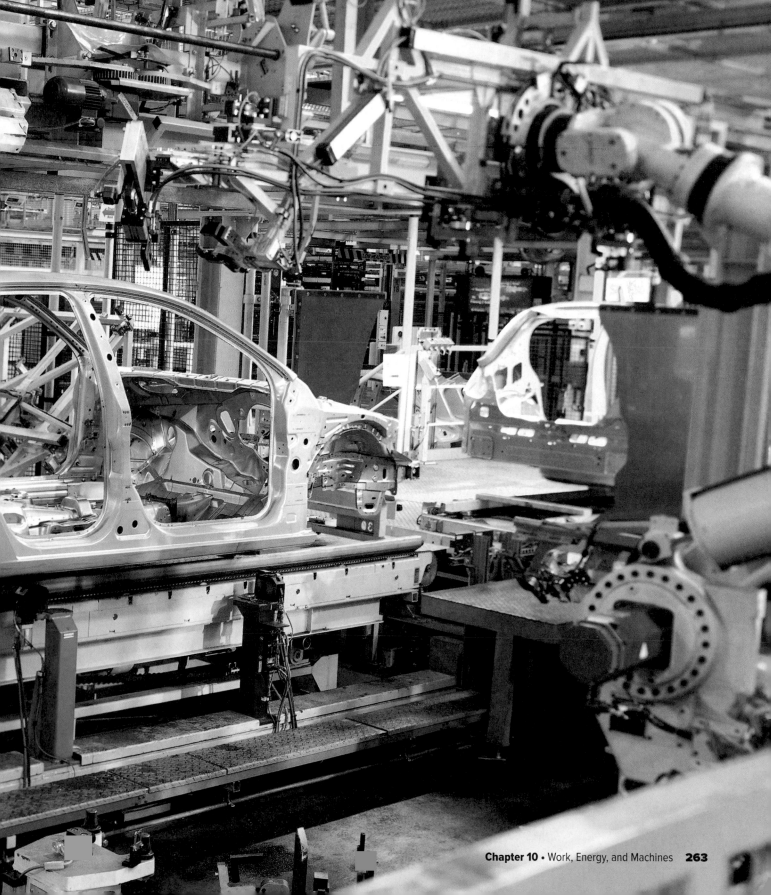

Work and Energy

PHYSICS 4 YOU

Exercising to stay healthy is sometimes called working out. Job related activities are referred to as work. How do scientists define the term *work?*

MAINIDEA

Work is the transfer of energy that occurs when a force is applied through a displacement.

Essential Questions

• What is work?

• What is energy?

• How are work and energy related?

• What is power, and how is it related to work and energy?

Review Vocabulary

law of conservation of momentum states that the momentum of any closed, isolated system does not change

New Vocabulary

work
joule
energy
work-energy theorem
kinetic energy
translational kinetic energy
power
watt

Work

Think about two cars colliding head-on and coming to an immediate stop. You know that momentum is conserved. But the cars were moving prior to the crash and come to a stop after the crash. Thus, it seems that there must be some other quantity that changed as a result of the force acting on each car.

Consider a force exerted on an object while the object moves a certain distance, such as the bookbag in **Figure 1.** There is a net force, so the object is accelerated, $a = \frac{F}{m}$, and its velocity changes. Recall from your study of motion that acceleration, velocity, and distance are related by the equation $v_f^2 = v_i^2 + 2ad$. This can be rewritten as $2ad = v_f^2 - v_i^2$. If you replace a with $\frac{F}{m}$ you obtain $2\left(\frac{F}{m}\right)d = v_f^2 - v_i^2$. Multiplying both sides by $\frac{m}{2}$ gives $Fd = \frac{1}{2}mv_f^2 - \frac{1}{2}mv_i^2$.

The left side of the equation describes an action that was done to the system by the external world. Recall that a system is the object or objects of interest and the external world is everything else. A force (F) was exerted on a system while the point of contact moved. When a force is applied through a displacement, **work** (W) is done on the system.

The SI unit of work is called a **joule** (J). One joule is equal to 1 N·m. One joule of work is done when a force of 1 N acts on a system over a displacement of 1 m. An apple weighs about 1 N, so it takes roughly 1 N of force to lift the apple at a constant velocity. Thus, when you lift an apple 1 m at a constant velocity, you do 1 J of work on it.

Figure 1 Work is done when a force is applied though a displacement.

Identify another example of when a force does work on an object.

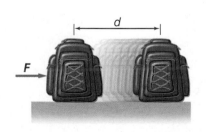

When a force (F) acts through a displacement (d) in the direction of an object's motion, work (W) is done.

The gravitational force does no work on a planet in a circular orbit because the force is perpendicular to the direction of motion.

Work done by a constant force In the book bag example, F is a constant force exerted in the direction in which the object is moving. In this case, work (W) is the product of the force and the magnitude of the system's displacement. That is, $W = Fd$.

What happens if the exerted force is perpendicular to the direction of motion? For example, for a planet in circular orbit, the force is always perpendicular to the direction of motion, as shown in **Figure 1**. Recall from your study of Newton's laws that a perpendicular force does not change the speed of a system, only its direction. The speed of the planet doesn't change and so the right side of the equation, $\frac{1}{2}mv_f^2 - \frac{1}{2}mv_i^2$, is zero. Therefore, the work done is also zero.

Constant force exerted at an angle What work does a force exerted at an angle do? For example, what work does the person pushing the car in **Figure 2** do? Recall that any force can be replaced by its components. If you use the coordinate system shown in **Figure 2**, the 125-N force (F) exerted in the direction of the person's arm has two components.

The magnitude of the horizontal component (F_x) is related to the magnitude of the applied force (F) by a cosine function: $\cos 25.0° = \frac{F_x}{F}$. By solving for F_x, you obtain

$$F_x = F \cos 25.0° = (125 \text{ N})(\cos 25.0°) = 113 \text{ N}.$$

Using the same method, the vertical component is

$$F_y = -F \sin 25.0° = -(125 \text{ N})(\sin 25.0°) = -52.8 \text{ N}.$$

The negative sign shows that the force is downward. Because the displacement is in the x direction, only the x-component does work. The y-component does no work. The work you do when you exert a force on a system at an angle to the direction of motion is equal to the component of the force in the direction of the displacement multiplied by the displacement. The magnitude of the component (F_x) force acting in the direction of displacement is found by multiplying the magnitude of force (F) by the cosine of the angle (θ) between the force and the direction of the displacement: $F_x = F \cos \theta$. Thus, the work done is represented by the following equation.

WORK
Work is equal to the product of the magnitude of the force and magnitude of displacement times the cosine of the angle between them.

$$W = Fd \cos \theta$$

☑ **READING CHECK** **Determine** the work you do when you exert a force of 3 N at an angle of 45° from the direction of motion for 1 m.

Notice that the equation above agrees with our expectations for constant forces exerted in the direction of displacement and for constant forces perpendicular to the displacement. In the book bag example, $\theta = 0°$ and $\cos 0° = 1$. Thus, $W = Fd(1) = Fd$, just as we found before. In the case of the orbiting planet, $\theta = 90°$ and $\cos 90° = 0$. Thus, $W = Fd(0) = 0$. This agrees with our previous conclusions.

Figure 2 Only the horizontal component of the force the man exerts on the car does work because the car's displacement is horizontal.

View an **animation about work**.

Concepts In Motion

+y

F_x

$\theta = 25.0°$ +x

F_y

$F = 125 \text{ N}$

MiniLAB

**FORCE APPLIED
AT AN ANGLE**

How does the direction of a force
affect the amount of work done on
an object?

Work done by many forces Suppose you are pushing a box on a frictionless surface while your friend is trying to prevent you from moving it, as shown in **Figure 3.** What forces are acting on the box? You are exerting a force to the right and your friend is exerting a force to the left. Earth's gravity exerts a downward force, and the ground exerts an upward normal force. How much work is done on the box?

When several forces are exerted on a system, calculate the work done by each force, and then add the results. For the box in **Figure 3,** the upward and downward forces (gravity and the normal force) are perpendicular ($\theta = 90°$) to the direction of motion and do no work. For these forces, $\theta = 90°$, which makes $\cos \theta = 0$, and thus, $W = 0$.

The force you exert ($F_{\text{on box by you}}$) is in the direction of the displacement, so the work you do is

$$W = F_{\text{on box by you}}d.$$

Your friend exerts a force (F_{friend}) in the direction opposite the displacement ($\theta = 180°$). Because $\cos 180° = -1$, your friend does negative work:

$$W = -F_{\text{on box by friend}}d$$

The total work done on the box would be

$$W = F_{\text{on box by you}}d - F_{\text{on box by friend}}d.$$

☑ **READING CHECK Explain** why you do positive work on the box and your friend does negative work on the box.

It is also important to consider each distance separately. For example, suppose you push a box 1 m with a force of 3 N. You then pull the box back 1 m to the starting point with a force of 3 N. You might think you did no work because the total displacement is zero.

This would be true if the force you exerted were constant, but your force changed direction. Your push was in the box's direction of motion for the first part, so you did 3 J of work. In the second part, both the force you exerted and the direction of motion reversed. Your pull and the box's motion were in the same direction, and you did 3 J of work on the box. Therefore, you did a total of 6 J of work on the box.

☑ **READING CHECK Describe** another scenario in which you do work on a system, and explain how much work is done on the system.

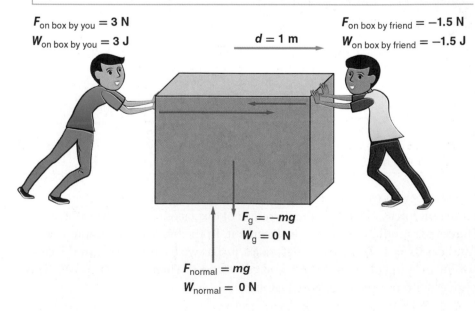

Figure 3 The total work done on a system is the sum of the work done by each agent that exerts a force on the system.

Describe the total work done on the box if your friend exerted a greater force than you did. Be sure to consider the direction of the displacement.

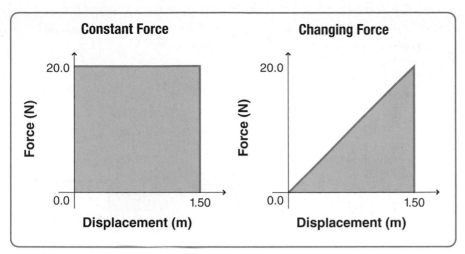

Constant Force

20.0

Force (N)

0.0
1.50

Displacement (m)

Changing Force

20.0

Force (N)

0.0
1.50

Displacement (m)

Figure 4 The area under a force-displacement graph is equal to the work.

Finding work done when forces change In the last example, the force changed, but we could determine the work done in each segment. But what if the force changes in a more complicated way? A graph of force versus displacement lets you determine the work done by a force. This graphical method can be used to solve problems in which the force is changing. The left panel of **Figure 4** shows the work done by a constant force of 20.0 N that is exerted to lift an object 1.50 m. The work done by this force is represented by $W = Fd = (20.0\text{ N})(1.50\text{ m}) = 30.0$ J. Note that the shaded area under the left graph is also equal to $(20.0\text{ N})(1.50\text{ m})$, or 30.0 J. The area under a force-displacement graph is equal to the work done by that force.

This is true even if the force changes. The right panel of **Figure 4** shows the force exerted by a spring, which varies linearly from 0.0 to 20.0 N as it is compressed 1.50 m. The work done by the force that compressed the spring is the area under the graph, which is the area of a triangle, $\left(\frac{1}{2}\right)(\text{base})(\text{altitude})$, or $W = \left(\frac{1}{2}\right)(20.0\text{ N})(1.50\text{ m}) = 15.0$ J. Use the problem-solving strategies below when you solve problems related to work.

PROBLEM-SOLVING STRATEGIES

WORK
When solving work-related problems, use the following strategies:

1. Identify and sketch the system. Show any forces doing work on the system.

2. Establish a coordinate system. Draw the displacement vectors of the system and each force vector doing work on the system.

3. Find the angle (θ) between each force and displacement.

4. Calculate the work done by each force using $W = Fd \cos \theta$.

5. Calculate the net work done.

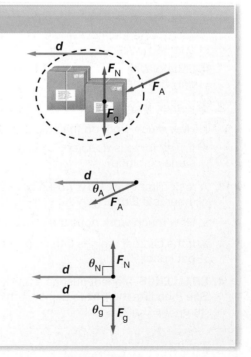

EXAMPLE PROBLEM

WORK A hockey player uses a stick to exert a constant 4.50-N force forward to a 105-g puck sliding on ice over a displacement of 0.150 m forward. How much work does the stick do on the puck? Assume friction is negligible.

1 ANALYZE AND SKETCH THE PROBLEM
- Identify the system and the force doing work on it.
- Sketch the situation showing initial conditions.
- Establish a coordinate system with +x to the right.
- Draw a vector diagram.

KNOWN	UNKNOWN
$m = 105$ g	$W = ?$
$F = 4.50$ N	
$d = 0.150$ m	
$\theta = 0°$	

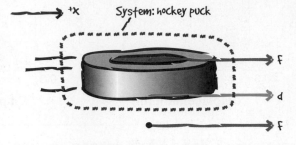

Player does work on hockey puck.

2 SOLVE FOR THE UNKNOWN
Use the definition for work.

$W = Fd \cos \theta$

$\quad = (4.50 \text{ N})(0.150 \text{ m})(1)$ ◀ Substitute $F = 4.50$ N, $d = 0.150$ m, $\cos 0° = 1$.

$\quad = 0.675$ N·m

$\quad = 0.675$ J ◀ 1 J = 1 N·m

3 EVALUATE THE ANSWER
- **Are the units correct?** Work is measured in joules.
- **Does the sign make sense?** The stick (external world) does work on the puck (the system), so the sign of work should be positive.

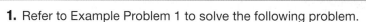

PRACTICE PROBLEMS

Do additional problems. Online Practice

1. Refer to Example Problem 1 to solve the following problem.

   a. If the hockey player exerted twice as much force (9.00 N) on the puck over the same distance, how would the amount of work the stick did on the puck be affected?

   b. If the player exerted a 9.00-N force, but the stick was in contact with the puck for only half the distance (0.075 m), how much work does the stick do on the puck?

2. Together, two students exert a force of 825 N in pushing a car a distance of 35 m.

   a. How much work do the students do on the car?

   b. If their force is doubled, how much work must they do on the car to push it the same distance?

3. A rock climber wears a 7.5-kg backpack while scaling a cliff. After 30.0 min, the climber is 8.2 m above the starting point.

   a. How much work does the climber do on the backpack?

   b. If the climber weighs 645 N, how much work does she do lifting herself and the backpack?

4. **CHALLENGE** Marisol pushes a 3.0-kg box 7.0 m across the floor with a force of 12 N. She then lifts the box to a shelf 1 m above the ground. How much work does Marisol do on the box?

EXAMPLE PROBLEM 2

Get help with **forces and work**. | Personal Tutor 👆

FORCE AND DISPLACEMENT AT AN ANGLE A sailor pulls a boat a distance of 30.0 m along a dock using a rope that makes a 25.0° angle with the horizontal. How much work does the rope do on the boat if its tension is 255 N?

1 ANALYZE AND SKETCH THE PROBLEM

- Identify the system and the force doing work on it.
- Establish coordinate axes.
- Sketch the situation showing the boat with initial conditions.
- Draw vectors showing the displacement, the force, and its component in the direction of the displacement.

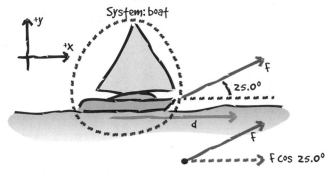

System: boat

Sailor does work on the boat.

KNOWN	UNKNOWN
$F = 255$ N $\theta = 25.0°$	$W = ?$
$d = 30.0$ m	

2 SOLVE FOR THE UNKNOWN

Use the definition of work.

$W = Fd \cos \theta$

$= (255 \text{ N})(30.0 \text{ m})(\cos 25.0°)$ ◀ Substitute $F = 255$ N, $d = 30.0$ m, $\theta = 25.0°$.

$= 6.93 \times 10^3$ J

3 EVALUATE THE ANSWER

- **Are the units correct?** Work is measured in joules.
- **Does the sign make sense?** The rope does work on the boat, which agrees with a positive sign for work.

PRACTICE PROBLEMS

Do additional problems. | Online Practice 👆

5. If the sailor in Example Problem 2 pulled with the same force and through the same displacement, but at an angle of 50.0°, how much work would be done on the boat by the rope?

6. Two people lift a heavy box a distance of 15 m. They use ropes, each of which makes an angle of 15° with the vertical. Each person exerts a force of 225 N. How much work do the ropes do?

7. An airplane passenger carries a 215-N suitcase up the stairs, a displacement of 4.20 m vertically and 4.60 m horizontally.

   a. How much work does the passenger do on the suitcase?

   b. The same passenger carries the same suitcase back down the same set of stairs. How much work does the passenger do on the suitcase to carry it down the stairs?

8. A rope is used to pull a metal box a distance of 15.0 m across the floor. The rope is held at an angle of 46.0° with the floor, and a force of 628 N is applied to the rope. How much work does the rope do on the box?

9. CHALLENGE A bicycle rider pushes a 13-kg bicycle up a steep hill. The incline is 25° and the road is 275 m long, as shown in **Figure 5**. The rider pushes the bike parallel to the road with a force of 25 N.

   a. How much work does the rider do on the bike?

   b. How much work is done by the force of gravity on the bike?

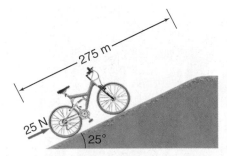

Figure 5

VOCABULARY
Science Usage v. Common Usage

Energy
• Science usage
the ability of a system to produce a change in itself or the world around it
The kinetic energy of the soccer ball decreased as it slowed down.

• Common usage
the capacity of acting or being active
The young students had a lot of energy during recess.

Watch a **BrainPOP video about kinetic energy.**

Figure 6 The bobsledders do work on the bobsled when they push it. The result is a change in the bobsled's kinetic energy.

Energy

Look again at the equation $W = \frac{1}{2}mv_f^2 - \frac{1}{2}mv_i^2$. What property of a system does $\frac{1}{2}mv^2$ describe? A massive, fast-moving vehicle can do damage to objects around it, and a baseball hit at high speed can rise high into the air. That is, a system with this property can produce a change in itself or the world around it. This ability of a system to produce a change in itself or the world around it is called **energy** and is represented by the symbol E.

The right side of the equation, $\frac{1}{2}mv_f^2 - \frac{1}{2}mv_i^2$, indicates a change in a specific kind of energy. That is, work causes a change in energy. This is the **work-energy theorem,** which states that when work is done on a system, the result is a change in the system's energy. The work-energy theorem can be represented by the following equation.

WORK-ENERGY THEOREM
Work done on a system is equal to the change in the system's energy.

$$W = \Delta E$$

Since work is measured in joules, energy must also be measured in joules. In fact, the unit gets its name from the nineteenth-century physicist James Prescott Joule, who established the relationship between work done and the change in energy. Recall that 1 joule equals 1 N·m and that 1 N equals 1 kg·m/s². Therefore, 1 joule equals 1 kg·m²/s². These are the same units $\frac{1}{2}mv^2$ has.

Through the process of doing work, energy can move between the external world and the system. The direction of energy transfer can be either way. If the external world does work on a system, then W is positive and the energy of the system increases. If a system does work on the external world, then W is negative and the energy of the system decreases. In summary, work is the transfer of energy that occurs when a force is applied through a displacement.

Changing kinetic energy So far, we have discussed the energy associated with a system's motion. For example, the bobsledders in **Figure 6** do work on their sled to get it moving at the beginning of a race. The energy associated with motion is called **kinetic energy** (KE).

In the examples we have considered, the object was changing position and its energy ($\frac{1}{2}mv^2$) was due to this motion. Energy due to changing position is called **translational kinetic energy** and can be represented by the following equation.

TRANSLATIONAL KINETIC ENERGY
A system's translational kinetic energy is equal to $\frac{1}{2}$ times the system's mass multiplied by the system's speed squared.

$$KE_{trans} = \frac{1}{2}mv^2$$

In the case of the bobsled, work resulted in a change in the object's translational kinetic energy. There are, however, many other forms of energy. Work can cause a change in these other forms, as well. Some of these forms, such as potential energy and thermal energy, will be explored in subsequent chapters.

Power

Suppose you had a stack of books to move from the floor to a shelf. You could lift the entire stack at once, or you could move the books one at a time. How would the amount of work compare between the two cases? Since the total force applied and the displacement are the same in both cases, the work is the same. However, the time needed is different. Recall that work causes a change in energy. The rate at which energy is transformed is **power.** Power is equal to the change in energy divided by the time required for the change.

POWER

Power is equal to the change in energy divided by the time required for the change.

$$P = \frac{\Delta E}{t}$$

When work causes the change in energy, power is equal to the work done divided by the time taken to do the work.

$$P = \frac{W}{t}$$

Consider the two forklifts in **Figure 7.** The left forklift raises the load in 5 seconds, and the right forklift raises the load in 10 seconds. The left forklift is more powerful than the right forklift. Even though the same work is accomplished by both, the left forklift accomplishes it in less time and thus develops more power.

Power is measured in watts (W). One **watt** is 1 J of energy transformed in 1 s. That is, 1 W = 1 J/s. A watt is a relatively small unit of power. For example, a glass of water weighs about 2 N. If you lift it 0.5 m to your mouth at a constant speed, you do 1 J of work. If you lift the glass in 1 s, you are doing work at the rate of 1 W. Because a watt is such a small unit, power often is measured in kilowatts (kW). One kilowatt is equal to 1000 W.

PhysicsLAB

STAIR CLIMBING AND POWER

What is the maximum power you can develop while climbing stairs?

Figure 7 The forklift on the left develops more power than the forklift on the right. It lifts the load at a faster rate.

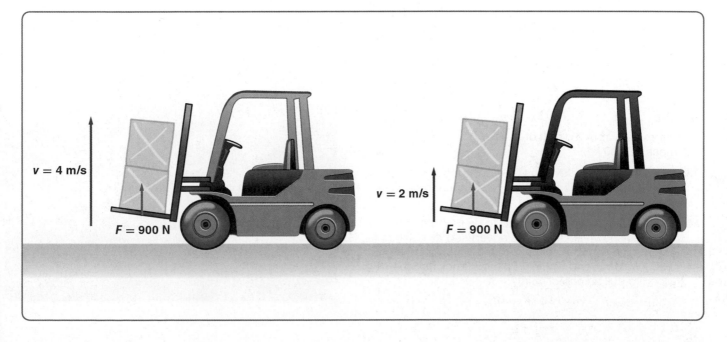

v = 4 m/s

F = 900 N

v = 2 m/s

F = 900 N

POWER An electric motor lifts an elevator 9.00 m in 15.0 s by exerting an upward force of 1.20×10^4 N. What power does the motor produce in kW?

1 ANALYZE AND SKETCH THE PROBLEM

- Sketch the situation showing the system as the elevator with its initial conditions.
- Establish a coordinate system with up as positive.
- Draw a vector diagram for the force and displacement.

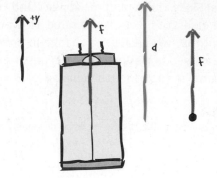

KNOWN

$d = 9.00$ m
$t = 15.0$ s
$F = 1.20 \times 10^4$ N

UNKNOWN

$P = ?$

2 SOLVE FOR THE UNKNOWN

Use the definition of power.

$$P = \frac{W}{t}$$

$$= \frac{Fd}{t} \qquad \blacktriangleleft \text{Substitute } W = Fd \cos 0° = Fd.$$

$$= \frac{(1.20 \times 10^4 \text{ N})(9.00 \text{ m})}{(15.0 \text{ s})} \qquad \blacktriangleleft \text{Substitute } F = 1.20 \times 10^4 \text{ N}, d = 9.00 \text{ m}, t = 15.0 \text{ s.}$$

$$= 7.20 \text{ kW}$$

3 EVALUATE THE ANSWER

- **Are the units correct?** Power is measured in joules per second or watts.
- **Does the sign make sense?** The positive sign agrees with the upward direction of the force.

PRACTICE PROBLEMS   Do additional problems. | Online Practice |

10. A 575-N box is lifted straight up a distance of 20.0 m by a cable attached to a motor. The box moves with a constant velocity and the job is done in 10.0 s. What power is developed by the motor in W and kW?

11. You push a wheelbarrow a distance of 60.0 m at a constant speed for 25.0 s by exerting a 145-N force horizontally.

    a. What power do you develop?

    b. If you move the wheelbarrow twice as fast, how much power is developed?

12. What power does a pump develop to lift 35 L of water per minute from a depth of 110 m? (1 L of water has a mass of 1.00 kg.)

13. An electric motor develops 65 kW of power as it lifts a loaded elevator 17.5 m in 35 s. How much force does the motor exert?

14. **CHALLENGE** A winch designed to be mounted on a truck, as shown in **Figure 8,** is advertised as being able to exert a 6.8×10^3-N force and to develop a power of 0.30 kW. How long would it take the truck and the winch to pull an object 15 m?

Figure 8

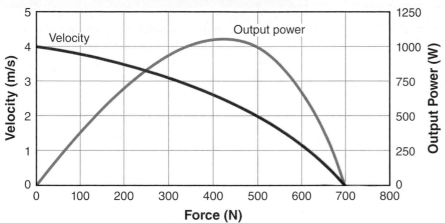

Power-Force/Velocity-Force Graph

Figure 9 When riding a multispeed bicycle, the output power depends on the force you exert and your speed.

Power and speed You might have noticed in Example Problem 3 that when the force has a component (F_x) in the same direction as the displacement, $P = \frac{F_x d}{t}$. However, because $v = \frac{d}{t}$, power also can be calculated using $P = F_x v$.

When you are riding a multispeed bicycle, how do you choose the correct gear? You want to get your body to deliver the largest amount of power. By considering the equation $P = Fv$, you can see that either zero force or zero speed results in no power delivered. The muscles cannot exert extremely large forces nor can they move very fast. Thus, some combination of moderate force and moderate speed will produce the largest amount of power. **Figure 9** shows that in this particular situation, the maximum power output is over 1000 W when the force is about 400 N and speed is about 2.6 m/s.

All engines—not just humans—have these limitations. Simple machines often are designed to match the force and the speed that the engine can deliver to the needs of the job. You will learn more about simple machines in the next section.

REAL-WORLD PHYSICS

TOUR DE FRANCE A bicyclist in the Tour de France rides at an average speed of about 8.94 m/s for more than 6 h a day. The power output of the racer is about 1 kW. One-fourth of that power goes into moving the bike against the resistance of the air, gears, and tires. Three-fourths of the power is used to cool the racer's body.

SECTION 1 REVIEW

Section Self-Check Check your understanding.

15. **MAIN IDEA** If the work done on an object doubles its kinetic energy, does it double its speed? If not, by what ratio does it change the speed?

16. **Work** Murimi pushes a 20-kg mass 10 m across a floor with a horizontal force of 80 N. Calculate the amount of work done by Murimi on the mass.

17. **Work** Suppose you are pushing a stalled car. As the car gets going, you need less and less force to keep it going. For the first 15 m, your force decreases at a constant rate from 210.0 N to 40.0 N. How much work did you do on the car? Draw a force-displacement graph to represent the work done during this period.

18. **Work** A mover loads a 185-kg refrigerator into a moving van by pushing it at a constant speed up a 10.0-m, friction-free ramp at an angle of inclination of 11°. How much work is done by the mover on the refrigerator?

19. **Work** A 0.180-kg ball falls 2.5 m. How much work does the force of gravity do on the ball?

20. **Work and Power** Does the work required to lift a book to a high shelf depend on how fast you raise it? Does the power required to lift the book depend on how fast you raise it? Explain.

21. **Power** An elevator lifts a total mass of 1.1×10^3 kg a distance of 40.0 m in 12.5 s. How much power does the elevator deliver?

22. **Mass** A forklift raises a box 1.2 m and does 7.0 kJ of work on it. What is the mass of the box?

23. **Work** You and a friend each carry identical boxes from the first floor of a building to a room located on the second floor, farther down the hall. You choose to carry the box first up the stairs, and then down the hall to the room. Your friend carries it down the hall on the first floor, then up a different stairwell to the second floor. How do the amounts of work done by the two of you on your boxes compare?

24. **Critical Thinking** Explain how to find the change in energy of a system if three agents exert forces on the system at once.

Machines

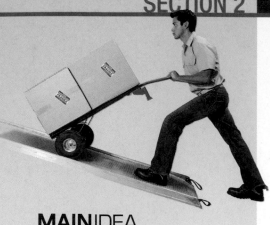

When you think of the word *machine,* you might think of vacuum cleaners, computers, or industrial equipment. However, ramps, screws, and crowbars are also considered machines.

MAINIDEA

Machines make tasks easier by changing the magnitude or the direction of the force exerted.

Essential Questions

- What is a machine, and how does it make tasks easier?

- How are mechanical advantage, the effort force, and the resistance force related?

- What is a machine's ideal mechanical advantage?

- What does the term efficiency mean?

Review Vocabulary

work a force applied through a distance

New Vocabulary

machine
effort force
resistance force
mechanical advantage
ideal mechanical advantage
efficiency
compound machine

Benefits of Machines

Machines, whether powered by engines or people, make tasks easier. A **machine** is a device that makes tasks easier by changing either the magnitude or the direction of the applied force. Consider the bottle opener in **Figure 10.** When you lift the handle, you do work on the opener. The opener lifts the cap, doing work on it. The work you do is called the input work (W_i). The work the machine does is called the output work (W_o).

Recall that work transfers energy. When you do work on the bottle opener, you transfer energy to the opener. The opener does work on the cap and transfers energy to it. The opener is not a source of energy. So, the cap cannot receive more energy than what you put into the opener. Thus, the output work can never be greater than the input work. The machine only aids in the transfer of energy from you to the bottle cap.

Mechanical advantage The force exerted by a user on a machine is called the **effort force** (F_e). The force exerted by the machine is called the **resistance force** (F_r). For the bottle opener in **Figure 10,** F_e is the upward force exerted by the person using the bottle opener and F_r is the upward force exerted by the bottle opener. The ratio of resistance force to effort force $\left(\dfrac{F_r}{F_e}\right)$ is called the **mechanical advantage** (*MA*) of the machine.

MECHANICAL ADVANTAGE
The mechanical advantage of a machine is equal to the resistance force divided by the effort force.

$$MA = \frac{F_r}{F_e}$$

Figure 10 This bottle opener makes opening a bottle easier. However, it does not lessen the energy required to do so.

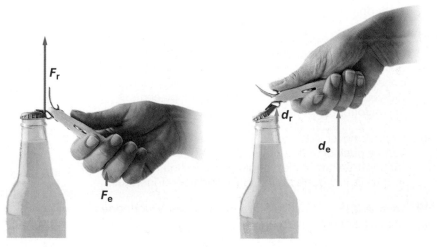

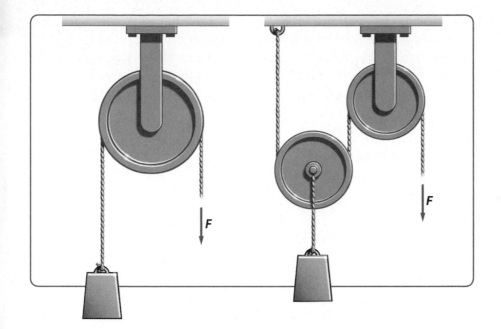

Figure 11 A fixed pulley has a mechanical advantage equal to 1 but is useful because it changes the direction of the force. A pulley system with a movable pulley has a mechanical advantage greater than 1, making the force applied to the object larger than the force originally exerted.

Watch a **BrainPOP video about pulleys.**

Video

For a fixed pulley, such as the one shown on the left in **Figure 11,** the effort force (F_e) and the resistance force (F_r) are equal. Thus, *MA* is 1. What is the advantage of this machine? The fixed pulley is useful, not because the effort force is lessened, but because the direction of the effort force is changed. Many machines, such as the bottle opener shown in **Figure 10** and the pulley system shown on the right in **Figure 11,** have a mechanical advantage greater than 1. When the mechanical advantage is greater than 1, the machine increases the force applied by a person.

☑ **READING CHECK** **Calculate** A machine has a mechanical advantage of 3. If the input force is 2 N what is the output force?

Ideal mechanical advantage A machine can increase force, but it cannot increase energy. An ideal machine transfers all the energy, so the output work equals the input work: $W_o = W_i$. The input work is the product of the effort force and the displacement the effort force acts through: $W_i = F_e d_e$. The output work is the product of the resistance force and the displacement the resistance force acts through: $W_o = F_r d_r$. Substituting these expressions into $W_o = W_i$ gives $F_r d_r = F_e d_e$. This equation can be rewritten $\frac{F_r}{F_e} = \frac{d_e}{d_r}$.

Recall that mechanical advantage is given by $MA = \frac{F_r}{F_e}$. Thus, for an ideal machine, **ideal mechanical advantage** (*IMA*) is equal to the displacement of the effort force divided by the displacement of the resistance force. *IMA* can be represented as follows.

IDEAL MECHANICAL ADVANTAGE
The ideal mechanical advantage of a machine is equal to the displacement of the effort force divided by the displacement of the resistance force.

$$IMA = \frac{d_e}{d_r}$$

Notice that you measure the displacements to calculate the ideal mechanical advantage, but you measure the forces exerted to find the actual mechanical advantage.

PhysicsLAB

LIFTING WITH PULLEYS
How does the arrangement of a pulley system affect its ideal mechanical advantage.

View an **animation on the benefits of machines.**

Concepts In Motion

Efficiency In a real machine, not all of the input work is available as output work. Energy removed from the system through heat or sound means that there is less output work from the machine. Consequently, the machine is less efficient at accomplishing the task. The **efficiency** of a machine (*e*) is defined as the ratio of output work to input work.

EFFICIENCY
The efficiency of a machine (in %) is equal to the output work, divided by the input work, multiplied by 100.

$$e = \frac{W_o}{W_i} \times 100$$

An ideal machine has equal output and input work, $\frac{W_o}{W_i} = 1$, and its efficiency is 100 percent. All real machines have efficiencies of less than 100 percent.

Efficiency can be expressed in terms of the mechanical advantage and ideal mechanical advantage. Efficiency, $e = \frac{W_o}{W_i}$, can be rewritten as follows:

$$e = \frac{W_o}{W_i} = \frac{F_r d_r}{F_e d_e}$$

Because $MA = \frac{F_r}{F_e}$ and $IMA = \frac{d_e}{d_r}$, the following expression can be written for efficiency.

EFFICIENCY
The efficiency of a machine (in %) is equal to its mechanical advantage, divided by the ideal mechanical advantage, multiplied by 100.

$$e = \left(\frac{MA}{IMA}\right) \times 100$$

A machine's design determines its ideal mechanical advantage. An efficient machine has an *MA* almost equal to its *IMA*. A less-efficient machine has a small *MA* relative to its *IMA*. To obtain the same resistance force, a greater force must be exerted in a machine of lower efficiency than in a machine of higher efficiency.

☑ **READING CHECK Determine** If a machine has an efficiency of 50 percent and an *MA* of 3, what is its *IMA*?

PHYSICS CHALLENGE

An electric pump pulls water at a rate of 0.25 m³/s from a well that is 25 m deep. The water leaves the pump at a speed of 8.5 m/s.

1. What power is needed to lift the water to the surface?

2. What power is needed to increase the water's kinetic energy?

3. If the pump's efficiency is 80 percent, how much power must be delivered to the pump?

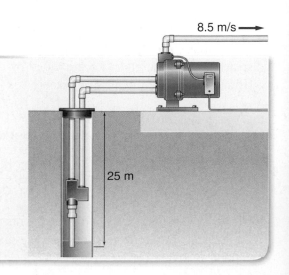

8.5 m/s →

25 m

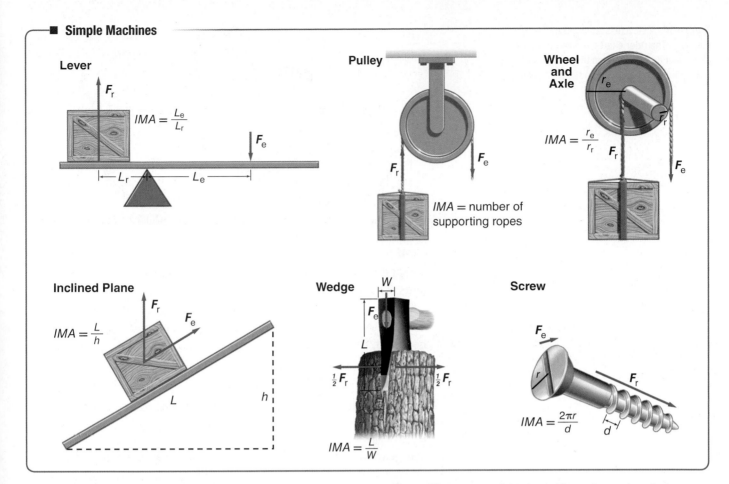

■ **Simple Machines**

Lever

$$IMA = \frac{L_e}{L_r}$$

Pulley

IMA = number of supporting ropes

Wheel and Axle

$$IMA = \frac{r_e}{r_r}$$

Inclined Plane

$$IMA = \frac{L}{h}$$

Wedge

$$IMA = \frac{L}{W}$$

Screw

$$IMA = \frac{2\pi r}{d}$$

Compound Machines

Most machines, no matter how complex, are combinations of one or more of the six simple machines: the lever, pulley, wheel and axle, inclined plane, wedge, and screw. These machines are shown in **Figure 12.**

The *IMA* of each machine shown in **Figure 12** is the ratio of the displacement of the effort force to the displacement of the resistance force. For machines such as the lever and the wheel and axle, this ratio can be replaced by the ratio of the displacements between the place where the force is applied and the pivot point. A common version of the wheel and axle is a steering wheel, such as the one shown in **Figure 13.** The *IMA* is the ratio of the radii of the wheel and axle.

A machine consisting of two or more simple machines linked in such a way that the resistance force of one machine becomes the effort force of the second is called a **compound machine.** Some examples of compound machines are scissors (wedges and levers) and wheelbarrows (lever and wheel and axle).

☑ **READING CHECK Compare and contrast** simple machines and compound machines.

Figure 12 Each type of simple machine makes work easier by changing the direction or magnitude of the force. Notice that the *IMA* of each machine is related to the properties of the machine. For example, the *IMA* of a lever is equal to the length of the effort arm (L_e) divided by the length of the resistance arm (L_r). Similarly, the *IMA* of a wedge is equal to its length (L) divided by its width (W).

Identify an everyday example of each simple machine.

Figure 13 A steering wheel is an example of a wheel and axle. Its *IMA* is $\frac{r_e}{r_r}$.

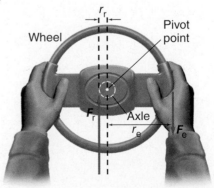

Figure 14 A bicycle is a compound machine. The rider exerts a force on the pedals, which exert a force on the chain. The chain then exerts a force on the rear gear.

$F_{\text{chain on gear}}$

$F_{\text{gear on chain}}$

$F_{\text{rider on road}}$

$F_{\text{rider on pedal}}$

MiniLAB

WHEEL AND AXLE

The gear mechanism on your bicycle multiplies the distance that you travel, but what does it do to the force you exert?

Mechanical advantage and bicycles A bicycle, such as the one shown in **Figure 14**, is a compound machine, consisting of two wheel-and-axle systems. The first is the pedal and front gear. Here, the effort force is the force that the rider exerts on the pedal, $F_{\text{rider on pedal}}$. The force of your foot is most effective when the force is exerted perpendicular to the arm of the pedal; that is, when the torque is largest. Therefore, we will assume that $F_{\text{rider on pedal}}$ is applied perpendicular to the pedal arm. The resistance is the force that the front gear exerts on the chain, $F_{\text{gear on chain}}$.

✓ **READING CHECK** **Identify** the effort force and the resistance force for the pedal and front gear.

The rear gear and the rear wheel act like another wheel and axle. The chain exerts an effort force on the rear gear, $F_{\text{chain on gear}}$. This force is equal to the force of the front gear on the chain. That is $F_{\text{gear on chain}} = F_{\text{chain on gear}}$. The resistance force is the force that the wheel exerts on the road, $F_{\text{wheel on road}}$.

The *MA* of a compound machine is the product of the *MA*s of the simple machines from which it is made. Therefore, the bicycle's mechanical advantage is given by the following equation:

$$MA_{\text{total}} = MA_{\text{pedal gear}} \times MA_{\text{rear wheel}}$$

$$MA_{\text{total}} = \left(\frac{F_{\text{gear on chain}}}{F_{\text{rider on pedal}}}\right)\left(\frac{F_{\text{wheel on road}}}{F_{\text{chain on gear}}}\right) = \frac{F_{\text{wheel on road}}}{F_{\text{rider on pedal}}}$$

Recall that $F_{\text{gear on chain}} = F_{\text{chain on gear}}$. Therefore, they cancel in the equation shown above.

Similarly, the *IMA* of the bicycle is $IMA = IMA_{\text{pedal gear}} \times IMA_{\text{rear wheel}}$. For each wheel and axle, the IMA is the ratio of the wheel to the axle.

For the pedal gear, $IMA = \dfrac{\text{pedal radius}}{\text{front gear radius}}$

For the rear wheel, $IMA = \dfrac{\text{rear gear radius}}{\text{wheel radius}}$
For the bicycle, then,

$$IMA = \left(\frac{\text{pedal radius}}{\text{front gear radius}}\right)\left(\frac{\text{rear gear radius}}{\text{wheel radius}}\right)$$

$$= \left(\frac{\text{rear gear radius}}{\text{front gear radius}}\right)\left(\frac{\text{pedal radius}}{\text{wheel radius}}\right)$$

✓ **READING CHECK** **Explain** What are the units of *MA* and *IMA* for the bicycle?

MECHANICAL ADVANTAGE You examine the rear wheel on your bicycle. It has a radius of 35.6 cm and has a gear with a radius of 4.00 cm. When the chain is pulled with a force of 155 N, the wheel rim moves 14.0 cm. The efficiency of this part of the bicycle is 95.0 percent.

a. What is the *IMA* of the wheel and gear?

b. What is the *MA* of the wheel and gear?

c. What is the resistance force?

d. How far was the chain pulled to move the rim 14.0 cm?

1 ANALYZE AND SKETCH THE PROBLEM
- Sketch the wheel and axle.
- Sketch the force vectors.

KNOWN		UNKNOWN	
$r_e = 4.00$ cm	$e = 95.0\%$	$IMA = ?$	$F_r = ?$
$r_r = 35.6$ cm	$d_r = 14.0$ cm	$MA = ?$	$d_e = ?$
$F_e = 155$ N			

2 SOLVE FOR THE UNKNOWN

a. Solve for *IMA*.

$$IMA = \frac{r_e}{r_r}$$

◀ For a wheel-and-axle machine, *IMA* is equal to the ratio of radii.

$$= \frac{4.00 \text{ cm}}{35.6 \text{ cm}} = 0.112$$

◀ Substitute $r_e = 4.00$ cm, $r_r = 35.6$ cm.

b. Solve for *MA*.

$$e = \frac{MA}{IMA} \times 100$$

$$MA = \left(\frac{e}{100}\right) \times IMA$$

$$= \left(\frac{95.0}{100}\right) \times 0.112 = 0.106$$

◀ Substitute $e = 95.0\%$, $IMA = 0.112$.

c. Solve for force.

$$MA = \frac{F_r}{F_e}$$

$$F_r = (MA)(F_e)$$

$$= (0.106)(155 \text{ N}) = 16.4 \text{ N}$$

◀ Substitute $MA = 0.106$, $F_e = 155$ N.

d. Solve for distance.

$$IMA = \frac{d_e}{d_r}$$

$$d_e = (IMA)(d_r)$$

$$= (0.112)(14.0 \text{ cm}) = 1.57 \text{ cm}$$

◀ Substitute $IMA = 0.112$, $d_r = 14.0$ cm.

3 EVALUATE THE ANSWER
- **Are the units correct?** Force is measured in newtons, and distance in centimeters.
- **Is the magnitude realistic?** *IMA* is low for a bicycle because a greater F_e is traded for a greater d_r. *MA* is always smaller than *IMA*. Because *MA* is low, F_r also will be low. The small distance the axle moves results in a large distance covered by the wheel. Thus, d_e should be very small.

25. If the gear radius of the bicycle in Example Problem 4 is doubled while the force exerted on the chain and the distance the wheel rim moves remain the same, what quantities change, and by how much?

26. A sledgehammer is used to drive a wedge into a log to split it. When the wedge is driven 0.20 m into the log, the log is separated a distance of 5.0 cm. A force of 1.7×10^4 N is needed to split the log, and the sledgehammer exerts a force of 1.1×10^4 N.

   a. What is the *IMA* of the wedge?

   b. What is the *MA* of the wedge?

   c. Calculate the efficiency of the wedge as a machine.

27. A worker uses a pulley to raise a 24.0-kg carton 16.5 m, as shown in **Figure 15**. A force of 129 N is exerted, and the rope is pulled 33.0 m.

   a. What is the *MA* of the pulley?

   b. What is the efficiency of the pulley?

28. A winch has a crank with a 45-cm radius. A rope is wrapped around a drum with a 7.5-cm radius. One revolution of the crank turns the drum one revolution.

   a. What is the ideal mechanical advantage of this machine?

   b. If, due to friction, the machine is only 75 percent efficient, how much force would have to be exerted on the handle of the crank to exert 750 N of force on the rope?

29. **CHALLENGE** You exert a force of 225 N on a lever to raise a 1.25×10^3-N rock a distance of 13 cm. If the efficiency of the lever is 88.7 percent, how far did you move your end of the lever?

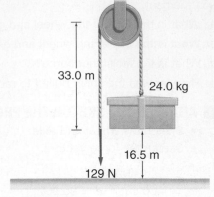

33.0 m

24.0 kg

16.5 m

129 N

Figure 15

Multi-gear bicycle Shifting gears on a bicycle is a way of adjusting the ratio of gear radii to obtain the desired *IMA*. On a multi-gear bicycle, the rider can change the *IMA* of the machine by choosing the size of one or both gears. **Figure 16** shows a rear gear with five different gear sizes. When accelerating or climbing a hill, the rider increases the ideal mechanical advantage to increase the force that the wheel exerts on the road.

To increase the *IMA*, the rider needs to make the rear gear radius large compared to the front gear radius (refer to the *IMA* equation earlier in the section). For the same force exerted by the rider, a larger force is exerted by the wheel on the road. However, the rider must rotate the pedals through more turns for each revolution of the wheel.

On the other hand, less force is needed to ride the bicycle at high speed on a level road. The rider needs a small rear gear and a large front gear, resulting in a smaller *IMA*. Thus, for the same force exerted by the rider, a smaller force is exerted by the wheel on the road. However, in return, the rider does not have to move the pedals as far for each revolution of the wheel.

An automobile transmission works in the same way. To accelerate a car from rest, large forces are needed and the transmission increases the *IMA* by increasing the gear ratio. At high speeds, however, the transmission reduces the gear ratio and the *IMA* because smaller forces are needed. Even though the speedometer shows a high speed, the tachometer indicates the engine's low angular speed.

☑ **READING CHECK** Explain why your car needs multiple gears.

Figure 16 A rider can change the *IMA* of the bicycle by shifting gears.

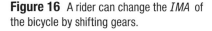

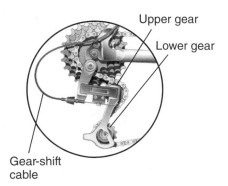

Upper gear

Lower gear

Gear-shift cable

The Human Walking Machine

▶ **CONNECTION TO BIOLOGY** Movement of the human body is explained by the same principles of force and work that describe all motion. Simple machines, in the form of levers, give humans the ability to walk, run, and perform many other activities. The lever systems of the human body are complex. However, each system has the following four basic parts:

1. a rigid bar (bone)
2. a source of force (muscle contraction)
3. a fulcrum or pivot (movable joints between bones)
4. a resistance (the weight of the body or an object being lifted or moved)

Figure 17 shows these parts in the lever system in a human leg.

Lever systems of the body are not very efficient, and mechanical advantages are low. This is why walking and jogging require energy (burn calories) and help people lose weight.

When a person walks, the hip acts as a fulcrum and moves through the arc of a circle, centered on the foot. The center of mass of the body moves as a resistance around the fulcrum in the same arc. The length of the radius of the circle is the length of the lever formed by the bones of the leg. Athletes in walking races increase their velocity by swinging their hips upward to increase this radius.

A tall person's body has lever systems with less mechanical advantage than a short person's does. Although tall people usually can walk faster than short people can, a tall person must apply a greater force to move the longer lever formed by the leg bones. How would a tall person do in a walking race? What are the factors that affect a tall person's performance? Walking races are usually 20 or 50 km long. Because of the inefficiency of their lever systems and the length of a walking race, very tall people rarely have the stamina to win.

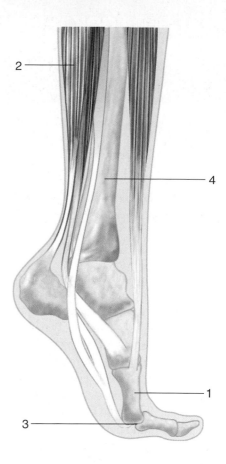

Figure 17 The human leg and foot function as a compound machine.

SECTION 2 REVIEW

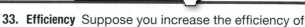
Section Self-Check — Check your understanding.

30. **MAIN IDEA** Classify each tool as a lever, a wheel and axle, an inclined plane, or a wedge. Describe how it changes the force to make the task easier.

 a. screwdriver   c. chisel
 b. pliers   d. nail puller

31. **IMA** A worker is testing a multiple pulley system to estimate the heaviest object that he could lift. The largest downward force he can exert is equal to his weight, 875 N. When the worker moves the rope 1.5 m, the object moves 0.25 m. What is the heaviest object that he could lift?

32. **Compound Machines** A winch has a crank on a 45-cm arm that turns a drum with a 7.5-cm radius through a set of gears. It takes three revolutions of the crank to rotate the drum through one revolution. What is the *IMA* of this compound machine?

33. **Efficiency** Suppose you increase the efficiency of a simple machine. Do the *MA* and *IMA* increase, decrease, or remain the same?

34. **Critical Thinking** The mechanical advantage of a multi-gear bicycle is changed by moving the chain to a suitable rear gear.

 a. To start out, you must accelerate the bicycle, so you want to have the bicycle exert the greatest possible force. Should you choose a small or large gear?

 b. As you reach your traveling speed, you want to rotate the pedals as few times as possible. Should you choose a small or large gear?

 c. Many bicycles also let you choose the size of the front gear. If you want even more force to accelerate while climbing a hill, would you move to a larger or smaller front gear?

*dis*ADVANTAGE?

Prosthetic Limbs In the past, losing both legs from the knee down might mean that an amputee could never again walk, let alone run. However, with modern running prosthetics—artificial legs with curved carbon-fiber blades—amputees can compete in world-class racing events.

Energy Runners with natural legs use their thighs, knees, calves, and ankles to absorb energy as each foot hits the ground and pushes off, as in **Figure 1.** With a prosthetic leg, some of the kinetic energy of the swinging blade is stored as the blade compresses when it hits the ground. The compressed blade acts like a spring, transforming the stored energy back into kinetic energy and pushing the runner forward with each stride, as shown in **Figure 2.**

Is running with prosthetics easier?
Many athletes think that prosthetics might even give amputees, who can wear technology that has been engineered to maximize their performance, an advantage over runners who have their natural legs. Scientists test prosthetics to determine how their performance measures up against natural legs and feet.

When it comes to running, prosthetic legs have disadvantages as well. At the starting block, runners with prosthetic legs lose more time launching into a sprint than do runners with natural legs. Furthermore, there is evidence that runners with prosthetics experience a disadvantage in force production, which could reduce their running speed.

FIGURE 1 The energy transformations for a natural leg are less efficient than those for a prosthetic leg.

FIGURE 2 When the foot of the prosthetic hits the ground, it compresses, storing energy from the impact.

GOING**FURTHER** >>>

Research the controversy surrounding athletic events in which some participants wear prosthetics. Write an opinion piece about the fairness of races in which runners that have both natural legs compete with double and single amputees.

STUDY GUIDE

BIGIDEA   Doing work on a system changes the system's energy.

VOCABULARY

- **work** (p. 264)
- **joule** (p. 264)
- **energy** (p. 270)
- **work-energy theorem** (p. 270)
- **kinetic energy** (p. 270)
- **translational kinetic energy** (p. 270)
- **power** (p. 271)
- **watt** (p. 271)

SECTION 1 **Work and Energy**

MAINIDEA   Work is the transfer of energy that occurs when a force is applied through a displacement.

- Work is done when a force is applied through a displacement. Work is the product of the force exerted on a system and the component of the distance through which the system moves that is parallel to the force.

$$W = Fd \cos \theta$$

The work done can be determined by calculating the area under a force-displacement graph.

- Energy is the ability of a system to produce a change in itself or its environment. A moving object has kinetic energy. Objects that are changing position have translational kinetic energy.

$$KE_{trans} = \frac{1}{2}mv^2$$

- The work done on a system is equal to the change in energy of the system. This is called the work-energy theorem.

$$W = \Delta E$$

- Power is the rate at which energy is transformed. When work causes the change in energy, power is equal to the rate of work done.

$$P = \frac{\Delta E}{t} = \frac{W}{t}$$

VOCABULARY

- **machine** (p. 274)
- **effort force** (p. 274)
- **resistance force** (p. 274)
- **mechanical advantage** (p. 274)
- **ideal mechanical advantage** (p. 275)
- **efficiency** (p. 276)
- **compound machine** (p. 277)

SECTION 2 **Machines**

MAINIDEA   Machines make tasks easier by changing the magnitude or the direction of the force exerted.

- Machines, whether powered by engines or humans, do not change the amount of work done, but they do make the task easier by changing the magnitude or direction of the effort force.

- The mechanical advantage (*MA*) is the ratio of resistance force to effort force.

$$MA = \frac{F_r}{F_e}$$

- The ideal mechanical advantage (IMA) is the ratio of the distances moved.

$$IMA = \frac{d_e}{d_r}$$

- The efficiency of a machine is the ratio of output work to input work.

$$e = \left(\frac{W_o}{W_i}\right) \times 100$$

The efficiency of a machine can be found from the real and ideal mechanical advantages. In all real machines, *MA* is less than *IMA*, and *e* is less than 100 percent.

$$e = \left(\frac{MA}{IMA}\right) \times 100$$

Games and Multilingual eGlossary

Vocabulary Practice

SECTION 1 **Work and Energy**

Mastering Concepts

35. In what units is work measured?

36. A satellite orbits Earth in a circular orbit. Does Earth's gravity do work on the satellite? Explain.

37. An object slides at constant speed on a frictionless surface. What forces act on the object? What is the work done by each force on the object?

38. Define *work* and *power*.

39. What is a watt equivalent to in terms of kilograms, meters, and seconds?

Mastering Problems

40. The third floor of a house is 8 m above street level. How much work must a pulley system do to lift a 150-kg oven at a constant speed to the third floor?

41. Haloke does 176 J of work lifting himself 0.300 m at a constant speed. What is Haloke's mass?

42. Tug-of-War During a tug-of-war, team A does 2.20×10^5 J of work in pulling team B 8.00 m. What average force did team A exert?

43. To travel at a constant velocity, a car exerts a 551-N force to balance air resistance. How much work does the car do on the air as it travels 161 km from Columbus to Cincinnati?

44. Cycling A cyclist exerts a 15.0-N force while riding 251 m in 30.0 s. What power does the cyclist develop?

45. A student librarian lifts a 2.2-kg book from the floor to a height of 1.25 m. He carries the book 8.0 m to the stacks and places the book on a shelf that is 0.35 m above the floor. How much work does he do on the book?

46. A horizontal force of 300.0 N is used to push a 145-kg mass 30.0 m horizontally in 3.00 s.

   a. Calculate the work done on the mass.

   b. Calculate the power developed.

47. Wagon A wagon is pulled by a force of 38.0 N exerted on the handle at an angle of 42.0° with the horizontal. If the wagon is pulled for 157 m, how much work is done on the wagon?

48. Lawn Mower To mow the yard, Shani pushes a lawn mower 1.2 km with a horizontal force of 66.0 N. Does all of the applied force do work on the mower, and how much work does Shani do on the mower?

49. A 17.0-kg crate is to be pulled a distance of 20.0 m, requiring 1210 J of work to be done on the crate. The job is done by attaching a rope and pulling with a force of 75.0 N. At what angle is the rope held?

50. Lawn Tractor The lawn tractor in **Figure 18** goes up a hill at a constant velocity in 2.5 s. Calculate the power that is developed by the tractor.

Figure 18

51. You slide a crate up a ramp at an angle of 30.0° to a vertical height of 1.15 m. You exert a 225-N force parallel to the ramp, and the crate moves at a constant speed. The coefficient of friction is 0.28. How much work do you do on the crate?

52. A 4.2×10^3-N piano is wheeled up a 3.5-m ramp at a constant speed. The ramp makes an angle of 30.0° with the horizontal. Find the work done by a man wheeling the piano up the ramp.

53. Sled Diego pulls a sled across level snow as shown in **Figure 19**. If the sled moves a distance of 65.3 m, how much work does Diego do on the sled?

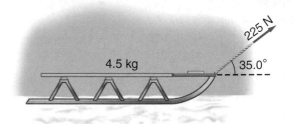

Figure 19

54. Escalator Sau-Lan's mass is 52 kg. She rides up the escalator at Ocean Park in Hong Kong. This is the world's longest escalator, with a length of 227 m and an average inclination of 31°.

   a. How much work does the escalator do on Sau-Lan?

   b. Sue's mass is 65 kg and she rides the escalator, too. How much work does the escalator do on Sue?

55. Lawn Roller A lawn roller is pushed across a lawn by a force of 115 N along the direction of the handle, which is 22.5° above the horizontal. If 64.6 W of power is developed for 90.0 s, what distance is the roller pushed?

56. Boat Engine An engine moves a boat through the water at a constant speed of 15 m/s. The engine must exert a force of 6.0 kN to balance the force that the water exerts against the hull. What power does the engine develop?

57. Maricruz slides a crate up an inclined ramp that is attached to a platform as shown in **Figure 20.** A 400.0-N force, parallel to the ramp, is needed to slide the crate up the ramp at a constant speed.

a. How much work does Maricruz do in sliding the crate up the ramp?

b. How much work would be done on the crate if Maricruz simply lifted the crate straight up from the floor to the platform at a constant speed?

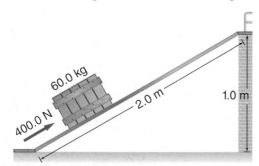

Figure 20

58. A worker pushes a 93-N crate up an inclined plane at a constant speed. As shown in **Figure 21,** the worker pushes parallel to the ground with a force of 85 N.

a. How much work does the worker do on the crate?

b. How much work is done by gravity on the crate? (Be careful with the signs you use.)

c. The coefficient of friction is $\mu = 0.20$. How much energy is transformed by friction? (Be careful with the signs you use.)

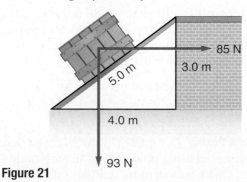

Figure 21

59. In **Figure 22,** the magnitude of the force necessary to stretch a spring is plotted against the distance the spring is stretched.

a. Calculate the slope of the graph (k), and show that $F = kd$, where $k = 25$ N/m.

b. Use the graph to find the work done in stretching the spring from 0.00 m to 0.20 m.

c. Show that the answer to part **b** can be calculated using the formula $W = \left(\frac{1}{2}\right)kd^2$, where W is the work, $k = 25$ N/m (the slope of the graph), and d is the distance the spring is stretched (0.20 m).

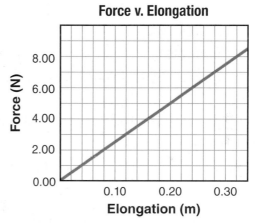

Figure 22

60. Use the graph in **Figure 22** to find the required work to stretch the spring from 0.12 m to 0.28 m.

61. The graph in **Figure 23** shows the force exerted on and displacement of an object being pulled.

a. Find the work done to pull the object 7.0 m.

b. Calculate the power that would be developed if the work was done in 2.0 s.

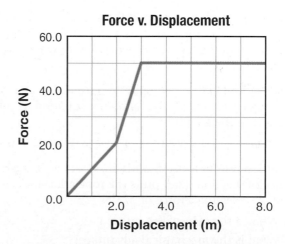

Figure 23

62. Oil Pump In 35.0 s, a pump delivers 0.550 m³ of oil into barrels on a platform 25.0 m above the intake pipe. The oil's density is 0.820 g/cm³.

   a. Calculate the work done by the pump on the oil.

   b. Calculate the power produced by the pump.

63. Conveyor Belt A 12.0-m-long conveyor belt, inclined at 30.0°, is used to transport bundles of newspapers from the mail room up to the cargo bay to be loaded onto delivery trucks. The mass of a newspaper is 1.0 kg, and each bundle has 25 newspapers. Find the power the conveyor develops if it delivers 15 bundles per minute.

SECTION 2 Machines

Mastering Concepts

64. Is it possible to get more work out of a machine than you put into it?

65. Explain how bicycle pedals are a simple machine.

Mastering Problems

66. Piano Takeshi raises a 1200-N piano a distance of 5.00 m using a set of pulleys. He pulls in 20.0 m of rope.

   a. How much effort force would Takeshi apply if this were an ideal machine?

   b. What force is used to balance the friction force if the actual effort is 340 N?

   c. What is the output work?

   d. What is the input work?

   e. What is the mechanical advantage?

67. Because there is very little friction, the lever is an extremely efficient simple machine. Using a 90.0-percent-efficient lever, what input work is needed to lift an 18.0-kg mass a distance of 0.50 m?

68. A student exerts a force of 250 N on a lever through a distance of 1.6 m as he lifts a 150-kg crate. If the efficiency of the lever is 90.0 percent, how far is the crate lifted?

69. Reverse Problem Write a physics problem with real-life objects for which the following equation would be part of the solution:

$$(12.5 \text{ N})d = \frac{1}{2}(6.0 \text{ kg})(1.10 \text{ m/s})^2 - \frac{1}{2}(6.0 \text{ kg})(0.05 \text{ m/s})^2$$

70. A pulley system is used to lift a 1345-N weight a distance of 0.975 m. Paul pulls the rope a distance of 3.90 m, exerting a force of 375 N.

   a. What is the IMA of the system?

   b. What is the mechanical advantage?

   c. How efficient is the system?

71. A force of 1.4 N is exerted on a rope in a pulley system. The force is exerted through a distance of 40.0 cm, lifting a 0.50-kg mass 10.0 cm. Calculate the following:

   a. the MA

   b. the IMA

   c. the efficiency

72. Use **Figure 24** to answer the following questions.

   a. What force, parallel to the ramp (F_A), is required to slide a 25-kg box at constant speed to the top of the ramp? Ignore friction.

   b. What is the IMA of the ramp?

   c. What are the actual MA and the efficiency of the ramp if a 75-N parallel force is needed?

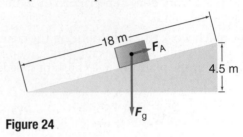

Figure 24

73. Bicycle Luisa pedals a bicycle with the wheel shown in **Figure 25**. If the wheel revolves once, what is the length of the chain that was used?

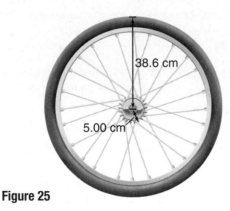

Figure 25

74. A motor with an efficiency of 88 percent runs a crane with an efficiency of 42 percent. The power supplied to the motor is 5.5 kW. At what constant speed does the crane lift a 410-kg crate?

75. What work is required to lift a 215-kg mass a distance of 5.65 m, using a machine that is 72.5 percent efficient?

76. Problem Posing Complete this problem so that it can be solved using power: "While rearranging furniture, Opa needs to move a 50-kg sofa …"

77. A compound machine is made by attaching a lever to a pulley system. Consider an ideal compound machine consisting of a lever with an *IMA* of 3.0 and a pulley system with an *IMA* of 2.0.

   a. Show that the compound machine's *IMA* is 6.0.

   b. If the compound machine is 60.0 percent efficient, how much effort must be applied to the lever to lift a 540-N box?

   c. If you move the effort side of the lever 12.0 cm, how far is the box lifted?

Applying Concepts

78. Which requires more work—carrying a 420-N backpack up a 200-m-high hill or carrying a 210-N backpack up a 400-m-high hill? Why?

79. Lifting You slowly lift a box of books from the floor and put it on a table. Earth's gravity exerts a force, magnitude *mg*, downward, and you exert a force, magnitude *mg*, upward. The two forces have equal magnitudes and opposite directions. It appears that no work is done, but you know that you did work. Explain what work was done.

80. You have an after-school job carrying cartons of new copy paper up a flight of stairs and then carrying recycled paper back down the stairs. The mass of the paper is the same in both cases. Your physics teacher says that you did no work, so you should not be paid. In what sense is the physics teacher correct? What arrangement of payments might you make to ensure that you are properly compensated?

81. Once downstairs, you carry the cartons of paper along a 15-m-long hallway. Are you doing work by carrying the boxes down the hall? Explain.

82. Climbing Stairs Two people of the same mass climb the same flight of stairs. The first person climbs the stairs in 25 s; the second person does so in 35 s.

   a. Which person does more work? Explain.

   b. Which person produces more power? Explain.

83. Show that power can be written as $P = Fv \cos \theta$.

84. How can you increase the ideal mechanical advantage of a machine?

85. BIGIDEA Orbits Explain why a planet orbiting the Sun does not violate the work-energy theorem.

86. Claw Hammer A standard claw hammer is used to pull a nail from a piece of wood. Where should you place your hand on the handle and where should the nail be located in the claw to make the effort force as small as possible?

87. Wedge How can you increase the mechanical advantage of a wedge without changing its ideal mechanical advantage?

Mixed Review

88. Ramps Isra has to get a piano onto a 2.0-m-high platform. She can use a 3.0-m-long frictionless ramp or a 4.0-m-long frictionless ramp. Which ramp should Isra use if she wants to do the least amount of work?

89. Brutus, a champion weight lifter, raises 240 kg of weights a distance of 2.35 m at a constant speed.

   a. Find the work Brutus does on the weights.

   b. How much work is done by Brutus on the weights holding the weights above his head?

   c. How much work is done by Brutus on the weights lowering them back to the ground?

   d. Does Brutus do work if he lets go of the weights and they fall back to the ground?

90. A 805-N horizontal force is needed to drag a crate across a horizontal floor at a constant speed. You drag the crate using a rope held at a 32° angle.

   a. What force do you exert on the rope?

   b. How much work do you do on the crate if you move it 22 m?

   c. If you complete the job in 8.0 s, what power is developed?

91. Dolly and Ramp A dolly is used to move a 115-kg refrigerator up a ramp into a house. The ramp is 2.10 m long and rises 0.850 m. The mover pulls the dolly with a force of 496 N parallel to the ramp. The dolly and ramp constitute a machine.

   a. What work does the mover do on the dolly?

   b. What is the work done on the refrigerator by the machine?

   c. What is the efficiency of the machine?

92. Sally does 11.4 kJ of work dragging a wooden crate 25.0 m across a floor at a constant speed. The rope she uses to pull the crate makes an angle of 48.0° with the horizontal.

   a. What force does the rope exert on the crate?

   b. Find the force of friction acting on the crate.

   c. How much energy is transformed by the force of friction between the floor and the crate?

93. Sledding An 845-N sled is pulled a distance of 185 m. The task requires 1.20×10^4 J of work and is done by pulling on a rope with a force of 125 N. At what angle is the rope held?

94. An electric winch pulls an 875-N crate up a 15° incline at 0.25 m/s. The coefficient of friction between the crate and incline is 0.45.

   a. What power does the winch develop?

   b. How much electrical power must be delivered to the winch if it is 85 percent efficient?

Thinking Critically

95. Apply Concepts A 75-kg sprinter runs the 50.0-m dash in 8.50 s. Assume the sprinter's acceleration is constant throughout the race.

   a. Find the sprinter's average power for the race.

   b. What is the maximum power the sprinter develops?

96. Apply Concepts The sprinter in the previous problem runs the 50.0-m dash again in 8.50 s. This time, however, the sprinter accelerates in the first second and runs the rest of the race at a constant velocity.

   a. Calculate the average power produced for that first second.

   b. What is the maximum power the sprinter now generates?

97. Analyze and Conclude You are carrying boxes to a storage loft that is 12 m above the ground. You need to move 30 boxes with a total mass of 150 kg as quickly as possible. You could carry more than one up at a time, but if you try to move too many at once, you will go very slowly and rest often. If you carry only one box at a time, most of the energy will go into raising your own body. The power that your body can develop over a long time depends on the mass that you carry, as shown in **Figure 26**. Find the number of boxes to carry on each trip that would minimize the time required. What time would you spend doing the job? Ignore the time needed to go back down the stairs and to lift and lower each box.

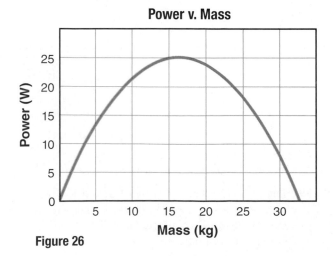

Power v. Mass

Figure 26

98. Ranking Task A 20-kg boy interacts with a bench as shown in **Figure 27**. Rank each interaction according to the work the boy does on the bench, from least to greatest. Clearly indicate any ties.

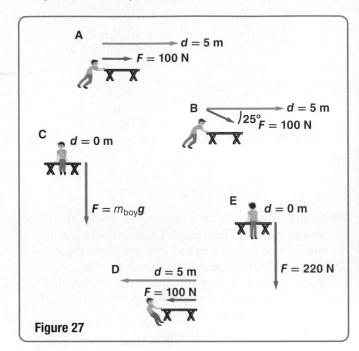

Figure 27

Writing in Physics

99. Just as a bicycle is a compound machine, so is an automobile. Find the efficiencies of the component parts of the power train (engine, transmission, wheels, and tires). Explore possible improvements in each of these efficiencies.

100. The terms *force, work, power,* and *energy* are often used as synonyms in everyday use. Obtain examples from radio, television, print media, and advertisements that illustrate meanings for these terms that differ from those used in physics.

Cumulative Review

101. You are gardening and fill a garbage can with soil and weeds. The 24-kg can is too heavy to lift so you push it across the yard. The coefficient of kinetic friction between the can and the muddy grass is 0.27, and the coefficient of static friction is 0.35. How hard must you push horizontally to get the can to just start moving?

102. Baseball A major league pitcher throws a fastball horizontally at a speed of 40.3 m/s (90 mph). How far has it dropped by the time it crosses home plate 18.4 m (60 ft, 6 in.) away?

MULTIPLE CHOICE

1. A 4-N soccer ball sits motionless on a field. A player's foot exerts a force of 5 N on the ball for a distance of 0.1 m, and the ball rolls a distance of 10 m. How much kinetic energy does the ball gain from the player?

   A. 0.5 J   **C.** 9 J

   B. 0.9 J   **D.** 50 J

2. A pulley system consists of two fixed pulleys and two movable pulleys that lift a rock that has a 300-N weight at a constant speed. If the effort force used to lift the rock is 100 N, what is the mechanical advantage of the system?

   A. $\frac{1}{3}$ **C.** 3

   B. $\frac{3}{4}$ **D.** 6

3. A compound machine used to raise heavy boxes consists of a ramp and a pulley. The efficiency of pulling a 100-kg box up the ramp is 50 percent. If the efficiency of the pulley is 90 percent, what is the overall efficiency of the compound machine?

   A. 40 percent   **C.** 50 percent

   B. 45 percent   **D.** 70 percent

4. A 20.0-N block is attached to the end of a rope, and the rope is looped around a pulley system. If you pull the opposite end of the rope a distance of 2.00 m, the pulley system raises the block a distance of 0.40 m. What is the pulley system's ideal mechanical advantage?

   A. 2.5   **C.** 5.0

   B. 4.0   **D.** 10.0

5. Two people carry identical 40.0-N boxes up a ramp. The ramp is 2.00 m long and rests on a platform that is 1.00 m high. One person walks up the ramp in 2.00 s, and the other person walks up the ramp in 4.00 s. What is the difference in power the two people use to carry the boxes up the ramp?

   A. 5.00 W   **C.** 20.0 W

   B. 10.0 W   **D.** 40.0 W

6. A skater with a mass of 50.0 kg slides across an icy pond with negligible friction. As he approaches a friend, both he and his friend hold out their hands, and the friend exerts a force in the direction opposite to the skater's movement, which slows the skater's speed from 2.0 m/s to 1.0 m/s. What is the change in the skater's kinetic energy?

   A. −25 J   **C.** −100 J

   B. −75 J   **D.** −150 J

7. The box in the diagram is being pushed up the ramp with a force of 100.0 N. What is the work done on the box?
(sin 30° = 0.50, cos 30° = 0.87, tan 30° = 0.58)

   A. 150 J   **C.** 450 J

   B. 260 J   **D.** 600 J

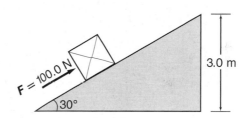

FREE RESPONSE

8. The diagram shows a box being pulled along a horizontal surface with a force of 200.0 N. Calculate the work done on the box and the power required to pull it a distance of 5.0 m in 10.0 s.
(sin 45° = cos 45° = 0.71)

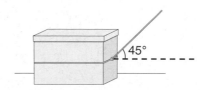

NEED EXTRA HELP?

If You Missed Question	1	2	3	4	5	6	7	8
Review Section	1	2	2	2	1	1	1	1

Online Test Practice

CHAPTER 11

Energy and Its Conservation

BIGIDEA   Within a closed, isolated system, energy can change form, but the total energy is constant.

SECTIONS

1 **The Many Forms of Energy**

2 **Conservation of Energy**

LaunchLAB

ENERGY OF A BOUNCING BALL

What determines how high a basketball bounces?

WATCH THIS!

Video

FUEL EFFICIENCY

The phrase *fuel-efficient* is often used as a selling point when describing the features of a car. But what does that phrase really mean? How is fuel-efficiency calculated? And how does it relate to the conservation of energy?

PHYSICS T.V.

(l)McGraw-Hill Education; (r)Tadao Yamamoto/amana images/Getty Images

The Many Forms of Energy

Water has several forms. It forms ice and water vapor. In a similar way, energy has different forms, such as energy due to motion or energy due to object interactions.

MAINIDEA

Kinetic energy is energy due to an object's motion, and potential energy is energy stored due to the interactions of two or more objects.

Essential Questions

- How is a system's motion related to its kinetic energy?
- What is gravitational potential energy?
- What is elastic potential energy?
- How are mass and energy related?

Review Vocabulary

system the object or objects of interest that can interact with each other and with the outside world

New Vocabulary

rotational kinetic energy
potential energy
gravitational potential energy
reference level
elastic potential energy
thermal energy

A Model of the Work-Energy Theorem

The word *energy* is used in many different ways in everyday speech. Some fruit-and-cereal bars are advertised as energy sources. Athletes use energy in sports. Companies that supply your home with electricity, natural gas, or heating fuel are called energy companies.

Scientists and engineers use the term *energy* more precisely. Recall that the work-energy theorem states that doing work on a system causes a change in the energy of that system. That is, work is the process that transfers energy between a system and the external world. Recall that a system is the object or objects of interest. When an agent performs work on a system, the system's energy increases. When the system does work on its surroundings, the system's energy decreases.

Modeling transformations In some ways, energy is like ice cream—it comes in different varieties. You can have vanilla, chocolate, or peach ice cream. These are different varieties, but they are all ice cream. However, unlike ice cream, energy can be changed from one form to another. A system of objects can possess energy in a variety of forms. In this chapter, you will learn how energy is transformed from one variety (or form) to another and how to keep track of the changes.

Keeping track of energy changes is much like keeping track of your savings account. Bar diagrams, such as those shown in **Figure 1,** can be used to track money or energy. The amount of money you have changes when you earn more or spend it. Similarly, energy can be stored or it can be used. In each case, the behavior of a system is affected.

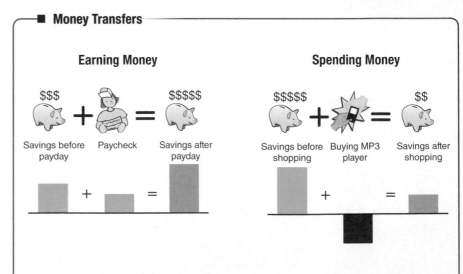

Figure 1 Energy is similar to a savings account. The amount can increase or decrease.

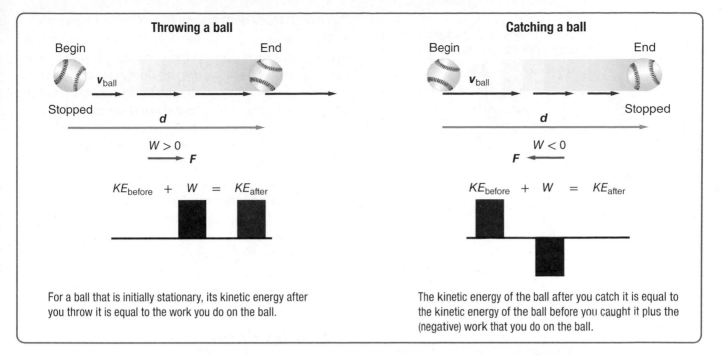

Throwing a ball		Catching a ball	
Begin	End	Begin	End

$KE_{before} + W = KE_{after}$

For a ball that is initially stationary, its kinetic energy after you throw it is equal to the work you do on the ball.

The kinetic energy of the ball after you catch it is equal to the kinetic energy of the ball before you caught it plus the (negative) work that you do on the ball.

Throwing a ball Gaining and losing energy can be illustrated by throwing and catching a ball. Recall that when you exert a constant force (F) on an object through a distance (d) in the direction of the force, you do an amount of work, represented by $W = Fd$. The work is positive because the force and the motion are in the same direction. The energy of the object increases by an amount equal to the work (W). Suppose the object is a stationary ball, and you exert a force to throw the ball. As a result of the work you do, the ball gains kinetic energy. This process is shown on the left in **Figure 2.** You can represent the process using an energy bar diagram. The height of the bar represents the amount of work (in joules). The kinetic energy after the work is done is equal to the sum of the initial kinetic energy plus the work done on the ball.

Catching a ball A similar process occurs when you catch the ball. Before the moving ball is caught, it has kinetic energy. In catching it, you exert a force on the ball in the direction *opposite* to its motion. Therefore, you do negative work on it, causing it to stop. Now that the ball is not moving, it has no kinetic energy. This process is represented in the energy bar diagram on the right in **Figure 2.** Kinetic energy is always positive, so the initial kinetic energy of the ball is positive. The work done on the ball is negative and the final kinetic energy is zero. Again, the kinetic energy after the ball has stopped equals the sum of the initial kinetic energy and the work done on the ball.

Figure 2 When you do work on an object, such as a ball, you change its energy.

Identify two other events in which work causes a change in energy.

Kinetic Energy

Recall that translational kinetic energy is due to an object's change in position. It is represented by the equation $KE_{trans} = \frac{1}{2}mv^2$, where m is the object's mass and v is the magnitude of the object's velocity or speed. Kinetic energy is proportional to the object's mass. For example, a 7.26-kg bowling ball thrown through the air has more kinetic energy than a 0.148-kg baseball with the same speed. An object's kinetic energy is also proportional to the square of the object's speed. A car moving at 20 m/s has four times the kinetic energy of the same car moving at 10 m/s.

Diver has translational and rotational kinetic energy.

Diver has translational kinetic energy.

Diving board does work on diver.

Figure 3 The diving board does work on the diver. This work increases the diver's kinetic energy.

Watch a **BrainPOP video about potential energy.**

Rotational Kinetic Energy Kinetic energy also can be due to rotational motion. If you spin a toy top in one spot, does it have kinetic energy? You might say that it does not because the top does not change its position. However, to make the top rotate, you had to do work on it. The top has **rotational kinetic energy,** which is energy due to rotational motion. Rotational kinetic energy can be calculated using $KE_{rot} = \frac{1}{2}I\omega^2$, where I is the object's moment of inertia and ω is the object's angular velocity.

In **Figure 3,** a diving board does work on a diver. This work transfers energy to the diver, including both translational and rotational kinetic energy. Her center of mass moves as she leaps, so she has translational kinetic energy. She rotates about her center of mass, so she has rotational kinetic energy. When she slices into the water, she has mostly translational kinetic energy.

Potential Energy

The money model that was discussed earlier also illustrates the transformation of energy from one form to another. Both money and energy can be in different forms. You can have one five-dollar bill, 20 quarters, or 500 pennies. In all of these forms, you still have five dollars. In the same way, energy can be kinetic energy, stored energy, or another form.

For an example of stored energy, consider a small group of boulders high on a mountain. These boulders have been moved away from Earth's center by geological processes against the gravitational force; thus, the system that includes both Earth and the boulders has stored energy. Energy that is stored due to interactions between objects in a system is called **potential energy.**

Not all potential energy is due to gravity. A spring-loaded toy, such as a jack-in-the-box, is a system that has potential energy, but the energy is stored due to a compressed spring, not gravity. Can you think of other ways that a system could have potential energy?

PRACTICE PROBLEMS

Do additional problems. Online Practice

1. A 52.0-kg skater moves at 2.5 m/s and glides to a stop over a distance of 24.0 m. Find the skater's initial kinetic energy. How much of her kinetic energy is transformed into other forms of energy by friction as she stops? How much work must she do to speed up to 2.5 m/s again?

2. An 875.0-kg car speeds up from 22.0 m/s to 44.0 m/s. What are the initial and final kinetic energies of the car? How much work is done on the car to increase its speed?

3. A comet with a mass of 7.85×10^{11} kg strikes Earth at a speed of 25.0 km/s. Find the kinetic energy of the comet in joules, and compare the work that is done by Earth in stopping the comet to the 4.2×10^{15} J of energy that was released by the largest nuclear weapon ever exploded.

4. **CHALLENGE** A 2-kg wheel rolls down the road with a linear speed of 15 m/s. Find its translational and rotational kinetic energies. (*Hint: $I = mr^2$*)

Gravitational Potential Energy

Look at the three oranges being juggled in **Figure 4.** If you consider the system to be only one orange, then it has several external forces acting on it. The force of the juggler's hand does work, giving the orange its initial kinetic energy. After the orange leaves the juggler's hand, only the gravitational force acts on it, assuming that air resistance is negligible. How much work does gravity do on the orange as its height changes?

Work done by gravity Let h represent the orange's height measured from the juggler's hand. On the way up, its displacement is upward, but the gravitational force on the orange (F_g) is downward, so the work done by gravity is negative: $W_g = -mgh$. On the way back down, the force and displacement are in the same direction, so the work done by gravity is positive: $W_g = mgh$. Thus, while the orange is moving upward, gravity does negative work, slowing the orange to a stop. On the way back down, gravity does positive work, increasing the orange's speed and thereby increasing its kinetic energy. The orange recovers all of its original kinetic energy when it returns to the same height as the juggler's hand, as shown in the energy bar diagrams at the bottom of the page. Notice in the diagram that the orange has equal amounts of gravitational potential energy and kinetic energy at $\frac{1}{2}h$.

☑ **READING CHECK** **Describe** the work done by gravity as the orange rises from the jugglers hand.

Recall that according to Newton's law of universal gravitation, there is a gravitational attraction between any pair of objects. This gravitational attraction between the objects is a force that performs work when one of the objects moves. Usually, we are most concerned with the gravitational potential energy between an object and Earth. For this particular example, we choose to discuss the object and Earth together as a system. If the object moves away from Earth, the system stores energy due to the gravitational force between the object and Earth. The stored energy due to gravity is called **gravitational potential energy,** represented by the symbol *GPE*. The height to which the object has risen is determined by using a **reference level,** the position where *GPE* is defined to be zero. In the example of the juggler, the reference level is the juggler's hand.

Figure 4 As the orange rises, its kinetic energy transforms into potential energy in the system composed of the orange and Earth. As it falls, the potential energy is transformed back into kinetic energy. In this process, Earth also gains and loses very small amounts of kinetic energy. The energy bar diagram at the bottom of the page illustrates how the energy changes as the orange moves along its path.

View an **animation about gravitational potential energy.**

Energy Bar Diagram

KE	GPE KE	GPE	GPE KE	KE
GPE		KE		GPE
Beginning Flight	**Rising** (at $\frac{1}{2}h$)	**Highest Point**	**Falling** (at $\frac{1}{2}h$)	**End of Flight**

Calculating gravitational potential energy

For an object with mass (m) that has risen to a height (h) above the reference level, the gravitational potential energy of the object-Earth system is represented by the following equation.

GRAVITATIONAL POTENTIAL ENERGY
The gravitational potential energy of an object-Earth system is equal to the product of the object's mass, the gravitational field strength, and the object's distance above the reference level.

$$GPE = mgh$$

In this equation, g is the gravitational field strength. Gravitational potential energy, like kinetic energy, is measured in joules.

Energy transformations Consider the energy of a system consisting of Earth and an orange being juggled. The energy of the system is in two forms: kinetic energy and gravitational potential energy. At the beginning of the orange's flight, all the energy is in the form of kinetic energy, as shown on the left in **Figure 5.** As the orange rises, it slows down and the system's energy changes from kinetic energy to gravitational potential energy. At the highest point of the orange's flight, the velocity is zero and all the energy is in the form of gravitational potential energy.

As the orange falls, gravitational potential energy changes back into kinetic energy. The sum of kinetic energy and gravitational potential energy is constant at all times because no work is done on the system by forces outside of the system.

Reference levels So far, we have set the reference level to be the juggler's hand. That is, the height of the orange is measured from where it left contact with the juggler's hand. Thus, at the juggler's hand, $h = 0$ m and $GPE = 0$ J. However, you can set the reference level at any height that is convenient for solving a given problem.

Suppose the reference level is set at the highest point of the orange's flight. Then, $h = 0$ m and the system's $GPE = 0$ J at that point, as illustrated on the right in **Figure 5.** The system's potential energy is negative at the beginning of the orange's flight, zero at the highest point, and negative at the end of the orange's flight. If you were to calculate the total energy of the system represented on the left in **Figure 5,** it would be different from the total energy of the system represented on the right in **Figure 5.** This is because the reference levels are different in each case. However, the total energy of the system in each situation is constant at all times during the flight of the orange.

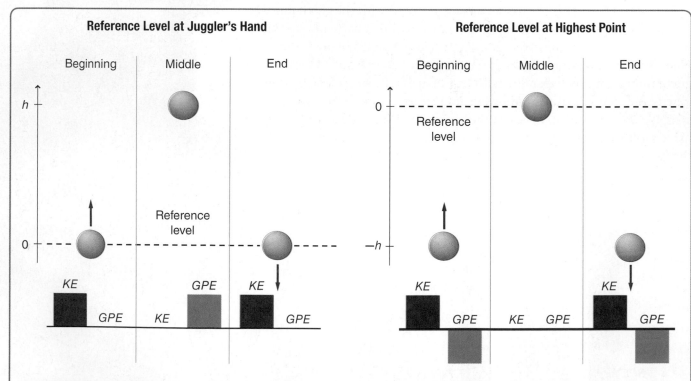

Figure 5 Where you put the reference level affects the system's gravitational potential energy. Where you put the reference level does not affect how the system's gravitational potential energy changes over time, however.
Analyze Where else might you place the reference level?

GRAVITATIONAL POTENTIAL ENERGY You lift a 7.30-kg bowling ball from the storage rack and hold it up to your shoulder. The storage rack is 0.61 m above the floor and your shoulder is 1.12 m above the floor.

a. When the bowling ball is at your shoulder, what is the ball-Earth system's gravitational potential energy relative to the floor?

b. When the bowling ball is at your shoulder, what is the ball-Earth system's gravitational potential energy relative to the rack?

c. How much work was done by gravity as you lifted the ball from the rack to shoulder level?

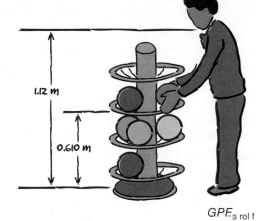

1 ANALYZE AND SKETCH THE PROBLEM

- Sketch the situation.
- Choose a reference level.
- Draw an energy bar diagram showing the gravitational potential energy with the floor as the reference level.

KNOWN	UNKNOWN
$m = 7.30$ kg $g = 9.8$ N/kg	$GPE_{s\,rel\,f} = ?$
$h_r = 0.61$ m (rack relative to the floor)	$GPE_{s\,rel\,r} = ?$
$h_s = 1.12$ m (shoulder relative to the floor)	$W = ?$

2 SOLVE FOR THE UNKNOWN

a. Set the reference level to be at the floor.

Determine the gravitational potential energy of the system when the ball is at shoulder level.

$GPE_{s\,rel\,f} = mgh_s = (7.30$ kg$)(9.8$ N/kg$)(1.12$ m$)$ ◀ Substitute $m = 7.30$ kg, $g = 9.8$ N/kg, $h_s = 1.12$ m

$= 8.0 \times 10^1$ J

b. Set the reference level to be at the rack height.

Determine the height of your shoulder relative to the rack.

$h = h_s - h_r$

Determine the gravitational potential energy of the system when the ball is at shoulder level.

$GPE_{s\,rel\,r} = mgh = mg(h_s - h_r)$ ◀ Substitute $h = h_s - h_r$

$= (7.30$ kg$)(9.8$ N/kg$)(1.12$ m $- 0.61$ m$)$ ◀ Substitute $m = 7.30$ kg, $g = 9.8$ N/kg, $h_s = 1.12$ m, $h_r = 0.61$ m

$= 36$ J ◀ This also is equal to the work done by you as you lifted the ball.

c. The work done by gravity is the weight of the ball times the distance the ball was lifted.

$W = Fd = -(mg)h = -(mg)(h_s - h_r)$ ◀ The weight opposes the motion of lifting, so the work is negative.

$= -(7.30$ kg$)(9.8$ N/kg$)(1.12$ m $- 0.61$ m$)$ ◀ Substitute $m = 7.30$ kg, $g = 9.8$ N/kg, $h_s = 1.12$ m, $h_r = 0.61$ m

$= -36$ J

3 EVALUATE THE ANSWER

- **Are the units correct?** Both potential energy and work are measured in joules.
- **Does the answer make sense?** The system should have a greater *GPE* measured relative to the floor than relative to the rack because the ball's distance above the floor level is greater than the ball's distance above the rack.

5. In Example Problem 1, what is the potential energy of the ball-Earth system when the bowling ball is on the floor? Use the rack as your reference level.

6. If you slowly lower a 20.0-kg bag of sand 1.20 m from the trunk of a car to the driveway, how much work do you do?

7. A boy lifts a 2.2-kg book from his desk, which is 0.80 m high, to a bookshelf that is 2.10 m high. What is the potential energy of the book-Earth system relative to the desk when the book is on the shelf?

8. If a 1.8-kg brick falls to the ground from a chimney that is 6.7 m high, what is the change in the potential energy of the brick-Earth system?

9. **CHALLENGE** A worker picks up a 10.1-kg box from the floor and sets it on a 1.1-m-high table. He slides the box 5.0 m along the table and then lowers it back to the floor. What were the changes in the box-Earth system's energy, and how did the system's total energy change? (Ignore friction.)

Elastic Potential Energy

When the string on the bow shown in **Figure 6** is pulled, work is done on the bow string and energy is transferred to the bow. If you identify the system as the bow, the arrow, and Earth, the energy of the system increases. When the string and the arrow are released, the stored energy is changed into kinetic energy. The stored energy due to the pulled string is elastic potential energy. **Elastic potential energy** is stored energy due to an object's change in shape. Systems that include springs, rubber bands, and trampolines often have elastic potential energy.

☑ **READING CHECK** **Define** the term *elastic potential energy*.

You can also store elastic potential energy in a system when you bend an elastic object in that system. For example, when stiff metal or bamboo poles were used in pole-vaulting, the poles did not bend easily. Little work was done on the poles, and the poles did not store much potential energy. Since flexible fiberglass poles were introduced, however, record pole-vaulting heights have soared.

Figure 6 The archer-bow-arrow system has maximum elastic potential energy before the string is released, as shown on the left. When the arrow and string disengage, the elastic potential energy is completely transformed into kinetic energy, as shown on the right.

Elastic Potential Energy

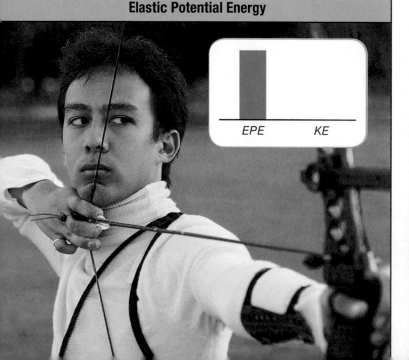

EPE   KE

Kinetic Energy

EPE   KE

A pole-vaulter runs with a flexible pole and plants its end into a socket in the ground. When the pole-vaulter bends the pole, as shown in **Figure 7,** some of the pole-vaulter's kinetic energy is transformed to elastic potential energy. When the pole straightens, the elastic potential energy is transformed to gravitational potential energy and kinetic energy as the pole-vaulter is lifted as high as 6 m above the ground. Unlike stiff metal poles or bamboo poles, fiberglass poles have an increased capacity for storing elastic potential energy. Thus, pole-vaulters can clear bars that are set very high.

Mass

Albert Einstein recognized yet another form of potential energy that is proportional to the object's mass. He demonstrated that mass represents a form of energy. This energy is called the rest energy (E_0) and can be calculated using the following formula.

REST ENERGY
An object's rest energy is equal to its mass times the speed of light squared.

$$E_0 = mc^2$$

According to this famous formula, mass is a form of energy that can be transformed into other forms of energy. Because the speed of light is very fast (300,000,000 m/s), even a small amount of mass is equivalent to a large amount of energy. For example, 1 g of mass is equivalent to 90 trillion J of energy. As a result, mass-energy equivalence is only apparent when large amounts of energy are involved, such as during nuclear explosions and particle physics experiments.

Other Forms of Energy

Think about all the forms and sources of energy you encounter every day. Gasoline provides energy to run your car. You eat food to obtain energy. Power plants harness the energy of wind, water, fossil fuels, and atoms and transform it into electrical energy.

Chemical and nuclear energy Recall that fossil fuels release chemical energy when they are burned. A similar process occurs when you digest food. Your body uses the energy from the chemical bonds in your food as an energy source. The bonds inside an atom's nucleus also store energy. This energy is called nuclear energy and is released when the structure of an atom's nucleus changes. You will learn more about nuclear energy in later chapters.

Figure 7 The pole-vaulter stores elastic potential energy in the vaulter-pole-track system. The elastic potential energy of this system then changes into kinetic energy and gravitational potential energy.

Figure 8 In your daily activities, you encounter many types of energy, such as chemical energy, nuclear energy, thermal energy, and electrical energy.

Thermal, electrical, and radiant energy Thermal energy, which is associated with temperature, is another form of energy. **Thermal energy** is the sum of the kinetic energy and potential energy of the particles in a system. When you warm your hands by rubbing them together, you transform kinetic energy into thermal energy. Thermal energy can also be transferred. For example, the stove in **Figure 8** transfers thermal energy to the pan.

Power plants, such as the one shown in **Figure 8,** transform various types of energy into electrical energy, which is energy associated with charged particles. Electrical energy can be transmitted through wires, such as those shown in **Figure 8,** to your home. Appliances transform this electrical energy into other forms of energy, such as radiant energy in lightbulbs or into kinetic energy of moving fan blades. Radiant energy is energy transferred by electromagnetic waves. You will learn more about these forms of energy in later chapters.

SECTION 1 **REVIEW**

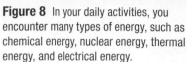 **Section Self-Check** Check your understanding.

10. **MAIN**IDEA How can you apply the work-energy theorem to lifting a bowling ball from a storage rack to your shoulder?

11. **Elastic Potential Energy** You get a spring-loaded jumping toy ready by compressing the spring. The toy then flies straight up. Draw bar graphs that describe the forms of energy present in the following instances. Assume the system includes the spring toy and Earth.

    a. The toy is pushed down thereby compressing the spring.

    b. The spring expands and the toy jumps.

    c. The toy reaches the top of its flight.

12. **Potential Energy** A 25.0-kg shell is shot from a cannon at Earth's surface. The reference level is Earth's surface.

    a. What is the shell-Earth system's gravitational potential energy when the shell's height is 425 m?

    b. What is the change in the system's potential energy when the shell falls to a height of 225 m?

13. **Potential Energy** A 90.0-kg rock climber climbs 45.0 m upward, then descends 85.0 m. The initial height is the reference level. Find the potential energy of the climber-Earth system at the top and at the bottom. Draw bar graphs for both situations.

14. **Rotational Kinetic Energy** On a playground, some children push a merry-go-round so that it turns twice as fast as it did before they pushed it. What are the relative changes in angular momentum and rotational kinetic energy of the merry-go-round?

15. **Critical Thinking** Karl uses an air hose to exert a constant horizontal force on a puck, which is on a frictionless air table. The force is constant as the puck moves a fixed distance.

    a. Explain what happens in terms of work and energy. Draw bar graphs.

    b. Suppose Karl uses a different puck with half the first one's mass. All other conditions remain the same. How should the kinetic energy and work done differ from those in the first situation?

    c. Describe what happened in parts a and b in terms of impulse and momentum.

Conservation of Energy

PHYSICS 4 YOU

If you count your money and find that you are $3 short, you would not assume that the money just disappeared. You might search for it. Just as money does not vanish into thin air, energy does not vanish.

MAIN IDEA

In a collision in a closed, isolated system, the total energy is conserved, but kinetic energy might not be conserved.

Essential Questions

- Under what conditions is energy conserved?
- What is mechanical energy, and when is it conserved?
- How are momentum and kinetic energy conserved or changed in a collision?

Review Vocabulary

closed system a system that does not gain or lose mass

New Vocabulary

law of conservation of energy
mechanical energy
elastic collision
inelastic collision

APPLYING PRACTICES

Develop and Use Models Go to the resources tab in ConnectED to find the Applying Practices worksheet *Modeling Changes in Energy*.

The Law of Conservation of Energy

Scientists look for energy that seems to be missing by determining where and how the energy was transferred. We certainly expect that the total amount of energy in a system remains constant as long as the system is closed and isolated from external forces.

The **law of conservation of energy** states that in a closed, isolated system, energy can neither be created nor destroyed; rather, energy is conserved. Under these conditions, energy can change form but the system's total energy in all of its forms remains constant.

Mechanical energy For many situations you focus on the energy that comes from the motions of and interactions between objects. The sum of the kinetic energy and potential energy of the objects in a system is the system's **mechanical energy** (*ME*). The kinetic energy includes both the translational and rotational kinetic energies of the objects in the system. Potential energy includes the gravitational and elastic potential energies of the system. In any system, mechanical energy is represented by the following equation.

MECHANICAL ENERGY OF A SYSTEM

The mechanical energy of a system is equal to the sum of the kinetic energy and potential energy of the system's objects.

$$ME = KE + PE$$

Imagine a system consisting of a 10.0-N bowling ball falling to Earth, as shown in **Figure 9.** For this case, the system's mechanical energy is 20 J.

Figure 9 When a bowling ball is dropped, mechanical energy is conserved.

Predict Suppose the ball started with an upward velocity. What would its energy graph look like?

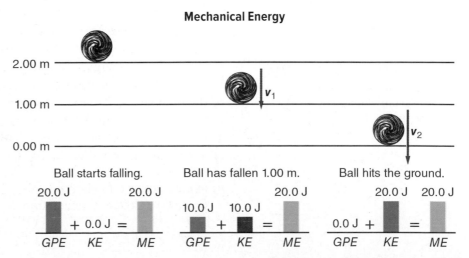

Mechanical Energy

Ball starts falling.
Ball has fallen 1.00 m.
Ball hits the ground.

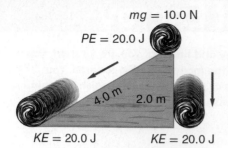

$mg = 10.0\ N$

$PE = 20.0\ J$

4.0 m   2.0 m

$KE = 20.0\ J$ $KE = 20.0\ J$

Figure 10 If friction does no work on the ball, the ball's final kinetic energy is equal to its initial gravitational potential energy regardless of which path it follows.

View an **animation of conservation of mechanical energy.**

Concepts In Motion

APPLYING PRACTICES

PBL Go to the resources tab in ConnectED to find the PBL *Earth Power*.

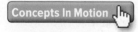

ENERGY EXCHANGE
How does energy transform when a ball is shot upward?

INTERRUPTED PENDULUM
How is the mechanical energy of a pendulum affected by an obstruction?

Conservation of mechanical energy Suppose that you let go of the bowling ball. As it falls, the gravitational potential energy of the ball-Earth system is transformed into kinetic energy. When the ball is 1.00 m above Earth's surface: $GPE = mgh = F_gh = (10.0\ N)(1.00\ m) = 10.0\ J$. What is the system's kinetic energy when the ball is at a height of 1.00 m? The system consisting of the ball and Earth is closed and isolated because no external forces are doing work upon it. Therefore, the total mechanical energy of the system remains constant at 20.0 J.

$$ME = KE + PE$$
$$KE = ME - PE$$
$$KE = 20.0\ J - 10.0\ J = 10.0\ J$$

When the ball reaches ground level, the system's potential energy is zero, and the system's kinetic energy is 20.0 J. The equation that describes conservation of mechanical energy can be written as follows.

CONSERVATION OF MECHANICAL ENERGY

When mechanical energy is conserved, the sum of the system's kinetic energy and potential energy before an event is equal to the sum of the system's kinetic energy and potential energy after that event.

$$KE_i + PE_i = KE_f + PE_f$$

What happens if the ball does not fall straight down but rolls down a ramp, as shown in **Figure 10**? If friction does not affect the system, then the system of the ball and Earth remains closed and isolated. The ball still moves down a vertical distance of 2.00 m, so the decrease in potential energy is 20.0 J. Therefore, the increase in kinetic energy is 20.0 J. As long as friction doesn't affect the system's energy, the path the ball takes does not matter. However, in the case with the ramp, the kinetic energy is split between translational kinetic energy (it is moving forward) and rotational kinetic energy (it is rolling). Three situations involving conservation of mechanical energy are shown in **Figure 11** on the next page.

Conservation and other forms of energy A swinging pendulum eventually stops, a bouncing ball comes to rest, and the heights of roller-coaster hills get lower and lower as the ride progresses. Where does the mechanical energy in such systems go? Objects moving through the air experience the forces of air resistance. In a roller coaster, there are also frictional forces between the wheels and the tracks. Each of these forces transforms mechanical energy into other forms of energy.

When a ball bounces from a surface, it compresses. Most of the kinetic energy transforms into elastic potential energy. Some, but not all, of this elastic potential energy transforms back into kinetic energy during the bounce. But as in the cases of the pendulum and the roller coaster, some of the original mechanical energy in the system transforms into another form of energy within the system or transmits to energy outside the system, as in air resistance. The bouncing ball becomes warmer (thermal energy), and we hear the sound of the bounce (sound energy).

✓ **READING CHECK Analyze** how the ball's final kinetic energy in **Figure 10** would be different if friction transformed some of the system's energy.

Figure 11 The conservation of mechanical energy is an important consideration in designing roller coasters, ski slopes, and the pendulums for grandfather clocks.

Investigate **roller coasters.**

Virtual Investigation

Roller Coasters The roller-coaster car is nearly at rest at the top of the first hill, and the total mechanical energy in the Earth-roller coaster car system is the system's gravitational potential energy at that point. If a hill farther along the track were higher than the first one, the roller-coaster car would not be able to climb the higher hill because the energy required to do so would be greater than the total mechanical energy of the system.

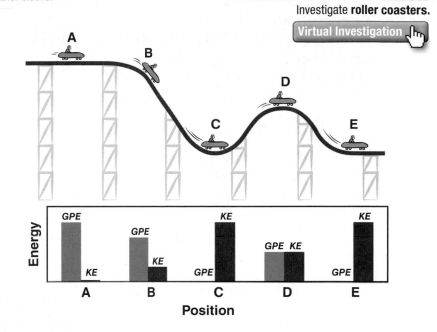

Skiing When you ski down a steep slope, you begin from rest at the top of the slope and the total mechanical energy is equal to the gravitational potential energy. Once you start skiing downhill, the gravitational potential energy is transformed to kinetic energy. As you ski down the slope, your speed increases as more potential energy is transformed to kinetic energy.

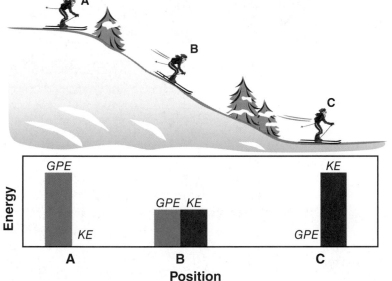

Pendulums The simple oscillation of a pendulum also demonstrates conservation of mechanical energy. The system is the pendulum bob and Earth. The reference level is chosen to be the height of the bob at the lowest point. When an external force pulls the bob to one side, the force does work that adds mechanical energy to the system. At the instant the bob is released, all the energy is in the form of potential energy, but as the bob swings downward, potential energy is transformed to kinetic energy. When the bob is at the lowest point, the gravitational potential energy is zero, and the kinetic energy is equal to the total mechanical energy.

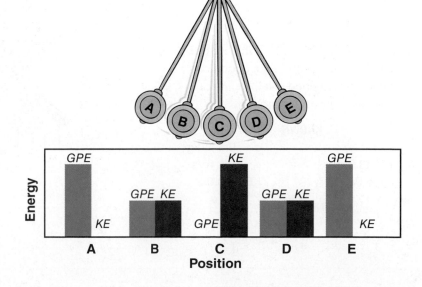

PROBLEM-SOLVING STRATEGIES

CONSERVATION OF ENERGY

Use the following strategies when solving problems about the conservation of energy.

1. Carefully identify the system. Determine whether the system is closed. In a closed system, no objects enter or leave the system.

2. Identify the forms of energy in the system. Identify which forms are part of the mechanical energy of the system.

3. Identify the initial and final states of the system.

4. Is the system isolated?

    a. If there are no external forces acting on the system, then the system is isolated and the total energy of the system is constant.
$$E_{initial} = E_{final}$$

    b. If there are external forces, then the final energy is the sum of the initial energy and the work done on the system. Remember that work can be negative.
$$E_{initial} + W = E_{final}$$

5. For an isolated system, identify the types of energy in the system before and after. If the only forms of energy are potential and kinetic, mechanical energy is conserved.
$$KE_i + PE_i = KE_f + PE_f$$

Decide on the reference level for gravitational potential energy. Draw energy bar diagrams showing initial and final energies like the diagram shown to the right.

Energy Bar Diagram

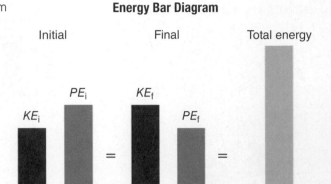

Find help with **square and cube roots.** Math Handbook

EXAMPLE PROBLEM 2

CONSERVATION OF MECHANICAL ENERGY A 22.0-kg tree limb is 13.3 m above the ground. During a hurricane, it falls on a roof that is 6.0 m above the ground.
a. Find the kinetic energy of the limb when it reaches the roof. Assume that the air does no work on the tree limb.
b. What is the limb's speed when it reaches the roof?

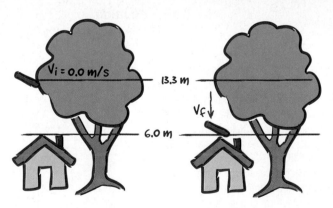

1 ANALYZE AND SKETCH THE PROBLEM

- Sketch the initial and final conditions.
- Choose a reference level.
- Draw an energy bar diagram.

KNOWN

$m = 22.0$ kg	$g = 9.8$ N/kg
$h_{limb} = 13.3$ m	$v_i = 0.0$ m/s
$h_{roof} = 6.0$ m	$KE_i = 0.0$ J

UNKNOWN

$GPE_i = ?$	$KE_f = ?$
$GPE_f = ?$	$v_f = ?$

2 SOLVE FOR THE UNKNOWN

a. Set the reference level as the height of the roof.
Find the initial height of the limb relative to the roof.

$h = h_{limb} - h_{roof} = 13.3 \text{ m} - 6.0 \text{ m} = 7.3 \text{ m}$ ◀ Substitute h_{limb} = 13.3 m, h_{roof} = 6.0 m

Determine the initial potential energy of the limb-Earth system.

$GPE_i = mgh = (22.0 \text{ kg})(9.8 \text{ N/kg})(7.3 \text{ m})$ ◀ Substitute m = 22.0 kg, g = 9.8 N/kg, h = 7.3 m
$= 1.6 \times 10^3 \text{ J}$

Identify the initial kinetic energy of the system.

$KE_i = 0.0 \text{ J}$ ◀ The tree limb is initially at rest.

Identify the final potential energy of the system.

$GPE_f = 0.0 \text{ J}$ ◀ h = 0.0 m at the roof.

Use the law of conservation of energy to find the KE_f.

$KE_f + GPE_f = KE_i + GPE_i$
$KE_f = KE_i + GPE_i - GPE_f$
$= 0.0 \text{ J} + 1.6 \times 10^3 \text{ J} - 0.0 \text{ J}$ ◀ Substitute KE_i = 0.0 J, GPE_i = 1.6×10³ J, and GPE_f = 0 J.
$= 1.6 \times 10^3 \text{ J}$

b. Determine the speed of the limb.

$KE_f = \frac{1}{2} m v_f^2$

$v_f = \sqrt{\frac{2KE_f}{m}} = \sqrt{\frac{2(1.6 \times 10^3 \text{ J})}{22.0 \text{ kg}}}$ ◀ Substitute KE_f = 1.6×10³ J, m = 22.0 kg

$= 12 \text{ m/s}$

3 EVALUATE THE ANSWER

- **Are the units correct?** The magnitude of velocity is measured in m/s, and energy is measured in kg·m²/s² = J.
- **Do the signs make sense?** KE and speed are both positive in this scenario.

PRACTICE PROBLEMS

Do additional problems. **Online Practice**

16. A bike rider approaches a hill at a speed of 8.5 m/s. The combined mass of the bike and the rider is 85.0 kg. Choose a suitable system. Find the initial kinetic energy of the system. The rider coasts up the hill. Assuming friction is negligible, at what height will the bike come to rest?

17. Suppose that the bike rider in the previous problem pedaled up the hill and never came to a stop. In what system is energy conserved? From what form of energy did the bike gain mechanical energy?

18. A skier starts from rest at the top of a 45.0-m-high hill, skis down a 30° incline into a valley, and continues up a 40.0-m-high hill. The heights of both hills are measured from the valley floor. Assume that friction is negligible and ignore the effect of the ski poles.

  a. How fast is the skier moving at the bottom of the valley?

  b. What is the skier's speed at the top of the second hill?

  c. Do the angles of the hills affect your answers?

19. In a belly-flop diving contest, the winner is the diver who makes the biggest splash upon hitting the water. The size of the splash depends not only on the diver's style, but also on the amount of kinetic energy the diver has. Consider a contest in which each diver jumps from a 3.00-m platform. One diver has a mass of 136 kg and simply steps off the platform. Another diver has a mass of 100 kg and leaps upward from the platform. How high would the second diver have to leap to make a competitive splash?

20. CHALLENGE The spring in a pinball machine exerts an average force of 2 N on a 0.08-kg pinball over 5 cm. As a result, the ball has both translational and rotational kinetic energy. If the ball is a uniform sphere $\left(I = \frac{5}{2} mr^2\right)$, what is its linear speed after leaving the spring? (Ignore the table's tilt.)

Figure 12 In a car crash, the kinetic energy decreases.

Explain What happened to the energy?

PhysicsLABs

CONSERVATION OF ENERGY
Is energy conserved when a ball rolls down a ramp?

IS ENERGY CONSERVED?
How does friction affect the energy of a system?

Analyzing Collisions

A collision between two objects, whether the objects include an automobile such as the one shown in **Figure 12,** hockey players, or subatomic particles, is one of the most common situations analyzed in physics. Because the details of a collision can be very complex during the collision itself, the strategy is to find the motion of the objects just before and just after the collision. What conservation laws can be used to analyze such a system? If the system is closed and isolated, then momentum and energy are conserved. However, the potential energy or thermal energy in the system might decrease, remain the same, or increase. Therefore, you cannot predict whether kinetic energy is conserved. **Figure 13** on the next page shows four different kinds of collisions. Look at the calculations for kinetic energy for each of the cases. In all the cases except case 1, there is a change in the amount of kinetic energy.

In case 1, the momentum of the system before and after the collision is represented by the following:

$$p_i = p_{Ai} + p_{Bi} = (1.00 \text{ kg})(1.00 \text{ m/s}) + (1.00 \text{ kg})(0.00 \text{ m/s})$$
$$= 1.00 \text{ kg·m/s}$$

$$p_f = p_{Af} + p_{Bf} = (1.00 \text{ kg})(0.00 \text{ m/s}) + (1.00 \text{ kg})(1.00 \text{ m/s})$$
$$= 1.00 \text{ kg·m/s}$$

Thus, in case 1, the momentum is conserved.

☑ **READING CHECK** **Calculate** the initial and final momentums for the remaining cases in **Figure 13.** Verify that momentum is conserved for each case.

Elastic and inelastic collisions Next, consider again the kinetic energy of the system in each of these cases. In case 1 the kinetic energy of the system before and after the collision was the same. A collision in which the kinetic energy does not change is called an **elastic collision.** Collisions between hard objects, such as those made of steel, glass, or hard plastic, often are called nearly elastic collisions.

Kinetic energy decreased in cases 2 and 3. Some of it may have been transformed to thermal energy. A collision in which kinetic energy decreases is called an **inelastic collision.** Objects made of soft, sticky materials, such as clay, act in this way. In case 2, the objects stuck together after colliding. A collision in which colliding objects stick together, such as in case 2, is called a perfectly inelastic collision.

In case 4, the kinetic energy of the system increased. If energy in the system is conserved, then one or more other forms of energy must have decreased. Perhaps potential energy from a compressed spring was released during the collision, adding kinetic energy to the system. This kind of collision is called a superelastic or explosive collision.

The four kinds of collisions can be represented using energy bar diagrams, also shown in **Figure 13.** The kinetic energies before and after the collisions can be calculated, but we do not always have detailed information about the other forms of energy involved. In automobile collisions, kinetic energy can be transformed into thermal energy and sound, but they are difficult to measure. We infer their amounts from the amount of kinetic energy that is transformed.

Figure 13 In a closed, isolated system, momentum is conserved when objects collide. Whether kinetic energy is conserved depends on whether the collision is elastic, inelastic, or superelastic.

Case 1: Elastic Collision

$$KE_f = KE_i$$

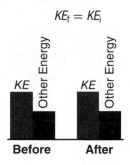

Before   After

Before collision

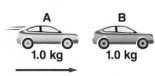

1.0 kg   1.0 kg

$v_{Ai} = 1.00$ m/s $v_{Bi} = 0.00$ m/s

$KE_i = KE_{Ai} + KE_{Bi}$
$= \frac{1}{2}(1.00 \text{ kg})(1.00 \text{ m/s})^2 + \frac{1}{2}(1.00 \text{ kg})(0.00 \text{ m/s})^2$
$= 0.50$ J

After collision

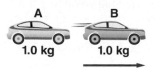

1.0 kg   1.0 kg

$v_{Af} = 0.00$ m/s $v_{Bf} = 1.00$ m/s

$KE_i = KE_{Ai} + KE_{Bi}$
$= \frac{1}{2}(1.00 \text{ kg})(0.00 \text{ m/s})^2 + \frac{1}{2}(1.00 \text{ kg})(1.00 \text{ m/s})^2$
$= 0.50$ J

Case 2: Perfectly Inelastic Collision

$$KE_f < KE_i$$

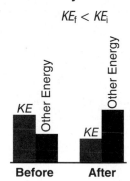

Before   After

Before collision

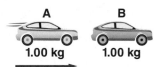

1.00 kg   1.00 kg

$v_{Ai} = 1.00$ m/s $v_{Bi} = 0.00$ m/s

$KE_i = KE_{Ai} + KE_{Bi}$
$= \frac{1}{2}(1.00 \text{ kg})(1.00 \text{ m/s})^2 + \frac{1}{2}(1.00 \text{ kg})(0.00 \text{ m/s})^2$
$= 0.50$ J

After collision

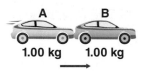

1.00 kg   1.00 kg

$v_{Af} = v_{Bf} = 0.50$ m/s

$KE_f = KE_{Af} + KE_{Bf}$
$= \frac{1}{2}(1.00 \text{ kg})(0.50 \text{ m/s})^2 + \frac{1}{2}(1.00 \text{ kg})(0.50 \text{ m/s})^2$
$= 0.25$ J

Case 3: Inelastic Collision

$$KE_f < KE_i$$

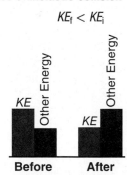

Before   After

Before collision

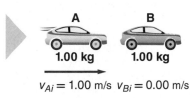

1.00 kg   1.00 kg

$v_{Ai} = 1.00$ m/s $v_{Bi} = 0.00$ m/s

$KE_i = KE_{Ai} + KE_{Bi}$
$= \frac{1}{2}(1.00 \text{ kg})(1.00 \text{ m/s})^2 + \frac{1}{2}(1.00 \text{ kg})(0.00 \text{ m/s})^2$
$= 0.50$ J

After collision

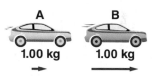

1.00 kg   1.00 kg

$v_{Af} = 0.25$ m/s $v_{Bf} = 0.75$ m/s

$KE_f = KE_{Af} + KE_{Bf}$
$= \frac{1}{2}(1.00 \text{ kg})(0.25 \text{ m/s})^2 + \frac{1}{2}(1.00 \text{ kg})(0.75 \text{ m/s})^2$
$= 0.31$ J

Case 4: Superelastic Collision

$$KE_f > KE_i$$

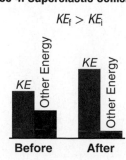

Before   After

Before collision

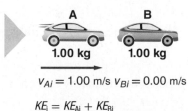

1.00 kg   1.00 kg

$v_{Ai} = 1.00$ m/s $v_{Bi} = 0.00$ m/s

$KE_i = KE_{Ai} + KE_{Bi}$
$= \frac{1}{2}(1.00 \text{ kg})(1.00 \text{ m/s})^2 + \frac{1}{2}(1.00 \text{ kg})(0.00 \text{ m/s})^2$
$= 0.50$ J

After collision

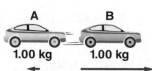

1.00 kg   1.00 kg

$v_{Af} = -0.20$ m/s $v_{Bf} = 1.20$ m/s

$KE_f = KE_{Af} + KE_{Bf}$
$= \frac{1}{2}(1.00 \text{ kg})(-0.20 \text{ m/s})^2 + \frac{1}{2}(1.00 \text{ kg})(0.00 \text{ m/s})^2$
$= 0.74$ J

EXAMPLE PROBLEM

INELASTIC COLLISION In an accident on a slippery road, a compact car with a mass of 1150 kg moving at 15.0 m/s smashes into the rear end of a car with mass of 1575 kg moving at 5.0 m/s in the same direction.
a. What is the final velocity if the wrecked cars lock together?
b. How much is the kinetic energy decreased in the collision?

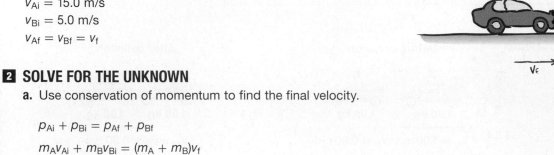

Before (initial)

$M_A v_{Ai}$ $M_B v_{Bi}$

v_{Ai} v_{Bi}

After (final)

$(M_A + M_B) v_f$

v_f

1 ANALYZE AND SKETCH THE PROBLEM

- Sketch the initial and final conditions.
- Sketch the momentum diagram.

KNOWN   **UNKNOWN**

$m_A = 1150$ kg $v_f = ?$
$m_B = 1575$ kg $\Delta KE = KE_f - KE_i = ?$
$v_{Ai} = 15.0$ m/s
$v_{Bi} = 5.0$ m/s
$v_{Af} = v_{Bf} = v_f$

2 SOLVE FOR THE UNKNOWN

a. Use conservation of momentum to find the final velocity.

$$p_{Ai} + p_{Bi} = p_{Af} + p_{Bf}$$

$$m_A v_{Ai} + m_B v_{Bi} = (m_A + m_B)v_f$$

$$v_f = \frac{m_A v_{Ai} + m_B v_{Bi}}{m_A + m_B}$$

$$= \frac{(1150 \text{ kg})(15.0 \text{ m/s}) + (1575 \text{ kg})(5.0 \text{ m/s})}{(1150 \text{ kg} + 1575 \text{ kg})}$$

◀ Substitute m_A = 1150 kg, v_{Ai} = 15.0 m/s, m_B = 1575 kg, v_{Bi} = 5.0 m/s

$$= 9.2 \text{ m/s, in the direction of the motion before the collision}$$

b. To determine the change in kinetic energy of the system, KE_f and KE_i are needed.

$$KE_f = \frac{1}{2} m v_f^2 = \frac{1}{2}(m_A + m_B)v_f^2$$

◀ Substitute $m = m_A + m_B$

$$= \frac{1}{2}(1150 \text{ kg} + 1575 \text{ kg})(9.2 \text{ m/s})^2$$

◀ Substitute m_A = 1150 kg, m_B = 1575 kg, v_f = 9.2 m/s

$$= 1.2 \times 10^5 \text{ J}$$

$$KE_i = KE_{Ai} + KE_{Bi}$$

$$= \frac{1}{2}(m_A v_{Ai}^2) + \frac{1}{2}(m_B v_{Bi}^2)$$

◀ Substitute $KE_{Ai} = \frac{1}{2}(m_A v_{Ai}^2)$, $KE_{Bi} = \frac{1}{2}(m_B v_{Bi}^2)$

$$= \frac{1}{2}(1150 \text{ kg})(15.0 \text{ m/s})^2 + \frac{1}{2}(1575 \text{ kg})(5.0 \text{ m/s})^2$$

◀ Substitute m_A = 1150 kg, m_B = 1575 kg, v_{Ai} = 15.0 m/s, v_{Bi} = 5.0 m/s

$$= 1.5 \times 10^5 \text{ J}$$

Solve for the change in kinetic energy of the system.

$$\Delta KE = KE_f - KE_i$$

$$= 1.2 \times 10^5 \text{ J} - 1.5 \times 10^5 \text{ J} = -3 \times 10^4 \text{ J}$$

◀ Substitute KE_f = 1.2×10⁶ J, KE_i = 1.5×10⁶ J

3 EVALUATE THE ANSWER

- **Are the units correct?** Velocity is measured in meters per second; energy is measured in joules.
- **Does the sign make sense?** The change in kinetic energy is negative, meaning that kinetic energy decreased as expected in an inelastic collision.

21. An 8.00-g bullet is fired horizontally into a 9.00-kg block of wood on an air table and is embedded in it. After the collision, the block and bullet slide along the frictionless surface together with a speed of 10.0 cm/s. What was the initial speed of the bullet?

22. A 91.0-kg hockey player is skating on ice at 5.50 m/s. Another hockey player of equal mass, moving at 8.1 m/s in the same direction, hits him from behind. They slide off together.

   a. What are the total mechanical energy and momentum of the system before the collision?

   b. What is the velocity of the two hockey players after the collision?

   c. How much was the system's kinetic energy decreased in the collision?

23. CHALLENGE A 0.73-kg magnetic target is suspended on a string. A 0.025-kg magnetic dart, shot horizontally, strikes the target head-on. The dart and the target together, acting like a pendulum, swing 12.0 cm above the initial level before instantaneously coming to rest.

   a. Sketch the situation and choose a system.

   b. Decide what is conserved in each step of the process and explain your decision.

   c. What was the initial velocity of the dart?

PHYSICS CHALLENGE

A bullet of mass m, moving at speed v_1, goes through a motionless wooden block and exits with speed v_2. After the collision, the block, which has mass m_B, is moving.

1. What is the final speed (v_B) of the block?

2. What was the change in the bullet's mechanical energy?

3. How much energy was lost to friction inside the block?

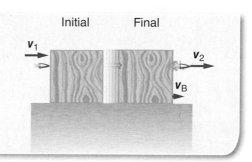

Initial   Final

v_1 v_2

v_B

SECTION 2 REVIEW

[Section Self-Check] Check your understanding.

24. MAINIDEA A child jumps on a trampoline. Draw energy bar diagrams to show the forms of energy present in the following situations.

   a. The child is at the highest point.

   b. The child is at the lowest point.

25. Closed Systems Is Earth a closed, isolated system? Support your answer.

26. Kinetic Energy Suppose a glob of chewing gum and a small, rubber ball collide head-on in midair and then rebound apart. Would you expect kinetic energy to be conserved? If not, what happens to the energy?

27. Kinetic Energy In table tennis, a very light but hard ball is hit with a hard rubber or wooden paddle. In tennis, a much softer ball is hit with a racket. Why are the two sets of equipment designed in this way? Can you think of other ball-paddle pairs in sports? How are they designed?

28. Potential Energy A rubber ball is dropped from a height of 8.0 m onto a hard concrete floor. The ball hits the floor and bounces repeatedly. Each time the ball hits the floor, the ball-Earth system loses 1/5 of its mechanical energy. How many times will the ball bounce before it bounces back up to a height of only about 4 m?

29. Energy As shown in **Figure 14,** a child slides down a playground slide. At the bottom of the slide, she is moving at 3.0 m/s. How much energy was transformed by friction as she slid down the slide?

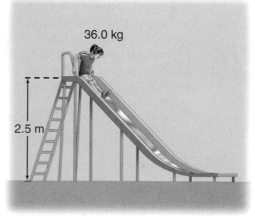

36.0 kg

2.5 m

Figure 14

30. Critical Thinking A ball drops 20 m. When it has fallen half the distance, or 10 m, half of the energy is potential and half is kinetic. When the ball has fallen for half the amount of time it takes to fall, will more, less, or exactly half of the energy be potential energy?

What's the **Alternative?**

Are you familiar with the terms *energy crisis* and *alternative energy*? An energy crisis can occur when the cost of energy rises too high or the supply of an energy resource dips too low. Scientists and engineers are searching for ways to provide alternative energy that is plentiful, less expensive, and green. Most of these sources have one thing in common—they can be used to turn a turbine attached to an electric generator that converts mechanical energy to electrical energy.

Energy from the wind
As wind spins the blades of wind turbines, some of the wind's kinetic energy turns a shaft connected to an electric generator. A group of wind turbines is called a wind farm.

Energy from water
In hydroelectric plants, the kinetic energy of falling water turns the turbines of electric generators. These plants are highly efficient and produce little pollution.

Energy from Earth
Geothermal plants harness steam produced in Earth's hot interior to turn turbines and generate electricity. In addition, hot springs water and ground-source heat pumps can be used to heat buildings.

Energy from biomass
For centuries, people burned wood for light and heat. Wood is a biomass energy source because it is an organic material that was living or recently living. Today, wood, garbage, waste products, landfill gases, and alcohol fuels are burned as biomass energy sources or produce other energy sources.

Energy from the Sun
A photovoltaic cell is a device that uses radiant energy to knock electrons loose from atoms. The flow of these electrons in an electric circuit is electricity. Solar panels use photovoltaic cells to convert sunlight directly into electricity.

Energy from nuclear reactions
Two nuclear reactions generate large amounts of thermal energy that can be used to heat water and produce steam that spins turbines—fission and fusion. Within the core of a nuclear power plant, fission splits uranium nuclei. Research is ongoing into fusion, the process of joining small nuclei to form a larger nucleus.

GOING**FURTHER** >>>

Class Debate Many factors must be considered when deciding which type of energy to use. Brainstorm a list of important factors, such as cost, availability, and efficiency. Research the pros and cons of various alternative energies, then debate the future of each.

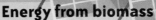

STUDY GUIDE

BIGIDEA   Within a closed, isolated system, energy can change form, but the total energy is constant.

VOCABULARY

- **rotational kinetic energy** (p. 294)
- **potential energy** (p. 294)
- **gravitational potential energy** (p. 295)
- **reference level** (p. 295)
- **elastic potential energy** (p. 298)
- **thermal energy** (p. 300)

SECTION 1 The Many Forms of Energy

MAINIDEA   Kinetic energy is energy due to an object's motion, and potential energy is energy stored due to the interactions of two or more objects.

- The translational kinetic energy of an object is proportional to its mass and the square of its velocity. The rotational kinetic energy of an object is proportional to the object's moment of inertia and the square of its angular velocity.

- The gravitational potential energy of an object-Earth system depends on the object's weight and that object's distance from the reference level. The reference level is the position where the gravitational potential energy is defined to be zero ($h = 0$).

$$GPE = mgh$$

- Elastic potential energy is energy stored due to stretching, compressing, or bending an object.

- Albert Einstein recognized that mass itself is a form of energy. This energy is called rest energy.

$$E_0 = mc^2$$

VOCABULARY

- **law of conservation of energy** (p. 301)
- **mechanical energy** (p. 301)
- **elastic collision** (p. 306)
- **inelastic collision** (p. 306)

SECTION 2 Conservation of Energy

MAINIDEA   In a collision in a closed, isolated system, the total energy is conserved, but kinetic energy might not be conserved.

- The total energy of a closed, isolated system is constant. Within the system, energy can change form, but the total amount of energy does not change. Thus, energy is conserved.

- The sum of kinetic and potential energy is called mechanical energy.

$$ME = KE + PE$$

In a closed, isolated system where the only forms of energy are kinetic energy and potential energy, mechanical energy is conserved. The mechanical energy before an event is the same as the mechanical energy after the event.

$$KE_{before} + PE_{before} = KE_{after} + PE_{after}$$

- Momentum is conserved in collisions if the external force is zero. The kinetic energy may be unchanged or decreased by the collision, depending on whether the collision is elastic or inelastic. The type of collision in which the kinetic energy before and after the collision is the same is called an elastic collision. The type of collision in which the kinetic energy after the collision is less than the kinetic energy before the collision is called an inelastic collision.

Games and Multilingual eGlossary

Vocabulary Practice

SECTION 1 The Many Forms of Energy

Mastering Concepts

Unless otherwise noted, air resistance does no work.

31. Explain how work and energy change are related.

32. Explain how force and energy change are related.

33. What form of energy does a system that contains a wound-up spring toy have? What form of energy does the toy have when it is going? When the toy runs down, what has happened to the energy?

34. A ball is dropped from the top of a building. You choose the top of the building to be the reference level, while your friend chooses the bottom. Explain whether the energy calculated using these two reference levels is the same or different for the following situations:

a. the ball-Earth system's potential energy

b. the change in the system's potential energy as a result of the fall

c. the kinetic energy of the system at any point

35. Can a baseball's kinetic energy ever be negative?

36. Can a baseball-Earth system ever have a negative gravitational potential energy? Explain without using a formula.

37. If a sprinter's velocity increases to three times the original velocity, by what factor does the kinetic energy increase?

38. What energy transformations take place when an athlete is pole-vaulting?

39. The sport of pole-vaulting was drastically changed when the stiff, wooden poles were replaced by flexible fiberglass poles. Explain why.

Mastering Problems

Unless otherwise noted, air resistance does no work.

40. A 1600-kg car travels at a speed of 12.5 m/s. What is its kinetic energy?

41. Shawn and his bike have a combined mass of 45.0 kg. Shawn rides his bike 1.80 km in 10.0 min at a constant velocity. What is the system's kinetic energy?

42. Tony has a mass of 45 kg and a speed of 10.0 m/s.

a. Find Tony's kinetic energy.

b. Tony's speed decreases to 5.0 m/s. Now what is his kinetic energy?

c. Find the ratio of the kinetic energies in parts a and b. Explain how this ratio relates to the change in speed.

43. A racing car has a mass of 1525 kg. What is its kinetic energy if it has a speed of 108 km/h?

44. Katia and Angela each have a mass of 45 kg and are moving together with a speed of 10.0 m/s.

a. What is their combined kinetic energy?

b. What is the ratio of their combined mass to Katia's mass?

c. What is the ratio of their combined kinetic energy to Katia's kinetic energy? Explain how this ratio relates to the ratio of their masses.

45. Train In the 1950s an experimental train with a mass of 2.50×10^4 kg was powered along 509 m of level track by a jet engine that produced a thrust of 5.00×10^5 N. Assume friction is negligible.

a. Find the work done on the train by the jet engine.

b. Find the change in kinetic energy.

c. Find the final kinetic energy of the train if it started from rest.

d. Find the final speed of the train.

46. Car Brakes The driver of the car in **Figure 15** suddenly applies the brakes, and the car slides to a stop. The average force between the tires and the road is 7100 N. How far will the car slide after the brakes are applied?

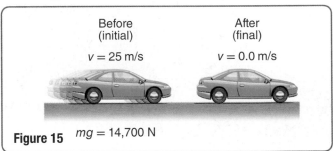

| Before (initial) | After (final) |
| v = 25 m/s | v = 0.0 m/s |

Figure 15   *mg* = 14,700 N

47. A 15.0-kg cart moves down a level hallway with a velocity of 7.50 m/s. A constant 10-N force acts on the cart, slowing its velocity to 3.20 m/s.

a. Find the change in the cart's kinetic energy.

b. How much work was done on the cart?

c. How far did the cart move while the force acted?

48. DeAnna, with a mass of 60.0 kg, climbs 3.5 m up a gymnasium rope. How much energy does a system containing DeAnna and Earth gain from this climb?

49. A 6.4-kg bowling ball is lifted 2.1 m to a shelf. Find the increase in the ball-Earth system's energy.

50. Mary weighs 505 N. She walks down a 5.50-m-high flight of stairs. What is the change in the potential energy of the Mary-Earth system?

51. Weight Lifting A weight lifter raises a 180-kg barbell to a height of 1.95 m. What is the increase in the potential energy of the barbell-Earth system?

52. A science museum display about energy has a small engine that pulls on a rope to lift a block 1.00 m. The display indicates that 1.00 J of work is done. What is the mass of the block?

53. Antwan raised a 12.0-N book from a table 75 cm above the floor to a shelf 2.15 m above the floor. Find the change in the system's potential energy.

54. Tennis A professional tennis player serves a ball. The 0.060-kg ball is in contact with the racket strings, as shown in **Figure 16,** for 0.030 s. If the ball starts at rest, what is its kinetic energy as it leaves the racket?

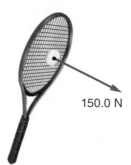

150.0 N

Figure 16

55. Pam has a mass of 45 kg. Her rocket pack supplies a constant force for 22.0 m, and Pam acquires a speed of 62.0 m/s as she moves on frictionless ice.

a. What is Pam's final kinetic energy?

b. What is the magnitude of the force?

56. Collision A 2.00×10^3-kg car has a speed of 12.0 m/s when it hits a tree. The tree doesn't move, and the car comes to rest, as shown in **Figure 17.**

a. Find the change in kinetic energy of the car.

b. Find the amount of work done by the tree on the car as the front of the car crashes into the tree.

c. Find the magnitude of the force that pushed in the front of the car by 50.0 cm.

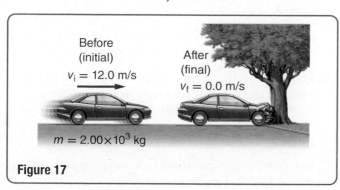

Before (initial)

v_i = 12.0 m/s

After (final)

v_f = 0.0 m/s

$m = 2.00 \times 10^3$ kg

Figure 17

Section 2 **Conservation of Energy**

Mastering Concepts

Unless otherwise noted, air resistance does no work.

57. You throw a clay ball at a hockey puck on ice. The smashed clay ball and the hockey puck stick together and move slowly.

a. Is momentum conserved in the collision? Explain.

b. Is kinetic energy conserved? Explain.

58. Draw energy bar diagrams for the following processes.

a. An ice cube, initially at rest, slides down a frictionless slope.

b. An ice cube, initially moving, slides up a frictionless slope and comes momentarily to rest.

59. BIGIDEA Describe the transformations from kinetic energy to potential energy and vice versa for a roller-coaster ride.

60. Describe how the kinetic energy and elastic potential energy of a bouncing rubber ball decreases. What happens to the ball's motion?

Mastering Problems

Unless otherwise noted, friction is negligible and air resistance does no work.

61. A 10.0-kg test rocket is fired vertically. When the engine stops firing, the rocket's kinetic energy is 1960 J. After the fuel is burned, to what additional height will the rocket rise?

62. A constant net force of 410 N is applied upward to a stone that weighs 32 N. The upward force is applied through a distance of 2.0 m, and the stone is then released. To what height, from the point of release, will the stone rise?

63. A 98.0-N sack of grain is hoisted to a storage room 50.0 m above the ground floor of a grain elevator.

a. How much work was done?

b. What is the increase in potential energy of a system containing the sack of grain and Earth?

c. The rope being used to lift the sack of grain breaks just as the sack reaches the storage room. What kinetic energy does the sack have just before it strikes the ground floor?

64. A 2.0-kg rock is initially at rest. The rock falls, and the potential energy of the rock-Earth system decreases by 407 J. How much kinetic energy does the system gain as the rock falls?

65. A rock sits on the edge of a cliff, as shown in **Figure 18**.

  a. What potential energy does the rock-Earth system possess relative to the base of the cliff?

  b. The rock falls without rolling from the cliff. What is its kinetic energy just before it strikes the ground?

  c. What is the rock's speed as it hits the ground?

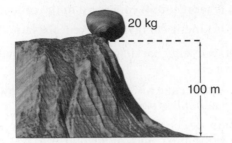

20 kg

100 m

Figure 18

66. Archery An archer fits a 0.30-kg arrow to the bow-string. He exerts an average force of 201 N to draw the string back 0.5 m.

  a. If all the energy goes into the arrow, with what speed does the arrow leave the bow?

  b. The arrow is shot straight up. What is the arrow's speed when it reaches a height of 10 m above the bow?

67. Railroad Car A 5.0×10^5-kg railroad car collides with a stationary railroad car of equal mass. After the collision, the two cars lock together and move off, as shown in **Figure 19.**

  a. Before the collision, the first railroad car was moving at 8.0 m/s. What was its momentum?

  b. What was the total momentum of the two cars after the collision?

  c. What were the kinetic energies of the two cars before and after the collision?

  d. Account for the change of kinetic energy.

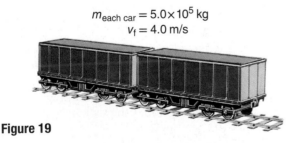

$m_{each\ car} = 5.0\times10^5$ kg
$v_f = 4.0$ m/s

Figure 19

68. Slide Lorena's mass is 28 kg. She climbs the 4.8-m ladder of a slide and reaches a velocity of 3.2 m/s at the bottom of the slide. How much mechanical energy did friction transform to other forms?

69. From what height would a compact car have to be dropped to have the same kinetic energy that it has when being driven at 1.00×10^2 km/h?

70. Problem Posing Complete this problem so that it can be solved using each concept listed below: "A cartoon character is holding a 50-kg anvil at the edge of a cliff…"

  a. conservation of mechanical energy

  b. Newton's second law

71. Kelli weighs 420 N, and she sits on a playground swing that hangs 0.40 m above the ground. Her mom pulls the swing back and releases it when the seat is 1.00 m above the ground.

  a. How fast is Kelli moving when the swing passes through its lowest position?

  b. If Kelli moves through the lowest point at 2.0 m/s, how much work was done on the swing by friction?

72. Hakeem throws a 10.0-g ball straight down from a height of 2.0 m. The ball strikes the floor at a speed of 7.5 m/s. Find the ball's initial speed.

73. A 635-N person climbs a ladder to a height of 5.0 m. Use the person and Earth as the system.

  a. Draw energy bar diagrams of the system before the person starts to climb and after the person stops at the top. Has the mechanical energy changed? If so, by how much?

  b. Where did this energy come from?

Applying Concepts

74. The driver of a speeding car applies the brakes and the car comes to a stop. The system includes the car but not the road. Apply the work-energy theorem to the following situations to describe the changes in energy of the system.

  a. The car's wheels do not skid.

  b. The brakes lock and the car's wheels skid.

75. A compact car and a trailer truck are traveling at the same velocity. Did the car engine or the truck engine do more work in accelerating its vehicle?

76. Catapults Medieval warriors used catapults to assault castles. Some catapults worked by using a tightly wound rope to turn the catapult arm. What forms of energy are involved in catapulting a rock into the castle wall?

77. Two cars collide and come to a complete stop. Where did all their kinetic energy go?

78. Skating Two skaters of unequal mass have the same speed and are moving in the same direction. If the ice exerts the same frictional force on each skater, how will the stopping distances of their bodies compare?

79. You swing a mass on the end of a string around your head in a horizontal circle at constant speed, as shown from above in **Figure 20.**

    a. How much work is done on the mass by the tension of the string in one revolution?

    b. Is your answer to part a in agreement with the work-energy theorem? Explain.

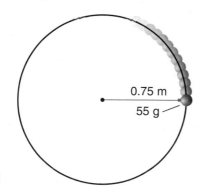

0.75 m

55 g

Figure 20

80. Roller Coaster You have been hired to make a roller coaster more exciting. The owners want the speed at the bottom of the first hill doubled. How much higher must the first hill be built?

81. Two identical balls are thrown from the top of a cliff, each with the same initial speed. One is thrown straight up, the other straight down. How do the kinetic energies and speeds of the balls compare as they strike the ground?

82. Give specific examples that illustrate the following processes.

    a. Work is done on a system, increasing kinetic energy with no change in potential energy.

    b. Potential energy is changed to kinetic energy with no work done on the system.

    c. Work is done on a system, increasing potential energy with no change in kinetic energy.

    d. Kinetic energy is reduced, but potential energy is unchanged. Work is done by the system.

Mixed Review

83. Suppose a chimpanzee swings through the jungle on vines. If it swings from a tree on a 13-m-long vine that starts at an angle of 45°, what is the chimp's speed when it reaches the ground?

84. An 0.80-kg cart rolls down a frictionless hill of height 0.32 m. At the bottom of the hill, the cart rolls on a flat surface, which exerts a frictional force of 2.0 N on the cart. How far does the cart roll on the flat surface before it comes to a stop?

85. A stuntwoman finds that she can safely break her fall from a one-story building by landing in a box filled to a 1-m depth with foam peanuts. In her next movie, the script calls for her to jump from a five-story building. How deep a box of foam peanuts should she prepare?

86. Football A 110-kg football linebacker collides head-on with a 150-kg defensive end. After they collide, they come to a complete stop. Before the collision, which player had the greater kinetic energy? Which player had the greater momentum?

87. A 2.0-kg lab cart and a 1.0-kg lab cart are held together by a compressed spring. The lab carts move at 2.1 m/s in one direction. The spring suddenly becomes uncompressed and pushes the two lab carts apart. The 2-kg lab cart comes to a stop, and the 1.0-kg lab cart moves ahead. How much energy did the spring add to the lab carts?

88. A 55.0-kg scientist roping through the treetops in the jungle sees a lion about to attack an antelope. She swings down from her 12.0-m-high perch and grabs the antelope (21.0 kg) as she swings. They barely swing back up to a tree limb out of the lion's reach. How high is the tree limb?

89. A cart travels down a hill as shown in **Figure 21.** The distance the cart must roll to the bottom of the hill is 0.50 m/sin 30.0° = 1.0 m. The surface of the hill exerts a frictional force of 5.0 N on the cart. Does the cart reach the bottom of the hill?

$m = 0.80$ kg

$F = 5.0$ N

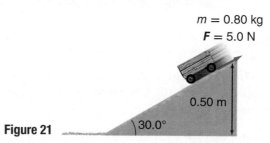

0.50 m

30.0°

Figure 21

90. Object A, sliding on a frictionless surface at 3.2 m/s, hits a 2.0-kg object, B, which is motionless. The collision of A and B is completely elastic. After the collision, A and B move away from each other in opposite directions at equal speeds. What is the mass of object A?

Thinking Critically

91. Apply Concepts A golf ball with a mass of 0.046 kg rests on a tee. It is struck by a golf club with an effective mass of 0.220 kg and a speed of 44 m/s. Assuming that the collision is elastic, find the speed of the ball when it leaves the tee.

92. Apply Concepts A fly hitting the windshield of a moving pickup truck is an example of a collision in which the mass of one of the objects is many times larger than the other. On the other hand, the collision of two billiard balls is one in which the masses of both objects are the same. How is energy transferred in these collisions? Consider an elastic collision in which billiard ball m_1 has velocity v_1 and ball m_2 is motionless.

   a. If $m_1 = m_2$, what fraction of the initial energy is transferred to m_2?

   b. If $m_1 \gg m_2$, what fraction of the initial energy is transferred to m_2?

   c. In a nuclear reactor, neutrons must be slowed down by causing them to collide with atoms. (A neutron is about as massive as a proton.) Would hydrogen, carbon, or iron atoms be more desirable to use for this purpose?

93. Reverse Problem Write a physics problem with real-life objects for which the energy bar diagram in **Figure 22** would be part of the solution:

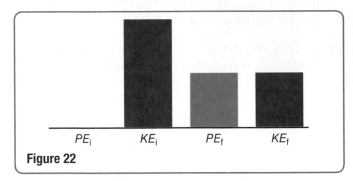

Figure 22

94. Reverse Problem Write a physics problem with real-life objects for which the following equation would be part of the solution:

$$\frac{1}{2}(6.0 \text{ kg})v^2 = (6.0 \text{ kg})(9.8 \text{ N/kg})(0.92 \text{ m})$$

95. Analyze and Conclude In a perfectly elastic collision, both momentum and kinetic energy are conserved. Two balls, with masses m_A and m_B, are moving toward each other with speeds v_A and v_B, respectively. Solve the appropriate equations to find the speeds of the two balls after the collision.

96. Ranking Task Five clay pots are dropped or thrown from the same rooftop as described below. Rank them according to their speed when they strike the ground, from least to greatest. Specifically indicate any ties.

   A. 1 kg, dropped from rest

   B. 1 kg, thrown downward with $v_i = 2$ m/s

   C. 1 kg, thrown upward with $v_i = 2$ m/s

   D. 1 kg, thrown horizontally with $v_i = 2$ m/s

   E. 2 kg, dropped from rest

97. Analyze and Conclude A 25-g ball is fired with an initial speed of v_1 toward a 125-g ball that is hanging motionless from a 1.25-m string. The balls have a perfectly elastic collision. As a result, the 125-g ball swings out until the string makes an angle of 37.0° with the vertical. What is v_1?

Writing in Physics

98. Most energy comes from the Sun and allows us to live and to operate our society. In what forms does this solar energy come to us? Research how the Sun's energy is turned into a form that we can use. After we use the Sun's energy, where does it go? Explain.

99. All forms of energy can be classified as either kinetic or potential energy. How would you describe nuclear, electrical, chemical, biological, solar, and light energy, and why? For each of these types of energy, research what objects are moving and how energy is stored due to the interactions between those objects.

Cumulative Review

100. A satellite is in a circular orbit with a radius of 1.0×10^7 m and a period of 9.9×10^3 s. Calculate the mass of Earth. *Hint: Gravity is the net force on such a satellite. Scientists have actually measured Earth's mass this way.*

101. A 5.00-g bullet is fired with a velocity of 100.0 m/s toward a 10.00-kg stationary solid block resting on a frictionless surface.

   a. What is the change in momentum of the bullet if it is embedded in the block?

   b. What is the change in momentum of the bullet if it ricochets in the opposite direction with a speed of 99 m/s?

   c. In which case does the block end up with a greater speed?

MULTIPLE CHOICE

1. You lift a 4.5-kg box from the floor and place it on a shelf that is 1.5 m above the ground. How much energy did you use in lifting the box?

A. 9.0 J

C. 11 J

B. 49 J

D. 66 J

2. A bicyclist increases her speed from 4.0 m/s to 6.0 m/s. The combined mass of the bicyclist and the bicycle is 55 kg. How much work did the bicyclist do in increasing her speed?

A. 11 J

C. 55 J

B. 28 J

D. 550 J

3. You move a 2.5-kg book from a shelf that is 1.2 m above the ground to a shelf that is 2.6 m above the ground. What is the change in potential energy of a system containing the book and Earth?

A. 1.4 J

C. 3.5 J

B. 25 J

D. 34 J

4. The illustration below shows a ball swinging on a rope. The mass of the ball is 4.0 kg. Assume that the mass of the rope is negligible. Assume that friction is negligible. What is the maximum speed of the ball as it swings back and forth?

A. 0.14 m/s

C. 7.0 m/s

B. 21 m/s

D. 49 m/s

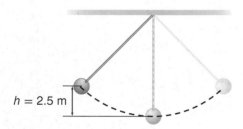

$h = 2.5$ m

5. A hockey puck of mass m slides along the ice at a speed of v_1. It strikes a wall and bounces back in the opposite direction. The energy of the puck after striking the wall is half its initial energy. Assuming friction is negligible, which expression gives the puck's new speed as a function of its initial speed?

A. $\frac{1}{2}v_1$

C. $\frac{\sqrt{2}}{2}(v_1)$

B. $\sqrt{2}(v_1)$

D. $2v_1$

Online Test Practice

6. The illustration below shows a box on a curved, frictionless track. The box starts with zero velocity at the top of the track. It then slides from the top of the track to the horizontal part at the ground. Its velocity just at the moment it reaches the ground is 14 m/s. What is the height (h) from the ground to the top of the track?

A. 7 m

C. 10 m

B. 14 m

D. 20 m

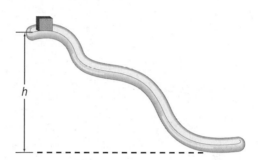

7. You drop a 6.0×10^{-2}-kg ball from a height of 1.0 m above a hard, flat surface. The ball strikes the surface and its energy decreases by 0.14 J. It then bounces back upward. How much kinetic energy does the ball have just after it bounces from the flat surface?

A. 0.20 J

C. 0.45 J

B. 0.59 J

D. 0.73 J

FREE RESPONSE

8. A box sits on a platform supported by a compressed spring. The box has a mass of 1.0 kg. When the spring is released, it does 4.9 J of work on the box, and the box flies upward. What will be the maximum height above the platform reached by the box before it begins to fall?

9. A 90.0-kg hockey player moves at 5.0 m/s and collides head-on with a 110-kg player moving at 3.0 m/s in the opposite direction. After the collision, they move off together at 1.0 m/s. Find the decrease in kinetic energy from the collision for the system containing both hockey players.

NEED EXTRA HELP?

If You Missed Question	1	2	3	4	5	6	7	8	9
Review Section	1	1	1	2	2	2	2	2	2

CHAPTER 12

Thermal Energy

BIGIDEA   Thermal energy is related to the motion of an object's particles and can be transferred and transformed.

SECTIONS

1 **Temperature, Heat, and Thermal Energy**

2 **Changes of State and Thermodynamics**

LaunchLAB

THERMAL ENERGY TRANSFER

How is thermal energy transferred between your hands and a glass of water?

WATCH THIS!

Video

WARM-BLOODED V. COLD-BLOODED

You might remember from biology that some organisms generate body heat internally, while others modify their behavior to regulate their body heat. What is the physics of body heat?

PHYSICS T.V.

(l)Ingram Publishing, (r)Jetta Productions/Dana Neely/Getty Images

Temperature, Heat, and Thermal Energy

Have you seen a balloon outside on a cold day? It was probably shrunken. But if you took it into a warm house, it would return to its normal size. Why does temperature affect the balloon's size?

MAINIDEA

Heat is a transfer of thermal energy that occurs spontaneously from a warmer object to a cooler object.

Essential Questions

• How are temperature and thermal energy related?

• How are thermal equilibrium and temperature related?

• How is thermal energy transferred?

• What is specific heat?

Review Vocabulary

thermal energy the sum of the kinetic and potential energies of the particles that make up an object

New Vocabulary

thermal conduction
thermal equilibrium
heat
convection
radiation
specific heat

Thermal Energy

You have studied how objects collide and trade kinetic energies. Every material is made of microscopic particles. The many particles present in a gas have linear and rotational kinetic energies. The particles also might have potential energy due to their internal bonds and interactions with each other. As gas particles collide with each other and with the walls of the container, they transfer energy. There are numerous molecules that make up the gas, resulting in many collisions. The energies of the particles become randomly distributed.

Thus, it is convenient to discuss the total energy of the particles that compose the gas and the average energy per particle in the gas. Recall that the sum of the particles' energies is the object's thermal energy. The average kinetic energy per particle is related to the temperature of the gas. The relationship between the particles' random motions and the bulk property of the material is described by kinetic theory.

Hot objects and cold objects What makes an object hot? Consider a helium-filled balloon. The balloon is kept inflated by the repeated pounding from helium atoms on the balloon wall. Each of the approximately 10^{22} helium atoms in the balloon collides with the balloon wall, bounces back, and hits the balloon wall again somewhere else. The size and the temperature of the balloon are affected by the average kinetic energy of the helium atoms, as shown in **Figure 1**.

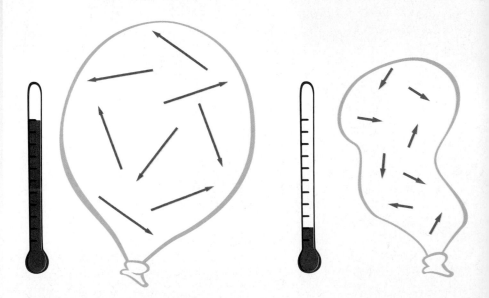

Figure 1 The temperature of an object is related to the average kinetic energy of its particles. The average kinetic energy of the particles that make up a hot object is greater than the average kinetic energy of particles that make up a cold object.

If you put a balloon in sunlight, energy absorbed from the sunlight makes each of the helium atoms move faster in random directions and bounce off the rubber walls of the balloon more often. Each atomic collision with the balloon wall puts a greater force on the balloon and stretches the rubber. Thus, the warmed balloon expands. On the other hand, if you refrigerate a balloon, you will find that it shrinks. It must do so because the particles are moving more slowly. The refrigeration has removed some of their thermal energy.

Thermal energy in solids The atoms or molecules in solids also have kinetic energy, but they are unable to move everywhere as gas atoms do. One way to illustrate the structure of a solid is to picture a number of atoms that are held in place next to each other by atomic forces that act like springs. The atoms cannot move freely, but they do bounce back and forth, with some bouncing more than others. Each atom has some kinetic energy and some potential energy. If a solid has N atoms, then the total thermal energy in the solid is equal to the average kinetic energy plus potential energy per atom times N.

Thermal Energy and Temperature

You have just seen that, on average, a particle in a hot object has more kinetic energy than a particle in a cold object. This does not mean that each of the particles that compose an object has the same amount of energy; they have a wide range of energies. The average kinetic energy of the particles that compose a hot object, however, is greater than the average kinetic energy of the particles that compose a cold object.

To understand this, consider the heights of students in a 9th-grade class and the heights of students in a 12th-grade class. The students' heights vary, as shown in **Figure 2,** but you can calculate the average height for each class. The average height of the 12th-grade class is greater than the average of height of the 9th-grade class, even though some 9th-grade students might be taller than some 12th-grade students.

Temperature depends only on the average kinetic energy of the particles in the object. It does not depend on the number of particles that compose the object. For example, consider the two muffins shown in **Figure 3.** They are at the same temperature, but the large muffin has ten times as many particles as the small muffin. Thus, the large muffin has ten times the thermal energy of the small muffin. The thermal energy of an object depends on both its temperature and the number of particles that make up that object.

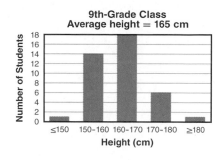

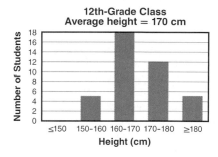

Figure 2 The average height in a 9th-grade class is less than the average height in a 12th-grade class. Similarly, the average kinetic energy of a hot object's particles is greater than the average kinetic energy of a cold object's particles.

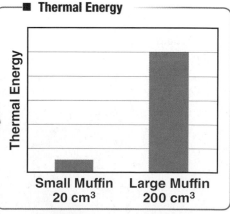

Figure 3 Two muffins at the same temperature can have different thermal energies.

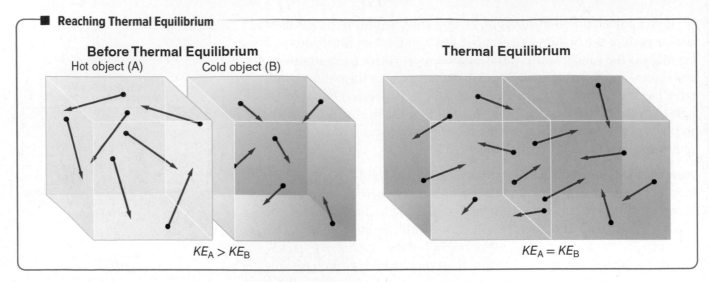

Before Thermal Equilibrium
Hot object (A)   Cold object (B)

$KE_A > KE_B$

Thermal Equilibrium

$KE_A = KE_B$

Figure 4 When a hot object and a cold object are in contact, there is a net transfer of thermal energy from the hot object to the cold object. When the two objects reach thermal equilibrium, the transfer of energy between the objects is equal and the objects are at the same temperature.

Equilibrium and Thermometers

How do you measure your body temperature? You might place a thermometer in your mouth and wait for a moment before checking the reading. Measuring your temperature involves random collisions and energy transfers between the particles of the thermometer and the particles of your body. Your body is hotter than the thermometer. That is, the average kinetic energy of the particles that compose your body is greater than the average kinetic energy of the thermometer's particles.

When the cool thermometer touches your skin, the particles of your skin collide with the particles of the thermometer. On average, the more energetic skin particles will transfer energy to the less energetic particles of the thermometer. **Thermal conduction** is the transfer of thermal energy that occurs when particles collide. As a result of these collisions, the thermal energy of the thermometer's particles increases. At the same time, the thermal energy of your skin's particles decreases.

Thermal equilibrium The thermometer's particles also transfer energy to your body's particles. As the thermometer's particles gain more energy, the amount of energy they give back to the skin increases. At some point, the rate of energy transfer from the thermometer to your body is equal to the rate of transfer in the other direction. At this point, your body and the thermometer have reached thermal equilibrium. **Thermal equilibrium** is the state in which the rates of thermal energy transfer between two objects are equal and the objects are at the same temperature. **Figure 4** shows two blocks reaching equilibrium.

☑ **READING CHECK Identify** a situation where two objects are in thermal equilibrium and a situation where two objects are not in thermal equilibrium.

Figure 5 Liquid crystal thermometers change color with temperature.

Summarize the process that occurs when this thermometer is placed on your forehead.

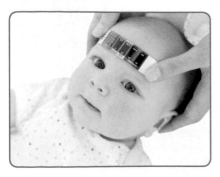

Thermometers Every thermometer has some useful property that changes with temperature. Household thermometers often contain colored alcohol that expands when heated. The hotter the thermometer, the more the alcohol expands and the higher it rises in the tube. The liquid crystal thermometer in **Figure 5** uses a variety of long molecules that rearrange and cause a color change at specific temperatures. Medical thermometers and the thermometers that monitor automobile engines use very small, temperature-sensitive electronic circuits to take rapid measurements.

Temperature limits You might say that a fire is hot and a freezer is cold. But the temperatures of everyday objects are only a small part of the wide range of temperatures present in the universe, as shown in **Figure 6.** Temperatures do not appear to have an upper limit. The interior of the Sun is at least $1.5 \times 10^7 \, °C$. Supernova cores are even hotter. On the other hand, liquefied gases can be very cold. For example, helium liquefies at $-269°C$. Even colder temperatures can be reached by making use of special properties of solids, helium isotopes, and atoms and lasers.

Temperatures do, however, have a lower limit. Generally, materials contract as they cool. If an ideal atomic gas in a balloon were cooled to $-273.15°C$, it would contract in such a way that it occupied a volume that is only the size of the atoms, and the atoms would become motionless. At this temperature, all the thermal energy that could be removed has been removed from the gas, and the temperature cannot be reduced any further. Therefore, there can be no temperature less than $-273.15°C$, which is called absolute zero.

✅ **READING CHECK Explain** why the term *absolute zero* is appropriate for the coldest temperature possible.

Figure 6 Temperatures in the universe range from just above absolute zero to more than 10^{10} K.

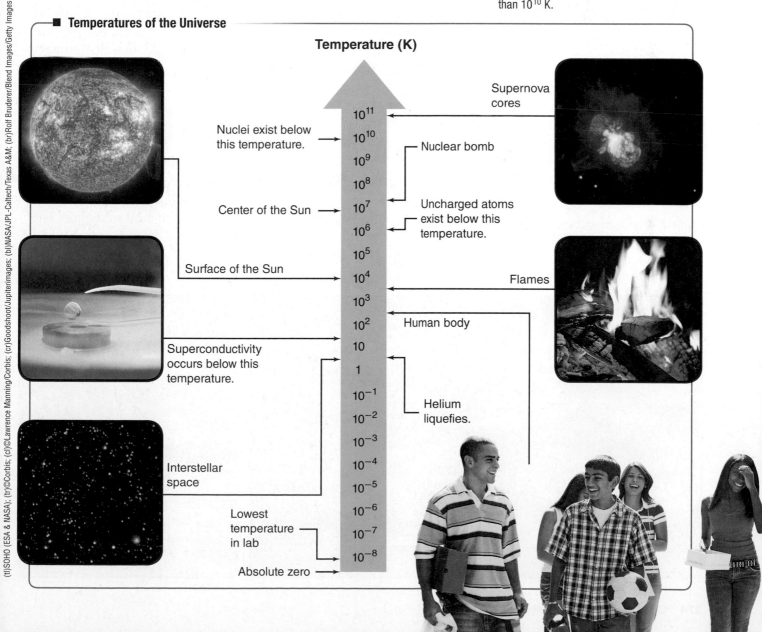

■ **Temperatures of the Universe**

Temperature (K)

Nuclei exist below this temperature. → 10^{11}
10^{10}
10^9 — Nuclear bomb
10^8
Center of the Sun → 10^7 ← Uncharged atoms exist below this temperature.
10^6
10^5
Surface of the Sun → 10^4
10^3 — Flames
10^2 — Human body
Superconductivity occurs below this temperature. → 10
1
10^{-1} — Helium liquefies.
10^{-2}
10^{-3}
Interstellar space → 10^{-4}
10^{-5}
Lowest temperature in lab → 10^{-6}
10^{-7}
Absolute zero → 10^{-8}

Supernova cores

Comparing Temperature Scales

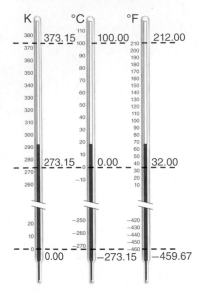

Figure 7 The Kelvin and Celsius scales are used by scientists. In the United States, the Fahrenheit scale is often used for weather reports and cooking.

Temperature scales In the United States, weather agencies report temperatures in degrees Fahrenheit. Scientists, however, use the Celsius and Kelvin scales. The Celsius scale is based on the properties of water and was devised in 1741 by Swedish physicist Anders Celsius. On the Celsius scale, the freezing point of pure water at sea level is defined to be 0°C. The boiling point of pure water at sea level is defined to be 100°C. The Celsius scale is useful for day-to-day measurements of temperature.

The Kelvin scale On the Celsius scale, temperatures can be negative. Negative temperatures suggest a particle could have negative kinetic energy. Because temperature represents average kinetic energy of the object's particles, it makes more sense to use a temperature scale whose zero temperature is where the particles' kinetic energy is also zero. Therefore the zero point of the Kelvin scale is defined to be absolute zero. On the Kelvin scale, the freezing point of water (0°C) is about 273 K and the boiling point of water is about 373 K. Each interval on this scale, called a kelvin, is equal to 1°C. Thus, $T_C + 273 = T_K$. **Figure 7** compares the Fahrenheit, Celsius, and Kelvin scales.

Heat and Thermal Energy Transfer

When two objects come in contact with each other, they redistribute their thermal energies. **Heat** (Q) is the transfer of thermal energy, which occurs spontaneously from a hotter object to a cooler object. Thermal energy cannot be transferred from a colder object to a hotter object without work being done. Like work and energy, heat is measured in joules. In the thermometer example, thermal energy was transferred from the warm skin to the cold thermometer because of the collisions of particles. If thermal energy has been absorbed by an object, Q is positive. If thermal energy is transferred from an object, Q is negative.

Conduction, convection, and radiation **Figure 8** shows the three types of heat—conduction, convection, and radiation. If you place one end of a metal rod in a flame, the hot gas conducts heat to the rod. The other end of the rod also becomes warm because the particles that make up the rod conduct thermal energy to their neighbors.

Figure 8 Thermal energy can be transferred by conduction, convection, or radiation.

Identify other common occurrences of conduction, convection, and radiation.

| Conduction | Convection | Radiation |

Have you ever noticed the motion on the surface of a pot of water just about to boil? The water at the bottom of the pot is heated by conduction, expands, and floats to the top, while the colder, denser water at the top sinks to the bottom. The motion of the hot water rapidly carries heat from the bottom of the pot to the top surface of the water. Heating caused by the motion of fluid in a liquid or gas due to temperature differences is called **convection.** Atmospheric turbulence is caused by convection of gases in the atmosphere. Thunderstorms and hurricanes are excellent examples of large-scale atmospheric convection. Convection also contributes to ocean currents that move water and materials over large distances.

The third method of thermal transfer, unlike the first two, does not depend on the presence of matter. The Sun warms Earth from over 150 million km away via **radiation,** which is the transfer of energy by electromagnetic waves. These waves carry the energy from the hot Sun through the vacuum of space to the much cooler Earth.

Specific Heat

Some objects are easier to warm up than others. On a bright summer day, the Sun radiates thermal energy to the sand on a beach and to the ocean water. The sand on the beach becomes quite hot, while the ocean water stays relatively cool. When an object is heated, its thermal energy increases and its temperature can increase. The amount of the increase in temperature depends on the size of the object and its composition.

The **specific heat** of a material is the amount of energy that must be added to a unit mass of the material to raise its temperature by one temperature unit. In SI units, specific heat (C) is measured in J/(kg·K).

☑ **READING CHECK Define** the term *specific heat.*

Table 1 provides values of specific heat for some common substances. For example, 897 J must be added to 1 kg of aluminum to raise its temperature by 1 K. The specific heat of aluminum is therefore 897 J/(kg·K). Materials with different specific heats are used for different purposes. Metals, such as those used to make the pans in **Figure 9,** have low specific heats and are good thermal conductors. Notice that liquid water has a high specific heat compared to other substances. Ice and water vapor also have relatively high specific heats. These high specific heats have had significant effects on our climate and our bodies.

Figure 9 These pans are made of stainless steel and have copper bottoms and plastic handles.

Explain how the selection of these materials is related to their specific heats.

Table 1 Specific Heat of Common Substances			
Material	**Specific Heat (J/(kg·K))**	**Material**	**Specific Heat (J/(kg·K))**
Aluminum	897	Lead	130
Brass	376	Methanol	2450
Carbon	710	Silver	235
Copper	385	Water Vapor	2020
Glass	840	Water	4180
Ice	2060	Zinc	388
Iron	450		

HEATING AND COOLING
How does a constant supply of
thermal energy affect the
temperature of water?

Measuring heat When a substance is heated, the substance's temperature can change. The change of temperature (ΔT) depends on heat (Q), the mass of the substance, and the specific heat of the substance. By using the following equation, you can calculate the heat (Q) required to change the temperature of an object.

HEAT

Heat is equal to the mass of an object times the specific heat of the object times the difference between the final and initial temperatures.

$$Q = mC\Delta T = mC(T_f - T_i)$$

For example, when the temperature of 10.0 kg of water is increased from 80 K to 85 K, the heat is $Q = (10.0 \text{ kg})(4180 \text{ J/(kg·K)})(85 \text{ K} - 80 \text{ K}) = 2.1 \times 10^5$ J. Remember that the temperature interval for the Kelvin scale is the same as that for the Celsius scale. For this reason, you can calculate ΔT on the Kelvin scale or on the Celsius scale.

EXAMPLE PROBLEM 1

Find help with **order of operations.** Math Handbook

HEAT TRANSFER A 5.10-kg cast-iron skillet is heated on the stove from 295 K to 373 K. How much thermal energy had to be transferred to the iron?

1 ANALYZE AND SKETCH THE PROBLEM

Sketch the thermal energy transfer into the skillet from the stove top.

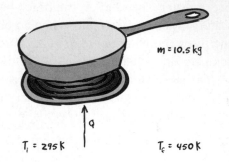

$m = 10.5 \text{ kg}$

$T_i = 295 \text{ K}$ $T_f = 450 \text{ K}$

KNOWN		UNKNOWN
$m = 5.10$ kg	$C = 450$ J/(kg·K)	$Q = ?$
$T_i = 295$ K	$T_f = 373$ K	

2 SOLVE FOR THE UNKNOWN

$Q = mC(T_i - T_f)$

$\quad = (5.10 \text{ kg})(450 \text{ J/(kg·K)})(373 \text{ K} - 295 \text{ K})$ ◀ Substitute m = 5.10 kg, C = 450 J/(kg·K), T_f = 373 K, T_i = 295 K.

$\quad = 1.8 \times 10^5$ J

3 EVALUATE THE ANSWER

- **Are the units correct?** Heat is measured in joules.

- **Does the sign make sense?** Temperature increased, so Q is positive.

PRACTICE PROBLEMS

Do additional problems. Online Practice

1. When you turn on the hot water to wash dishes, the water pipes heat up. How much thermal energy is absorbed by a copper water pipe with a mass of 2.3 kg when its temperature is raised from 20.0°C to 80.0°C?

2. Electrical power companies sell electrical energy by the kilowatt-hour, where 1 kWh = 3.6×10^6 J. Suppose that it costs $0.15 per kWh to run your electric water heater. How much does it cost to heat 75 kg of water from 15°C to 43°C to fill a bathtub?

3. **CHALLENGE** A car engine's cooling system contains 20.0 L of water (1 L of water has a mass of 1 kg).

   a. What is the change in the temperature of the water if 836.0 kJ of thermal energy is added?

   b. Suppose that it is winter, and the car's cooling system is filled with methanol. The density of methanol is 0.80 g/cm³. What would be the increase in temperature of the methanol if it absorbed 836.0 kJ of thermal energy?

   c. Which coolant, water or methanol, would better remove thermal energy from a car's engine? Explain.

EXAMPLE PROBLEM

PRACTICE PROBLEMS

Measuring Specific Heat

A calorimeter, such as the simple one shown in **Figure 10,** is a device that measures changes in thermal energy. A calorimeter is carefully insulated so that thermal energy transfer to the external world is kept to a minimum. A measured mass of a substance that has been heated to a high temperature (T_A) is placed in the calorimeter. The calorimeter also contains a known mass of cold water at a measured temperature (T_B). Thermal energy is transferred from the warmer substance to the cooler water until they come to an equilibrium temperature (T_f). By measuring these three temperatures, the specific heat of the unknown substance can be calculated.

Energy conservation The operation of a calorimeter depends on the conservation of energy in an isolated, closed system composed of the water and the substance being measured. Energy can neither enter nor leave this system but can be transferred from one part of the system to another. Therefore, if the thermal energy of the test substance changes by an amount (ΔE_A) then the change in thermal energy of the water (ΔE_B) must be related by the equation $\Delta E_A + \Delta E_B = 0$. This can be rearranged to form the equation:

$$\Delta E_A = -\Delta E_B$$

The change in energy of the cold water is positive, and the change in energy of the hot test substance is negative. A positive change in energy indicates a rise in temperature, and a negative change in energy indicates a fall in temperature.

In an isolated, closed system, no work is done, so the change in thermal energy for each substance is equal to the heat and can be expressed by the following equation:

$$\Delta E = Q = mC\Delta T = mC(T_f - T_i)$$

Combining this equation with $\Delta E_A = -\Delta E_B$ gives:

$$m_A C_A(T_f - T_A) = -m_B C_B(T_f - T_B)$$

The final temperatures of the two substances are equal because they are in thermal equilibrium. Solving for the unknown specific heat (C_A) gives the equation:

$$C_A = \frac{-m_B C_B \Delta T_B}{m_A \Delta T_A}$$

Calorimeter

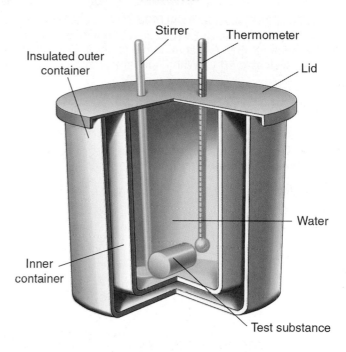

Figure 10 In a simple calorimeter, a hot test substance and a known volume of cold water are placed in an isolated system and allowed to come to thermal equilibrium. The ideal calorimeter has perfect insulation and does not transfer thermal energy to or from the outside. More sophisticated types of calorimeters are used to measure chemical reactions and the energy content of various foods.

View an **animation about using calorimetry to determine specific heat.**

 Concepts In Motion

APPLYING PRACTICES

Plan and Conduct an Investigation Go to the resources tab in ConnectED to find the Applying Practices worksheet *Coffee Cup Calorimetry*.

PhysicsLAB

HOW MANY CALORIES ARE THERE?
FORENSIC LAB How can a calorimeter be used to determine energy transfers?

EXAMPLE PROBLEM 2

Find help with **operations with significant figures.** Math Handbook

TRANSFERRING HEAT IN A CALORIMETER A calorimeter contains 0.50 kg of water at 15°C. A 0.10-kg block of an unknown substance at 62°C is placed in the water. The final temperature of the system is 16°C. What is the substance?

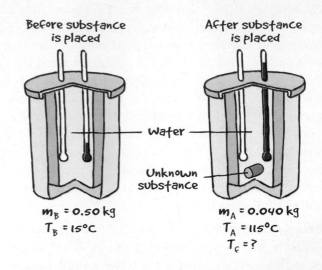

Before substance is placed

After substance is placed

water

Unknown substance

$m_B = 0.50$ kg
$T_B = 15°C$

$m_A = 0.040$ kg
$T_A = 115°C$
$T_f = ?$

1 ANALYZE AND SKETCH THE PROBLEM

- Let the unknown be sample A and water be sample B.
- Sketch the transfer of thermal energy from the hotter unknown sample to the cooler water.

KNOWN

$m_A = 0.10$ kg
$T_A = 62°C$
$m_B = 0.50$ kg
$C_B = 4180$ J/(kg·K)
$T_B = 15°C$
$T_f = 16°C$

UNKNOWN

$C_A = ?$

2 SOLVE FOR THE UNKNOWN

Determine the final temperature using the following equation. Beware of the minus signs.

$$C_A = \frac{-m_B C_B \Delta T_B}{m_A \Delta T_A}$$

◀ Substitute $m_A = 0.10$ kg, $T_A = 62°C$, $m_B = 0.50$ kg, $C_B = 4180$ J/(kg·K), $T_B = 15°C$, $T_f = 16°C$.

$$= \frac{-(0.50 \text{ kg})(4180 \text{ J/(kg·K)})(16°C - 15°C)}{(0.10 \text{ kg})(16°C - 62°C)}$$

$$= 450 \text{ J/(kg·K)}$$

According to **Table 1**, the specific heat of the unknown substance equals that of iron.

3 EVALUATE THE ANSWER

- **Are the units correct?** Specific heat is measured in J/(kg·K).
- **Is the magnitude realistic?** The answer is of the same magnitude as most metals listed in **Table 1**.

PRACTICE PROBLEMS

Do additional problems. Online Practice

4. A 1.00×10^2-g aluminum block at 100.0°C is placed in 1.00×10^2 g of water at 10.0°C. The final temperature of the mixture is 26.0°C. What is the specific heat of the aluminum?

5. Three metal fishing weights, each with a mass of 1.00×10^2 g and at a temperature of 100.0°C, are placed in 1.00×10^2 g of water at 35.0°C. The final temperature of the mixture is 45.0°C. What is the specific heat of the metal in the weights?

6. A 2.00×10^2-g sample of water at 80.0°C is mixed with 2.00×10^2 g of water at 10.0°C in a calorimeter. What is the final temperature of the mixture?

7. A 1.50×10^2-g piece of glass at a temperature of 70.0°C is placed in a container with 1.00×10^2 g of water initially at a temperature of 16.0°C. What is the equilibrium temperature of the water?

8. **CHALLENGE** A 4.00×10^2-g sample of water at 15.0°C is mixed with 4.00×10^2 g of water at 85.0°C. After the system reaches thermal equilibrium, 4.00×10^2 g of methanol at 15°C is added. Assume there is no thermal energy lost to the surroundings. What is the final temperature of the mixture?

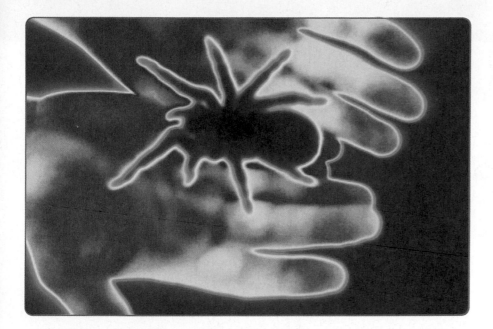

Figure 11 Cold-blooded animals depend on external sources of thermal energy to maintain their body temperatures. In contrast, warm-blooded animals maintain their body temperatures internally. In this thermal image, the spider is at the same temperature as the surrounding air, but the person's hands are significantly warmer than the surrounding air.

Animals and Thermal Energy

▶ **CONNECTION TO BIOLOGY** Animals can be divided into two groups based on how they control their body temperatures. Most, such as the spider in **Figure 11,** are cold-blooded animals. Their body temperatures depend on the environment. A cold-blooded animal regulates the transfer of thermal energy by its behavior, such as hiding under a rock to keep cool or sunning itself to keep warm.

The others are warm-blooded animals whose body temperatures are controlled internally. That is, a warm-blooded animal's body temperature remains stable regardless of the temperature of the environment. For example, humans are warm-blooded and have body temperatures close 37°C. To regulate its body temperature, a warm-blooded animal relies on bodily responses initiated by the brain, such as shivering and sweating, to counteract a rise or fall in body temperature.

SECTION 1 REVIEW

Section Self-Check Check your understanding.

9. **MAIN IDEA** The hard tile floor of a bathroom always feels cold to bare feet even though the rest of the room is warm. Is the floor colder than the rest of the room?

10. **Temperature** Make the following conversions:

   a. 5°C to kelvins

   b. 34 K to degrees Celsius

   c. 212°C to kelvins

   d. 316 K to degrees Celsius

11. **Units** Are the units the same for heat (Q) and specific heat (C)? Explain.

12. **Types of Energy** Describe the mechanical energy and the thermal energy of a bouncing basketball.

13. **Thermal Energy** Could the thermal energy of a bowl of hot water equal that of a bowl of cold water? Explain your answer.

14. **Cooling** On a dinner plate, a baked potato always stays hot longer than any other food. Why?

15. **Heat and Food** It takes much longer to bake a whole potato than potatoes that have been cut into pieces. Why?

16. **Cooking** Stovetop pans are made from metals such as copper, iron, and aluminum. Why are these materials used?

17. **Specific Heat** If you take a plastic spoon out of a cup of hot cocoa and put it in your mouth, you are not likely to burn your tongue. However, you could very easily burn your tongue if you put the hot cocoa in your mouth. Why?

18. **Critical Thinking** As water heats in a pot on a stove, it might produce some mist above its surface right before the water begins to roll. What is happening?

Changes of State and Thermodynamics

PHYSICS 4 YOU

You might have heard of perpetual motion machines. These machines, which would theoretically continue to move forever once started, do not actually work. If they did, they would violate the laws of thermodynamics.

MAINIDEA

When thermal energy is transferred, energy is conserved and the total entropy of the universe will increase.

Essential Questions

- How are the heats of fusion and vaporization related to changes in state?
- What is the first law of thermodynamics?
- How do engines, heat pumps, and refrigerators demonstrate the first law of thermodynamics?
- What is the second law of thermodynamics?

Review Vocabulary

joule (J) the SI unit of work and energy; 1 J of work is done when a force of 1 N acts on an object over a displacement of 1 m

New Vocabulary

heat of fusion
heat of vaporization
first law of thermodynamics
heat engine
second law of thermodynamics
entropy

Changes of State

In a steam engine, heat turns liquid water into water vapor. The water vapor pushes a piston to turn the engine, and then the water vapor cools and condenses into a liquid again. Adding thermal energy to water can change the water's structure as well as its temperature.

The three most common states of matter on Earth are solid, liquid, and gas. As the temperature of a solid rises, that solid usually changes to a liquid. At even higher temperatures, it becomes a gas. If the gas cools, it returns to the liquid state. If the cooling continues, the liquid will return to the solid state. How can these changes be explained? Recall that when the thermal energy of a material changes, the motion of its particles also changes, as does the temperature.

Figure 12 diagrams the changes of state as thermal energy is added to 1.0 kg of water starting at 243 K (ice) and continuing until that water reaches 473 K (water vapor). Between points A and B, the ice is warmed to 273 K. At this point, the added thermal energy gives the water molecules enough energy to partially overcome the forces holding them together. The particles are still touching each other, but they have more freedom of movement. Eventually, the molecules become free enough to slide past each other.

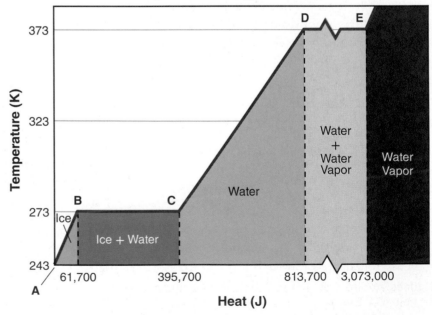

Adding Thermal Energy to Water

Figure 12 Thermal energy added to a substance can raise the temperature or cause a change in state. Note that the scale is broken between points D and E.

View an **animation of changes of state.**

Concepts In Motion

Melting point At this point, water changes from a solid to a liquid, just as the snowman in **Figure 13** does. The temperature at which this change occurs is the melting point of a substance. When a substance melts, the addition of thermal energy allows the particles to move, rotate, and vibrate in ways not available in the solid. Each of these new motions might add new modes of kinetic or potential energy. The added thermal energy does not change the temperature of the material. This can be observed between points B and C in **Figure 12**, where the added thermal energy melts all the ice at a constant 273 K.

Boiling point Once the ice is completely melted, adding more thermal energy increases the motion of the water molecules again. The temperature rises as shown between points C and D in **Figure 12**. As the temperature increases further, some of the particles that make up the liquid acquire enough energy to break free from the other particles.

At a specific temperature, known as the boiling point, adding more energy to a substance causes it to undergo another change of state. All the added thermal energy converts it from the liquid state to the gas state. As in melting, the temperature does not rise while a liquid boils, as shown between points D and E in **Figure 12**. After the water is entirely converted to gas, any added thermal energy again increases the motion of the molecules, and the temperature rises. After point E, the water vapor is heated to temperatures greater than 373 K.

Heat of fusion and heat of vaporization The amount of thermal energy needed to melt 1 kg of a substance is called the substance's **heat of fusion** (H_f). For ice the heat of fusion is 3.34×10^5 J/kg. If 1 kg of ice at its melting point, 273 K, absorbs 3.34×10^5 J, the ice becomes 1 kg of water at the same temperature, 273 K. The added energy causes a change in state but not in temperature. The horizontal distance in **Figure 12** from point B to point C represents the heat of fusion.

The thermal energy needed to vaporize 1 kg of a liquid is called the **heat of vaporization** (H_v). Water's heat of vaporization is 2.26×10^6 J/kg. The horizontal distance from point D to point E in **Figure 12** represents the heat of vaporization. Every material has a characteristic heat of fusion and heat of vaporization. The values of some heats of fusion (H_f) and heats of vaporization (H_v) are shown in **Table 2**.

Figure 13 Thermal energy is transferred from the warmer air to the snowman, causing the snowman to melt.

Watch a **BrainPOP video on how matter changes states.**

Video

MiniLAB

MELTING
How does heating affect the temperature and state of water?

Table 2 Heats of Fusion and Vaporization of Common Substances		
Material	**Heat of Fusion H_f (J/kg)**	**Heat of Vaporization H_v (J/kg)**
Copper	2.05×10^5	5.07×10^6
Mercury	1.15×10^4	2.72×10^5
Gold	6.30×10^4	1.64×10^6
Methanol	1.09×10^5	8.78×10^5
Iron	2.66×10^5	6.29×10^6
Silver	1.04×10^5	2.36×10^6
Lead	2.04×10^4	8.64×10^5
Water (ice)	3.34×10^5	2.26×10^6

Figure 14 One way to measure the absorption of energy by a material is to add energy from a constant source of thermal energy and measure the change in temperature over time. The plot of temperature v. time is called a heating curve. For this figure, a beaker of cold water was placed on a hot plate. The resulting heating curve is graphed.

Explain why thermal energy must be added at a constant rate to calculate the specific heat of water from this graph.

APPLYING PRACTICES

Develop and Use Models Go to the resources tab in ConnectED to find the Applying Practices worksheet *Modeling Energy at Different Scales*.

Investigate **phase changes.**

Virtual Investigation

PhysicsLAB

HEAT OF FUSION

How can you measure the heat of fusion of ice?

APPLYING PRACTICES

Develop and Use Models Go to the resources tab in ConnectED to find the Applying Practices worksheet *Modeling Changes in Energy*.

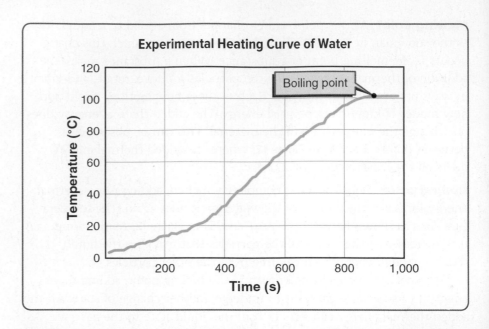

Experimental Heating Curve of Water

Boiling point

Energy and changes of state There is a definite slope to the graph in **Figure 14** between about 300 s and 800 s. Heat is added at a constant rate, so this slope is proportional to the reciprocal of the specific heat of water. The slope between points A and B in **Figure 12** is proportional to the reciprocal of the specific heat of ice, and the slope above point E is proportional to the reciprocal of the specific heat of water vapor. The slope for water is less than those of both ice and water vapor. This is because water has a greater specific heat than does ice or water vapor. The heat (Q) required to melt a solid of mass (m) is given by the following equation.

HEAT REQUIRED TO MELT A SOLID
The heat required to melt a solid is equal to the mass of the solid times the heat of fusion of the solid.

$$Q = mH_f$$

Similarly, the heat (Q) required to vaporize a mass (m) of liquid is given by the following equation.

HEAT REQUIRED TO VAPORIZE A LIQUID
The heat required to vaporize a liquid is equal to the mass of the liquid times the heat of vaporization of the liquid.

$$Q = mH_v$$

When a liquid freezes, an amount of thermal energy ($Q = -mH_f$) must be removed from the liquid to turn it into a solid. The negative sign indicates that the thermal energy is transferred from the sample to the external world. In the same way, when a vapor condenses to a liquid, an amount of thermal energy ($Q = -mH_v$) must be removed from the vapor.

Water absorbs significant amounts of thermal energy when it melts or evaporates. Every day you use the large heats of fusion and vaporization of water. Each gram of sweat that evaporates from your skin carries off about 2.3 kJ of thermal energy. This is one cooling process that many warm-blooded animals use to regulate their body temperatures. Similarly the melting of a 24-g cube of ice absorbs enough thermal energy, 8 kJ, to cool a glass of water by about 30°C.

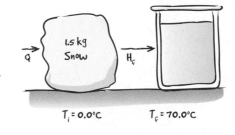

HEAT Suppose that you are camping in the mountains. You need to melt 1.50 kg of snow at 0.0°C and heat it to 70.0°C to make hot cocoa. How much heat will you need?

1 ANALYZE AND SKETCH THE PROBLEM

- Sketch the relationship between heat and water in its solid and liquid states.
- Sketch the transfer of heat as the temperature of the water increases.

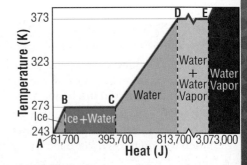

KNOWN		UNKNOWN
m = 1.50 kg	H_f = 3.34×10⁵ J/kg	$Q_{melt\ ice}$ = ?
T_i = 0.0°C	T_f = 70.0°C	$Q_{heat\ liquid}$ = ?
C = 4180 J/(kg·K)		Q_{total} = ?

2 SOLVE FOR THE UNKNOWN

Calculate the heat needed to melt ice.

$Q_{melt\ ice} = mH_f$
$= (1.50\ kg)(3.34×10^5\ J/kg)$
$= 5.01×10^5\ J = 5.01×10^2\ kJ$ ◀ **Substitute** m = 1.50 kg, H_f = 3.34×10⁵ J/kg.

Calculate the temperature change.

$\Delta T = T_f - T_i$
$= 70.0°C - 0.0°C = 70.0°C$ ◀ **Substitute** T_f = 70.0°C, T_i = 0.0°C.

Calculate the heat needed to raise the water temperature.

$Q_{heat\ liquid} = mC\Delta T$
$= (1.50\ kg)(4180\ J/(kg·K))(70.0°C)$
$= 4.39×10^5\ J = 4.39×10^2\ kJ$ ◀ **Substitute** m = 1.50 kg, C = 4180 J/(kg·K), ΔT = 70.0°C.

Calculate the total amount of heat needed.

$Q_{total} = Q_{melt\ ice} + Q_{heat\ liquid}$
$= 5.01×10^2\ kJ + 4.39×10^2\ kJ$ ◀ **Substitute** $Q_{melt\ ice}$ = 5.01×10² kJ, $Q_{heat\ liquid}$ = 4.39×10² kJ.
$= 9.40×10^2\ kJ$

3 EVALUATE THE ANSWER

- **Are the units correct?** Energy units are in joules.
- **Does the sign make sense?** Q is positive when thermal energy is absorbed.
- **Is the magnitude realistic?** To check the magnitude, perform a quick estimation:

$Q = (1.5\ kg)(300,000\ J/kg) + (1.5\ kg)(4000\ J/(kg·K))(70\ K) = 9×10^2\ kJ.$

PRACTICE PROBLEMS   Do additional problems. **Online Practice**

19. How much thermal energy is absorbed by $1.00×10^2$ g of ice at −20.0°C to become water at 0.0°C?

20. A $2.00×10^2$-g sample of water at 60.0°C is heated to water vapor at 140.0°C. How much thermal energy is absorbed?

21. Use the graph in **Figure 15** to calculate the heat of fusion and heat of vaporization of water in joules per kilogram.

22. A steel plant operator wishes to change 100 kg of 25°C iron into molten iron (melting point = 1538°C). How much thermal energy must be added?

23. **CHALLENGE** How much thermal energy is needed to change $3.00×10^2$ g of ice at −30.0°C to water vapor at 130.0°C?

Figure 15

Figure 16 Steam engines change thermal energy into useful mechanical energy.

The First Law of Thermodynamics

The first steam engines were built in the eighteenth century and used to power trains and factories. Steam engines, such as the one in **Figure 16**, change thermal energy into mechanical energy. The invention of the steam engine contributed greatly to the Industrial Revolution and to the study of how heat is related to work. The study of how thermal energy is transformed into other forms of energy is called thermodynamics.

Before 1900, scientists did not realize that the concepts of thermodynamics were linked to the motions of particles in an object and they considered thermodynamics to be a separate topic from mechanics. Today, engineers routinely apply the concepts of thermodynamics to develop higher performance refrigerators, automobile engines, aircraft engines, and numerous other machines.

The first law developed for thermodynamics was a statement about what thermal energy is and where it can go. As you know, you can raise the temperature of a glass of cold water by placing it on a hot plate by stirring it. That is, you can increase the water's thermal energy by heating or by doing work on it. If we consider the system to be the water, the work the system does on you is equal to the negative of the work you do on the system.

The **first law of thermodynamics** states that the change in thermal energy (ΔU) of an object is equal to the heat (Q) that is added to the object minus the work (W) done by the object. Note that ΔU, Q, and W are all measured in joules, the unit of energy.

THE FIRST LAW OF THERMODYNAMICS
The change in thermal energy of an object is equal to the heat added to the object minus the work done by the object.

$$\Delta U = Q - W$$

The first law of thermodynamics is merely a restatement of the law of conservation of energy, which states that energy is neither created nor destroyed but can be changed into other forms.

Another example of changing the amount of thermal energy in a system is a hand pump used to inflate a bicycle tire. As a person pumps, the air and the hand pump become warm. The mechanical energy in the moving piston is converted into thermal energy of the gas. Similarly, other forms of energy, such as light, sound, and electrical energy, can be changed into thermal energy. For example, a toaster converts electrical energy into radiant energy when it toasts bread. You can probably think of many more examples from your everyday life.

MiniLAB

CONVERT ENERGY
How are work and thermal energy related?

Do additional problems. **Online Practice**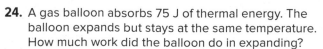

PRACTICE PROBLEMS

24. A gas balloon absorbs 75 J of thermal energy. The balloon expands but stays at the same temperature. How much work did the balloon do in expanding?

25. A drill bores a small hole in a 0.40-kg block of aluminum and heats the aluminum by 5.0°C. How much work did the drill do in boring the hole?

26. How many times would you have to drop a 0.50-kg bag of lead shot from a height of 1.5 m to heat the shot by 1.0°C?

27. When you stir a cup of tea, you do about 0.050 J of work each time you circle the spoon in the cup. How many times would you have to stir the spoon to heat a 0.15-kg cup of tea by 2.0°C?

28. CHALLENGE An expansion valve does work on 100 g of water. The system is isolated, and all of the work is used to convert the 90°C water into water vapor at 110°C. How much work does the expansion valve do on the water?

Heat engines When you rub your hands together to warm them, you convert mechanical energy into thermal energy. This energy conversion is easy to produce. However, the reverse process, the conversion of thermal energy into mechanical energy, is more difficult. A **heat engine** is a device that is able to continuously convert thermal energy to mechanical energy. A heat engine requires a high-temperature source, a low-temperature receptacle, called a sink, and a way to convert the thermal energy into work. **Figure 17** shows that some thermal energy from the source is used to do work and some is transferred to the sink.

Internal combustion engines An automobile's internal combustion engine, such as the one shown in **Figure 18,** is one example of a heat engine. In the engine, input heat (Q_H) is transferred from a high-temperature flame to a mixture of air and gas vapor in the cylinder. The hot air expands and pushes on a piston, thereby changing thermal energy into mechanical energy. The heated air is expelled, the piston returns to the top of the cylinder, and the cycle repeats. Car engines repeat this cycle many times each minute. The thermal energy from the flame is converted into mechanical energy, which propels the car.

Waste heat Not all thermal energy from the flame is converted into mechanical energy. When the engine is running, the gases and the engine parts become hot. The exhaust comes in contact with outside air and heats the air. In addition, thermal energy from the engine is transferred to a radiator. Outside air passes through the radiator and becomes warmer. All of this energy (Q_C) transferred out of the automobile engine is called waste heat. When the engine is working continuously, its internal energy does not change. That is, $\Delta U = 0 = Q - W$. The net heat going into the engine is $Q = Q_H - Q_C$. Thus, the work done by the engine is $W = Q_H - Q_C$. All heat engines generate waste heat, and therefore no engine can ever convert all the energy into useful kinetic energy.

Energy Diagram of a Heat Engine

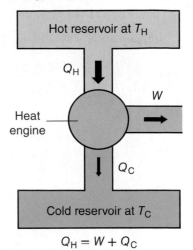

$$Q_H = W + Q_C$$

Figure 17 Heat engines transform thermal energy into mechanical energy and waste heat. This schematic shows the energy transfers and transformations.

Watch an **animation of a heat engine.**

Concepts In Motion

Figure 18 Internal combustion engines are one type of heat engine. They are used in automobiles.

■ **Internal Combustion Engine**

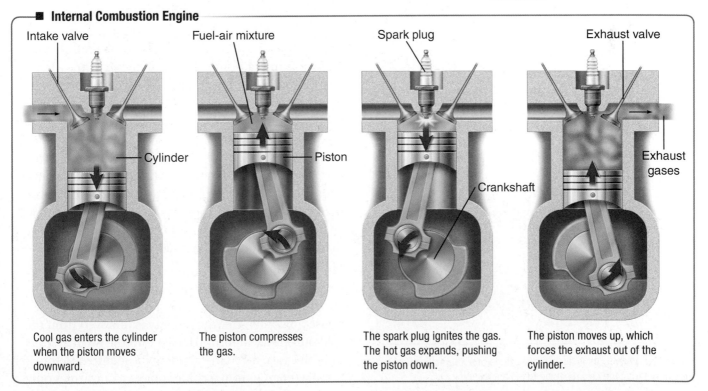

Intake valve | Fuel-air mixture | Spark plug | Exhaust valve

Cylinder | Piston | Crankshaft | Exhaust gases

Cool gas enters the cylinder when the piston moves downward.

The piston compresses the gas.

The spark plug ignites the gas. The hot gas expands, pushing the piston down.

The piston moves up, which forces the exhaust out of the cylinder.

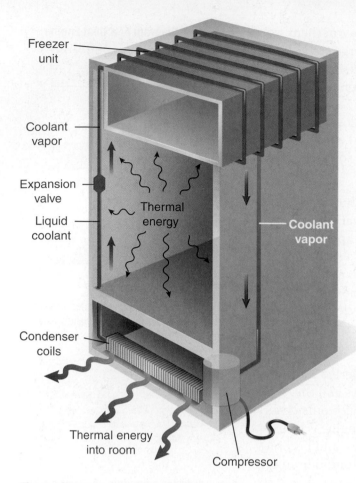

Freezer unit

Coolant vapor

Expansion valve

Liquid coolant

Thermal energy

Coolant vapor

Condenser coils

Thermal energy into room

Compressor

Figure 19 Liquid coolant is pumped into an expansion valve, where it absorbs energy from its surroundings and becomes a gas. The gas then heats up as it absorbs thermal energy from inside the refrigerator. A compressor does work on the gas to cool it to a liquid, and the cycle begins again.

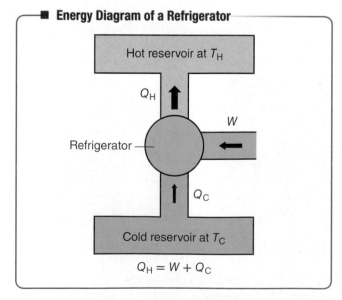

■ Energy Diagram of a Refrigerator

Hot reservoir at T_H

Q_H

W

Refrigerator

Q_C

Cold reservoir at T_C

$$Q_H = W + Q_C$$

Figure 20 When work is done on the refrigerator, thermal energy is transferred from the cold reservoir to the hot reservoir.

Efficiency Engineers and car salespeople often discuss the fuel efficiencies of automobile engines. They are referring to the amount of the input heat (Q_H) that is turned into useful work (W). The actual efficiency of an engine is given by the ratio W/Q_H. If all the input heat could be turned into useful work by the engine, the engine would have an efficiency of 100 percent. Because there is always waste heat (Q_C), even the most efficient engines fall short of 100 percent efficiency.

☑ **READING CHECK Infer** Would a more efficient engine burn more fuel or less fuel for the same amount of work than a less efficient engine?

In fact, most heat engines are significantly less than 100 percent efficient. For example, even the most efficient automobile gasoline engines have an efficiency of less than 40 percent. A typical gasoline engine in an automobile has an efficiency that is closer to 20 percent.

A considerable amount of thermal energy transfers from a hot automobile engine to the cooler surroundings. Does any device transfer thermal energy from a cold object to warmer surroundings?

Refrigerators Thermal energy is transferred from a warm object to a cold object spontaneously. But it is also possible to remove thermal energy from a colder object and add it to a warmer object if work is done. A refrigerator, such as the one in **Figure 19,** is a common example of a device that accomplishes this transfer. Electrical energy runs a motor that does work on a gas and compresses it.

The gas draws thermal energy from the interior of the refrigerator, passes from the compressor through the condenser coils on the outside of the refrigerator, and cools into a liquid. Thermal energy is transferred into the air in the room. The liquid reenters the interior, vaporizes, and absorbs thermal energy from its surroundings. The gas returns to the compressor, and the process is repeated. The overall change in the thermal energy of the gas is zero. Thus, according to the first law of thermodynamics, the sum of the thermal energy removed from the refrigerator's contents and the work done by the motor is equal to the thermal energy expelled. These energy transfers and transformations are shown in **Figure 20.**

Heat pumps A heat pump is a refrigerator that can be run in two directions. In summer, the pump removes thermal energy from a house and cools the house. In winter, thermal energy is removed from the cold outside air and transferred into the warmer house. In both cases, mechanical energy is required to transfer thermal energy from a colder object to a warmer one.

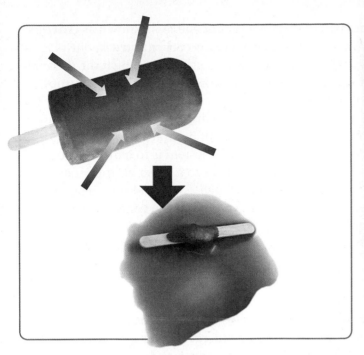

(t)Photodisc/Getty Images; (t)Robbie Augspurger/Flickr/Getty Images; (b)Martin Poole/Getty Images; (b)Barry Wong/Getty Images

The Second Law of Thermodynamics

We take for granted that many daily events occur spontaneously only in one direction. You would be shocked if the reverse of the same events occurred spontaneously. For example, you are not surprised when a metal spoon, heated at one end, soon becomes uniformly hot. Consider your reaction, however, if a spoon lying on a table suddenly, on its own, became red hot at one end and icy cold at the other. This reverse process would not violate the first law of thermodynamics—the thermal energy of the spoon would remain the same. Many processes that are consistent with the first law of thermodynamics have never been observed to occur spontaneously. However, there is more to modeling thermal events than making sure energy is conserved.

Energy spreads out Examine the melting ice pop and the cooling pizza in **Figure 21.** The first law of thermodynamics does not prohibit net thermal energy transfers from the cold ice pop to the air or from the air to the hot pizza. This does not occur, however, because of the second law of thermodynamics. When a hot object is placed in contact with cooler surroundings, the thermal energy in the hot object has the opportunity to disperse, or spread out more. Some of the thermal energy moves into the cold object, warming it and therefore cooling the originally hotter object. The **second law of thermodynamics** states that whenever there is an opportunity for energy dispersal, the energy always spreads out.

Consider the cooling pizza. The particles in the pizza have a greater average kinetic energy than the particles in the air. Some of the pizza's original thermal energy disperses into the air. As a result, the pizza's temperature decreases and the air temperature increases a small amount. When the pizza and the air reach the same temperature, the average kinetic energy of the particles in the pizza and the air will be the same. That is, the energy spreads out among the particles. Similarly, if you leave the ice pop sitting on the counter, thermal energy from the air will be dispersed to the ice pop. The ice pop will heat up and melt, while the air will experience a small temperature decrease.

Figure 21 According to the second law of thermodynamics, thermal energy always spreads out. The red arrows represent thermal energy flow. Thermal energy spontaneously flows from a warmer object to a colder object.

Entropy The measure of this dispersal of energy is known as **entropy** (S). A system in which the thermal energy is concentrated in one place is referred to as a system with low entropy. A system in which the thermal energy is spread throughout the system is a system with high entropy.

Another way of stating the second law of thermodynamics is that natural processes go in a direction that maintains or increases the total entropy of the universe. That is, energy will naturally disperse unless some action is taken to localize it. Once a system is in a high-entropy state, it is highly unlikely that it will return to a lower entropy state on its own. Events that occur spontaneously, such as the melting ice pop or the cooling pizza, are events in which the entropy of the system increases. Processes that would decrease the entropy of a system do not tend to occur spontaneously but require work done by an external agent.

☑ **READING CHECK State** the second law of thermodynamics using the term *entropy*.

Entropy and heat engines How does entropy relate to heat engines? If heat engines completely converted thermal energy into mechanical energy with no waste heat, energy would still be conserved, and so the first law of thermodynamics would be obeyed. However, waste heat is always generated, dispersing thermal energy beyond the engine. In the nineteenth century, French engineer Sadi Carnot studied the ability of engines to convert thermal energy into mechanical energy. He developed a logical proof that even an ideal engine would generate some waste heat. Carnot's result was one of the first formal analyses leading to the development of the concept of entropy.

Changes in entropy Like energy, entropy is a property of a system. If thermal energy is added to a system, the entropy increases. If thermal energy is removed from a system, its entropy decreases. If a system does work on its surroundings without any transfer of thermal energy, the entropy does not change. For a reversible process, the change in entropy (ΔS) is expressed by the following equation, in which entropy has units of J/K and the temperature is constant and measured in kelvins.

CHANGE IN ENTROPY
For a reversible process, the change in entropy of a system is equal to the heat added to the system divided by the temperature of the system in kelvins.

$$\Delta S = \frac{Q}{T}$$

Investigate **kinetic theory and how energy spreads out.**

Virtual Investigation 👆

PHYSICS CHALLENGE

Entropy has some interesting properties. Calculate the change in entropy for the following situations. Explain how and why these changes in entropy are different from each other. For these small temperature changes, you can use the original temperature to find the change in entropy.

1. Heating 1.0 kg of water from 273 K to 274 K.

2. Heating 1.0 kg of water from 353 K to 354 K.

3. Heating 1.0 kg of lead from 273 K to 274 K.

4. Completely melting 1.0 kg of ice at 273 K.

Entropy and the energy crisis The second law of thermodynamics and the increase in entropy also give new meaning to what has been commonly called the energy crisis. The energy crisis refers to the continued use of limited resources such as natural gas and petroleum. When you use a resource, you do not use up the energy in the resource. For example, when you drive a car, such as those in **Figure 22,** the gas ignites and the chemical energy contained in the molecules of the gas is converted into the kinetic energy that runs the car and the thermal energy that heats the engine. Even if friction converts the car's kinetic energy into thermal energy, the energy is not lost or used up.

The entropy, however, has increased. The chemical energy of the unburned gas has dispersed into many more objects contained within a much larger volume. While it is mathematically possible for all of the dispersed energy to be brought back together in some one object, the probability of this happening is very near zero. For this reason, entropy often is used as a measure of the unavailability of useful energy. The energy in the warmed air in a home is not as available to do mechanical work or to transfer thermal energy to other objects as the energy in the original gas molecules was. The lack of usable energy is actually a surplus of entropy.

Figure 22 Burning gasoline uses up natural resources and increases entropy but does not use up energy. The energy is no longer in a useful form.

SECTION 2 REVIEW

Section Self-Check ✋ Check your understanding.

29. **MAIN**IDEA Describe the energy transformations and transfers made by a heat engine, and explain why operating a heat engine causes an increase in entropy.

30. **Heat of Vaporization** Old heating systems sent water vapor into radiators in each room of a house. In the radiators, the water vapor condensed to water. Analyze this process and explain how it heated a room.

31. **Heat of Fusion** How much thermal energy is needed to change 50.0 g of ice at −20.0°C to water at 10.0°C?

32. **Heat of Vaporization** How much energy is needed to heat 1.0 kg of mercury metal from 10.0°C to its boiling point (357°C) and vaporize it completely? For mercury, $C = 140$ J/kg·°C and $H_v = 3.06 \times 10^5$ J/kg.

33. **Mechanical Energy and Thermal Energy** A man uses a 320-kg hammer moving at 5.0 m/s to smash a 3.0-kg block of lead against a 450-kg rock. When he measured the temperature of the lead block he found that it had increased by 5.0°C. Explain how this happened.

34. **Mechanical Energy and Thermal Energy** James Joule carefully measured the difference in temperature of water at the top and the bottom of a waterfall. Why did he expect a difference?

35. **Mechanical Energy and Thermal Energy** For the waterfall in **Figure 23,** calculate the temperature difference between the water at the top and the bottom of the fall. Assume that the potential energy of the water is all converted to thermal energy.

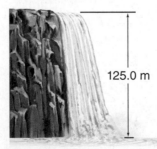

125.0 m

Figure 23

36. **Entropy** Evaluate why heating a home with natural gas results in increased entropy.

37. **Critical Thinking** Many outdoor amusement parks and zoos have systems that spray a fine mist of water, which evaporates quickly. Explain why this process cools the surrounding air.

2's Company, 3's a CROWD...

PHYSICS MODELS & PEDESTRIAN DYNAMICS

Have you ever been part of an extremely dense crowd at a concert or sports event? The forces can change direction quickly, pushing you along. You might feel helpless to move in any particular direction. Uncontrolled crowding leads to thousands of deaths every year, notably in sports stadiums and the annual Hajj pilgrimages to Mecca. In an effort to prevent such accidents, stadium managers, public safety officials, and architects are working with physicists to understand how crowds move. Understanding how crowds behave can lead to improvements in building design and crowd safety.

People as particles Physicists noticed that crowds often behave like many-particle systems. When the crowd's density is low, pedestrians can move freely and their motion resembles the behavior of gases. At medium and high densities, crowds resemble liquids. In addition, crowds pass through narrow openings in much the same way that granular materials, such as sand and salt, do. People bunched up near exits have an archlike structure characteristic of granular materials passing though an opening. By applying the same equations used to describe particle movement in homogenous gases, physicist found that they could describe and predict crowd movement.

Social forces Physicists realized, however, that a realistic model for pedestrians must include interactions not governed by Newton's laws. These interactions involve people's internal motivations and are called social forces. Even though human behavior often seems chaotic and unpredictable, behavioral conventions exist. For example, an individual's desire for personal space acts as a repulsive force. People also try to move at the same speed as the crowd and tend to form lanes. Including social forces can greatly improve a pedestrian model.

Future applications Currently, researchers in Germany are developing an evacuation assistant that uses live crowd data. This computerized pedestrian model can help stadium managers plan efficient and safe evacuations routes.

GOING**FURTHER** >>>

Think Critically Study your school's evacuation plan. Suggest any ways it can be improved.

STUDY GUIDE

BIGIDEA Thermal energy is related to the motion of an object's particles and can be transferred and transformed.

SECTION 1 Temperature, Heat, and Thermal Energy

MAINIDEA Heat is a transfer of thermal energy that occurs spontaneously from a warmer object to a cooler object.

- Thermal energy is the sum of the kinetic and potential energies of an object's particles. An object's temperature is a measure of the average kinetic energy of its particles.

- When two objects are in thermal equilibrium, there is no net transfer of thermal energy between the objects and the two objects are at the same temperature. A thermometer measures temperature by reaching thermal equilibrium with its surroundings. When an object's temperature is at absolute zero, the average kinetic energy of its particles is zero and that object cannot transfer thermal energy.

- Heat is the transfer of thermal energy. Thermal energy is spontaneously transferred from a warm object to a cool object. Thermal energy is transferred by three processes: conduction, convection, and radiation.

- Substances heat differently, based on their specific heats. Specific heat (C) is the heat required to raise the temperature of 1 kg of a substance 1 K.

$$Q = mC\Delta T = mC(T_\text{f} - T_\text{i})$$

A calorimeter is a closed system used to measure changes in thermal energies. Specific heat is calculated by using measurements from a calorimeter.

SECTION 2 Changes of State and Thermodynamics

MAINIDEA When thermal energy is transferred, energy is conserved and the total entropy of the universe will increase.

- Thermal energy transferred during a change of state does not change the temperature of a substance. The heat of fusion is the quantity of heat needed to change 1 kg of a substance from a solid state to a liquid state at its melting point.

$$Q = mH_\text{f}$$

The heat of vaporization is the quantity of heat needed to change 1 kg of a substance from a liquid state to a gaseous state at its boiling point.

$$Q = mH_\text{v}$$

- The first law of thermodynamics states that the change in the thermal energy of an object is equal to the heat added to the object minus the work done by the object.

$$\Delta U = Q - W$$

- A heat engine converts thermal energy to mechanical energy. A heat pump and a refrigerator use mechanical energy to transfer thermal energy from a region of lower temperature to one of higher temperature.

- The second law of thermodynamics states that whenever there is an opportunity for energy dispersal, the energy always spreads out. Entropy (S) is a measure of the energy dispersal of a system. The second law of thermodynamics indicates that natural processes go in a direction that maintains or increases the total entropy of the universe. The change in entropy of an object is defined as the heat added to the object divided by the object's temperature.

Games and Multilingual eGlossary

Vocabulary Practice

SECTION 1
Temperature, Heat, and Thermal Energy
Mastering Concepts

38. BIGIDEA Explain the differences among the mechanical energy of a ball, its thermal energy, and its temperature.

39. Can temperature be assigned to a vacuum? Explain.

40. Do all the molecules or atoms in a liquid have the same speed?

41. Is your body a good judge of temperature? On a cold winter day, a metal doorknob feels much colder to your hand than a wooden door does. Explain why this is true.

42. When thermal energy is transferred from a warmer object to a colder object it is in contact with, do the two have the same temperature changes?

Mastering Problems

43. How much thermal energy is needed to raise the temperature of 50.0 g of water from 4.5°C to 83.0°C?

44. Coffee Cup A glass coffee cup is at room temperature. It is then plunged into hot dishwater, as shown in **Figure 24.** If the temperature of the cup reaches that of the dishwater, how much heat does the cup absorb? Assume that the mass of the dishwater is large enough so that its temperature does not change appreciably.

20.0°C

80.0°C

4.00×10^2 g

Figure 24

45. A 1.00×10^2-g mass of tungsten at 100.0°C is placed in 2.00×10^2 g of water at 20.0°C. The mixture reaches equilibrium at 21.2°C. Calculate the specific heat of tungsten.

46. A 6.0×10^2-g sample of water at 90.0°C is mixed with 4.00×10^2 g of water at 22.0°C. Assume that there is no thermal energy lost to the container or surroundings. What is the final temperature of the mixture?

47. A 5.00×10^2-g block of metal absorbs 5016 J of thermal energy when its temperature changes from 20.0°C to 30.0°C. Calculate the specific heat of the metal.

48. The kinetic energy of a compact car moving at 100 km/h is 2.9×10^5 J. To get an idea of the amount of energy needed to heat water, how many liters of water would 2.9×10^5 J of energy warm from room temperature (20.0°C) to boiling (100.0°C)?

49. Car Engine A 2.50×10^2-kg cast-iron car engine contains water as a coolant. Suppose that the engine's temperature is 35.0°C when it is shut off, and the air temperature is 10.0°C. The thermal energy given off by the engine and water in it as they cool to air temperature is 4.40×10^6 J. What mass of water is used to cool the engine?

50. Water Heater An electric immersion heater is used to heat a cup of water, as shown in **Figure 25.** The cup is made of glass and contains 250 g of water at 15°C. How much time is needed to bring the water to the boiling point? Assume that the temperature of the cup is the same as the temperature of the water at all times and that no thermal energy is lost to the air.

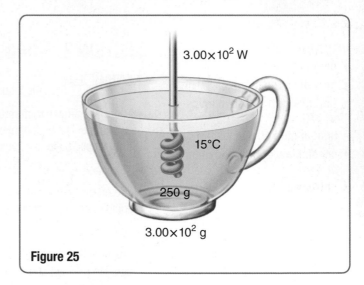
3.00×10^2 W

15°C

250 g

3.00×10^2 g

Figure 25

51. Ranking Task The following materials are each placed in identical containers holding equal amounts of room-temperature methanol. Rank the materials according to the amount of thermal energy they transfer to the methanol, from least to greatest. Specifically indicate any ties.

A. 50 g of aluminum at 30°C

B. 60 g of aluminum at 40°C

C. 50 g of glass at 30°C

D. 50 g of silver at 30°C

E. 50 g of zinc at 30°C

52. A piece of zinc at 71.0°C is placed in a container of water, as shown in **Figure 26.** What is the final temperature of the water and the zinc?

20.0 kg

10.0 kg

10.0°C

Figure 26

SECTION 2
Changes of State and Thermodynamics
Mastering Concepts

53. Can you add thermal energy to an object without increasing its temperature? Explain.

54. When wax freezes, does it absorb or release energy?

55. Explain why water in a canteen that is surrounded by dry air stays cooler if it has a canvas cover that is kept wet.

56. Which process occurs at the coils of a running air conditioner inside a house, vaporization or condensation? Explain.

Mastering Problems

57. Years ago, a block of ice with a mass of about 20.0 kg was used daily in a home icebox. The temperature of the ice was 0.0°C when it was delivered. As it melted, how much thermal energy did the ice absorb?

58. A 40.0-g sample of chloroform is condensed from a vapor to a liquid at 61.6°C. It releases 9870 J of thermal energy. What is the heat of vaporization of chloroform?

59. A 750-kg car moving at 23 m/s brakes to a stop. Assume that all the kinetic energy is transformed into thermal energy. The brakes contain about 15 kg of iron, which absorbs the energy. What is the increase in temperature of the brakes?

60. How much thermal energy is added to 10.0 g of ice at −20.0°C to convert it to water vapor at 120.0°C?

61. A 4.2-g lead bullet moving at 275 m/s strikes a steel plate and comes to a stop. If all its kinetic energy is converted to thermal energy and none leaves the bullet, what is its temperature change?

62. Soft Drink A soft drink from Australia is labeled *Low-Joule Cola.* The label says "100 mL yields 1.7 kJ." The can contains 375 mL of cola. Chandra drinks the cola and then wants to offset this input of food energy by climbing stairs. How high must Chandra climb if her mass is 65.0 kg?

Applying Concepts

63. Cooking Sally is cooking pasta in a pot of boiling water. Will the pasta cook faster if the water is boiling vigorously or if it is boiling gently?

64. Which liquid would an ice cube cool faster, water or methanol? Explain.

65. Equal masses of aluminum and lead are heated to the same temperature. The pieces of metal are placed on a block of ice. Which metal melts more ice? Explain.

66. Why do easily vaporized liquids, such as acetone and methanol, feel cool to the skin?

67. Explain why fruit growers spray their trees with water to protect the fruit from freezing when frost is expected.

68. Two blocks of lead have the same temperature. Block A has twice the mass of block B. They are dropped into identical cups of water of equal temperatures. Will the two cups of water have equal temperatures after equilibrium is achieved? Explain.

69. Windows Often architects design most windows of a house on the north side. How does this affect the heating and cooling of the house?

Mixed Review

70. What is the efficiency of an engine that outputs 1800 J/s while burning gasoline to produce 5300 J/s? How much waste heat does the engine produce per second?

71. Stamping Press A metal stamping machine in a factory does 2100 J of work each time it stamps out a piece of metal. Assume that the work changes only the metal's thermal energy. Each stamped piece is then dipped in a 32.0-kg vat of water for cooling. By how many degrees does the vat heat up each time a piece of stamped metal is dipped into it?

72. Problem Posing Complete this problem so that it must be solved using the concepts listed below: "A beaker of water has a temperature of 35°C ... ."

a. specific heat

b. entropy

73. A 1500-kg automobile comes to a stop from 25 m/s. All the energy of the automobile is deposited in the brakes. Assuming that the brakes are about 45 kg of aluminum, what is the change in temperature of the brakes?

74. Iced Tea To make iced tea, you brew the tea with hot water and then add ice. If you start with 1.0 L of 90°C tea, how much ice is needed to cool it to 0°C? Would it be better to let the tea cool to room temperature before adding the ice?

75. A block of copper comes in contact with a block of aluminum, and they come to thermal equilibrium, as shown in **Figure 27**. What are the relative masses of the blocks?

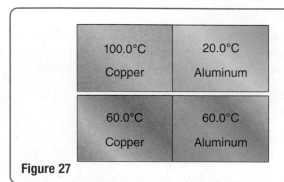

100.0°C	20.0°C
Copper	Aluminum
60.0°C	60.0°C
Copper	Aluminum

Figure 27

76. Two copper blocks, each with a mass of 0.35 kg, slide toward each other at the same speed and collide. The two blocks come to a stop together after the collision. Their temperatures increase by 0.20°C as a result of the collision. Assume that all kinetic energy is transformed into thermal energy. What was their speed before the collision?

77. A 2.2-kg block of ice slides across a rough floor. Its initial velocity is 2.5 m/s, and its final velocity is 0.50 m/s. How much of the ice block melted as a result of the work done by friction?

Thinking Critically

78. Analyze and Conclude Chemists use calorimeters to measure the heat produced by chemical reactions. Suppose a chemist dissolves 1.0×10^{22} molecules of a powdered substance into a calorimeter containing 0.50 kg of water. The molecules break up and release their binding energy to the water. The water temperature increases by 2.3°C. What is the binding energy per molecule for this substance?

79. Reverse Problem Write a physics problem with real-life objects for which the following equation would be part of the solution:

$$75 \text{ J/K} = \frac{m(4180 \text{ J/(kg·K)})(260 \text{ K} - 250 \text{ K})}{250 \text{ K}}$$

80. Analyze and Conclude A certain heat engine removes 50.0 J of thermal energy from a hot reservoir at temperature $T_H = 545$ K and expels 40.0 J of thermal energy to a colder reservoir at temperature $T_C = 325$ K. In the process, it also transfers entropy from one reservoir to the other.

a. Find the total entropy change of the reservoirs.

b. What would be the total entropy change in the reservoirs if $T_C = 205$ K?

81. Analyze and Conclude During a game, the metabolism of basketball players often increases by as much as 30.0 W. How much perspiration must a player vaporize per hour to dissipate this extra thermal energy?

82. Apply Concepts All energy on Earth comes from the Sun. The Sun's surface temperature is approximately 10^4 K. What would be the effect on Earth if the Sun's surface temperature were 10^3 K?

Writing in Physics

83. Our understanding of the relationship between heat and energy was influenced by a brewer named James Prescott Joule and a soldier named Benjamin Thompson, Count Rumford. Investigate what experiments they did and evaluate whether it is fair that the unit of energy is called the joule and not the thompson.

84. Water has an unusually large specific heat and large heats of fusion and vaporization. The weather and ecosystems depend upon water in all three states. How would the world be different if water's thermodynamic properties were like other materials, such as methanol?

Cumulative Review

85. A rope is wound around a drum with a radius of 0.250 m and a moment of inertia of 2.25 kg·m². The rope is connected to a 4.00-kg block. Find the linear acceleration of the block. Find the angular acceleration of the drum. Find the tension (F_T) in the rope. Find the angular velocity of the drum after the block has fallen 5.00 m.

86. A weight lifter raises a 180-kg barbell to a height of 1.95 m. How much work is done by the weight lifter in lifting the barbell?

87. In a Greek myth, Sisyphus is fated to forever roll a huge rock up a hill. Each time he reaches the top, the rock rolls back to the bottom. If the rock has a mass of 215 kg, the hill is 33 m in height, and Sisyphus produces an average power of 0.2 kW, how many times in 1 h can he roll the rock up the hill?

MULTIPLE CHOICE

Use the following information as needed.

$C_{ice} = 2060$ J/(kg·K)
$C_{water} = 4180$ J/(kg·K)
$C_{water\,vapor} = 2020$ J/(kg·K)
$H_{f\,water} = 3.34×10^5$ J/kg
$H_{v\,water} = 2.26×10^6$ J/kg

1. Which temperature conversion is incorrect?

A. $-273°C = 0$ K   **C.** 298 K $= 571°C$

B. $273°C = 546$ K   **D.** 88 K $= -185°C$

2. What are the units of entropy?

A. J/K   **C.** J

B. K/J   **D.** kJ

3. Which statement about two objects in thermal equilibrium is false?

A. Energy exchange between the objects continues to occur.

B. The net flow of energy between the objects is zero.

C. The objects are at the same temperature.

D. There is a net flow of energy from one object to the other.

4. How much heat is needed to warm 363 mL of water from 24°C to 38°C?

A. 21 kJ   **C.** 121 kJ

B. 36 kJ   **D.** 820 kJ

5. In the figure below, 81 g of ice melts and warms to 10°C. How much thermal energy is absorbed from the surroundings when this occurs?

A. 0.34 kJ   **C.** 30 kJ

B. 27 kJ   **D.** 190 kJ

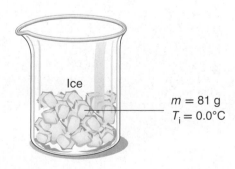

Ice

$m = 81$ g
$T_i = 0.0°C$

6. How much heat is required to heat 87 g of ice at 14 K to water vapor at 140°C?

A. 45 kJ   **C.** 315 kJ

B. 58 kJ   **D.** 280 kJ

7. You do 0.050 J of work on the coffee in your cup each time you stir it. What is the increase in entropy in 125 mL of coffee at 65°C when you stir it 85 times?

A. 0.013 J/K   **C.** 0.095 J/K

B. 0.050 J   **D.** 4.2 J

8. Why is there always some waste heat in a heat engine?

A. The entropy decreases at each stage.

B. The engine is not as efficient as it could be.

C. The entropy increases at each stage.

D. The energy is being used up.

9. Which statement about entropy and energy is true?

A. When ice freezes, it gives off energy and its entropy increases.

B. When ice freezes, it gives off energy and its entropy decreases.

C. When ice freezes, it absorbs energy and its entropy increases.

D. When ice freezes, it absorbs energy and its entropy decreases.

FREE RESPONSE

10. What is the difference in heat required to melt 454 g of ice at 0.00°C and to turn 454 g of water at 100.0°C into water vapor? Is the amount of this difference greater or less than the amount of energy required to heat the 454 g of water from 0.00°C to 100.0°C?

NEED EXTRA HELP?

If You Missed Question	1	2	3	4	5	6	7	8	9	10
Review Section	1	2	1	1	2	2	2	2	2	2

Online Test Practice

CHAPTER 13

States of Matter

BIGIDEA   The thermal energy of a material and the forces between that material's particles determine its properties.

SECTIONS

1 Properties of Fluids

2 Forces Within Liquids

3 Fluids at Rest and in Motion

4 Solids

LaunchLAB

MEASURING BUOYANCY

How does the density of an object affect how it floats?

WATCH THIS!

Video

PLASMA

You've probably explored the three main states of matter, but how much experience do you have with plasma? You might be surprised!

PHYSICS T.V.

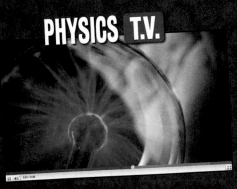

Properties of Fluids

PHYSICS 4 YOU

You might not notice it very often, but gases in the atmosphere are exerting pressure on your body. If you've ever ridden in an elevator in a tall building, been to the top of a mountain, or flown in an airplane, you might have had your ears pop. Your ears pop to help balance the changing pressure between the inside and the outside of your ear.

MAINIDEA
Fluids flow, have no definite shape, and include liquids, gases, and plasmas.

Essential Questions
• What is a fluid?
• What are the relationships among the pressure, volume, and temperature of a gas?
• What is the ideal gas law?
• What is plasma?

Review Vocabulary
linear relationship a relationship in which the dependent variable varies linearly with the independent variable

New Vocabulary
fluid
pressure
pascal
combined gas law
ideal gas law
thermal expansion
plasma

Liquids and Gases

Water and air are two of the most common substances in the everyday lives of people. We feel their effects when we drink, when we bathe, and literally with every breath we take. Although you might not think about it every day, water and air have a great deal in common. Both water and air flow, and unlike solids, neither one of them has a definite shape. Gases and liquids are two states of matter in which atoms and molecules have great freedom to move. In this chapter, you will learn about the principles that explain how liquids and gases respond to changes in temperature and pressure, how hydraulic systems can multiply applied forces, and how huge metallic ships can float on water.

Fluids Study the ice chunks in the lake in **Figure 1.** Like ice cubes, these chunks have a certain mass and a certain shape, and neither of these quantities depends on the size or shape of the lake basin. What happens, however, when the ice melts? Its mass remains the same, but its shape changes. The water flows to take the shape of the basin and forms a definite, flat, upper surface. As evaporation occurs, the liquid water changes into a gas in the form of water vapor. Like liquid water, water vapor flows and does not have any definite shape. Both liquids and gases are fluids. **Fluids** are materials that can flow and have no definite shape of their own.

Figure 1 The chunks of ice in this lake, which are solids, have definite shapes. However, the liquid water, a fluid, takes the shape of the lake basin.

Identify What fluid is filling the space above the water?

Pressure

When considering fluids (as well as solids), it is often useful to think about pressure as well as force. You have probably heard people talk about water pressure and air pressure, and you might already have a sense that pressure and force are related. Pressure and force are not the same, however. **Pressure** is the perpendicular component of a force on a surface divided by the area of the surface. Since pressure is force exerted on a surface, anything that exerts pressure is capable of producing change and doing work. In **Figure 2,** both the astronaut and the legs of the landing module are exerting pressure on the Moon's surface.

PRESSURE

Pressure equals the perpendicular component of the force divided by the surface area to which it is applied.

$$P = \frac{F}{A}$$

Pressure is a scalar. In the SI system, the unit of pressure is the **pascal** (Pa), which is 1 N/m^2. One pascal is a very small amount of pressure, about equal to the pressure that a flat dollar bill exerts on a tabletop. Thus the kilopascal (kPa), equal to 1000 Pa, is usually used. **Table 1** shows pressures in various locations.

Table 1 Some Typical Pressures	
Location	**Pressure (kPa)**
The center of Earth	4×10^8
The deepest ocean trench	1.1×10^5
Standard atmosphere	1.01325×10^2
Blood pressure	1.6×10^1
Air pressure on top of Mt. Everest	3×10^1
The best vacuum	1×10^{-10}

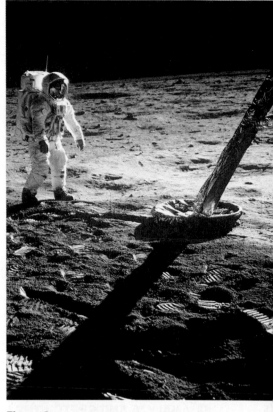

Figure 2 The landing module and the astronaut both exert pressure on the surface of the Moon.

Calculate If the lunar module weighed approximately 12,000 N and rested on four pads that were each 91 cm in diameter, what pressure did it exert on the Moon's surface? How could you estimate the pressure exerted by the astronaut?

Solids, liquids, and pressure Imagine you are standing on the surface of a frozen lake. The forces your feet exert on the ice are spread over the area of your shoes, resulting in pressure on the ice. Ice is a solid that is made up of vibrating water molecules, and the forces that hold the water molecules in place cause the ice to exert upward forces on your feet that equal your weight. If the ice melted, most of the bonds between the water molecules would be weakened. Although the molecules would continue to vibrate and remain close to each other, they also would slide past one another, and you would break through the surface. The moving water molecules would continue to exert forces on your body.

Gases and pressure The pressure exerted by a gas can be understood by applying the kinetic-molecular theory of gases, which explains the properties of an ideal gas. In this model, particles are treated as taking up no space and having no intermolecular attractive forces. In spite of the fact that particles of a real gas take up space and exert attractive forces, an ideal gas is an accurate model of a real gas under most conditions.

According to the kinetic-molecular theory, the particles in a gas are in random motion at high speeds and colliding elastically with each other. When a gas particle hits a container's surface, it rebounds, which changes its momentum. The impulses exerted by these collisions result in gas pressure on the surface.

Atmospheric pressure At sea level, gases of the atmosphere exert a force in all directions of approximately 10 N, about the weight of a 1-kg object, on each centimeter of surface area. This atmospheric pressure on your body is so well balanced by your body's outward forces that you seldom notice it. You probably become aware of this pressure only when your ears pop as the result of pressure changes, as when you ride an elevator in a tall building or fly in an airplane. Atmospheric pressure is about 10 N per 1 cm^2 (10^{-4} m^2), which is about 1.0×10^5 N/m^2, or 100 kPa. Other planets in our solar system also have atmospheres that exert pressure. For example, the pressure at the surface of Venus is about 92 times that at Earth's surface.

Find help with **dimensional calculations.** Math Handbook

EXAMPLE PROBLEM 1

EXAMPLE PROBLEM

CALCULATING PRESSURE A child weighs 364 N and sits on a three-legged stool, which weighs 41 N. The bottoms of the stool's legs touch the ground over a total area of 19.3 cm^2.

a. What is the average pressure that the child and the stool exert on the ground?

b. How does the pressure change when the child leans over so that only two legs of the stool touch the floor?

1 ANALYZE AND SKETCH THE PROBLEM

- Sketch the child and the stool, labeling the total force that they exert on the ground.

- List the variables, including the force that the child and the stool exert on the ground and the areas for parts **a** and **b**.

KNOWN		UNKNOWN
$F_{g\ child} = 364$ N	$A_a = 19.3$ cm^2	$P_a = ?$
$F_{g\ stool} = 41$ N	$A_b = \frac{2}{3} \times 19.3$ cm^2	$P_b = ?$
$F_{g\ total} = F_{g\ child} + F_{g\ stool}$	$= 12.9$ cm^2	
$= 364$ N $+ 41$ N		
$= 405$ N		

$f_g = 405$ N

2 SOLVE FOR THE UNKNOWN

Find each pressure.

$$P = \frac{F}{A}$$

a. $P_a = \left(\dfrac{405\ N}{19.3\ cm^2}\right)\left(\dfrac{(100\ cm)^2}{(1\ m)^2}\right)$ ◄ Substitute $F = F_{g\ total} = 405$ N, $A = A_A = 19.3$ cm^2

$= 2.10\times10^2$ kPa

b. $P_b = \left(\dfrac{405\ N}{12.9\ cm^2}\right)\left(\dfrac{(100\ cm)^2}{(1\ m)^2}\right)$ ◄ Substitute $F = F_{g\ total} = 405$ N, $A = A_B = 12.9$ cm^2

$= 3.14\times10^2$ kPa

3 EVALUATE THE ANSWER

Are the units correct? The units for pressure should be Pa, and 1 N/m^2 = 1 Pa.

1. The atmospheric pressure at sea level is about 1.0×10^5 Pa. What is the force at sea level that air exerts on the top of a desk that is 152 cm long and 76 cm wide?

2. A car tire makes contact with the ground on a rectangular area of 12 cm by 18 cm. If the car's mass is 925 kg, what pressure does the car exert on the ground as it rests on all four tires?

3. A lead brick, 5.0 cm × 10.0 cm × 20.0 cm, rests on the ground on its smallest face. Lead has a density of 11.8 g/cm³. What pressure does the brick exert on the ground?

4. Suppose that during a storm, the atmospheric pressure suddenly drops by 15 percent outside. What net force would be exerted on a front door to a house that is 195 cm high and 91 cm wide? In what direction would this force be exerted?

5. **CHALLENGE** Large pieces of industrial equipment are placed on wide steel plates that spread the weight of the equipment over larger areas. If an engineer plans to install a 454-kg device on a floor that is rated to withstand additional pressure of 5.0×10^4 Pa, how large should the steel support plate be?

The Gas Laws

Think about a container of gas that is held at a constant temperature. If you reduced the volume, what would happen to the pressure of the gas? There would be more collisions between the particles and the container's walls, and so the pressure would increase. Similarly, if you increased the volume, there would be fewer collisions, decreasing the pressure. This inverse relationship was found by seventeenth-century chemist and physicist Robert Boyle. Because the product of inversely related variables is a constant, Boyle's law can be written $PV = $ constant, or $P_1V_1 = P_2V_2$. The subscripts that you see in the gas laws help keep track of different variables, such as pressure and volume, as they change throughout a problem. The relationship between the pressure and the volume of a gas is critical to the scuba diver in **Figure 3.**

About 100 years after Boyle's work, Jacques Charles cooled a gas and found that the volume shrank by $\frac{1}{273}$ of its original volume for every degree cooled, which is a linear relationship. At the time, Charles could not cool gases to the extremely low temperatures achieved in modern laboratories. In order to see what lower limits might be possible, he extended, or extrapolated, the graph of his data to these temperatures. This extrapolation suggested that if the temperature were reduced to −273°C, a gas would have zero volume. The temperature at which a gas would have zero volume is now called absolute zero, which is represented by the zero of the Kelvin temperature scale.

These experiments indicated that under constant pressure, the volume of a sample of gas varies directly with its Kelvin temperature, a result that is now called Charles's law. Charles's law can be written $\frac{V}{T} = $ constant, or $\frac{V_1}{T_1} = \frac{V_2}{T_2}$.

Combined gas law Combining Boyle's law and Charles's law relates the pressure, temperature, and volume of a fixed amount of ideal gas, which leads to the **combined gas law.**

COMBINED GAS LAW
For a fixed amount of an ideal gas, the pressure times the volume, divided by the Kelvin temperature equals a constant.

$$\frac{P_1V_1}{T_1} = \frac{P_2V_2}{T_2} = \text{constant}$$

Watch a **video on Charles's law.**

Video

Figure 3 The gas in the tank on the diver's back is at high pressure. The regulator in the diver's mouth reduces the pressure, making the pressure of the gas the diver breathes equal to that of the water pressure. Bubbles are emitted from the regulator as the diver exhales.

Figure 4 The combined gas law shows the relationship among pressure, temperature, and volume of a fixed amount of an ideal gas. Boyle's law and Charles's law can each be derived from the combined gas law under certain conditions.

Explain What happens if you hold volume constant?

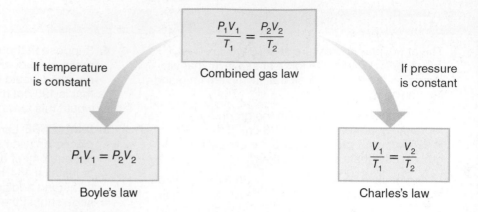

If temperature is constant

$$\frac{P_1V_1}{T_1} = \frac{P_2V_2}{T_2}$$

Combined gas law

If pressure is constant

$$P_1V_1 = P_2V_2$$

Boyle's law

$$\frac{V_1}{T_1} = \frac{V_2}{T_2}$$

Charles's law

As shown in **Figure 4,** the combined gas law reduces to Boyle's law under conditions of constant temperature and to Charles's law under conditions of constant pressure.

The ideal gas law What might the constant in the combined gas law depend on? Suppose the volume and temperature of an ideal gas are held constant while the number of particles (N) is increased. What happens to the pressure? The particles will have more collisions with the sides of the container, thus increasing the pressure. Removing particles decreases the number of collisions, and thus, decreases the pressure. Therefore, the constant in the combined gas law equation is proportional to N.

$$\frac{PV}{T} = kN$$

The constant (k) is called Boltzmann's constant, and its value is 1.38×10^{-23} Pa·m³/K. In any practical application, the number of particles (N) is very large. Instead of using N, scientists often use a unit called a mole. One mole (abbreviated mol) is similar to one dozen, except that instead of representing 12 items, one mole represents 6.022×10^{23} particles. This number is called Avogadro's number, after Italian scientist Amedeo Avogadro.

Avogadro's number is numerically equal to the number of particles in a sample of matter whose mass equals the molar mass of the substance. You can use this relationship to convert between mass and n, the number of moles present. Using moles instead of the number of particles, however, changes Boltzmann's constant. This new constant is abbreviated R, and it has the value 8.31 Pa·m³/(mol·K). Rearranging, you can write the ideal gas law in its most familiar form. The **ideal gas law** states that for an ideal gas, the pressure times the volume is equal to the number of moles multiplied by the constant R and the Kelvin temperature.

IDEAL GAS LAW

For an ideal gas, the pressure times the volume is equal to the number of moles times the constant R times the temperature.

$$PV = nRT$$

Note that with the given value of R, volume must be expressed in cubic meters, temperature in kelvin, and pressure in pascals. In practice, the ideal gas law predicts the behavior of gases remarkably well, except under conditions of high pressures or low temperatures.

View an **animation of the ideal gas law.**

Concepts In Motion

MiniLAB

PRESSURE

How much pressure do you exert when standing on one foot?

GAS LAWS A 20.0-L sample of argon gas at 273 K is at atmospheric pressure (101.3 kPa). The temperature is lowered to 120 K, and the pressure is increased to 145 kPa.

a. What is the new volume of the argon sample?

b. Find the number of moles of argon atoms in the argon sample.

c. Find the mass of the argon sample. The molar mass (M) of argon is 39.9 g/mol.

1 ANALYZE AND SKETCH THE PROBLEM

- Sketch the situation. Indicate the conditions in the container of argon before and after the change in temperature and pressure.
- List the known and unknown variables.

KNOWN	UNKNOWN
$V_1 = 20.0$ L	$V_2 = ?$
$P_1 = 101.3$ kPa	moles of argon $= ?$
$T_1 = 273$ K	mass of argon sample $= ?$
$P_2 = 145$ kPa	
$T_2 = 120$ K	
$R = 8.31$ Pa·m^3/(mol·K)	
$M_{argon} = 39.9$ g/mol	

$T_1 = 273$ K $T_2 = 120$ K
$P_1 = 101.3$ kPa $P_2 = 145$ kPa
$V_1 = 20.0$ L $V_2 = ?$

2 SOLVE FOR THE UNKNOWN

a. Use the combined gas law and solve for V_2.

$$\frac{P_1V_1}{T_1} = \frac{P_2V_2}{T_2}$$

$$V_2 = \frac{P_1V_1T_2}{P_2T_1}$$

$$= \frac{(101.3 \text{ kPa})(20.0 \text{ L})(120 \text{ K})}{(145 \text{ kPa})(273 \text{ K})}$$

◀ Substitute $P_1 = 101.3$ kPa, $P_2 = 145$ kPa, $V_1 = 20.0$ L, $T_1 = 273$ K, $T_2 = 120$ K.

$$= 6.1 \text{ L}$$

b. Use the ideal gas law, and solve for n.

$$PV = nRT$$

$$n = \frac{PV}{RT}$$

$$= \frac{(101.3 \times 10^3 \text{ Pa})(0.0200 \text{ m}^3)}{(8.31 \text{ Pa·m}^3/(\text{mol·K}))(273 \text{ K})}$$

◀ Substitute $P = 101.3 \times 10^3$ Pa, $V = 0.0200$ m^3, $R = 8.31$ m^3/(mol·K), $T = 273$ K.

$$= 0.893 \text{ mol}$$

c. Use the molar mass to convert from moles of argon in the sample to mass of the sample.

$$m = Mn$$

$$m_{argon\ sample} = (39.9 \text{ g/mol})(0.893 \text{ mol})$$ ◀ Substitute $M = 39.9$ g/mol, $n = 0.893$ mol.

$$= 35.6 \text{ g}$$

3 EVALUATE THE ANSWER

- **Are the units correct?** The volume (V_2) is in liters, and the mass of the sample is in grams.

- **Is the magnitude realistic?** The change in volume is consistent with an increase in pressure and a decrease in temperature. The calculated mass of the argon sample is reasonable.

6. A tank of helium gas used to inflate toy balloons is at a pressure of 15.5×10^6 Pa and a temperature of 293 K. The tank's volume is 0.020 m³. How large a balloon would it fill at 1.00 atmosphere and 323 K?

7. What is the mass of the helium gas in the previous problem? The molar mass of helium gas is 4.00 g/mol.

8. A tank containing 200.0 L of hydrogen gas at 0.0°C is kept at 156 kPa. The temperature is raised to 95°C, and the volume is decreased to 175 L. What is the new pressure of the gas?

9. CHALLENGE The average molar mass of the components of air (mainly diatomic nitrogen gas and diatomic oxygen gas) is about 29 g/mol. What is the volume of 1.0 kg of air at atmospheric pressure and 20.0°C?

Thermal Expansion

As you applied the combined gas law, you discovered how gases expand as their temperatures increase. **Thermal expansion** is a property of all forms of matter that causes the matter to expand, becoming less dense, when heated. Thermal expansion has many useful applications, such as circulating air in a room.

Convection currents Figure 5 shows that when the air near the radiator of a room is heated, it becomes less dense and, therefore rises. Gravity pulls the denser, colder air near the ceiling down. The cold air is subsequently warmed by the radiator, and air continues to circulate. This circulation of air within a room is called a convection current. Convection currents also occur in a pot of hot, but not boiling, water on a stove. When the pot is heated from the bottom, the colder and denser water sinks to the bottom where it is warmed and then pushed up by the continuous flow of cooler water from the top.

This thermal expansion occurs in most fluids. A good model for all liquids does not exist, but it is useful to think of a liquid as a finely ground solid. Groups of two, three, or more particles move together as if they were tiny pieces of a solid. When a liquid is heated, particle motion causes these groups to expand in the same way that particles in a solid are pushed apart. The spaces between groups increase. As a result, the whole liquid expands. With an equal change in temperature, liquids expand considerably more than solids but not as much as gases.

☑ **READING CHECK** **Explain** the role of thermal expansion in the formation of a convection current.

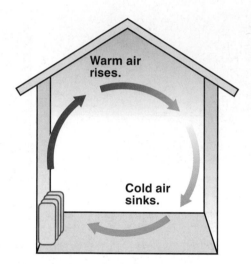

Figure 5 Convection currents occur as warmer, less dense air rises and cooler, denser air sinks.

Watch a **video on water properties.**

[Video]

Why ice floats Because matter expands as it is heated, you might predict that ice would be more dense than water, and therefore, it should sink. However, when water is heated from 0°C to 4°C, instead of expanding, it contracts as the forces between particles increase and the ice crystals collapse. These forces between water molecules are strong, and the crystals that make up ice have a very open structure. Even when ice melts, tiny crystals remain. These remaining crystals are melting, and the volume of the water decreases until the temperature reaches 4°C. However, once the temperature of water moves above 4°C, its volume increases because of greater molecular motion. The practical result is that water is most dense at 4°C and ice floats. This unique property of water is very important to our lives and environment. If ice sank, lakes would freeze from the bottom each winter and many would never melt completely in the summer.

Figure 6 Plasma emits light as it conducts electricity. The color produced by glowing plasma depends on the gas inside the tube.

Plasma

If you heat a solid, it melts to form a liquid. Further heating results in a gas. What happens if you increase the temperature still further? Collisions between the particles become violent enough to tear the electrons off the atoms, thereby producing positively charged ions. The gaslike state of negatively charged electrons and positively charged ions is called **plasma.** Plasma is considered to be another state of matter.

The plasma state may seem to be uncommon, but plasma is actually the most common state of matter in the universe. Stars consist mostly of plasma at extremely high temperatures. Much of the matter between stars and galaxies consists of energetic hydrogen that has no electrons. This hydrogen is in the plasma state. The primary difference between gas and plasma is that plasmas can conduct an electric current, whereas gases cannot. Lightning bolts are in the plasma state. Neon signs, such as the one shown in **Figure 6,** contain plasma. The fluorescent bulbs that probably light your school also contain plasma.

SECTION 1 REVIEW

 **Section Self-Check** Check your understanding.

10. **MAIN**IDEA Compare and contrast liquids, gases, and plasmas.

11. **Pressure and Force** Two boxes are each suspended by thin strings in midair. One is 20 cm × 20 cm × 20 cm. The other is 20 cm × 20 cm × 40 cm.

   a. How does the pressure of the air on the outside of the two boxes compare?

   b. How does the magnitude of the total force of the air on the two boxes compare?

12. **Meteorology** A weather balloon used by meteorologists is made of a flexible bag that allows the gas inside to freely expand. If a weather balloon containing 25.0 m^3 of helium gas is released from sea level, what is the volume of gas when the balloon reaches a height of 2100 m, where the pressure is 0.82×10^5 Pa? Assume the temperature is unchanged.

13. **Density and Temperature** Starting at 0°C, how will the density of water change as it is heated to 4°C? To 8°C?

14. **Gas Compression** In a certain internal-combustion engine, 0.0021 m^3 of air at atmospheric pressure and 303 K is rapidly compressed to a pressure of 20.1×10^5 Pa and a volume of 0.0003 m^3. What is the final temperature of the compressed gas?

15. **The Standard Molar Volume** What is the volume of 1.00 mol of a gas at atmospheric pressure and a temperature of 273 K?

16. **The Air in a Refrigerator** How many moles of air are in a refrigerator with a volume of 0.635 m^3 at a temperature of 2.00°C? If the average molar mass of air is 29 g/mol, what is the mass of the air in the refrigerator?

17. **Critical Thinking** Compared to the particles that make up carbon dioxide gas, the particles that make up helium gas are very small. What can you conclude about the number of particles in a 2.0-L sample of carbon dioxide gas compared to the number of particles in a 2.0-L sample of helium gas if both samples are at the same temperature and pressure?

Forces Within Liquids

PHYSICS 4 YOU

During exercise or on a hot day, your body perspires to cool itself. As sweat evaporates from the skin's surface, particles with higher-than-average kinetic energy escape from the liquid perspiration. The average kinetic energy of the remaining particles decreases, which leads to a decrease in temperature.

MAIN IDEA

Cohesive forces occur between the particles of a substance, while adhesive forces occur between particles of different substances.

Essential Questions

• What is surface tension?

• What is capillary action?

• How do clouds form?

Review Vocabulary

net force the vector sum of all the forces on an object

New Vocabulary

cohesive forces
adhesive forces

Cohesive Forces

Figure 7 shows a water strider walking across the surface of a pond. This lightweight insect can do this because of surface tension—the tendency of the surface of a liquid to contract to the smallest possible area. Surface tension results from the cohesive forces among the particles of a liquid. **Cohesive forces** are the forces of attraction that like particles exert on one another. Notice that beneath the liquid's surface in **Figure 7,** each particle of the liquid is attracted equally in all directions by neighboring particles. As a result, no net force acts on any of the particles beneath the surface. At the surface, however, the particles are attracted downward and to the sides but not upward. There is a net downward force, which acts on the top layers and causes the surface layer to be slightly compressed. The surface layer acts like a tightly stretched sheet that is strong enough to support the weight of very light objects, such as the water strider.

You might have seen beaded water droplets on a freshly washed and waxed car. Why do these spherical drops form? The force pulling the surface particles into a liquid causes the surface to become as small as possible, and the shape that has the least surface for a given volume is a sphere. The higher the surface tension of the liquid, the more resistant the liquid is to having its surface broken. For example, liquid mercury has much stronger cohesive forces than water does. Liquid mercury forms spherical drops, even when it is placed on a smooth surface. A drop of water flattens out on a smooth surface.

Figure 7 A water strider can walk on water because molecules at the surface experience a net downward force. Below the surface, each particle of liquid is equally attracted in all directions.

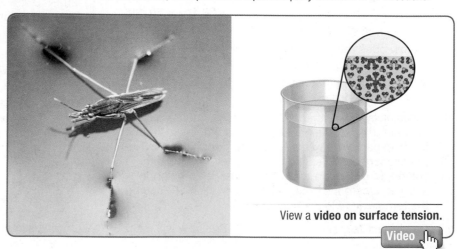

View a **video on surface tension.**

Video

Viscosity In nonideal fluids, the cohesive forces and collisions between fluid molecules cause internal friction that slows the fluid flow and dissipates mechanical energy. The measure of this internal friction is called the viscosity of the liquid. Water is not very viscous, but motor oil is very viscous. As a result of its viscosity, motor oil flows slowly over the parts of an engine to coat the metal and reduce rubbing. Lava, molten rock that flows from a volcano or vent in Earth's surface, is one of the most viscous fluids. There are several types of lava, and the viscosity of each type varies with composition and temperature.

Adhesive Forces

Similar to cohesive forces, **adhesive forces** are attractive forces that act between particles of different substances. When a glass tube is placed in a beaker of water, the surface of the water climbs the outside of the tube, as shown in **Figure 8.** The adhesive forces between the particles that make up the glass and the water molecules are greater than the cohesive forces between the water molecules. In contrast, the cohesive forces between mercury atoms are greater than the adhesive forces between the mercury and the glass, so the liquid does not climb the tube. These forces also cause the mercury's surface to depress, as shown in **Figure 8.**

If a glass tube with a small inner diameter is placed in water, the water rises inside the tube. This happens because the adhesive forces between glass and water molecules are stronger than the cohesive forces between water molecules. The water continues to rise until the weight of the water that is lifted balances the total adhesive force between the glass and water molecules. If the radius of the tube increases, the volume and the weight of the water will increase proportionally faster than the surface area of the tube. Thus, water is lifted higher in a narrow tube than in a wider one. This phenomenon is called capillary action. It causes molten wax to rise in a candle's wick and water to move up through the soil and into the roots of plants.

Evaporation and Condensation

Why does a puddle of water disappear on a hot, dry day? As you have previously read, the particles in a liquid are moving at random speeds. If a fast-moving particle can break through the surface layer, it will escape from the liquid. Because there is a net downward cohesive force at the surface only the most energetic particles escape. This escape of particles is called evaporation.

Water

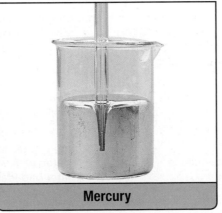

Mercury

Figure 8 Due to adhesive forces, water climbs the outside wall of the glass tube. In the mercury, however, the forces of attraction between mercury atoms are stronger than any adhesive forces between the mercury and the glass. Therefore, the mercury is depressed by the tube.

Figure 9 Warm, moist, surface air rises. Clouds form when the air cools and the water vapor condenses.

PhysicsLAB

EVAPORATIVE COOLING
How can you infer the relationship between cohesive forces and evaporation rates?

Watch a **BrainPOP** video on humidity.

Evaporative cooling Evaporation has a cooling effect. On a hot day, your body perspires, and the evaporation of your sweat cools you. In a puddle of water, evaporation causes the remaining liquid to cool. Each time a particle with higher-than-average kinetic energy escapes from the water, the average kinetic energy of the remaining particles decreases. As you learned earlier, a decrease in average kinetic energy is a decrease in temperature. Rubbing alcohol has a noticeable cooling effect when it evaporates from a person's skin. Alcohol molecules evaporate easily because they have weak cohesive forces. A liquid that evaporates quickly is called a volatile liquid.

Have you ever wondered why humid days feel warmer than dry days at the same temperature? On a humid day, the water vapor content of the air is high. Because there are already many water molecules in the air, the water molecules in perspiration are less likely to evaporate from the skin. Evaporation is the body's primary cooling mechanism, so the body is not able to cool itself as effectively on a humid day.

☑ **READING CHECK** **Explain** why evaporation has a cooling effect.

Condensation Particles of liquid that have evaporated into the air can also return to the liquid phase if the kinetic energy or temperature decreases, a process called condensation. What happens if you bring a cold glass into a hot, humid area? The outside of the glass soon becomes coated with condensed water. Water molecules moving randomly in the air surrounding the glass strike the cold surface, and if they lose enough energy, the cohesive forces become strong enough to prevent their escape.

The air above any body of water, as shown in **Figure 9,** contains evaporated water vapor, which is water in the form of gas. If the temperature is reduced, the water vapor condenses around tiny dust particles in the air and produces droplets only 0.01 mm in diameter. A cloud of these droplets that forms at Earth's surface is called fog. Fog often forms when moist air is chilled by the cold ground. Fog also occurs briefly when a carbonated drink is opened. The sudden decrease in pressure causes the temperature of the gas in the container to drop, which condenses the water vapor dissolved in that gas.

SECTION 2 REVIEW

Section Self-Check | Check your understanding.

18. **MAIN**IDEA The English language includes the term *adhesive tape* and the phrase *working as a cohesive group.* In these examples, are *adhesive* and *cohesive* being used in the same context as their meanings in physics? Explain your answer.

19. **Surface Tension** A paper clip, which has a density greater than that of water, can be made to stay on the surface of water. What procedures must you follow for this to happen? Explain.

20. **Floating** How can you tell that the paper clip in the previous problem was not floating?

21. **Adhesion and Cohesion** In terms of adhesion and cohesion, explain why alcohol clings to the surface of a glass rod but mercury does not.

22. **Evaporation and Cooling** In the past when a baby had a high fever, the doctor might have suggested gently sponging off the baby with a liquid that evaporates easily. Why would this help?

23. **Critical Thinking** On a hot, humid day, Beth sat outside with a glass of cold water. The outside of the glass was coated with water. Her friend, Sally, suggested that the water had leaked out through the glass. Design an experiment for Beth to show Sally where the water came from.

Fluids at Rest and in Motion

A 2.5-g penny will sink in a glass of water, but a canoe with several passengers can float on a lake or river. Why does the heavier item float but the lighter item sink? What might happen if the canoe were filled with water?

Fluids at Rest

If you have ever dived deep into a swimming pool or a lake, you likely felt pressure on your ears. You might have noticed that the pressure you felt did not depend on whether your head was upright or tilted, but that if you swam deeper, the pressure increased.

Pascal's principle Blaise Pascal, a French physician, found that the pressure at a point in a fluid depends on its depth in the fluid and is unrelated to the shape of the fluid's container. He also noted that any change in pressure applied at any point on a confined fluid is transferred undiminished throughout the fluid, a fact that is now known as **Pascal's principle.** Every time you squeeze an open tube of toothpaste, you demonstrate Pascal's principle. The pressure that your fingers exert at the bottom of the tube is transmitted through the toothpaste and forces the paste out at the top. Likewise, if you squeeze one end of a fluid-filled balloon, the other end of the balloon expands.

One application of Pascal's principal is using fluids in machines to multiply forces. In the hydraulic system shown in **Figure 10,** a fluid is confined to two connecting chambers. Each chamber has a piston that is free to move, and the pistons have different surface areas. Recall that if a force (F_1) is exerted on the first piston with a surface area of A_1, the pressure (P_1) exerted on the fluid is $P_1 = \dfrac{F_1}{A_1}$. The pressure exerted by the fluid on the second piston, with a surface area A_2, is $P_2 = \dfrac{F_2}{A_2}$.

MAINIDEA

Hydraulic lifts, floating objects, and carburetors rely on the forces exerted by fluids.

Essential Questions

- What is Pascal's principle?
- How does Archimedes' principle apply to buoyancy?
- What is the role of Bernoulli's principle in airflows?

Review Vocabulary

pressure the perpendicular component of a force on a surface divided by the area of the surface

New Vocabulary

Pascal's principle
buoyant force
Archimedes' principle
Bernoulli's principle
streamlines

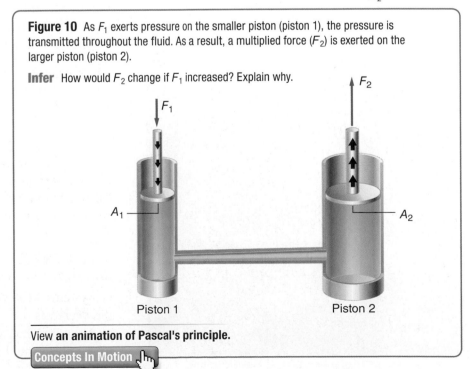

Figure 10 As F_1 exerts pressure on the smaller piston (piston 1), the pressure is transmitted throughout the fluid. As a result, a multiplied force (F_2) is exerted on the larger piston (piston 2).

Infer How would F_2 change if F_1 increased? Explain why.

F_1

F_2

A_1

A_2

Piston 1

Piston 2

View **an animation of Pascal's principle.**

Concepts In Motion

Online Practice

24. Dentists' chairs are examples of hydraulic-lift systems. If a chair weighs 1600 N and rests on a piston with a cross-sectional area of 1440 cm², what force must be applied to the smaller piston, with a cross-sectional area of 72 cm², to lift the chair?

25. A mechanic exerts a force of 55 N on a 0.015 m² hydraulic piston to lift a small automobile. The piston the automobile sits on has an area of 2.4 m². What is the weight of the automobile?

26. **CHALLENGE** By multiplying a force, a hydraulic system serves the same purpose as a lever or a seesaw. If a 400-N child standing on one piston is balanced by a 1100-N adult standing on another piston, what is the ratio of the areas of their pistons?

PhysicsLAB

UNDER PRESSURE
What causes a scuba diver's ears to hurt?

Figure 11 Submersibles are built to withstand the crushing pressure exerted by the water column.

According to Pascal's principle, pressure is transmitted without change throughout a fluid, so pressure P_2 is equal in value to P_1. You can determine the force exerted by the second piston by setting the pressures equal to each other and solving for F_2.

FORCE EXERTED BY A HYDRAULIC LIFT
The force exerted by the second piston is equal to the force exerted by the first piston multiplied by the ratio of the area of the second piston to the area of the first piston.

$$F_2 = F_1 \frac{A_2}{A_1}$$

Swimming Under Pressure

When you are swimming, you feel the pressure of the water increase as you dive deeper. This pressure is a result of gravity; it is related to the weight of the water above you. The deeper you go, the more water there is above you, and the greater the pressure. The pressure of the water is equal to the weight (F_g) of the column of water above you divided by the column's cross-sectional area (A). Even though gravity pulls only in the downward direction, the fluid transmits the pressure in all directions: up, down, and to the sides. As before, the pressure of the water is $P = \frac{F_g}{A}$.

The weight of the column of fluid is $F_g = mg$, and the mass is equal to the density (ρ) of the fluid times its volume, $m = \rho V$. You also know that the volume of the fluid is the area of the base of the column times its height, $V = Ah$. Therefore, $F_g = \rho Ahg$. Substituting ρAhg for F_g gives $P = \frac{F_g}{A} = \frac{\rho Ahg}{A}$. Divide A from the numerator and denominator to arrive at the simplest form of the equation for the pressure exerted by a column of fluid on a submerged body.

PRESSURE OF FLUID ON A BODY
The pressure a column of fluid exerts on a body is equal to the density of the fluid times the height of the column times the free-fall acceleration.

$$P = \rho hg$$

The pressure of a fluid on a body depends on the density of the fluid, its depth, and g. As shown in **Figure 11,** submersibles have explored the deepest ocean trenches and encountered pressures in excess of 1000 times standard air pressure.

▶ **CONNECTION TO BIOLOGY** Scientists use submersibles to learn more about deep ocean ecosystems. In 1977 the first hydrothermal vents were discovered as the crewed submersible ALVIN cruised over the Pacific Ocean floor. Hydrothermal vents form when superheated water flows up from cracks in the seafloor.

Because these vents are located thousands of meters below the ocean surface, the fluid pressure can be over one hundred times the standard atmospheric pressure. Despite the high pressure and the fact that sunlight does not reach them, hydrothermal vents thrive with life. Giant tube worms harbor bacteria in their tissues. The bacteria use hydrogen sulfide from the vent water to produce sugar, which provides the energy that supports the entire ecosystem. Other organisms that live at hydrothermal vents include fish, mussels, shrimp, clams, and octopuses. Submersibles have been used to explore hydrothermal vents in the Atlantic, Indian, and Arctic Oceans.

Buoyancy What produces the upward force that allows you to swim? The increase in pressure with increasing depth creates an upward force called the **buoyant force.** By comparing the buoyant force on an object with its weight, you can predict whether the object will sink or float.

Suppose that a box is immersed in water. It has a height of l, and its top and bottom each have a surface area of A. Its volume, then, is $V = lA$. Water pressure exerts forces on all sides, as shown in **Figure 12.** Will the box sink or float? As you know, the pressure on the box depends on its depth (h). To find out whether the box will float in water, you will need to analyze the forces acting on it, which are its weight and the forces on each side due to the pressure of the fluid. Compare these two equations:

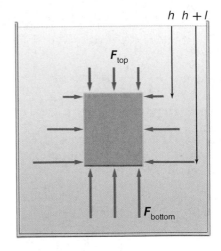

Figure 12 A fluid exerts a greater upward force on the bottom of an immersed object than the downward force on the top of the object. The net upward force is called the buoyant force.

View an **animation of buoyancy.**

Concepts In Motion

$$F_{\text{top}} = P_{\text{top}} A = \rho hgA$$
$$F_{\text{bottom}} = P_{\text{bottom}} A = \rho(l + h)gA$$

On the four vertical sides, the forces are equal in all directions, so there is no net horizontal force. The upward force on the bottom is larger than the downward force on the top, so there is a net upward force. The buoyant force can now be determined.

$$\begin{aligned} F_{\text{buoyant}} &= F_{\text{bottom}} - F_{\text{top}} \\ &= \rho(l + h)gA - \rho hgA \\ &= \rho lgA = \rho Vg \end{aligned}$$

These calculations show the net upward force to be proportional to the volume of the box. This volume equals the volume of the fluid displaced, or pushed out of the way, by the box. Therefore, the magnitude of the buoyant force (ρVg) equals the weight of the fluid displaced by the object.

BUOYANT FORCE
The buoyant force on an object is equal to the weight of the fluid displaced by the object, which is equal to the density of the fluid in which the object is immersed multiplied by the object's volume and the free-fall acceleration.

$$F_{\text{buoyant}} = \rho_{\text{fluid}} Vg$$

Archimedes' principle The relationship between buoyant force and the weight of the fluid displaced by an object was discovered in the third century B.C. by Greek scientist and mathematician Archimedes. **Archimedes' principle** states that an object immersed in a fluid has an upward force on it that is equal to the weight of the fluid displaced by the object. The force does not depend on the weight of the object, only on the weight of the displaced fluid.

THE BUOYANT FORCE OF WATER

Why does a rock feel lighter in water?

Sink or float? If you want to know whether an object sinks or floats, you have to take into account all of the forces acting on the object. The buoyant force pushes up, but the weight of the object pulls it down. The difference between the buoyant force and the object's weight determines whether an object sinks or floats.

Suppose you submerge three objects in a tank of water ($\rho_{water} = 1.00 \times 10^3$ kg/m³). Each object has a volume of 400 cm³, or 4.00×10^{-4} m³. The first object is a steel block with a mass of 3.60 kg. The second is a can of soda with a mass of 0.40 kg. The third is an ice block with a mass of 0.36 kg. How will each item move when it is immersed in water and released? Since the three objects have the same volume, they displace the same volume of water, and the upward force on all three objects is the same, as shown in **Figure 13**. This buoyant force can be calculated as follows.

$$F_{buoyant} = \rho_{water} V g$$
$$= (1.00 \times 10^3 \text{ kg/m}^3)(4.00 \times 10^{-4} \text{ m}^3)(9.8 \text{ N/kg})$$
$$= 3.9 \text{ N}$$

Figure 13 All of the forces on an object must be accounted for to determine whether an object will sink or float.

Describe the circumstances under which an object will float.

■ **Buoyant Force**

Sinking The weight of the block of steel is 8.8 N, much greater than the buoyant force. There is a net downward force, so the block will sink to the bottom of the tank. The net downward force is less than the object's weight. All objects in a liquid, even those that sink, experience a net force that is less than the net force when the object is in air.

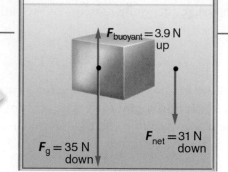

Neutral The weight of the soda can is 0.98 N, the same as the weight of the water displaced. There is, therefore, no net force, and the can will remain wherever it is placed in the water. It is said to have neutral buoyancy.

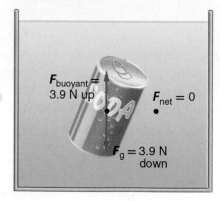

Floating The weight of the ice cube is 0.88 N, less than the buoyant force, so there is a net upward force, and the ice cube will rise. An object will float if its density is less than the density of the fluid in which it is placed.

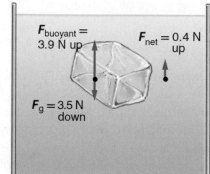

Note: Force vectors are not drawn to scale.

ARCHIMEDES' PRINCIPLE A cubic decimeter (1.00×10^{-3} m³) of a granite building block is submerged in water. The density of granite is 2.70×10^3 kg/m³.

a. What is the magnitude of the buoyant force acting on the block?

b. What is the net force on the block?

1 ANALYZE AND SKETCH THE PROBLEM

- Sketch the cubic decimeter of granite immersed in water.
- Show the upward buoyant force and the downward force due to gravity acting on the granite.

KNOWN

$V = 1.00\times10^{-3}$ m³

$\rho_{granite} = 2.70\times10^3$ kg/m³

$\rho_{water} = 1.00\times10^3$ kg/m³

UNKNOWN

$F_{buoyant} = ?$

$F_{net} = ?$

2 SOLVE FOR THE UNKNOWN

a. Calculate the buoyant force on the granite block.

$F_{buoyant} = \rho_{water}Vg$

　　　$= (1.00\times10^3$ kg/m³$)(1.00\times10^{-3}$ m³$)(9.8$ N/kg$)$

　　　$= 9.8$ N

◀ Substitute $\rho_{water} = 1.00\times10^3$ kg/m³, $V = 1.00\times10^{-3}$ m³, $g = 9.8$ N/kg.

b. Calculate the granite's weight, and then find its net force.

$F_g = mg = \rho_{granite}Vg$

　　$= (2.70\times10^3$ kg/m³$)(1.00\times10^{-3}$ m³$)(9.8$ N/kg$)$

　　$= 26.5$ N

◀ Substitute $\rho_{granite} = 2.70\times10^3$ kg/m³, $V = 1.00\times10^{-3}$ m³, $g = 9.8$ N/kg.

$F_{net} = F_g - F_{buoyant}$

　　　$= 26.5$ N $- 9.8$ N

　　　$= 17$ N

◀ Substitute $F_g = 26.5$ N, $F_{buoyant} = 9.8$ N.

3 EVALUATE THE ANSWER

- **Are the units correct?** The forces and the weight are in newtons, as expected.
- **Is the magnitude realistic?** The buoyant force is about one-third the weight of the granite, a sensible answer because the density of water is about one-third that of granite.

27. Common brick is about 1.8 times denser than water. What is the net force on a 0.20 m³ block of bricks under water?

28. A girl is floating in a freshwater lake with her head just above the water. If she weighs 610 N, what is the volume of the submerged part of her body?

29. What is the tension in a wire supporting a 1250-N camera submerged in water? The volume of the camera is 16.5×10^{-3} m³.

30. Plastic foam is about 0.10 times as dense as water. What weight of bricks could you stack on a 1.0-m × 1.0-m × 0.10-m slab of foam so that the slab of foam floats in water and is barely submerged, leaving the bricks dry?

31. CHALLENGE Canoes often have plastic foam blocks mounted under the seats for flotation in case the canoe fills with water. What is the approximate minimum volume of foam needed for flotation for a 480-N canoe?

THE FIRST FORENSIC SCIENTIST

Forensics Lab Was Archimedes the first forensic scientist?

Figure 14 You can demonstrate Bernoulli's principle by narrowing the opening of the hose as water flows out. As the velocity of the water increases, the pressure it exerts decreases.

Ships How can ships can be made of steel and still float? You can investigate this by making a small boat out of folded aluminum foil. The boat should float easily. Add a cargo of paper clips or some pennies, and it will ride lower in the water. Crumple the foil into a tight ball, and it will sink. When the boat is hollow and large enough so that its average density is less than the density of water, it floats. As you add cargo, the density increases, and more of the boat is submerged. The crumpled boat has a density greater than the water's and sinks.

Other examples of Archimedes' principle include submarines and fishes. Submarines take advantage of Archimedes' principle as water is pumped into or out of chambers to change the submarine's average density, causing it to rise or sink. Fishes that have swim bladders also use Archimedes' principle to control their depths. To move upward in the water, a fish expands its swim bladder by filling it with gas to displace more water and increase the buoyant force. The fish moves downward by contracting the volume of its swim bladder.

Bernoulli's Principle

Study the flow of water from the hose in **Figure 14**. In the photo on the top, the water flows from the hose unobstructed. In the photo on the bottom, the hose opening has been narrowed by a person's thumb being placed over it. Notice that the stream of water on the bottom looks different than it does on the top. The velocity of the water stream in the bottom photo is greater compared to the velocity of the stream in the top photo. What you can't see from the photographs is that pressure exerted by the water in the bottom photo decreased. The relationship between the velocity and pressure exerted by a moving fluid is named for Swiss scientist Daniel Bernoulli. **Bernoulli's principle** states that as the velocity of a fluid increases, the pressure exerted by that fluid decreases. This principle is a statement of work and energy conservation as applied to fluids.

Another instance in which the velocity of water can change is in a stream. You might have seen the water in a stream speed up as it passed through narrowed sections of the stream bed. As the opening of the hose and the stream channel become wider or narrower, the velocity of the fluid changes to maintain the overall flow of water. The pressure of blood in our circulatory systems depends partly on Bernoulli's principle. Bernoulli's principle also helps explain how smoke is pulled up a fireplace chimney.

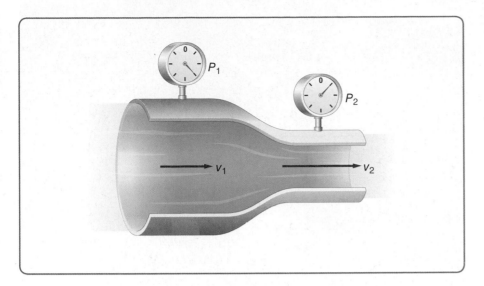

Figure 15 The fluid flowing through this pipe also demonstrates Bernoulli's principle. As the velocity of the fluid increases (v_2 is greater than v_1), the pressure it exerts decreases (P_2 is less than P_1).

Consider a horizontal pipe completely filled with a smoothly flowing ideal fluid. If a certain mass of the fluid enters one end of the pipe, then an equal mass must come out the other end. What happens if the cross section becomes narrower, as shown in **Figure 15?** To keep the same mass of fluid moving through the narrow section in a fixed amount of time, the velocity of the fluid in the tube must increase. As the fluid's velocity increases, so does its kinetic energy. This means that net work has been done on the swifter fluid. This net work comes from the difference between the work that was done to move the mass of fluid into the pipe and the work that was done by the fluid pushing the same mass out of the pipe. The work is proportional to the force on the fluid, which, in turn, depends on the pressure. If the net work is positive, the pressure at the input end of the section, where the velocity is lower, must be larger than the pressure at the output end, where the velocity is higher.

☑ **READING CHECK** **Describe** the relationship between the velocity of a fluid and the pressure it exerts according to Bernoulli's principle.

Applications of Bernoulli's principle There are many common applications of Bernoulli's principle, such as paint sprayers and sprayers attached to garden hoses to apply fertilizers and pesticides to lawns and gardens. In a hose-end sprayer, a strawlike tube is sunk into the chemical solution in the sprayer. The sprayer is attached to a hose. A trigger on the sprayer allows water from the hose to flow at a high speed, producing an area of low pressure above the tube. The solution is then sucked up through the tube and into the stream of water.

A gasoline engine's carburetor, which is where air and gasoline are mixed, is another common application of Bernoulli's principle. Part of the carburetor is a tube with a constriction, as shown in the diagram in **Figure 16.** The pressure on the gasoline in the fuel supply is the same as the pressure in the thicker part of the tube. Air flowing through the narrow section of the tube, which is attached to the fuel supply, is at a lower pressure, so fuel is forced into the air flow. By regulating the flow of air in the tube, the amount of fuel mixed into the air can be varied. Carburetors are used in motorcycles, stock car race cars, and the motors of small gasoline-powered machines, such as lawn mowers.

Figure 16 In a carburetor, low pressure in the narrow part of the tube draws fuel into the air flow.

Carburetor

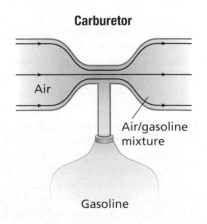

Air

Air/gasoline mixture

Gasoline

Figure 17 The streamline shows the air flowing above a cyclist pedaling in a wind tunnel.

Streamlines Automobile and aircraft manufacturers spend a great deal of time and money testing new designs in wind tunnels to ensure the greatest efficiency of movement through air. The flow of fluids around objects is represented by **streamlines,** as shown in **Figure 17.** Objects require less energy to move through a smooth streamlined flow.

Streamlines can best be illustrated by a simple demonstration. Imagine carefully squeezing tiny drops of food coloring into a smoothly flowing fluid. If the colored lines that form stay thin and well defined, the flow is said to be streamlined. Notice that if the flow narrows, the streamlines move closer together. Closely spaced streamlines indicate greater velocity and, therefore, reduced pressure. If streamlines swirl and become diffused, the flow of the fluid is said to be turbulent. Bernoulli's principle does not apply to turbulent flow.

SECTION 3 REVIEW

Section Self-Check Check your understanding.

32. **MAIN**IDEA All soda cans contain the same volume of liquid, 354 mL, and displace the same volume of water. What might be a difference between a can that sinks and one that floats? *Hint: Place a full can of regular soda and a full can of diet soda in water.*

33. **Transmission of Pressure** A toy rocket launcher is designed so that a child stomps on a rubber cylinder, which increases the air pressure in a launching tube and pushes a foam rocket into the sky. If the child stomps with a force of 150 N on a 2.5×10^{-3}-m^2 area piston, what is the additional force transmitted to the 4.0×10^{-4}-m^2 launch tube?

34. **Floating in Air** A helium balloon rises because of the buoyant force of the air lifting it. The density of helium is 0.18 kg/m^3, and the density of air is 1.3 kg/m^3. How large a volume would a helium balloon need to lift the lead brick shown in **Figure 18?**

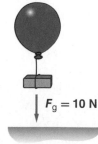

$F_g = 10$ N

Figure 18

35. **Floating and Density** A fishing bobber made of cork floats with one-tenth of its volume below the water's surface. What is the density of cork?

36. **Pressure and Force** An automobile weighing 2.3×10^4 N is lifted by a hydraulic cylinder with an area of 0.15 m^2.

   a. What is the pressure in the hydraulic cylinder?

   b. The pressure in the lifting cylinder is produced by pushing on a 0.0082-m^2 cylinder. What force must be exerted on this small cylinder to lift the automobile?

37. Which displaces more water when placed in a pool?

   a. a 1.0-kg block of aluminum or a 1.0-kg block of lead

   b. a 10-cm^3 block of aluminum or a 10-cm^3 block of lead

38. **Critical Thinking** A tornado passing over a house sometimes makes the house explode from the inside out. How might Bernoulli's principle explain this phenomenon? What could be done to reduce the danger of a door or window exploding outward?

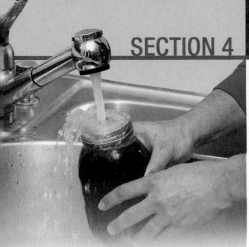

SECTION 4 | Solids

If a glass jar with a metal lid will not open, placing it in warm water will often help loosen the lid. This is because, when heated, the metal lid expands more than the glass jar. What would happen if both the lid and the jar were made of the same material?

MAINIDEA

Solids usually expand when heated.

Essential Questions

- How do a solid's properties relate to that solid's structure?
- Why do solids expand and contract when the temperature changes?
- Why is thermal expansion important?

Review Vocabulary

cohesive force the force of attraction that particles within a substance exert on one another

New Vocabulary

crystal lattice
amorphous solid
coefficient of linear expansion
coefficient of volume expansion

Solid Bodies

How do solids and liquids differ? Solids are stiff, they can be cut in pieces, and they retain their shapes. You can push on solids. Liquids flow and do not retain their shapes. If you push your finger on water, your finger will move through it. Under certain conditions, however, solids and liquids are not easily distinguished. As bottle glass is heated to a molten state, the change from solid to liquid is so gradual that, at times during the process, it is difficult to tell which is which.

When the temperature of a liquid is lowered, the average kinetic energy of the particles decreases. As the particles slow down, the cohesive forces have more effect, and for many solids, the particles become frozen into a fixed pattern called a **crystal lattice,** shown in **Figure 19.** Although the cohesive forces hold the particles in place, the particles in a crystalline solid do not stop moving completely. Rather, they vibrate around their fixed positions. In other materials, such as butter and glass, the particles do not form a fixed crystalline pattern. A substance that has no regular crystal structure but does have a definite volume and shape is called an **amorphous solid.**

Pressure and freezing As a liquid becomes a solid, its particles usually fit more closely together than in the liquid state, making solids more dense than liquids. However, water is an exception because it is most dense at 4°C. Water is also an exception to another general rule. For most liquids, an increase in the pressure on the surface of the liquid increases its freezing point. Because water expands as it freezes, an increase in pressure forces the molecules closer together and opposes the freezing. Therefore, higher pressure lowers the freezing point of water very slightly.

Figure 19 As the temperature of water is lowered and it changes from a liquid to a solid, the particles are frozen in a pattern called a crystal lattice.

| Solid Ice | Crystal Lattice |

● = O
● = H

It has been hypothesized that the drop in water's freezing point caused by the pressure of an ice-skater's blades produces a thin film of liquid between the ice and the blades. Recent measurements have shown that the friction between the blade and the ice generates enough thermal energy to melt the ice and create a thin layer of water. This explanation is supported by measurements of the spray of ice particles, which are considerably warmer than the ice itself. The same process of melting occurs during snow skiing.

Elasticity of solids External forces applied to a solid object may twist, stretch, or bend it out of shape. The ability of a solid object to return to its original form when the external forces are removed is called the elasticity of the solid. If too much deformation occurs, the object will not return to its original shape because its elastic limit has been exceeded. Elasticity is a property of each substance and depends on the forces holding its particles together. Malleability and ductility are two properties that depend on the structure and elasticity of a substance. Because gold can be flattened and shaped into thin sheets, it is said to be malleable. Copper is a ductile metal because it can be pulled into thin strands of wire.

Thermal Expansion of Solids

It is standard practice for engineers to design small gaps, called expansion joints, into concrete-and-steel highway bridges to allow for the expansion of parts in the heat of summer. Expansion joints are shown in **Figure 20.** Objects expand only a small amount when they are heated, but that small amount could be several centimeters in a 100-m-long bridge. If expansion gaps were not present, the bridge could buckle, or parts of it could break. Some materials, such as the ovenproof glass used for laboratory experiments and cooking, are designed to have the least possible thermal expansion. Large telescope mirrors are made of a ceramic material that is designed to undergo essentially no thermal expansion.

To understand the expansion of heated solids, picture a solid as a collection of particles connected by springs that represent the attractive forces between the particles. When the particles get too close, the springs push them apart. When a solid is heated, the kinetic energy of the particles increases, and they vibrate rapidly and move farther apart, weakening the attractive forces between the particles. As a result, when the particles vibrate more violently with increased temperature, their average separation increases and the solid expands.

Figure 20 Expansion joints are included when bridges, roads, and train tracks are built.

Infer If there were no expansion joints in this road, what might happen during the summer?

The change in length of a solid is proportional to the change in temperature, as shown in **Figure 21.** A solid will expand in length twice as much when its temperature is increased by 20°C than when it is increased by 10°C. The expansion also is proportional to its length. This means that a 2-m bar will expand twice as much as a 1-m bar with the same change in temperature. The length (L_2) of a solid at temperature T_2 can be found using the following relationship, where L_1 is the length at temperature T_1 and alpha (α) is the coefficient of linear expansion. The **coefficient of linear expansion** is equal to the change in length divided by the original length and the change in temperature.

$$L_2 = L_1 + \alpha L_1 (T_2 - T_1)$$

With some algebra, you can solve for α.

$$L_2 - L_1 = \alpha L_1 (T_2 - T_1)$$
$$\Delta L = \alpha L_1 \Delta T$$

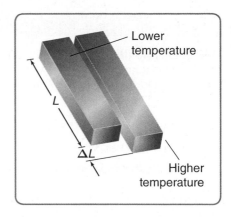

Figure 21 The change in length of a material is proportional to its original length and the change in temperature.

COEFFICIENT OF LINEAR EXPANSION
The coefficient of linear expansion is equal to the change in length divided by the product of the original length and the change in temperature.

$$\alpha = \frac{\Delta L}{L_1 \Delta T}$$

The unit for the coefficient of linear expansion is the reciprocal of degrees Celsius (which can be written as 1/°C or °C^{-1}). The **coefficient of volume expansion** is equal to the change in volume divided by the original volume and the change in temperature. The coefficient of volume expansion (β) is generally about three times the coefficient of linear expansion.

COEFFICIENT OF VOLUME EXPANSION
The coefficient of volume expansion is equal to the change in volume divided by the product of the original volume and the change in temperature.

$$\beta = \frac{\Delta V}{V_1 \Delta T}$$

The unit for β is also 1/°C (°C^{-1}). The two coefficients of thermal expansion for a variety of materials are given in **Table 2.**

Table 2 Coefficients of Thermal Expansion at 20°C		
Material	**Coefficient of Linear Expansion, α (°C^{-1})**	**Coefficient of Volume Expansion, β (°C^{-1})**
Solids		
Aluminum	23×10^{-6}	69×10^{-6}
Glass (soft)	9×10^{-6}	27×10^{-6}
Glass (ovenproof)	3×10^{-6}	9×10^{-6}
Concrete	12×10^{-6}	36×10^{-6}
Copper	17×10^{-6}	51×10^{-6}
Liquids		
Methanol	Not Applicable	1200×10^{-6}
Gasoline	Not Applicable	950×10^{-6}
Water	Not Applicable	210×10^{-6}

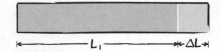

Get help with **operations with scientific notation.**   Math Handbook

LINEAR EXPANSION A metal bar is 1.60 m long at room temperature (21°C). The bar is put into an oven and heated to a temperature of 84°C. It is then measured and found to be 1.7 mm longer. What is the coefficient of linear expansion of this material?

1 ANALYZE AND SKETCH THE PROBLEM

- Sketch the bar, which is 1.7 mm longer at 84°C than at 21°C.
- Identify the initial length of the bar (L_1) and the change in length (ΔL).

KNOWN

$L_1 = 1.60$ m
$\Delta L = 1.7 \times 10^{-3}$ m
$T_1 = 21°C$
$T_2 = 84°C$

UNKNOWN

$\alpha = ?$

2 SOLVE FOR THE UNKNOWN

Calculate the coefficient of linear expansion using the relationship among known length, change in length, and change in temperature.

$$\alpha = \frac{\Delta L}{L_1 \Delta T}$$

$$= \frac{1.7 \times 10^{-3} \text{ m}}{(1.60 \text{ m})(84°C - 21°C)}$$

◀ Substitute $\Delta L = 1.7 \times 10^{-3}$ m, $L_1 = 1.60$ m, $\Delta T = (T_2 - T_1) = 84°C - 21°C$

$$= 1.7 \times 10^{-5} \text{ °C}^{-1}$$

3 EVALUATE THE ANSWER

- **Are the units correct?** The units are correctly expressed in °C^{-1}.
- **Is the magnitude realistic?** The magnitude of the coefficient is close to the accepted value for copper.

Do additional problems.   Online Practice

39. A piece of aluminum house siding is 3.66 m long on a cold winter day of −28°C. How much longer is it on the hot summer day shown in **Figure 22?**

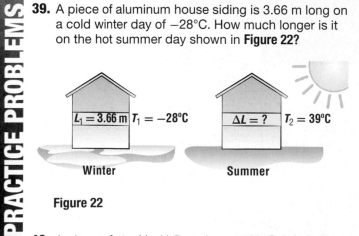

Winter   Summer

Figure 22

40. A piece of steel is 11.5 cm long at 22°C. It is heated to 1221°C, close to its melting temperature. How long is it?

41. A 400-mL glass beaker at room temperature is filled to the brim with cold water at 4.4°C. When the water warms up to 30.0°C, how much water will spill from the beaker?

42. A tank truck takes on a load of 45,725 L of gasoline in Houston, where the temperature is 28.0°C. The truck delivers its load in Minneapolis, where the temperature is −12.0°C.

a. How many liters of gasoline does the truck deliver?

b. What happened to the gasoline?

43. A hole with a diameter of 0.85 cm is drilled into a steel plate. At 30.0°C, the hole exactly accommodates an aluminum rod of the same diameter. What is the spacing between the plate and the rod when they are cooled to 0.0°C?

44. CHALLENGE A steel ruler is marked in millimeters so that the ruler is absolutely correct at 30.0°C. By what percentage would the ruler be incorrect at −30.0°C?

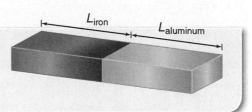
Applications of thermal expansion Engineers take the thermal expansion of materials into consideration as they design structures. You've already read about the expansion joints that are installed on concrete highways and bridges. The regular gaps between slabs of concrete in sidewalks also help keep sidewalks from buckling when the concrete expands during hot weather. Different materials expand at different rates, as indicated by the different coefficients of expansion given in **Table 2.** Engineers also consider different expansion rates when designing systems. Steel bars are often used to reinforce concrete. The steel and concrete must have the same expansion coefficient. Otherwise, the structure could crack on a hot day. Similarly, filling materials used to repair teeth must expand and contract at the same rate as tooth enamel.

Different rates of expansion have useful applications. For example, engineers have taken advantage of these differences to construct a useful device called a bimetallic strip, which is used in thermostats. A bimetallic strip consists of two strips of different metals welded or riveted together. Usually, one strip is brass and the other is iron. When heated, brass expands more than iron does. Thus, when the bimetallic strip of brass and iron is heated, the brass part of the strip becomes longer than the iron part. As a result, the bimetallic strip bends with the brass on the outside of the curve. If the bimetallic strip is cooled, it bends in the opposite direction. The brass is then on the inside of the curve.

In a home thermostat the bimetallic strip is installed so that it bends toward an electric contact as the room cools. When the room cools below the setting on the thermostat, the bimetallic strip bends enough to make electric contact with the switch, which turns on the heater. As the room warms, the bimetallic strip bends in the other direction. When the room's temperature reaches the setting on the thermostat, the electric circuit is broken and the heater switches off.

MiniLAB

JUMPERS
Can heat change the shape of a bimetallic disk?

SECTION 4 REVIEW

Section Self-Check — Check your understanding.

45. MAINIDEA On a hot day, you are installing an aluminum screen door in a concrete doorframe. You want the door to fit well on a cold winter day. Should you make the door fit tightly in the frame or leave extra room?

46. Types of Solids What is the difference between the structure of candle wax and that of ice?

47. Thermal Expansion Can you heat a piece of copper enough to double its length?

48. States of Matter Does **Table 2** provide a way to distinguish between solids and liquids?

49. Solids and Liquids A solid can be defined as a material that can be bent and will resist bending. Explain how these properties relate to the binding of atoms in a solid but do not apply to a liquid.

50. Critical Thinking The iron ring in **Figure 23** was made by cutting a small piece from a solid ring. If the ring in the figure is heated, will the gap become wider or narrower? Explain your answer.

Figure 23

BENDING THE RULES!

In a match prior to the 1998 World Cup, Brazilian soccer player Roberto Carlos amazed fans and players by bending a free kick around a wall of defenders by at least a meter and clipping the inside of the post for a goal. The only onlookers who weren't mystified were the physicists in the crowd, as they could easily explain the shot's strange trajectory!

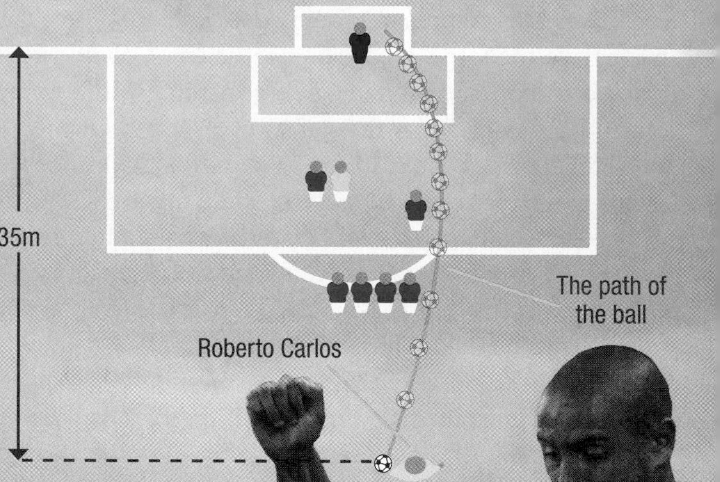

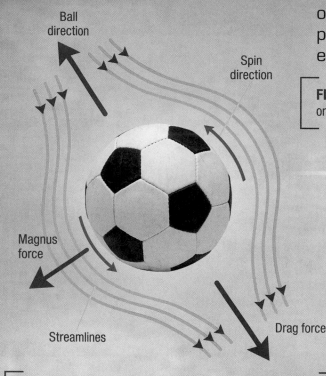

Ball direction

Spin direction

Magnus force

Streamlines

Drag force

FIGURE 1 A bird's eye view of a soccer ball spinning on its axis perpendicular to the flow of air across it.

35m

The path of the ball

Roberto Carlos

FIGURE 2 The curving path of a rotating ball will spiral exponentially smaller, provided the rotation remains constant.

Bend it like Carlos The path of Carlos's curving shot is shown in **Figure 1.** In a curving shot, the drag force is not the only force that will act on the ball from the air. Another force, called the Magnus force, also acts on the ball. The Magnus force was first explained by Gustav Magnus in 1852. He was trying to determine why spinning artillery shells and bullets deflect to one side. The Magnus force acts on spinning soccer balls too.

In **Figure 2** the left side of the ball is spinning in the same direction as the flow of air around the ball as it moves. As a result, the pressure on the left side of the ball is reduced. Notice that the right side of the ball is spinning in the opposite direction of the airflow. On this side of the ball the drag force is increased. Due to this imbalance of forces, the ball curves to the left. A slow-moving ball with a lot of spin will experience a larger sideways force than a fast-moving ball with the same spin. As the ball slows at the end of its trajectory, the curve becomes more pronounced. Because Carlos had practiced this shot countless times, he knew precisely where the ball would bend in its flight toward the goal.

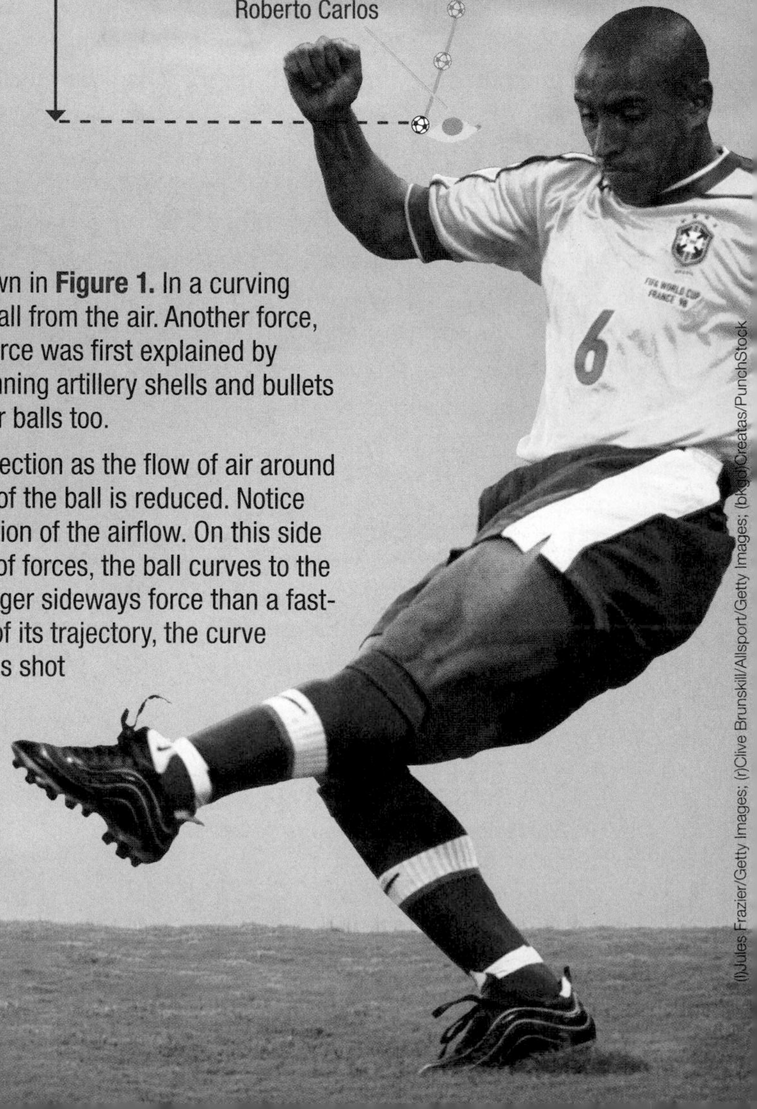

GOING**FURTHER** >>>

Research how different projectiles, such as baseballs, hockey pucks, and flying disks, are designed to travel and how this affects the games in which they are used.

STUDY GUIDE

BIGIDEA   The thermal energy of a material and the forces between the material's particles determine its properties.

VOCABULARY
- **fluid** *(p. 348)*
- **pressure** *(p. 349)*
- **pascal** *(p. 349)*
- **combined gas law** *(p. 351)*
- **ideal gas law** *(p. 352)*
- **thermal expansion** *(p. 354)*
- **plasma** *(p. 355)*

SECTION 1 Properties of Fluids

MAINIDEA   Fluids flow, have no definite shape, and include liquids, gases, and plasmas.

- Matter in the fluid state flows and has no definite shape of its own.
- The combined gas law represents the relationships among the pressure, volume, and temperature of gases.

$$\frac{P_1 V_1}{T_1} = \frac{P_2 V_2}{T_2}$$

- The ideal gas law represents the relationship among pressure, volume, temperature, and the number of moles of a gas.
- Plasma is a gaslike state of negatively charged electrons and positively charged ions.

VOCABULARY
- **cohesive forces** *(p. 356)*
- **adhesive forces** *(p. 357)*

SECTION 2 Forces Within Liquids

MAINIDEA   Cohesive forces occur between the particles of a substance, while adhesive forces occur between particles of different substances.

- Surface tension is the tendency of the surface of a liquid to contract to the smallest possible area. Surface tension results from the attractive forces that like particles exert on one another.
- Capillary action occurs when a liquid rises in a thin tube because the adhesive forces between the tube and the liquid are stronger than the cohesive forces between liquid's molecules.
- Clouds form when water vapor in the atmosphere cools and condenses, forming droplets around dust particles.

VOCABULARY
- **Pascal's principle** *(p. 359)*
- **buoyant force** *(p. 361)*
- **Archimedes' principle** *(p. 361)*
- **Bernoulli's principle** *(p. 364)*
- **streamlines** *(p. 366)*

SECTION 3 Fluids at Rest and in Motion

MAINIDEA   Hydraulic lifts, floating objects, and carburetors rely on the forces exerted by fluids.

- Pascal's principle states that an applied pressure change is transmitted undiminished throughout a fluid.
- According to Archimedes' principle, the buoyant force equals the weight of the fluid displaced by an object.
- Bernoulli's principle states that the pressure exerted by a fluid decreases as its velocity increases.

VOCABULARY
- **crystal lattice** *(p. 367)*
- **amorphous solid** *(p. 367)*
- **coefficient of linear expansion** *(p. 369)*
- **coefficient of volume expansion** *(p. 369)*

SECTION 4 Solids

MAINIDEA   Solids usually expand when heated.

- A crystalline solid has a regular pattern of particles, and an amorphous solid has an irregular pattern of particles. Malleability and ductility depend on structure type.
- As the temperature of a solid changes, the kinetic energy of its particles changes accordingly. As the vibrations of the particles change, a solid expands as temperature increases and contracts as temperature decreases.
- Expansion rates of different materials must be considered when designing structures.

Games and Multilingual eGlossary

Vocabulary Practice 🖱

Chapter Self-Check

SECTION 1 Properties of Fluids

Mastering Concepts

51. How are force and pressure different?

52. A gas is placed in a sealed container, and some liquid is placed in a container of the same size. The gas and liquid both have definite volume. How do they differ?

53. What characteristics do gases and plasmas have in common? How do gases and plasma differ?

54. The Sun is made of plasma. How is this plasma different from the plasmas on Earth?

Mastering Problems

55. Textbooks A 0.85-kg physics book with dimensions of 24.0 cm × 20.0 cm is at rest on a table.

   a. What force does the book apply to the table?

   b. What pressure does the book apply to the table?

56. Ranking Task Rank the following situations according to the pressure, from least to greatest. Specifically indicate any ties.

   A. 20 N exerted over a surface of 0.35 m^2

   B. 20 N exerted over a surface of 0.65 m^2

   C. 50 N exerted over a surface of 0.05 m^2

   D. 50 N exerted over a surface of 0.35 m^2

   E. 60 N exerted over a surface of 0.55 m^2

57. As shown in **Figure 24,** a constant-pressure thermometer is made with a cylinder containing a piston that can move freely inside the cylinder. The pressure and the amount of gas enclosed in the cylinder are kept constant. As the temperature increases or decreases, the piston moves up or down in the cylinder. At 0°C, the height of the piston is 20 cm. What is the height of the piston at 100°C?

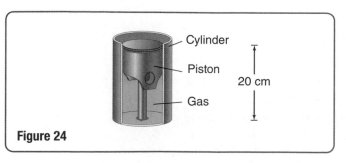

Figure 24

58. Soft Drinks Sodas are made fizzy by the carbon dioxide (CO_2) dissolved in the liquid. An amount of carbon dioxide equal to about 8.0 L of carbon dioxide gas at atmospheric pressure and 300.0 K can be dissolved in a 2-L bottle of soda. The molar mass of CO_2 is 44 g/mol.

   a. How many moles of carbon dioxide are in the 2-L bottle? (1 L = 0.001 m^3)

   b. What is the mass of the carbon dioxide in the 2-L bottle of soda?

59. A piston with an area of 0.015 m^2 encloses a constant amount of gas in a cylinder with a volume of 0.23 m^3. The initial pressure of the gas equals 1.5×10^5 Pa. A 150-kg mass is placed on the piston, and the piston moves downward to a new position, as shown in **Figure 25.** If the temperature is constant, what is the new volume of the gas in the cylinder?

Volume = 0.23 m3
Piston area = 0.015 m2

Volume = ?

Figure 25

60. Automobiles A certain automobile tire is specified to be used at a gauge pressure of 30.0 psi (psi is pounds per square inch). (One pound per square inch equals 6.90×10^3 Pa.) The term *gauge pressure* means the pressure above atmospheric pressure. Thus, the actual pressure in the tire is 1.01×10^5 Pa + (30.0 psi)(6.90×10^3 Pa/psi) = 3.08×10^5 Pa. As the car is driven, the tire's temperature increases, and the volume and pressure increase. Suppose you filled a car's tire to a volume of 0.55 m^3 at a temperature of 280 K. The initial pressure was 30.0 psi, but during the drive, the tire's temperature increased to 310 K and the tire's volume increased to 0.58 m^3.

   a. What is the new pressure in the tire?

   b. What is the new gauge pressure?

SECTION 2 **Forces Within Liquids**

Mastering Concepts

61. Lakes A frozen lake melts in the spring. What effect does this have on the temperature of the air above the lake?

62. Hiking Canteens used by hikers often are covered with canvas bags. If you wet the canvas bag covering a canteen, the water in the canteen will be cooled. Explain.

SECTION 3 **Fluids at Rest and in Motion**

Mastering Concepts

63. What do the equilibrium tubes in **Figure 26** tell you about the pressure exerted by a liquid?

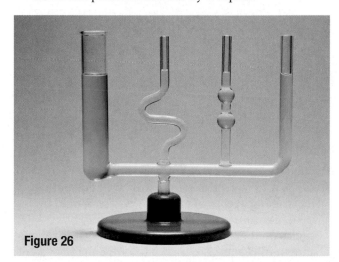

Figure 26

64. According to Pascal's principle, what happens to the pressure at the top of a container if the pressure at the bottom is increased?

65. How does the water pressure 1 m below the surface of a small pond compare with the water pressure the same distance below the surface of a lake?

66. Does Archimedes' principle apply to an object inside a flask that is inside a spaceship in orbit?

67. A stream of water goes through a garden hose into a nozzle. As the water speeds up, what happens to the water pressure?

Mastering Problems

68. Reservoirs A reservoir behind a dam is 17 m deep. What is the pressure of the water at the following locations?

a. the base of the dam

b. 4.0 m from the top of the dam

69. A test tube standing vertically in a test-tube rack contains 2.5 cm of oil ($\rho = 0.81$ g/cm^3) and 6.5 cm of water. What is the pressure exerted by the two liquids on the bottom of the test tube?

70. Antiques An antique yellow metal statuette of a bird is suspended from a spring scale. The scale reads 11.81 N when the statuette is suspended in air, and it reads 11.19 N when the statuette is completely submerged in water.

a. Find the volume of the statuette.

b. Is the bird more likely made of gold ($\rho = 19.3 \times 10^3$ kg/cm^3) or gold-plated aluminum ($\rho = 2.7 \times 10^3$ kg/cm^3)?

71. During an ecology experiment, an aquarium half-filled with water is placed on a scale. The scale shows a weight of 195 N.

a. A rock weighing 8 N is added to the aquarium. If the rock sinks to the bottom of the aquarium, what will the scale read?

b. The rock is removed from the aquarium, and the amount of water is adjusted until the scale again reads 195 N. A fish weighing 2 N is added to the aquarium. What is the scale reading with the fish in the aquarium?

72. Oceanography As shown in **Figure 27,** a large buoy used to support an oceanographic research instrument is made of a cylindrical, hollow iron tank. The tank is 2.1 m in height and 0.33 m in diameter. The total mass of the buoy and the research instrument is about 120 kg. The buoy must float so that one end is above the water to support a radio transmitter. Assuming that the mass of the buoy is evenly distributed, how much of the buoy will be above the waterline when it is floating?

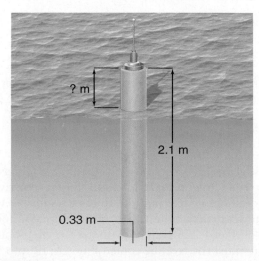

Figure 27

73. What is the magnitude of the buoyant force on a 26.0-N ball that is floating in fresh water?

74. What is the net force on a rock submerged in water if the rock weighs 45 N in air and has a volume of 2.1×10^{-3} m^3?

75. What is the maximum weight that a balloon filled with 1.00 m^3 of helium can lift in air? Assume that the density of air is 1.20 kg/m^3 and that of helium is 0.177 kg/m^3. Neglect the mass of the balloon.

76. If a rock weighs 54 N in air and experiences a net force of 46 N when submerged in a liquid with a density twice that of water, what will be the net force on it when it is submerged in water?

SECTION 4 Solids

Mastering Concepts

77. How does the arrangement of atoms in a crystalline solid differ from that in an amorphous solid?

78. Does the coefficient of linear expansion depend on the unit of length used? Explain.

Mastering Problems

79. A bar of an unknown metal has a length of 0.975 m at 45°C and a length of 0.972 m at 23°C. What is its coefficient of linear expansion?

80. An inventor constructs a thermometer from an aluminum bar that is 0.500 m in length at 273 K. He measures the temperature by measuring the length of the aluminum bar. If the inventor wants to measure a 1.0-K change in temperature, how precisely must he measure the length of the bar?

81. Bridges How much longer will a 300-m steel bridge be on a 30°C day in August than on a −10°C night in January?

82. What is the change in length of a 2.00-m copper pipe if its temperature is raised from 23°C to 978°C?

83. What is the change in volume of a 1.0-m^3 concrete block if its temperature is raised 45°C?

84. Bridge builders often use rivets that are larger than the rivet hole to make the joint tighter. The rivet is cooled before it is put into the hole. Suppose that a builder drills a hole 1.2230 cm in diameter for a steel rivet 1.2250 cm in diameter. To what temperature must the rivet be cooled if it is to fit into the rivet hole, which is at 20.0°C?

85. A steel tank filled with methanol is 2.000 m in diameter and 5.000 m in height. It is completely filled at 10.0°C. If the temperature rises to 40.0°C, how much methanol (in liters) will flow out of the tank, given that both the tank and the methanol will expand?

86. An aluminum sphere is heated from 11°C to 580°C. If the volume of the sphere is 1.78 cm^3 at 11°C, what is the increase in sphere volume at 580°C?

87. The volume of a copper sphere is 2.56 cm^3 after being heated from 12°C to 984°C. What was the volume of the copper sphere at 12°C?

88. A square iron plate that is 0.3300 m on each side is heated from 0°C to 95°C.

 a. What is the change in the length of the sides of the square?

 b. What is the change in area of the square?

89. An aluminum cube with a volume of 0.350 m^3 at 350.0 K is cooled to 270.0 K.

 a. What is its volume at 270.0 K?

 b. What is the length of a side of the cube at 270.0 K?

90. Industry A machinist builds a rectangular mechanical part for a special refrigerator system from two rectangular pieces of steel and two rectangular pieces of aluminum. At 293 K, the part is a perfect square, but at 170 K, the part becomes warped, as shown in **Figure 28**. Which parts were made of steel and which were made of aluminum?

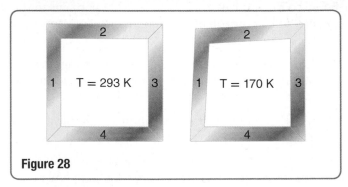

Figure 28

Applying Concepts

91. A rectangular box with its largest surface resting on a table is rotated so that its smallest surface is now on the table. Has the pressure on the table increased, decreased, or remained the same?

92. Show that a pascal is equivalent to a kg/m·s^2.

93. Shipping Cargo Compared to an identical empty ship, would a ship filled with table-tennis balls sink deeper into the water or rise in the water? Explain.

94. How deep would a water container have to be to have the same pressure at the bottom as that found at the bottom of a 10.0-cm-deep beaker of mercury, which is 13.55 times as dense as water?

95. Drops of mercury, water, ethanol, and acetone are placed on a smooth, flat surface, as shown in **Figure 29**. From this figure, what can you conclude about the cohesive forces in these liquids?

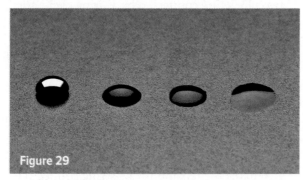

Figure 29

96. Alcohol evaporates more quickly than water does at the same temperature. What does this observation allow you to conclude about the properties of the particles in the two liquids?

97. Five objects with the following densities are put into a tank of water.

a. 85 g/cm³   **d.** 1.15 g/cm³

b. 0.95 g/cm³   **e.** 1.25 g/cm³

c. 1.05 g/cm³

The density of water is 1.00 g/cm³. The diagram in **Figure 30** shows six possible positions of these objects. Select a position, from 1 to 6, for each of the five objects. Not all positions need to be selected.

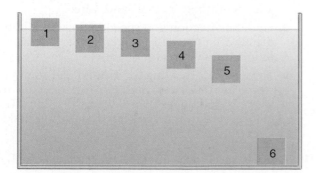

Figure 30

98. Equal volumes of water are heated in two tubes that are identical, except that tube A is made of soft glass and tube B is made of ovenproof glass. As the temperature increases, the water level rises higher in tube B than in tube A. Give a possible explanation.

99. A platinum wire easily can be sealed in a glass tube, but a copper wire does not form a tight seal with the glass. Explain.

Mixed Review

100. What is the pressure on the hull of a submarine at a depth of 65 m?

101. Scuba Diving A scuba diver swimming at a depth of 5.0 m under water exhales a 4.2×10^{-6} m³ bubble of air. What is the volume of that bubble just before it reaches the surface of the water?

102. Reverse Problem Write a physics problem with real-life objects for which the following equation would be part of the solution:

$$T_1 = \frac{(61.2 \text{ kPa})(28.0 \text{ L})(273 \text{ K})}{(77.0 \text{ kPa})(25.0 \text{ L})}$$

103. BIGIDEA An aluminum bar is floating in a bowl of mercury. When the temperature is increased, does the aluminum float higher or sink deeper into the mercury?

104. There is 100.0 mL of water in an 800.0-mL soft-glass beaker at 15.0°C. How much will the water level have dropped or risen when the beaker and water are heated to 50.0°C?

105. Auto Maintenance A hydraulic jack used to lift cars for repairs is called a 3-ton jack. The large piston is 22 mm in diameter, and the small one is 6.3 mm in diameter. Assume that a force of 3 tons is 3.0×10^4 N.

a. What force must be exerted on the small piston to lift a 3-ton weight?

b. Most jacks use a lever to reduce the force needed on the small piston. If the resistance arm is 3.0 cm, how long must the effort arm of an ideal lever be to reduce the force to 100.0 N?

106. Ballooning A hot-air balloon contains a fixed volume of gas. When the gas is heated, it expands and some gas escapes out the open end. As a result, the mass of the gas in the balloon is reduced. Why would the air in a balloon have to be hotter to lift the same number of people above Vail, Colorado, which has an altitude of 2400 m, than above Norfolk, Virginia, which has an altitude of 3 m?

ASSESSMENT

Thinking Critically

107. Problem Posing Complete this problem so that it can be solved using buoyant force and Newton's second law: "A block of metal has a volume of 2.4 cm³ and mass of 0.56 kg ...."

108. Analyze and Conclude One method of measuring the percentage of body fat is based on the fact that fatty tissue is less dense than muscle tissue. How can a person's average density be assessed with a scale and a swimming pool? What measurements does a physician need to record to find a person's average percentage of body fat?

109. Analyze and Conclude A downward force of 700 N is required to fully submerge a plastic foam sphere, as shown in **Figure 31**. The density of the foam is 95 kg/m³.

   a. What percentage of the sphere would be submerged if the sphere were released to float freely?

   b. What is the weight of the sphere in air?

   c. What is the volume of the sphere?

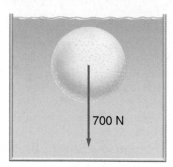

700 N

Figure 31

110. Apply Concepts Tropical fish for aquariums are often transported home from pet shops in transparent plastic bags filled mostly with water. If you placed a fish in its unopened transport bag in a home aquarium, which of the cases in **Figure 32** best represents what would happen? Explain your reasoning.

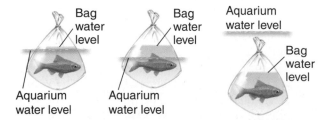

Figure 32

Writing in Physics

111. Some solid materials expand when they are cooled. Water between 4° and 0°C is the most common example, but rubber bands also expand in length when cooled. Research what causes this expansion.

112. Research Joseph Louis Gay-Lussac's contributions to the gas laws. How did Gay-Lussac's work lead to the discovery of the formula for water?

Cumulative Review

113. Two blocks are connected by a string over a frictionless, massless pulley such that one is resting on an inclined plane and the other is hanging over the top edge of the plane, as shown in **Figure 33**. The hanging block has a mass of 3.0 kg and the block on the plane has a mass of 2.0 kg. The coefficient of kinetic friction between the block and the inclined plane is 0.19. Answer the following questions assuming the blocks are released from rest.

   a. What is the acceleration of the blocks?

   b. What is the tension in the string connecting the blocks?

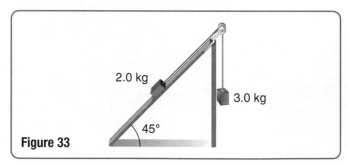

2.0 kg

3.0 kg

45°

Figure 33

114. A compact car with a mass of 875 kg, moving south at 15 m/s, is struck by a full-sized car with a mass of 1584 kg, moving east at 12 m/s. The two cars stick together, and momentum is conserved.

   a. Sketch the situation, assigning coordinate axes and identifying "before" and "after."

   b. Find the direction and speed of the wreck immediately after the collision, remembering that momentum is a vector quantity.

   c. The wreck skids along the ground and comes to a stop. The coefficient of kinetic friction while the wreck is skidding is 0.55. Assume that the acceleration is constant. How far does the wreck skid after impact?

115. A 188-W motor will lift a load at the rate (speed) of 6.50 cm/s. How great a load can the motor lift at this rate?

MULTIPLE CHOICE

1. Gas with a volume of 10.0 L is trapped in an expandable cylinder. If the pressure is tripled and the temperature is increased by 80.0 percent (as measured on the Kelvin scale), what will be the new volume of the gas?

A. 2.70 L

B. 6.00 L

C. 16.7 L

D. 54.0 L

2. Nitrogen gas at standard atmospheric pressure, 101.3 kPa, has a volume of 0.080 m³. If there are 3.6 mol of the gas, what is the temperature?

A. 0.27 K

B. 270 K

C. 0.27°C

D. 270°C

3. As diagrammed below, an operator applies a force of 200.0 N to the first piston of a hydraulic lift, which has an area of 5.4 cm². What is the pressure applied to the hydraulic fluid?

A. 3.7×10^1 Pa

B. 2.0×10^3 Pa

C. 3.7×10^3 Pa

D. 3.7×10^5 Pa

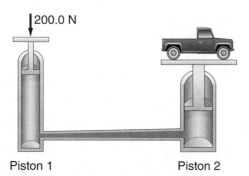

Piston 1   Piston 2

4. If the second piston in the lift diagrammed above exerts a force of 41,000 N, what is the area of the second piston?

A. 0.0049 m²

B. 0.026 m²

C. 0.11 m²

D. 11 m²

5. The density of cocobolo wood from Costa Rica is 1.10 g/cm³. What is the net force on a cocobolo wood figurine that displaces 786 mL when submerged in a freshwater lake?

A. 0.770 N

B. 0.865 N

C. 7.70 N

D. 8.47 N

Online Test Practice

6. What is the buoyant force on a 17-kg object that displaces 85 L of water?

A. 1.7×10^2 N

B. 8.3×10^2 N

C. 1.7×10^5 N

D. 8.3×10^5 N

7. Which one of the following items does not contain matter in the plasma state?

A. neon lighting

B. stars

C. lightning

D. incandescent lighting

8. Suppose you use a hole punch to make a circular hole in a piece of aluminum foil. If you heat the foil, what will happen to the size of the hole?

A. It will decrease.

B. It will increase.

C. It will decrease, then increase.

D. It will increase, then decrease.

FREE RESPONSE

9. A balloon has a volume of 125 mL of air at standard atmospheric pressure, 101.3 kPa. If the balloon is anchored 1.27 m under the surface of a swimming pool, as illustrated in the diagram below, what is the new volume of the balloon?

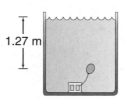

1.27 m

NEED EXTRA HELP?

If you Missed Question	1	2	3	4	5	6	7	8	9
Review Section	1	1	3	3	3	3	1	4	3

Vibrations and Waves

BIGIDEA   Waves and simple harmonic motion are examples of periodic motion.

SECTIONS

LaunchLAB

WAVE INTERACTION

What types of waves can you make on a spring?

WATCH THIS!

PLAYGROUND PHYSICS?

You probably remember playing on swings when you were younger. What did you have to do in order to keep going? How do swings compare to the periodic motion of a pendulum?

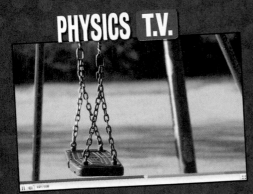

(l)Andrew Ward/Life File/Getty Images, (r)JVP photography/Flickr/Getty Images

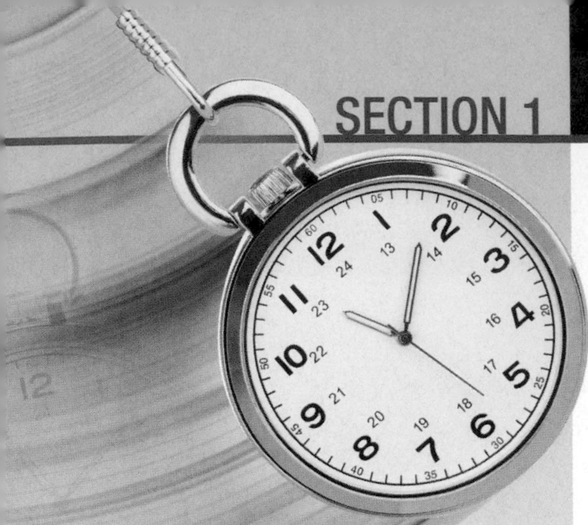

SECTION 1 Periodic Motion

PHYSICS 4 YOU **You might have seen cartoons in which a character is hypnotized using a pocket watch. The watch swings back and forth rhythmically, similar to the pendulum of a grandfather clock. Both the watch and the clock's pendulum undergo periodic motion.**

MAIN IDEA

Periodic motion repeats in a regular cycle.

Essential Questions

- What is simple harmonic motion?
- How much energy is stored in a spring?
- What affects a pendulum's period?

Review Vocabulary

gravitational field (*g*) force per unit mass resulting from the influence of Earth's gravity; near Earth's surface, *g* is 9.8 N/kg toward the center of Earth

New Vocabulary

periodic motion
period
amplitude
simple harmonic motion
Hooke's law
simple pendulum
resonance

Mass on a Spring

The motion of a mass bobbing up and down on a spring is similar to that of a pendulum. The motion repeats, following the same path during the same amount of time. Other similar examples include a vibrating guitar string and a tree swaying in the wind. These motions, which all repeat in a regular cycle, are examples of **periodic motion.**

In each example above, at one position the net force on the object is zero. At that position, the object is in equilibrium. Whenever the object moves away from its equilibrium position, the net force on the system becomes nonzero. This net force acts on the object to bring it back toward equilibrium. The **period** (T) is the time needed for an object to repeat one complete cycle of the motion. The **amplitude** of the motion is the maximum distance the object moves from the equilibrium position.

Simple harmonic motion In **Figure 1,** the force exerted by the spring is directly proportional to the distance the spring is stretched. If you pull the mass down and release it, the mass will bounce up and down through the equilibrium position. Any system in which the force acting to restore an object to its equilibrium position is directly proportional to the displacement of the object shows **simple harmonic motion.**

☑ **READING CHECK** **State** the requirement for simple harmonic motion.

■ **Mass on a Spring**

Figure 1 The force exerted on the mass by the spring is directly proportional to the mass's displacement.

Determine the displacement if the mass is 0.5 *mg*.

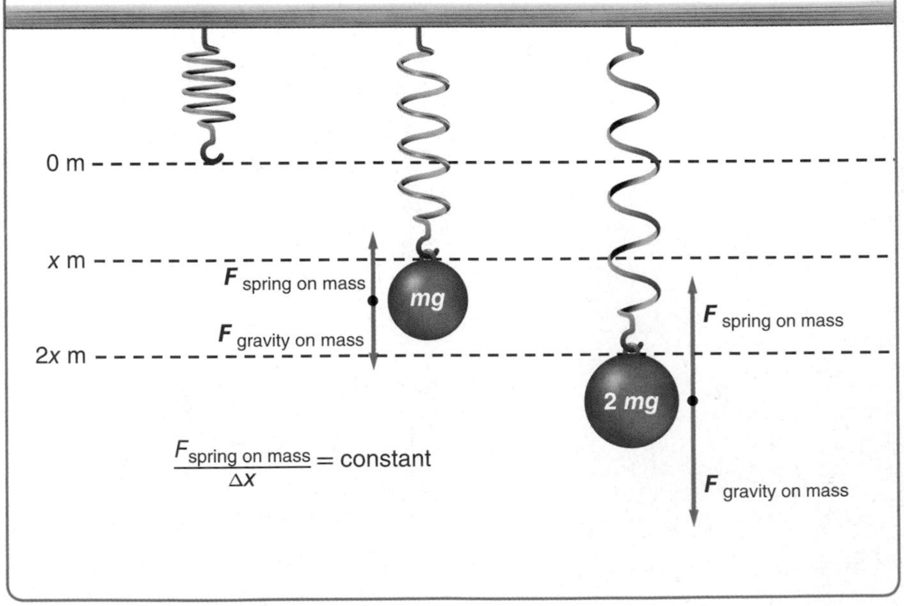

Table 1
Force Magnitude-Stretch Distance in a Spring

Stretch Distance (m)	Magnitude of Force Exerted by Spring (N)
0.0	0.0
0.030	1.9
0.060	3.7
0.090	6.3
0.12	7.8

Finding the spring constant

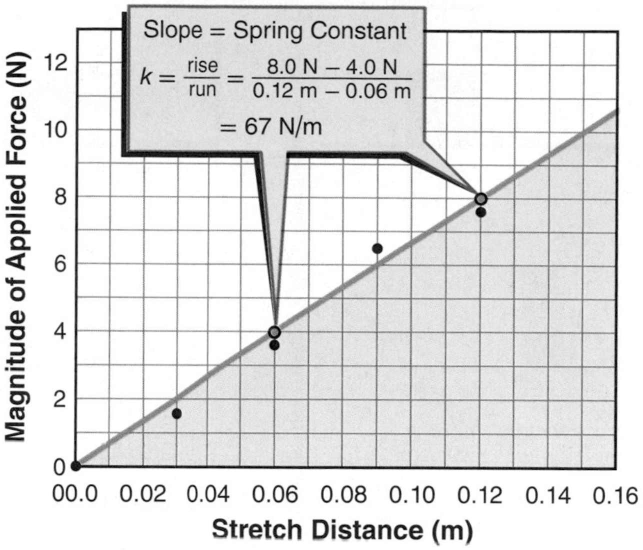

Slope = Spring Constant

$$k = \frac{rise}{run} = \frac{8.0\ N - 4.0\ N}{0.12\ m - 0.06\ m}$$

$$= 67\ N/m$$

Figure 2 The spring constant can be determined from the slope of the force magnitude-distance graph. The area under the curve is equal to the potential energy stored in the spring.

Hooke's Law

Table 1 shows the results of an investigation into the relationship between the magnitude of the force exerted by a spring and the distance the spring stretches. **Figure 2** is a graph of the data with the line of best fit. The linear relationship shown in the graph indicates that the magnitude of the force exerted by the spring is directly proportional to the amount the spring is stretched. A spring that exerts a force directly proportional to the distance stretched obeys **Hooke's law.**

HOOKE'S LAW
The magnitude of the force exerted by a spring is equal to the spring constant times the distance the spring is stretched or compressed from its equilibrium position.

$$F = -kx$$

In this equation, k is the spring constant, which depends on the stiffness and other properties of the spring, and x is the distance the spring is stretched from its equilibrium position. Notice that k is the slope of the line in the magnitude of the force v. stretch distance graph. A steeper slope, and thus a larger k, indicates that the spring is harder to stretch. The spring constant has the same units as the slope, newtons/meter (N/m).

The negative sign in Hooke's law indicates that the force is in the direction opposite the stretch or compression direction. The force exerted by the spring on the mass is always directed toward the spring's equilibrium position.

☑ **READING CHECK Explain** the negative sign in Hooke's law.

Hooke's law and real springs Not all springs obey Hooke's law. For example, rubber bands do not. Those that do obey Hooke's law are called elastic springs. Even for elastic springs, Hooke's law only applies over a limited range of distances. If a spring is stretched too far, that spring can become so deformed that the force is not proportional to the displacement. We say the spring has exceeded its elastic limit.

Potential energy When you stretch a spring you transfer energy to the spring, giving it elastic potential energy. The work done by an applied force is equal to the area under a force v. distance graph. How much work is done to stretch a spring? In **Figure 2,** the area under the curve represents the work done to stretch the spring. This work is equal to the elastic potential energy stored in the spring. To calculate the potential energy stored in the spring, find the area of the triangle by multiplying one-half the base of the triangle, which is x, by the height of the triangle. According to Hooke's law, the height of the triangle, which is the magnitude of the force, is equal to kx.

POTENTIAL ENERGY IN A SPRING
The potential energy in a spring is equal to one-half times the product of the spring constant and the square of the displacement.

$$PE_{spring} = \frac{1}{2}kx^2$$

As shown in **Figure 3** on the next page, during completely horizontal simple harmonic motion the spring's elastic potential energy is converted to kinetic energy and then back to potential energy.

■ **Simple Harmonic Motion**

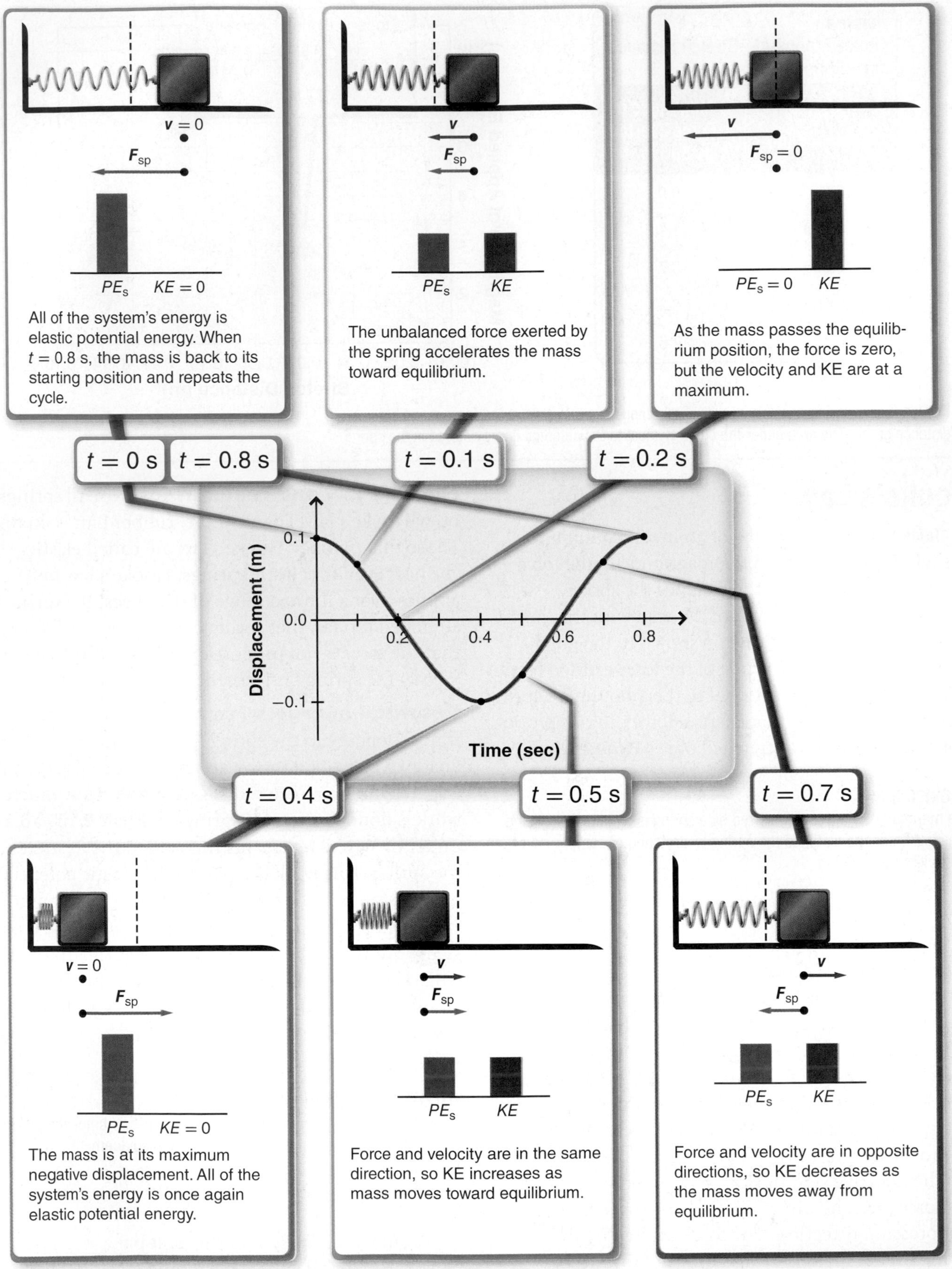

v = 0
F sp
PE s
KE = 0
All of the system's energy is elastic potential energy. When t = 0.8 s, the mass is back to its starting position and repeats the cycle.
v
F sp
PE s
KE
The unbalanced force exerted by the spring accelerates the mass toward equilibrium.
v
F sp = 0
PE s = 0
KE
As the mass passes the equilibrium position, the force is zero, but the velocity and KE are at a maximum.
t = 0 s
t = 0.8 s
t = 0.1 s
t = 0.2 s
Displacement (m)
0.1
0.0
−0.1
0.2
0.4
0.6
0.8
Time (sec)
t = 0.4 s
t = 0.5 s
t = 0.7 s
v = 0
F sp
PE s
KE = 0
The mass is at its maximum negative displacement. All of the system's energy is once again elastic potential energy.
v
F sp
PE s
KE
Force and velocity are in the same direction, so KE increases as mass moves toward equilibrium.
v
F sp
PE s
KE
Force and velocity are in opposite directions, so KE decreases as the mass moves away from equilibrium.

Figure 3 The total mechanical energy of the system is constant throughout the oscillation.

Identify At what position does the mass-and-spring system have the greatest potential energy?

View an **animation of a mass on a spring.**

Concepts In Motion

THE SPRING CONSTANT AND THE ENERGY OF A SPRING A spring stretches by 18 cm when a bag of potatoes weighing 56 N is suspended from its end.

a. Determine the spring constant.

b. How much elastic potential energy does the spring have when it is stretched this far?

1 ANALYZE AND SKETCH THE PROBLEM

- Sketch the situation.
- Show and label the distance the spring has stretched and its equilibrium position.

KNOWN	UNKNOWN
$x = 18$ cm	$k = ?$
$F = -56$ N	$PE_{sp} = ?$

2 SOLVE FOR THE UNKNOWN

a. Use Hooke's law and isolate k.

$$k = -\frac{F}{x}$$

$$= -\frac{-56 \text{ N}}{0.18 \text{ m}}$$ ◀ Substitute $F = -56$ N, $x = 0.18$ m. The force is negative because it is in the opposite direction of x.

$$= 310 \text{ N/m}$$

b. $PE_{sp} = \frac{1}{2}kx^2$

$$= \frac{1}{2}(310 \text{ N/m})(0.18 \text{ m})^2$$ ◀ Substitute $k = 310$ N/m, $x = 0.18$ m.

$$= 5.0 \text{ J}$$

3 EVALUATE THE ANSWER

- **Are the units correct?** N/m is the correct unit for the spring constant. $(\text{N/m})(\text{m}^2) = \text{N} \cdot \text{m} = \text{J}$, which is the correct unit for energy.

- **Is the magnitude realistic?** The average magnitude of the force the spring exerts is the average of 0 and 56 N. The work done is $W = Fx = (28 \text{ N})(0.18 \text{ m}) = 5.0$ J.

PRACTICE PROBLEMS

Do additional problems. **Online Practice**

1. What is the spring constant of a spring that stretches 12 cm when an object weighing 24 N is hung from it?

2. A spring with $k = 144$ N/m is compressed by 16.5 cm. What is the spring's elastic potential energy?

3. A spring has a spring constant of 56 N/m. How far will it stretch when a block weighing 18 N is hung from its end?

4. **CHALLENGE** A spring has a spring constant of 256 N/m. How far must it be stretched to give it an elastic potential energy of 48 J?

PHYSICS CHALLENGE

A car of mass m rests at the top of a hill of height h before rolling without friction into a crash barrier located at the bottom of the hill. The crash barrier contains a spring with a spring constant k, which is designed to bring the car to rest with minimum damage.

1. Determine, in terms of m, h, k, and g, the maximum distance (x) the spring will be compressed when the car hits it.

2. If the car rolls down a hill that is twice as high, by what factor will the spring compression increase?

3. What will happen after the car has been brought to rest?

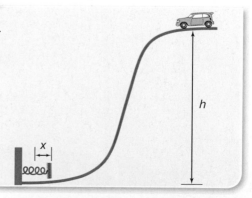

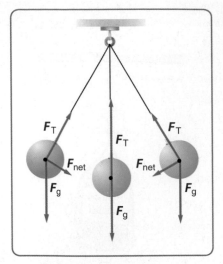

Figure 4 The pendulum's motion is an example of simple harmonic motion because the restoring force is directly proportional to the displacement from equilibrium.

Pendulums

Simple harmonic motion also occurs in the swing of a pendulum. A **simple pendulum** consists of a massive object, called the bob, suspended by a string or a light rod of length ℓ. The bob swings back and forth, as shown in **Figure 4.** The string or rod exerts a tension force (F_T), and gravity exerts a force (F_g) on the bob. Throughout the pendulum's path, the component of the gravitational force in the direction of the pendulum's circular path is a restoring force. At the left and right positions, the restoring force is at a maximum and the velocity is zero. At the equilibrium position, the restoring force is zero and the velocity is maximum.

For small angles (less than about 15°), the restoring force is proportional to the displacement from equilibrium. Similar to the motion of the mass on a spring discussed earlier, the motion of the pendulum is simple harmonic motion. The period of a pendulum is given by the following equation.

PERIOD OF A PENDULUM
The period of a pendulum is equal to 2π times the square root of the length of the pendulum divided by the gravitational field.

$$T = 2\pi \sqrt{\frac{\ell}{g}}$$

Notice that the period depends only on the length of the pendulum and the gravitational field, not on the mass of the bob or the amplitude of oscillation. One practical use of the pendulum is to measure *g*, which can vary slightly at different locations on Earth.

☑ **READING CHECK Describe** the relationship between the period of a simple pendulum and the mass of the pendulum's bob.

Figure 5 Wind around the Tacoma Narrows Bridge helped set the bridge in motion. Once in motion, a complex interaction between forces within the bridge's structure and forces applied by the wind caused the bridge's collapse.

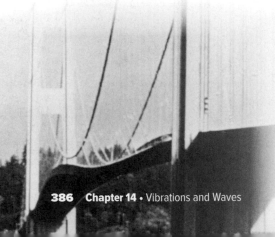

Resonance

To get a playground swing going, you can "pump" it by leaning back and pulling the chains at the same point in each swing. Another option is to have a friend give you repeated pushes at just the right times. **Resonance** occurs when forces are applied to a vibrating or oscillating object at time intervals equal to the period of oscillation. As a result, the amplitude of the vibration increases. Other familiar examples of resonance include rocking a car to free it from a snow bank and jumping rhythmically on a trampoline or a diving board to go higher.

Resonance in simple harmonic motion systems causes a larger and larger displacement as energy is added in small increments. The large-amplitude oscillations caused by resonance can produce both useful and catastrophic results. Resonance is used in musical instruments to amplify sounds and in clocks to increase accuracy.

In a particularly dramatic and famous instance, forces generated by the wind contributed to the destruction of the original Tacoma Narrows Bridge, shown in **Figure 5.** Varying forces from of the wind added energy to the bridge's oscillation. These oscillations grew so large that the bridge collapsed into the water below. Architects and engineers must take great care when designing and constructing buildings to insure resonance will not harm the structure.

EXAMPLE PROBLEM 2

Find help with **isolating a variable.** | Math Handbook

FINDING g USING A PENDULUM A pendulum with a length of 36.9 cm has a period of 1.22 s. What is the gravitational field at the pendulum's location?

1 ANALYZE AND SKETCH THE PROBLEM

- Sketch the situation.
- Label the length of the pendulum.

KNOWN

$\ell = 36.9$ cm

$T = 1.22$ s

UNKNOWN

$g = ?$

36.9 cm

2 SOLVE FOR THE UNKNOWN

$T = 2\pi\sqrt{\dfrac{\ell}{g}}$ ◀ Solve for g.

$g = \dfrac{(2\pi)^2\ell}{T^2}$

$= \dfrac{4\pi^2(0.369 \text{ m})}{(1.22 \text{ s})^2}$ ◀ Substitute $l = 0.369$ s, $T = 1.22$ s.

$= 9.78 \text{ m/s}^2 = 9.78 \text{ N/kg}$

3 EVALUATE THE ANSWER

- **Are the units correct?** N/kg is the correct unit for gravitational field.
- **Is the magnitude realistic?** The calculated value of g is quite close to the accepted value of g, 9.8 N/kg. This pendulum could be at a higher elevation above sea level.

PRACTICE PROBLEMS

Do additional problems. | Online Practice

5. What is the period on Earth of a pendulum with a length of 1.0 m?

6. How long must a pendulum be on the Moon, where $g = 1.6$ N/kg, to have a period of 2.0 s?

7. **CHALLENGE** On a planet with an unknown value of g, the period of a 0.75-m-long pendulum is 1.8 s. What is g for this planet?

SECTION 1 REVIEW

Section Self-Check | Check your understanding.

8. **MAIN IDEA** Explain why a pendulum is an example of periodic motion.

9. **Energy of a Spring** The springs shown in **Figure 6** are identical. Contrast the potential energies of the bottom two springs.

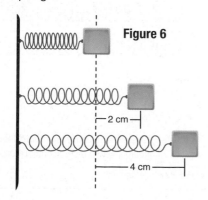

Figure 6

2 cm

4 cm

10. **Hooke's Law** Two springs look alike but have different spring constants. How could you determine which one has the larger spring constant?

11. **Hooke's Law** Objects of various weights are hung from a rubber band that is suspended from a hook. The weights of the objects are plotted on a graph against the stretch of the rubber band. How can you tell from the graph whether the rubber band obeys Hooke's law?

12. **Pendulum** How must the length of a pendulum be changed to double its period? How must the length be changed to halve the period?

13. **Resonance** If a car's wheel is out of balance, the car will shake strongly at a specific speed but not at a higher or lower speed. Explain.

14. **Critical Thinking** How is uniform circular motion similar to simple harmonic motion? How are they different?

Wave Properties

PHYSICS 4 YOU

Imagine tossing a ball to a friend. The ball moves from you and carries kinetic energy. If, however, you and your friend hold the ends of a rope and you give your end a quick shake, the rope remains in your hand. Even though no matter is transferred, the rope carries energy to your friend.

MAIN IDEA

Waves transfer energy without transferring matter.

Essential Questions

• What are waves?

• How do transverse and longitudinal waves compare?

• What is the relationship between wave speed, wavelength, and frequency?

Review Vocabulary

period in any periodic motion, the amount of time required for an object to repeat one complete cycle of motion

New Vocabulary

wave
wave pulse
transverse wave
periodic wave
longitudinal wave
surface wave
trough
crest
wavelength
frequency

APPLYING PRACTICES

Use Mathematical Representations Go to the resources tab in ConnectED to find the Applying Practices worksheet *Wave Characteristics*.

Mechanical Waves

A **wave** is a disturbance that carries energy through matter or space without transferring matter. You have learned how Newton's laws of motion and the law of conservation of energy govern the behavior of particles. These laws also govern the behavior of waves. Water waves, sound waves, and the waves that travel along a rope or a spring are mechanical waves. Mechanical waves travel through a physical medium, such as water, air, or a rope.

Transverse waves A **wave pulse** is a single bump or disturbance that travels through a medium. In the left panel of **Figure 7,** the wave pulse disturbs the rope in the vertical direction, but the pulse travels horizontally. A wave that disturbs the particles in the medium perpendicular to the direction of the wave's travel is called a **transverse wave.** If the disturbances continue at a constant rate, a **periodic wave** is generated.

Longitudinal waves In a coiled spring toy, you can produce another type of wave. If you squeeze together several turns of the coiled spring toy and then suddenly release them, pulses will move away in both directions. The result is called a **longitudinal wave** because the disturbance is parallel to the direction of the wave's travel. Sound waves are longitudinal waves in which the molecules are alternately compressed or decompressed along the path of the wave.

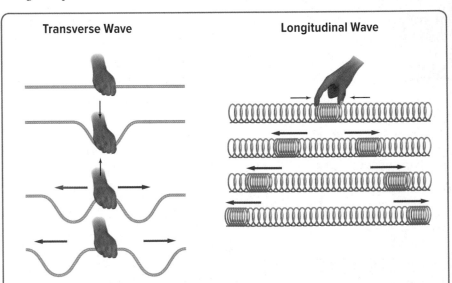

Figure 7 Shaking a rope up and down produces transverse wave pulses traveling in both directions. Squeezing and releasing the coils of a spring produces longitudinal wave pulses in both directions.

Explain the difference between transverse and longitudinal waves.

Crest

Wave motion →

Trough

Surface waves Waves that are deep in a lake or an ocean are longitudinal. In a **surface wave,** however, the medium's particles follow a circular path that is at times parallel to the direction of travel and at other times perpendicular to the direction of wave travel, as shown in **Figure 8.** Surface waves set particles in the medium, in this case water, moving in a circular pattern. At the top and bottom of the circular path, particles are moving parallel to the direction of the wave's travel. This is similar to a longitudinal wave. At the left and right sides of each circle, particles are moving up or down. This up-and-down motion is perpendicular to the wave's direction, similar to a transverse wave.

Wave Properties

Many types of waves share a common set of wave properties. Some wave properties depend on how the wave is produced, whereas others depend on the medium through which the wave travels.

Amplitude How does the pulse generated by gently shaking a rope differ from the pulse produced by a violent shake? The difference is similar to the difference between a ripple in a pond and an ocean breaker—they have different amplitudes. You read earlier that the amplitude of periodic motion is the greatest distance from equilibrium. Similarly, as shown in **Figure 9,** a transverse wave's amplitude is the maximum distance of the wave from equilibrium. Since amplitude is a distance, it is always positive. You will learn more about measuring the amplitude of longitudinal waves when you study sound.

Energy of a wave A wave's amplitude depends on how the wave is generated. More energy must be added to the system to generate a wave with a greater amplitude. For example, strong winds produce larger water waves than those formed by gentle breezes. Waves with greater amplitudes transfer more energy. Whereas a wave with a small amplitude might move sand on a beach a few centimeters, a giant wave can uproot and move a tree.

For waves that move at the same speed, the rate at which energy is transferred is proportional to the square of the amplitude. Thus, doubling the amplitude of a wave increases the amount of energy that wave transfers each second by a factor of four.

☑ **READING CHECK Predict** the factor by which the energy transfer per unit time increases if the amplitude of a wave is tripled.

Figure 8 Surface waves in water cause movement both parallel and perpendicular to the direction of wave motion. When these waves interact with the shore, the regular, circular motion is disrupted and the waves break on the beach.

REAL-WORLD PHYSICS
● ● ● ● ● ● ● ● ● ●

Tsunamis On March 11, 2011, a wall of water estimated to be ten meters high hit areas on the East coast of Japan —tsunami! A tsunami is a series of ocean waves that can have wavelengths over 100 km, periods of one hour, and wave speeds of 500–1000 km/h.

Figure 9 A wave's amplitude is measured from the equilibrium position to the highest or lowest point on the wave.

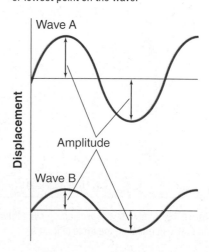

Wave A

Displacement

Amplitude

Wave B

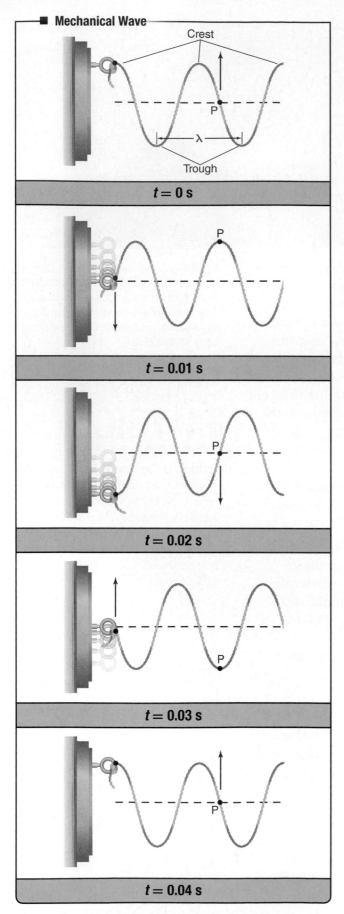

Mechanical Wave

Crest

P

λ

Trough

$t = 0$ s

$t = 0.01$ s

$t = 0.02$ s

$t = 0.03$ s

$t = 0.04$ s

Figure 10 A mechanical oscillator moves the left end of the rope up and down, completing the cycle in 0.04 s.

Wavelength Rather than focusing on one point on a wave, imagine taking a snapshot of the wave so you can see the whole wave at one instant in time. The top image in **Figure 10** shows each low point on a transverse wave, called a **trough,** and each high point on a transverse wave, called a **crest,** of a wave. The shortest distance between points where the wave pattern repeats itself is called the **wavelength.** Crests are spaced by one wavelength. Each trough also is one wavelength from the next. The Greek letter lambda (λ) represents wavelength.

Speed How fast does a wave travel? The speed of a wave pulse can be found in the same way the speed of a moving car is determined. First, measure the displacement of one of the wave's crests or compressions (Δd), then divide this by the time interval (Δt) to find the speed.

$$v = \frac{\Delta d}{\Delta t}$$

For most mechanical waves, both transverse and longitudinal, except water surface waves, wave speed does not depend on amplitude, frequency, or wavelength. Speed depends only on the medium through which the waves move.

☑ **READING CHECK Summarize** how changing a wave's amplitude, frequency, or wavelength affects the wave's speed.

Phase Any two points on a wave that are one or more whole wavelengths apart are said to be in phase. Particles in the medium are in phase with one another when they have the same displacement from the equilibrium position and the same velocity. Particles with opposite displacements from the equilibrium position and opposite velocities are 180° out of phase. A crest and a trough, for example, are 180° out of phase with each other. Two particles in a wave medium can be anywhere from 0° to 360° out of phase with each other.

Period and frequency Although wave speed and amplitude can describe both wave pulses and periodic waves, period (T) applies only to periodic waves. You have learned that the period of simple harmonic motion, such as the motion of a simple pendulum, is the time it takes for the motion to complete one cycle. Such motion is usually the source, or cause, of a periodic wave. The period of a wave is equal to the period of the source. In **Figure 10** the period (T) equals 0.04 s, which is the time it takes the source to complete one cycle. The same time is taken by P, a point on the rope, to return to its initial position and velocity.

Calculating frequency The **frequency** of a wave (f) is the number of complete oscillations a point on that wave makes each second. Frequency is measured in hertz (Hz). One hertz is one oscillation per second and is equal to 1/s or s^{-1}. The frequency and the period of a wave are related by the following equation.

FREQUENCY OF A WAVE
The frequency of a wave is equal to the reciprocal of the period.

$$f = \frac{1}{T}$$

Both the period and the frequency of a wave depend only on the wave's source. They do not depend on the wave's speed or the medium.

Calculating wavelength You can directly measure a wave's wavelength by measuring the distance between adjacent crests or troughs. You can also calculate it because the wavelength depends on both the frequency of the oscillator and the speed of the wave. In the time interval of one period, a wave moves one wavelength. Therefore, the wavelength of a wave is the speed multiplied by the period, $\lambda = vT$. Using the relation that $f = \frac{1}{T}$, the wavelength equation is very often written in the following way.

View a **BrainPOP video on waves.**
Video

WAVELENGTH
The wavelength of a wave is equal to the velocity divided by the frequency.

$$\lambda = \frac{v}{f}$$

Graphing waves If you took a snapshot of a transverse wave on a coiled spring toy, it might look like one of the waves shown in **Figure 10.** This snapshot could be placed on a graph grid to show more information about the wave, as in the left panel of **Figure 11.** Measuring from peak to peak or trough to trough on such a snapshot provides the wavelength. Now consider recording the motion of a single particle, such as point P in **Figure 10.** That motion can be plotted on a displacement-versus-time graph, as in the right graph in **Figure 11.** Measuring from peak to peak or trough to trough in this graph provides the wave's period.

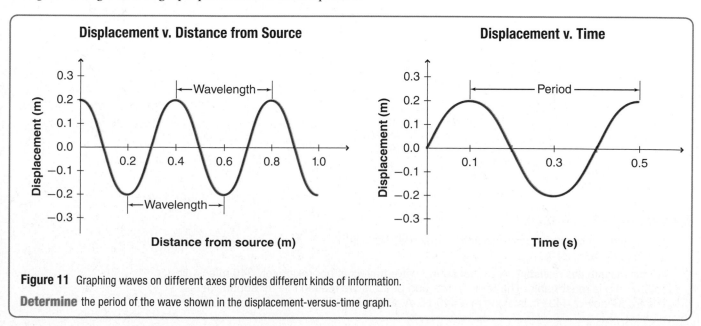

Figure 11 Graphing waves on different axes provides different kinds of information.

Determine the period of the wave shown in the displacement-versus-time graph.

EXAMPLE PROBLEM

CHARACTERISTICS OF A WAVE A sound wave has a frequency of 192 Hz and travels the length of a football field, 91.4 m, in 0.271 s.

a. What is the speed of the wave?

b. What is the wavelength of the wave?

c. What is the period of the wave?

d. If the frequency were changed to 442 Hz, what would be the new wavelength and period?

1 ANALYZE AND SKETCH THE PROBLEM

- Draw a diagram of the wave.

- Draw a velocity vector for the wave.

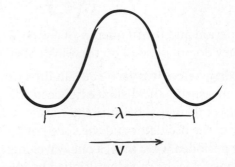

KNOWN	UNKNOWN
$f = 192$ Hz	$v = ?$
$\Delta d = 91.4$ m	$\lambda = ?$
$\Delta t = 0.271$ s	$T = ?$

2 SOLVE FOR THE UNKNOWN

a. Use the definition of velocity.

$$v = \frac{\Delta d}{\Delta t}$$ ◀ Substitute Δd = 91.4 m, Δt = 0.271 s

$$= \frac{91.4 \text{ m}}{0.271 \text{ s}}$$

$$= 337 \text{ m/s}$$

b. Use the relationship between wave velocity, wavelength, and frequency.

$$\lambda = \frac{v}{f}$$ ◀ Substitute v = 337 m/s, f = 192 Hz.

$$= \frac{337 \text{ m/s}}{192 \text{ Hz}}$$

$$= 1.76 \text{ m}$$

c. Use the relationship between period and frequency.

$$T = \frac{1}{f}$$ ◀ Substitute f = 192 Hz.

$$= \frac{1}{192 \text{ Hz}}$$

$$= 0.00521 \text{ s}$$

d. $\lambda = \frac{v}{f}$ ◀ Substitute v = 337 m/s, f = 442 Hz.

$$= \frac{337 \text{ m/s}}{442 \text{ Hz}}$$

$$= 0.762 \text{ m}$$

$$T = \frac{1}{f}$$ ◀ Substitute f = 442 Hz.

$$= \frac{1}{442 \text{ Hz}}$$

$$= 0.00226 \text{ s}$$

3 EVALUATE THE ANSWER

Are the units correct? Hz has the unit s^{-1}, so $(\frac{m}{s})/Hz = (\frac{m}{s}) \cdot s = m$, which is correct for λ.

Are the magnitudes realistic? A typical sound wave travels at approximately 340 m/s in air, so 337 m/s is reasonable. The frequencies and periods are reasonable for sound waves. The frequency of 442 Hz is close to a 440-Hz A, which is A above middle-C on a piano.

15. A sound wave produced by a clock chime is heard 515 m away 1.50 s later.

   a. Based on these measurements, what is the speed of sound in air?

   b. The sound wave has a frequency of 436 Hz. What is the period of the wave?

   c. What is its wavelength?

16. If you want to increase the wavelength of waves in a rope, should you shake it at a higher or lower frequency?

17. What is the speed of a periodic wave disturbance that has a frequency of 3.50 Hz and a wavelength of 0.700 m?

18. How does increasing the wavelength by 50 percent affect the frequency of a wave on a rope?

19. The speed of a transverse wave in a string is 15.0 m/s. If a source produces a disturbance that has a frequency of 6.00 Hz, what is its wavelength?

20. Five wavelengths are generated every 0.100 s in a tank of water. What is the speed of the wave if the wavelength of the surface wave is 1.20 cm?

21. A periodic longitudinal wave that has a frequency of 20.0 Hz travels along a coiled spring toy. If the distance between successive compressions is 0.600 m, what is the speed of the wave?

22. How does the frequency of a wave change when the period of the wave is doubled?

23. Describe the change in the wavelength of a wave when the period is reduced by one-half.

24. If the speed of a wave increases to 1.5 times its original speed while the frequency remains constant, how does the wavelength change?

25. **CHALLENGE** A hiker shouts toward a vertical cliff as shown in **Figure 12.** The echo is heard 2.75 s later.

   a. What is the speed of sound of the hiker's voice in air?

   b. The wavelength of the sound is 0.750 m. What is its frequency?

   c. What is the period of the wave?

Figure 12

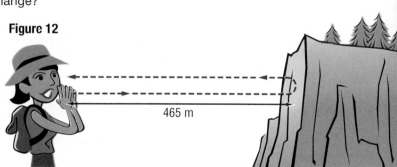

465 m

SECTION 2 **REVIEW**

Section Self-Check 🖑 Check your understanding.

26. **MAIN**IDEA Suppose you and your lab partner are asked to demonstrate that a transverse wave transports energy without transferring matter. How could you do it?

27. **Wave Characteristics** You are creating transverse waves on a rope by shaking your hand from side to side. Without changing the distance your hand moves, you begin to shake it faster and faster. What happens to the amplitude, wavelength, frequency, period, and velocity of the wave?

28. **Longitudinal Waves** Describe longitudinal waves. What types of mediums transmit longitudinal waves?

29. **Speeds in Different Mediums** If you pull on one end of a coiled spring toy, does the pulse reach the other end instantaneously? What happens if you pull on a rope? What happens if you hit the end of a metal rod? Compare the pulses traveling through these three materials.

30. **Critical Thinking** If a raindrop falls into a pool, it produces waves with small amplitudes. If a swimmer jumps into a pool, he or she produces waves with large amplitudes. Why doesn't the heavy rain in a thunderstorm produce large waves?

Wave Behavior

PHYSICS 4 YOU You have probably seen the outward ripples that occur when a rock is dropped into a calm lake. But what happens when two rocks are dropped into the lake near each other? How do the ripples from the rocks interact?

MAINIDEA

Waves can interfere with other waves.

Essential Questions

- How are waves reflected and refracted at boundaries between mediums?
- How does the principle of superposition apply to the phenomenon of interference?

Review Vocabulary

tension the specific name for the force exerted by a rope or a string

New Vocabulary

incident wave
reflected wave
principle of superposition
interference
node
antinode
standing wave
wavefront
ray
normal
law of reflection
refraction

Waves at Boundaries

Recall from Section 2 that the speed of a mechanical wave depends only on the properties of the medium it passes through, not on the wave's amplitude or frequency. For water waves, the depth of the water affects wave speed. For sound waves through air, the temperature affects wave speed. For waves on a spring, the speed depends on the spring's tension and mass per unit length.

Examine what happens when a wave travels across a boundary from one medium into another, such as two springs of different thicknesses joined end to end. **Figure 13** shows a wave pulse traveling from a larger spring into a smaller one. The pulse that strikes the boundary is called the **incident wave.** One pulse from the larger spring continues in the smaller spring, but the speed of the pulse is different in the smaller spring. Note that this transmitted wave pulse remains upward.

Some of the energy of the incident wave's pulse is reflected backward into the larger spring. This returning wave is called the **reflected wave.** Whether the reflected wave is upright or inverted depends on the characteristics of the two springs. For example, if the waves in the smaller spring have a greater speed because the spring is stiffer, then the reflected wave will be inverted.

■ **Wave at a Boundary**

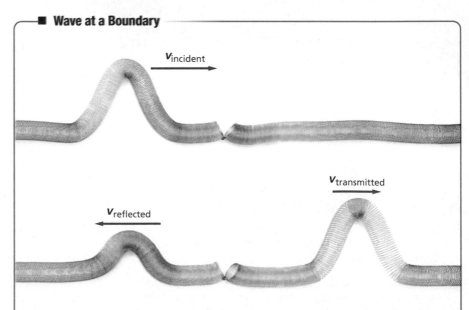

$v_{incident}$

$v_{transmitted}$

$v_{reflected}$

Figure 13 When the wave pulse meets the boundary between the two springs, a transmitted wave pulse and a reflected wave form.

Compare the energy of the incident wave to the energy of the reflected wave.

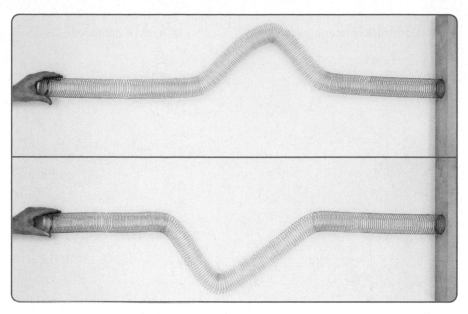

MiniLAB

WAVE REFLECTION
Does reflection change a wave's speed?

Rigid boundaries When a wave pulse hits a rigid boundary, the energy is reflected back, as shown in **Figure 14.** The wall is the boundary of a new medium through which the wave attempts to pass. Instead of passing through, the pulse is reflected from the wall with almost exactly the same amplitude as the pulse of the incident wave. Thus, almost all the wave's energy is reflected back. Very little energy is transmitted into the wall. Also note that the pulse is inverted.

Superposition of Waves

Suppose a pulse traveling along a spring meets a reflected pulse that is coming back from a boundary, as shown in **Figure 15.** In this case, two waves exist in the sam e place in the medium at the same time. Each wave affects the medium independently. The **principle of superposition** states that the displacement of a medium caused by two or more waves is the algebraic sum of the displacements caused by the individual waves. In other words, two or more waves can combine to form a new wave. If the waves move in the same medium, they can cancel or form a new wave of lesser or greater amplitude. The result of the superposition of two or more waves is called **interference.**

VOCABULARY
Science Usage v. Common Usage

Interference
• **Science usage**
the result of superposition of two or more waves
The amplitude of the interference of several waves was much larger than the amplitude of the individual waves.

• **Common usage**
the act of coming between in a way that hinders or impedes
Ehud was ejected from the game for an interference foul.

■ **Interference of Waves**

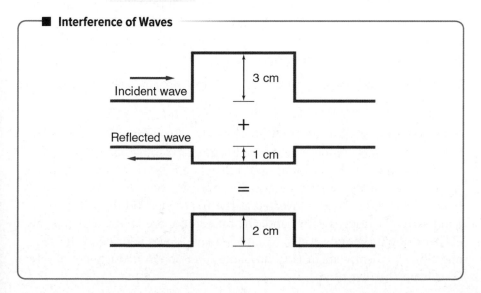

Figure 15 Waves add algebraically during superposition.

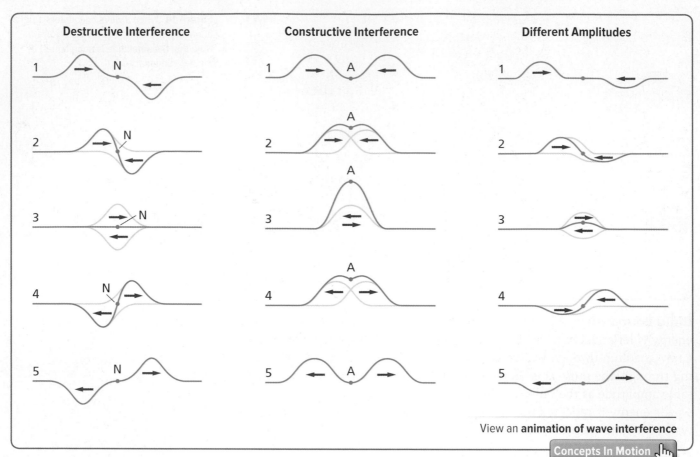

Destructive Interference	Constructive Interference	Different Amplitudes
1 N	1 A	1
2 N	2 A	2
3 N	3 A	3
4 N	4 A	4
5 N	5 A	5

View an **animation of wave interference**

Concepts In Motion

Figure 16 When waves add algebraically, the resulting combined waves can be quite different from the individual waves.

PhysicsLAB

INTERFERENCE AND DIFFRACTION
How do waves interfere with one another?

MiniLAB

WAVE INTERACTION
What happens when waves traveling in different directions meet?

Wave interference Wave interference can be either constructive or destructive. The first panel in **Figure 16** shows the superposition of waves with equal but opposite displacements, causing destructive interference. When the pulses meet and are in the same location, the displacement is zero. Point N, which does not move at all, is called a **node.** The pulses travel horizontally and eventually resume their original form.

Constructive interference occurs when wave displacements are in the same direction. The result is a wave that has an amplitude greater than those of the individual waves. A larger pulse appears at point A when the two waves meet. Point A has the largest displacement and is called the **antinode.** The two pulses pass through each other without changing their shapes or sizes. Even if the pulses have unequal amplitudes, the resultant pulse at the overlap is still the algebraic sum of the two pulses, as shown in the final panel of **Figure 16.**

☑ **READING CHECK** **Compare** the wave medium's displacement at a node and at an antinode.

Two reflections You can apply the concept of superimposed waves to the control of large-amplitude waves. Imagine attaching one end of a rope to a fixed point, such as a doorknob, a distance L away. When you vibrate the free end, the wave leaves your hand, travels along the rope toward the fixed end, is reflected and inverted at the fixed end, and returns to your hand. When it reaches your hand, the reflected wave is inverted and travels back down the rope. Thus, when the wave leaves your hand the second time, its displacement is in the same direction as it was when it left your hand the first time.

Standing waves Suppose you adjust the motion of your hand so that the period of vibration equals the time needed for the wave to make one round-trip from your hand to the door and back. Then, the displacement given by your hand to the rope each time will add to the displacement of the reflected wave. As a result, the amplitude of oscillation of the rope will be much greater than the motion of your hand. This large-amplitude oscillation is an example of mechanical resonance.

The ends of the rope are nodes and an antinode is in the middle, as shown in the top photo in **Figure 17.** Thus, the wave appears to be standing still and is called a **standing wave.** You should note, however, that the standing wave is the interference of waves traveling in opposite directions. If you double the frequency of vibration, you can produce one more node and one more antinode on the rope. Then it appears to vibrate in two segments. When you further increase the vibration frequency, it produces even more nodes and antinodes, as shown in the bottom photo in **Figure 17.**

Waves in Two Dimensions

You have studied waves on a rope and on a spring reflecting from rigid supports. During some of these interactions, the amplitude of the waves is forced to be zero by destructive interference. These mechanical waves travel in only one dimension. Waves on the surface of water, however, travel in two dimensions, and sound waves and electromagnetic waves will later be shown to travel in three dimensions. How can two-dimensional waves be represented?

Picturing waves in two dimensions When you throw a small stone into a calm pool of water, you see the circular crests and troughs of the resulting waves spreading out in all directions. You can sketch those waves by drawing circles to represent the wave crests. If you repeatedly dip your finger into water with a constant frequency, the resulting sketch would be a series of concentric circles, called wavefronts, centered on your finger. A **wavefront** is a line that represents the crest of a wave in two dimensions. Wavefronts can be used to show two-dimensional waves of any shape, including circular waves. The photo in **Figure 18** shows circular waves in water. The circles drawn on the diagram show the wavefronts that represent those water waves.

Whatever their shape, two-dimensional waves always travel in a direction that is perpendicular to their wavefronts. That direction can be represented by a **ray,** which is a line drawn at a right angle to the wavefront. When all you want to show is the direction in which a wave is traveling, it is convenient to draw rays instead of wavefronts. The red arrows in **Figure 18** are rays that show the water waves' direction of motion. One advantage of drawing wavefronts is when wavefronts are drawn to scale, they show the wave's wavelengths. In **Figure 18,** the wavelength equals the distance from one circle to the next.

☑ **READING CHECK** **Identify** the relationship between wavefronts and rays.

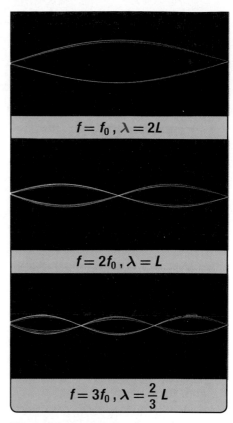

$f = f_0, \lambda = 2L$

$f = 2f_0, \lambda = L$

$f = 3f_0, \lambda = \frac{2}{3}L$

Figure 17 Interference produces standing waves only at certain frequencies.

Predict the wavelength if the frequency is four times the lowest frequency.

View an **animation of standing waves.**

Concepts In Motion

Figure 18 Waves spread out in a circular pattern from the oscillating source.

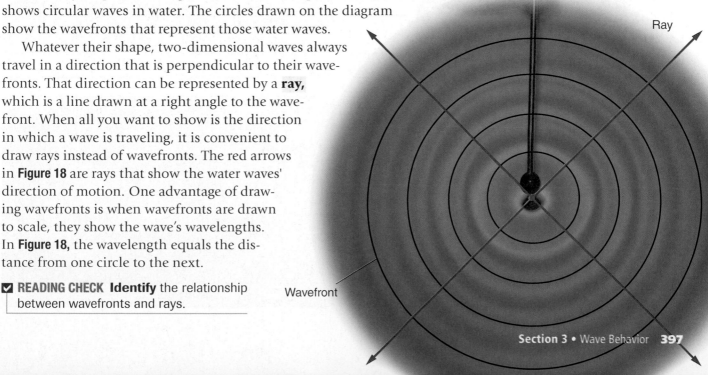

Ray

Wavefront

Creating two-dimensional waves A ripple tank is a piece of laboratory equipment that is used to investigate the properties of two-dimensional waves. The main portion of the ripple tank shown in **Figure 19** is a shallow tank that contains a thin layer of water. A board attached to a mechanical oscillator produces waves with long, straight wavefronts. A lamp above the tank produces shadows below the tank that show the locations of the crests of the waves. The top photo in **Figure 19** shows a wave traveling through the ripple tank. The direction the wave travels is modeled by a ray diagram. For clarity, the wavefronts are not extended the entire length of the wave.

Reflection of two-dimensional waves The bottom row of pictures in **Figure 19** shows an incident ray encountering a rigid barrier placed at an angle to the ray's path. The orientation of the barrier is shown by a line, called the **normal,** which is drawn perpendicular to the barrier. The angle between the incident ray and the normal is called the angle of incidence and is labeled θ_i in the diagram. The angle between the normal and the reflected ray is called the angle of reflection and is labeled θ_r. The **law of reflection** states that the angle of incidence is equal to the angle of reflection. The law of reflection applies to many different kinds of waves, not just the waves in a ripple tank.

☑ **READING CHECK** **Explain** how the angle of incidence and the angle of reflection are measured.

Figure 19 The ripple tank produces uniform waves that are useful for modeling wave behavior.

View a **simulation of a ripple tank.**

Concepts In Motion

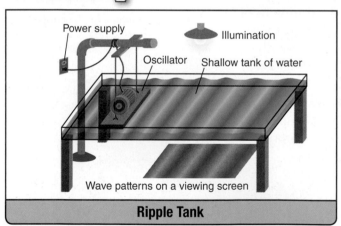
Ripple Tank

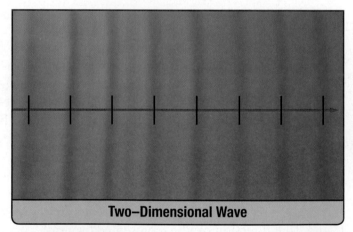
Two–Dimensional Wave

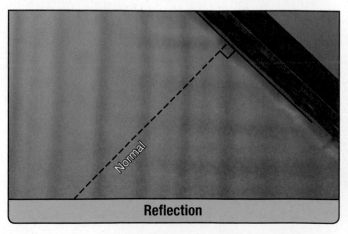
Reflection

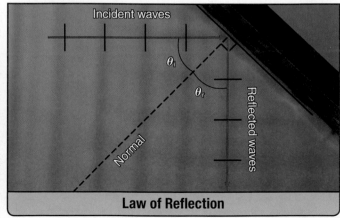
Law of Reflection

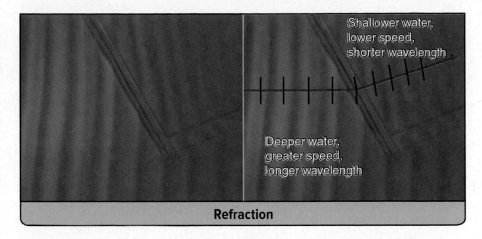

Figure 20 Waves in the ripple tank change direction as they enter shallower water.

Describe how the wavelength changes as the wave travels into the shallow water.

Refraction of waves in two dimensions A ripple tank can also model the behavior of waves as they travel from one medium into another. **Figure 20** shows a glass plate placed under the water in a ripple tank. The water above the plate is shallower than the water in the rest of the tank. As the waves travel from deep to shallow water, their speeds decrease and the direction of the waves changes. Such changes in speed are common when waves travel from one medium to another.

The waves in the shallow water are connected to the waves in the deep water. As a result, the frequency of the waves in the two mediums is the same. Based on the equation $\lambda = \frac{v}{f}$, the decrease in the speed of the waves means the wavelength is shorter in the shallower water. The change in the direction of waves at the boundary between two different mediums is known as **refraction**. **Figure 20** shows a wavefront and ray model of refraction. Part of the wave will refract through the boundary, and part will be reflected from the boundary.

✓ **READING CHECK Predict** the factor by which the wavelength changes if the speed of the refracted wave is half that of the incident wave.

Reflection and refraction occur for many different types of waves. Echoes are an example of reflection of sound waves by hard surfaces, such as the walls of a large gymnasium or a distant cliff face. Rainbows are the result of the reflection and refraction of light. As light from the Sun passes through a raindrop, reflection and refraction separate the light into its individual colors, producing a rainbow. The behavior of sound and light is discussed in more detail in other chapters.

PhysicsLAB

REFLECTION AND REFRACTION
How do waves behave at a barrier?

SECTION 3 REVIEW

Section Self-Check Check your understanding.

31. **MAIN**IDEA Which characteristics remain unchanged when a wave crosses a boundary into a different medium: frequency, amplitude, wavelength, velocity, and/or direction?

32. **Superposition of Waves** Sketch two wave pulses whose interference produces a pulse with an amplitude greater than either of the individual waves.

33. **Refraction of Waves** In **Figure 20,** the wave changes direction as it passes from one medium to another. Can two-dimensional waves cross a boundary between two mediums without changing direction? Explain.

34. **Standing Waves** In a standing wave on a string fixed at both ends, how is the number of nodes related to the number of antinodes?

35. **Critical Thinking** As another way to understand wave reflection, cover the right-hand side of each drawing in the left panel in **Figure 16** with a piece of paper. The edge of the paper should be at point N, the node. Now, concentrate on the resultant wave, shown in darker blue. Note that it acts as a wave reflected from a boundary. Is the boundary a rigid wall? Repeat this exercise for the middle panel in **Figure 16.**

Events of
MAGNITUDE

Japan experiences frequent and sometimes serious earthquakes, such as the Tohoku earthquake of March 2011. As a result, engineers have designed and built skyscrapers in Japan that can withstand even the worst earthquakes with only minimal damage. The 73-floor Landmark Tower in Yokohama is one such building.

1 A computer-guided pendulum system with a 170-ton mobile mass is hidden on the 71st floor of the Landmark Tower. This pendulum system absorbs oscillations from seismic waves and reduces swaying by up to 40 percent during earthquakes.

2 A dual tube structure is used for the Landmark Tower's frame. The outside tube runs along the building's outside wall. The inside tube surrounds the building's core. This structure absorbs sideways oscillations from seismic seismic waves.

3 The Landmark Tower has a wide base and a larger proportion of its mass near that base compared to other buildings of similar height. This construction helps minimize the motion of the foundation during an earthquake.

GOING**FURTHER** >>>

Research other earthquake-resistant buildings in Japan. Which technologies and designs were most effective at minimizing damage during the Tohoku earthquake in March 2011?

STUDY GUIDE

BIGIDEA   Waves and simple harmonic motion are examples of periodic motion.

SECTION 1 **Periodic Motion**

MAINIDEA   Periodic motion repeats in a regular cycle.

- Simple harmonic motion results when the restoring force on an object is directly proportional to the object's displacement from equilibrium.

- The elastic potential energy of a spring that obeys Hooke's law is expressed by the following equation:

$$PE_{sp} = \frac{1}{2}kx^2$$

- The period of a pendulum depends on the pendulum's length and the gravitational field strength at the pendulum's location. The period can be found using the following equation:

$$T = 2\pi\sqrt{\frac{\ell}{g}}$$

SECTION 2 **Wave Properties**

MAINIDEA   Waves transfer energy without transferring matter.

- Waves are disturbances that transfer energy without transferring matter.

- In transverse waves, the displacement of the medium is perpendicular to the direction the wave travels. In longitudinal waves, the displacement is parallel to the direction the wave travels.

- The velocity of a continuous wave is equal to the wave's frequency times its wavelength.

$$v = f\lambda$$

SECTION 3 **Wave Behavior**

MAINIDEA   Waves can interfere with other waves.

- When two-dimensional waves are reflected from boundaries, the angles of incidence and reflection are equal. The change in direction of waves at the boundary between two different mediums is called refraction.

- Interference occurs when two or more waves travel through the same medium at the same time. The principle of superposition states that the displacement of a medium resulting from two or more waves is the algebraic sum of the displacements of the individual waves.

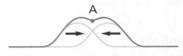

SECTION 1 Periodic Motion

Mastering Concepts

36. **BIGIDEA** What is periodic motion? Give three examples of periodic motion.

37. What is the difference between frequency and period? How are they related?

38. What is simple harmonic motion? Give an example of simple harmonic motion.

39. If a spring obeys Hooke's law, how does it behave?

40. How can the spring constant of a spring be determined from a graph of force magnitude versus displacement?

41. How can a spring's potential energy be determined from a graph of force magnitude versus displacement?

42. Does the period of a pendulum depend on the mass of the bob? The length of the string? The amplitude of oscillation? What else does the period depend on?

43. What conditions are necessary for resonance to occur?

Mastering Problems

44. A spring stretches 0.12 m when some apples are suspended from it, as shown in **Figure 21.** What is the spring constant of the spring?

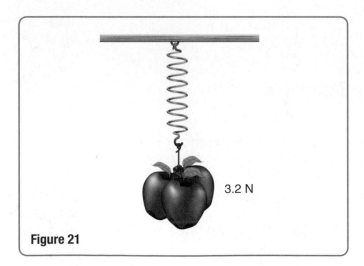

3.2 N

Figure 21

45. **Car Shocks** Each of the coil springs of a car has a spring constant of 25,000 N/m. How much is each spring compressed if it supports one-fourth of the car's 12,000-N weight?

46. A spring with a spring constant of 27 N/m is stretched 16 cm. What is the spring's potential energy?

47. **Rocket Launcher** A toy rocket launcher contains a spring with a spring constant of 35 N/m. How far must the spring be compressed to store 1.5 J of energy?

48. Force magnitude-versus-length data for a spring are plotted on the graph in **Figure 22.**

   a. What is the spring constant of the spring?

   b. What is the spring's potential energy when it is stretched to a length of 0.50 m?

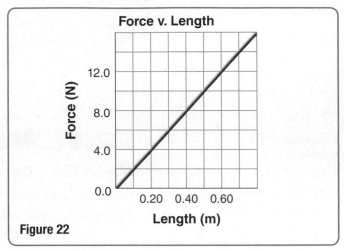

Force v. Length

Force (N) vs Length (m)

Figure 22

49. How long must a pendulum be to have a period of 2.3 s on the Moon, where $g = 1.6$ N/kg?

50. **Ranking Task** Rank the following pendulums according to period, from least to greatest. Specifically indicate any ties.

   A. 10 cm long, mass = 0.25 kg

   B. 10 cm long, mass = 0.35 kg

   C. 20 cm long, mass = 0.25 kg

   D. 20 cm long, mass = 0.35 kg

SECTION 2 Wave Properties

Mastering Concepts

51. Explain how the process of energy transfer associated with throwing a ball is different from the process of energy transfer associated with a mechanical wave.

52. What are the differences among transverse, longitudinal, and surface waves?

53. Waves are sent along a spring of fixed length.

   a. Can the speed of the waves in the spring be changed? Explain.

   b. Can the frequency of a wave in the spring be changed? Explain.

54. What is the wavelength of a wave?

55. Suppose you send a pulse along a rope. How does the position of a point on the rope before the pulse arrives compare to the point's position after the pulse has passed?

56. What is the difference between a wave pulse and a periodic wave?

57. Describe the difference between wave frequency and wave velocity.

58. Suppose you produce a transverse wave by shaking one end of a spring from side to side. How does the frequency of your hand compare with the frequency of the wave?

59. When are points on a wave in phase with each other? When are they out of phase? Give an example of each.

60. Describe the relationship between the amplitude of a wave and the energy it carries.

Mastering Problems

61. Building Motion The Willis Tower in Chicago sways back and forth in the wind with a frequency of about 0.12 Hz. What is its period of vibration?

62. Ocean Waves An ocean wave has a length of 12.0 m. A wave passes a fixed location every 3.0 s. What is the speed of the wave?

63. The wavelength of water waves in a shallow dish is 6.0 cm. The water moves up and down at a rate of 4.8 oscillations/s.

  a. What is the speed of the waves?

  b. What is the period of the waves?

64. Water waves in a lake travel 3.4 m in 1.8 s. The period of oscillation is 1.1 s.

  a. What is the speed of the water waves?

  b. What is their wavelength?

65. Sonar A sonar signal of frequency 1.00×10^6 Hz has a wavelength of 1.50 mm in water.

  a. What is the speed of the signal in water?

  b. What is its period in water?

66. The speed of sound in water is 1498 m/s. A sonar signal is sent straight down from a ship at a point just below the water surface, and 1.80 s later the reflected signal is detected. How deep is the water?

67. A sound wave of wavelength 0.60 m and a velocity of 330 m/s is produced for 0.50 s.

  a. What is the frequency of the wave?

  b. How many complete waves are emitted in this time interval?

  c. After 0.50 s, how far is the front of the wave from the source of the sound?

68. Pepe and Alfredo are resting on an offshore platform after a swim. They estimate that 3.0 m separates a trough and an adjacent crest of each surface wave on the lake. They count 12 crests in 20.0 s. Calculate how fast the waves are moving.

69. Earthquakes The velocity of the transverse waves produced by an earthquake is 8.9 km/s, and that of the longitudinal waves is 5.1 km/s. A seismograph records the arrival of the transverse waves 68 s before the arrival of the longitudinal waves. How far away is the earthquake?

SECTION 3 Wave Behavior
Mastering Concepts

70. When a wave crosses a boundary between a thin rope and a thick rope, as shown in **Figure 23**, its wavelength and speed change, but its frequency does not. Explain why the frequency is constant.

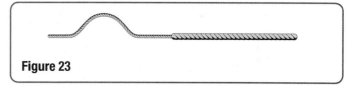

Figure 23

71. How does a wave pulse reflected from a rigid wall differ from the incident pulse?

72. Describe the motion of the particles of a medium located at the nodes of a standing wave.

73. Standing Waves A metal plate is held fixed in the center and sprinkled with sugar. With a violin bow, the plate is stroked along one edge and made to vibrate. The sugar begins to collect in certain areas and move away from others. Describe these regions in terms of standing waves.

74. If a string is vibrating in four parts, there are points where it can be touched without disturbing its motion. Explain. How many of these points exist?

75. Wavefronts pass at an angle from one medium into a second medium, where they travel with a different speed. Describe two changes in the wavefronts. What does not change?

ASSESSMENT

Chapter Self-Check

Mastering Problems

76. Sketch the result for each of the three cases shown in **Figure 24,** when the centers of the two approaching wave pulses lie on the dashed line so that the pulses exactly overlap.

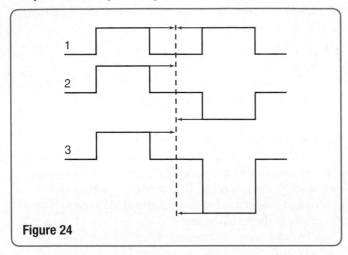

Figure 24

77. Guitars The wave speed in a guitar string is 265 m/s. The length of the string is 63 cm. You pluck the center of the string by pulling it up and letting go. Pulses move in both directions and are reflected off the ends of the string.

  a. How long does it take for the pulse to move to the string end and return to the center?

  b. When the pulses return, is the string above or below its resting location?

  c. If you plucked the string 15 cm from one end of the string, where would the two pulses meet?

78. Standing waves are created in the four strings shown in **Figure 25.** All strings have the same mass per unit length and are under the same tension. The lengths of the strings (L) are given. Rank the frequencies of the oscillations, from largest to smallest.

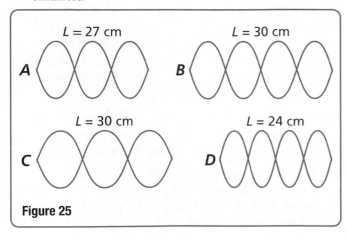

Figure 25

Applying Concepts

79. A ball bounces up and down on the end of a spring. Describe the energy changes that take place during one complete cycle. Does the total mechanical energy change?

80. Can a pendulum clock be used in the orbiting *International Space Station?* Explain.

81. Suppose you hold a 1-m metal bar in your hand and hit its end with a hammer, first, in a direction parallel to its length, and second, in a direction at right angles to its length. Describe the waves produced in the two cases.

82. Suppose you repeatedly dip your finger into a sink full of water to make circular waves. What happens to the wavelength as you move your finger faster?

83. What happens to the period of a wave as the frequency increases?

84. What happens to the wavelength of a wave as the frequency increases?

85. Suppose you make a single pulse on a stretched spring. How much energy is required to make a pulse with twice the amplitude?

86. You can make water slosh back and forth in a shallow pan only if you shake the pan with the correct frequency. Explain.

Mixed Review

87. What is the period of a 1.4-m pendulum?

88. Radio Wave AM-radio signals are broadcast at frequencies between 550 kHz (kilohertz) and 1600 kHz and travel at 3.0×10^8 m/s.

  a. What is the signals' range of wavelengths?

  b. FM frequencies range between 88 MHz (megahertz) and 108 MHz and travel at the same speed. What is the range of FM wavelengths?

89. The time needed for a water wave to change from the equilibrium level to the crest is 0.18 s.

  a. What fraction of a wavelength is this?

  b. What is the period of the wave?

  c. What is the frequency of the wave?

90. When a 225-g mass is hung from a spring, the spring stretches 9.4 cm. The spring and mass then are pulled 8.0 cm from this new equilibrium position and released. Find the spring constant of the spring.

91. You are floating offshore at the beach. Even though the wave pulses travel toward the beach, you don't move much closer to the beach.

   a. What type of wave are you experiencing as you float in the water?

   b. Explain why the wave does not move you closer to shore.

   c. In the course of 15 s, you count 10 waves that pass you. What is the period of the waves?

   d. What is the frequency of the waves?

   e. You estimate that the wave crests are 3 m apart. What is the velocity of the waves?

   f. After returning to the beach, you learn that the waves are moving at 1.8 m/s. What is the actual wavelength of the waves?

92. Bungee Jumper A 68-kg bungee jumper jumps from a hot-air balloon using a 540-m bungee cord. When the jump is complete and the jumper is suspended unmoving from the cord, it is 1710 m long. What is the spring constant of the bungee cord?

93. You have a mechanical fish scale that is made with a spring that compresses when weight is added to a hook attached below the scale. Unfortunately, the calibrations have completely worn off the scale. However, you have one known mass of 500.0 g that compresses the spring 2.0 cm.

   a. What is the spring constant for the spring?

   b. If a fish compresses the spring 4.5 cm, what is the mass of the fish?

94. Bridge Swinging In the summer over the New River in West Virginia, several teens swing from bridges with ropes, then drop into the river after a few swings back and forth.

   a. If Pam is using a 10.0-m length of rope, how long will it take her to complete one full swing?

   b. If Mike has a mass that is 20 kg more than Pam's, how would you expect the period of his swing to differ from Pam's?

   c. At what point in the swing is KE at a maximum?

   d. At what point in the swing is PE at a maximum?

   e. At what point in the swing is KE at a minimum?

   f. At what point in the swing is PE at a minimum?

95. Car Springs When you add a 45-kg load to the trunk of a new small car, the two rear springs compress an additional 1.0 cm.

   a. What is the spring constant for each of the springs?

   b. What is the increase in each of the springs' potential energy after loading the trunk?

96. Amusement Ride You notice that your favorite amusement-park ride seems bigger. The ride consists of a carriage that is attached to a structure so it swings like a pendulum. You remember that the carriage used to swing from one position to another and back again eight times in exactly 1 min. Now it only swings six times in 1 min. Give your answers to the following questions to two significant digits.

   a. What was the original period of the ride?

   b. What is the new period of the ride?

   c. What is the new frequency?

   d. How much longer is the arm supporting the carriage on the larger ride?

   e. If the park owners wanted to double the period of the ride, what percentage increase would need to be made to the length of the pendulum?

97. Sketch the result for each of the four cases shown in **Figure 26,** when the centers of each of the two wave pulses lie on the dashed line so that the pulses exactly overlap.

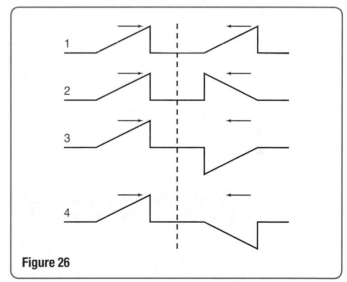

Figure 26

98. Clocks The speed at which a grandfather clock runs is controlled by a swinging pendulum.

   a. If you find that the clock loses time each day, what adjustment would you need to make to the pendulum so it would keep better time?

   b. If the pendulum currently is 15.0 cm long, by how much would you need to change the length to make the period shorter by 0.0400 s?

   c. Another clock has a pendulum 77.5 cm long. Over one day it loses 5.00 min. How much should its length be changed so that it keeps the correct time?

99. The velocity of a wave on a string depends on how tightly the string is stretched and on the mass per unit length of the string. If F_T is the tension in the string, and μ is the mass/unit length, then the velocity (v) can be determined by the following equation.

$$v = \sqrt{\frac{F_T}{\mu}}$$

A piece of string 5.30 m long has a mass of 15.0 g. What must the tension in the string be to make the wavelength of a 125-Hz wave 120.0 cm?

Thinking Critically

100. Reverse Problem Write a physics problem with real-life objects for which the following equation would be part of the solution:

$$(1.65 \text{ kg})(9.8 \text{ N/kg}) = k\,(0.15 \text{ m})$$

101. Problem Posing Complete this problem so that it must be solved using the concept listed below: "A block is suspended vertically from a spring with a spring constant of 200 N/m…"

a. conservation of energy

b. Newton's second law

c. frequency

102. Make and Use Graphs Several weights were suspended from a spring, and the resulting extensions of the spring were measured. **Table 2** shows the collected data.

Table 2 Weights on a Spring	
Force magnitude, F (N)	Extension, x (m)
2.5	0.12
5.0	0.26
7.5	0.35
10.0	0.50
12.5	0.60
15.0	0.71

a. Make a graph of the magnitude of the force applied to the spring versus the spring length. Plot the force on the y-axis.

b. Determine the spring constant from the graph.

c. Using the graph, find the elastic potential energy stored in the spring when it is stretched to 0.50 m.

103. Analyze and Conclude A 20-N force is required to stretch a spring by 0.5 m.

a. What is the spring constant?

b. How much energy does the spring store?

c. Why isn't the work done to stretch the spring equal to the force times the distance, or 10 J?

104. Apply Concepts Gravel roads often develop regularly spaced ridges that are perpendicular to the road, as shown in **Figure 27**. This effect, called washboarding, occurs because most cars travel at about the same speed, and the springs that connect the wheels to the cars oscillate at about the same frequency. If the ridges on a road are 1.5 m apart and cars travel on it at about 5 m/s, what is the frequency of the springs' oscillation?

Figure 27

Writing In Physics

105. Research Christiaan Huygens' work on waves and the controversy between him and Newton over the nature of light. Compare and contrast their explanations of such phenomena as reflection and refraction. Whose model would you choose as the best explanation? Explain why.

Cumulative Review

106. A 1400-kg drag racer automobile can complete a 402-m course in 9.8 s. The final speed of the automobile is 112 m/s.

a. What is the final kinetic energy of the automobile?

b. What is the minimum amount of work that was done by its engine? Why can't you calculate the total amount of work done?

c. What was the average acceleration of the automobile?

107. How much water would a steam engine have to evaporate in 1 s to produce 1 kW of power? Assume that the engine is 20 percent efficient.

MULTIPLE CHOICE

1. What is the value of the spring constant of a spring with a potential energy of 8.67 J when it's stretched 247 mm?

A. 70.2 N/m   **C.** 142 N/m

B. 71.1 N/m   **D.** 284 N/m

2. What is the magnitude of the force acting on a spring with a spring constant of 275 N/m that is stretched 14.3 cm?

A. 2.81 N   **C.** 39.3 N

B. 19.2 N   **D.** 3.93×10^{30} N

3. A mass stretches a spring as it hangs from the spring as shown in the figure below. What is the spring constant?

A. 0.25 N/m   **C.** 26 N/m

B. 0.35 N/m   **D.** 3.5×10^2 N/m

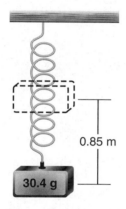

0.85 m

30.4 g

4. A spring with a spring constant of 350 N/m pulls a door closed. How much work is done as the spring pulls the door at a constant velocity from an 85.0-cm stretch to a 5.0-cm stretch?

A. 110 N·m   **C.** 220 N·m

B. 130 J   **D.** 1.1×10^3 J

5. What is the length of a pendulum that has a period of 4.89 s?

A. 5.94 m   **C.** 24.0 m

B. 11.9 m   **D.** 37.3 m

6. What is the frequency of a wave with a period of 3 s?

A. 0.3 Hz   **C.** $\frac{\pi}{3}$ Hz

B. $\frac{3}{\pi}$ Hz   **D.** 3 Hz

7. What is the correct rearrangement of the formula for the period of a pendulum to find the length of the pendulum?

A. $\ell = \dfrac{4\pi^2 g}{T^2}$ **C.** $\ell = \dfrac{T^2 g}{(2\pi)^2}$

B. $\ell = \dfrac{gT}{4\pi^2}$ **D.** $\ell = \dfrac{Tg}{2\pi}$

8. Which option describes a standing wave?

	Waves	Direction	Medium
a	Identical	Same	Same
b	Nonidentical	Opposite	Different
c	Identical	Opposite	Same
d	Nonidentical	Same	Different

9. What is the name given to the wave behavior in which a wave changes direction as it moves from one medium to another medium?

A. interference   **C.** reflection

B. rarefaction   **D.** refraction

10. The wave shown in the figure below travels 11.2 m to a wall and back again in 4 s. What is the wave's frequency?

A. 0.2 Hz   **C.** 5 Hz

B. 2 Hz   **D.** 9 Hz

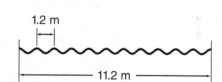

1.2 m

11.2 m

FREE RESPONSE

11. Use dimensional analysis of the equation $kx = mg$ to derive the units of k.

NEED EXTRA HELP?

If You Missed Question	1	2	3	4	5	6	7	8	9	10	11
Review Section	1	1	1	1	1	2	1	3	3	2	1

CHAPTER 15

Sound

BIGIDEA   Sound waves are pressure variations, and many can be detected by the human ear.

SECTIONS

1 **Properties and Detection of Sound**

2 **The Physics of Music**

LaunchLAB

PRODUCING MUSICAL NOTES

How can musical notes be produced using stemmed drinking glasses?

WATCH THIS!

MAKING MUSIC

From woodwinds to brass to percussion, every instrument makes a different kind of sound. What is the physics of making music?

PHYSICS T.V.

(l)Mary-Ella Keith/Alamy, (r)Robert Beck/Sports Illustrated/Getty Images

Properties and Detection of Sound

PHYSICS 4 YOU

You have probably noticed that an approaching truck sounds different than the same truck moving away from you. This effect, called the Doppler effect, was first studied in 1845. Dutch scientist C.H.D. Buys Ballot observed the changes in sound as a train carrying horn players went by.

MAIN IDEA

Our perception of a sound wave depends on that wave's physical properties.

Essential Questions

- What properties does sound share with other waves?
- How do the physical properties of sound waves relate to our perception of sound?
- What is the Doppler effect?
- What are some applications of the Doppler effect?

Review Vocabulary

wave a disturbance that carries energy through matter or space and transfers energy without transferring matter

New Vocabulary

sound wave
pitch
loudness
sound level
decibel
Doppler effect

Sound Waves

Sound is an important part of existence for many living things. For humans, the sound of a siren can heighten our awareness of our surroundings, while the sound of music can relax us. You already are familiar with several of the characteristics of sound, including volume, tone, and pitch, from your everyday experiences. Without thinking about it, you can use these, and other characteristics, to categorize many of the sounds that you hear. For example, some sound patterns are characteristic of speech, while others are characteristic of a musical group. In this chapter, you will study the physical principles of sound, which is a type of wave. You will use your knowledge of waves to describe some of sound's properties and interactions.

Pressure variations Put your fingers against your throat as you hum or speak. Can you feel the vibrations? **Figure 1** shows a vibrating bell that can represent your vocal cords, a loudspeaker, or any other sound source. As it moves back and forth, the edge of the bell strikes the particles in the air. When the edge moves forward, air particles are driven forward; that is, the air particles bounce off the bell with a greater velocity than they would if the bell were not moving. When the edge moves backward, air particles bounce off the bell with a lower velocity.

Figure 1 When the bell is at rest (top) the surrounding air is at average pressure. When the bell is struck (bottom) the vibrating edge creates regions of high and low pressure. For better understanding, the diagram shows the pressure regions moving in one direction. In reality, the waves move out from the bell in all directions.

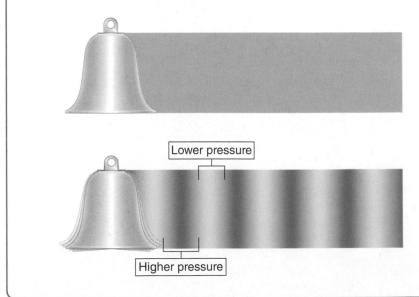

Lower pressure

Higher pressure

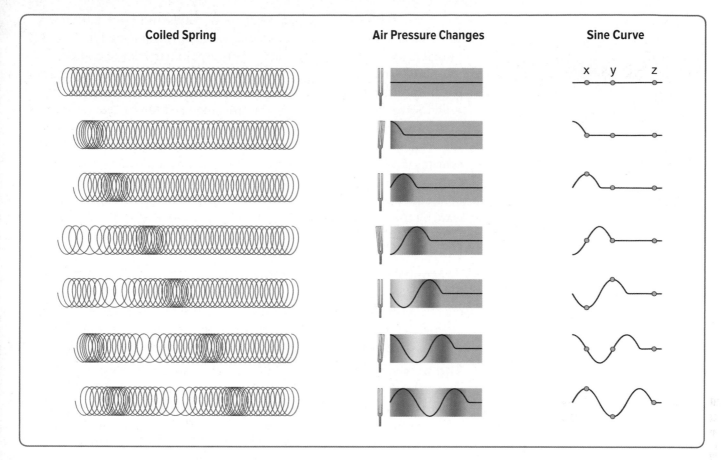

Coiled Spring	Air Pressure Changes	Sine Curve

The result of these velocity changes is that the forward motion of the bell produces a region where the air pressure is slightly higher than average, as shown in **Figure 1.** The backward motion produces slightly below-average pressure. Collisions of the air particles cause the pressure variations to move away from the bell in all directions. If you were to focus on one spot, you would see the value of the air pressure rise and fall. In this way, the pressure variations are transmitted through matter.

Describing sound A pressure oscillation that is transmitted through matter is a **sound wave.** Sound waves travel through air because a vibrating source produces regular variations, or oscillations, in air pressure. The air particles collide, transmitting the pressure variations away from the source of the sound. The pressure of the air oscillates about the mean air pressure, as shown in **Figure 2.** The frequency of the wave is the number of oscillations in pressure each second. The wavelength is the distance between successive regions of high or low pressure. Because the motion of the particles in air is parallel to the direction of the wave's motion, sound is a longitudinal wave.

Just like any other wave, the speed of sound depends on the medium through which it travels. In air, the speed depends on the temperature, increasing by about 0.6 m/s for each 1°C increase in air temperature. At room temperature (20°C) and sea level, the speed of sound is 343 m/s. For the problems in this book, you may assume these conditions unless otherwise stated.

☑ **READING CHECK Estimate** the speed of sound through air at sea level if the temperature is 25ºC.

Figure 2 A coiled spring models the oscillations created by a sound wave. As the sound wave travels through the air, the air pressure rises and falls. The changes in the sine curves correspond to the changes in air pressure. Note that the positions of x, y, and z show that the wave, not the matter, travels forward.

PhysicsLAB

SOUND TRAVELS THROUGH AIR
PROBEWARE LAB How fast does sound move in air?

Table 1
Speed of Sound in Various Media

Medium	m/s
Air (0°)	331
Air (20°)	343
Helium (0°)	965
Water (25°)	1497
Seawater (25°)	1535
Copper (20°)	4760
Iron (20°)	4994

In general, the speed of sound is greater in solids and liquids than in gases. **Table 1** lists the speeds of sound waves in various media. Sound cannot travel in a vacuum because there are no particles to collide.

Sound waves share the general properties of other waves. For example, they reflect off hard objects, such as the walls of a room. Reflected sound waves are called echoes. The time required for an echo to return to the source of the sound can be used to find the distance between the source and the reflective object. This principle is used by bats, by some cameras, and by ships that employ sonar. Two sound waves can interfere, causing dead spots at nodes where little sound can be heard. Recall that the frequency and wavelength of a wave are related to the speed of the wave by the equation $\lambda = v/f$.

Detection of Pressure Waves

Sound detectors transform sound energy—the kinetic energy of the vibrating particles of the transmitting medium—into another form of energy. A common detector is a microphone, which transforms sound energy into electrical energy. A microphone consists of a thin disk that vibrates in response to sound waves and produces an electrical signal.

The human ear As shown in **Figure 3,** the human ear is a sound detector that receives pressure waves and converts them to electrical impulses. The tympanic membrane, also called the eardrum, vibrates when sound waves enter the auditory canal. Three tiny bones in the middle ear then transfer these vibrations to fluid in the cochlea. Tiny hairs lining the spiral-shaped cochlea detect certain frequencies in the vibrating fluid. These hairs stimulate nerve cells, which send impulses to the brain and produce the sensation of sound.

The ear detects sound waves over a wide range of frequencies and is sensitive to an enormous range of amplitudes. In addition, human hearing can distinguish many different qualities of sound. Knowledge of both physics and biology is required to understand the complexities of the ear. The interpretation of sounds by the brain is even more complex, and is not totally understood.

Figure 3 The human ear is a sense organ that translates sound vibrations from the external environment into nerve impulses that are sent to the brain for interpretation. The eardrum vibrates when sound waves enter the auditory canal. The bones in the middle ear—the malleus, the incus, and the stapes—move as a result of the vibrations. The vibrations are then transmitted to the inner ear, where they trigger nerve impulses to the brain.

■ **The Human Ear**

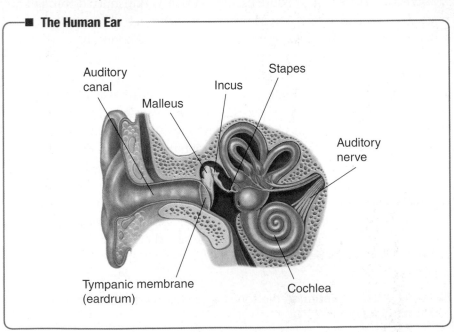

Perceiving Sound

How humans perceive sound depends partly on the physical characteristics of sound waves, such as frequency and amplitude.

Pitch Marin Mersenne and Galileo first determined that the pitch we hear depends on the frequency of vibration. **Pitch** is the highness or lowness of a sound, and it can be given a name on the musical scale. For instance, the note known as middle C has a frequency of 262 Hz. The highest note on a piano has a frequency of 4186 Hz. The human ear is not equally sensitive to all frequencies. Most people cannot hear sounds with frequencies below 20 Hz or above 16,000 Hz. Many animals, such as dogs, cats, elephants, and bats, are capable of hearing sounds at frequencies that humans cannot hear.

Loudness Frequency and wavelength are two physical characteristics of sound waves. Another physical characteristic of sound waves is amplitude. Amplitude is the measure of the variation in pressure in a wave. The **loudness** of a sound is the intensity of the sound as perceived by the ear and interpreted by the brain. This intensity depends primarily on the amplitude of the pressure wave.

The human ear is extremely sensitive to variations in the intensity of sound waves. Recall that 1 atmosphere of pressure equals $1.01{\times}10^5$ Pa. The ear can detect pressure-wave amplitudes of less than one-billionth of an atmosphere, or $2{\times}10^{-5}$ Pa. At the other end of the audible range, pressure variations of approximately 20 Pa or greater cause pain. It is important to remember that the ear detects pressure variations only at certain frequencies. Driving over a mountain pass changes the pressure on your ears by thousands of pascals, but this change does not take place at audible frequencies.

Because humans can detect a wide range of intensities, it is convenient to measure these intensities on a logarithmic scale called the **sound level.** The most common unit of measurement for sound level is the **decibel** (dB). The sound level depends on the ratio of the intensity of a given sound wave to that of the most faintly heard sound. This faintest sound is measured at 0 dB. A sound that is ten times more intense registers 20 dB. A sound that is another ten times more intense is 40 dB. Most people perceive a 10-dB increase in sound level as about twice as loud as the original level. **Figure 4** shows the sound level for a variety of sounds. In addition to intensity, pressure variations and the power of sound waves can be described by decibel scales.

VOCABULARY
Science Usage v. Common Usage

Pitch
• **Science usage**
the highness or lowness of a sound, which depends on the frequency of vibration
The flute and the tuba produce very different pitches.

• **Common usage**
the delivery of a ball by a pitcher to a batter
Sharon hit the pitch over the fence for a home run.

MiniLAB

SOUND CHARACTERISTICS
What physical factors determine sound?

Figure 4 This decibel scale shows the sound level for a variety of sounds.

Infer About how many times louder does an alarm clock sound than heavy traffic?

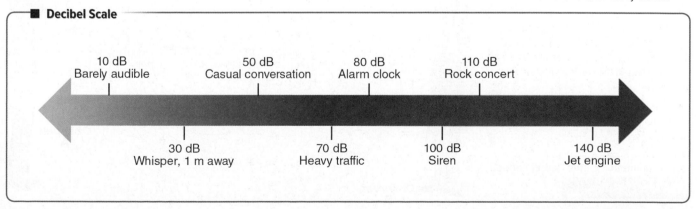

◼ Decibel Scale

10 dB
Barely audible

50 dB
Casual conversation

80 dB
Alarm clock

110 dB
Rock concert

30 dB
Whisper, 1 m away

70 dB
Heavy traffic

100 dB
Siren

140 dB
Jet engine

Figure 5 Hearing loss can occur with continuous exposure to loud sounds. Workers in many occupations, such as construction, wear ear protection. The jackhammer this worker is operating has a sound level of 130 dB.

The ear can lose its sensitivity, especially to high frequencies, after exposure to loud sounds in the form of noise or music. The longer a person is exposed to loud sounds, the greater the effect. A person can recover from short-term exposure in a period of hours, but the effects of long-term exposure can last for days or weeks. Long exposure to 100-dB or greater sound levels can produce permanent damage. Hearing loss also can result from loud music being transmitted to stereo headphones from personal radios and CD players. In some cases, the listeners are unaware of just how high the sound levels really are. Cotton earplugs reduce the sound level only by about 10 dB. Special ear inserts can provide a 25-dB reduction. Specifically designed earmuffs and inserts, as shown in **Figure 5,** can reduce the sound level by up to 45 dB.

The Doppler Effect

Have you ever noticed that the pitch of a fast car changed as the vehicle sped past you? The pitch was higher when the vehicle was moving toward you, then it dropped to a lower pitch as the vehicle moved away. The change in frequency of sound caused by the movement of either the source, the detector, or both is called the **Doppler effect.** The Doppler effect is illustrated in **Figure 6.** The sound source (S) is moving to the right with a speed of v_s. The waves that the source emits spread in circles centered on the source at the time it produced the waves. As the source moves toward the sound detector, observer A in **Figure 6,** more waves are crowded into the space between them. The wavelength is shortened to λ_A. Because the speed of sound is not changed, more crests reach the ear each second, which means that the frequency of the detected sound increases. When the source is moving away from the detector, observer B in **Figure 6,** the wavelength is lengthened to λ_B, fewer crests reach the ear each second, and the detected frequency is lower.

A Doppler shift also occurs if the detector is moving and the source is stationary. As the detector approaches a stationary source, it encounters more wave crests each second than if it were still, and a higher frequency is detected. If the detector recedes from the source, fewer crests reach it each second, resulting in a lower detected frequency.

Figure 6 As a sound-producing source moves toward observer A, the wavelength is shortened to λ_A. As the source moves away from observer B, the wavelength is lengthened to λ_B.

Describe What is the relative difference in the frequency of the detected sound for each observer?

View an **animation on the Doppler effect.**

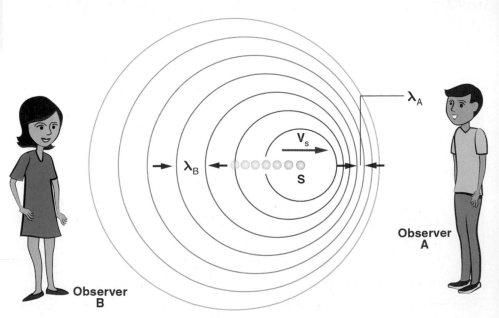

For any combination of moving source and moving observer, the frequency that the observer hears can be found using the relationship below.

DOPPLER EFFECT
The frequency perceived by a detector is equal to the velocity of the detector relative to the velocity of the wave, divided by the velocity of the source relative to the velocity of the wave, multiplied by the wave's frequency.

$$f_d = f_s \frac{v - v_d}{v - v_s}$$

Investigate the **Doppler effect.**

Virtual Investigation

In the Doppler effect equation, v is the velocity of the sound wave, v_s is the velocity of the sound's source, and v_d is the velocity of the observer of interest, who is detecting the sound. The subscript d is used instead of the letter o to avoid confusion with the number zero. The same subscripts are used to denote the corresponding frequencies.

Defining the coordinate system As you solve problems using the above equation, be sure to define the coordinate system so that the positive direction is from the source to the detector. The sound waves will be approaching the detector from the source, so the velocity of sound is always positive. Try drawing diagrams to confirm that the term $\frac{v - v_d}{v - v_s}$ behaves as you would predict based on what you have learned about the Doppler effect. Notice that for a source moving toward the detector (positive direction, which results in a smaller denominator compared to a stationary source) and for a detector moving toward the source (negative direction and increased numerator compared to a stationary detector), the detected frequency (f_d) increases. Similarly, if the source moves away from the detector or if the detector moves away from the source, then f_d decreases. Read the Connecting Math to Physics feature below to see how the Doppler effect equation reduces when the source or observer is stationary.

CONNECTING MATH TO PHYSICS

Reducing Equations When an element in an equation is equal to zero, the equation might reduce to a form that is easier to use.

Stationary detector, source in motion: $v_d = 0$	Stationary source, detector in motion: $v_s = 0$
$f_d = f_s \dfrac{v - v_d}{v - v_s}$	$f_d = f_s \dfrac{v - v_d}{v - v_s}$
$= f_s \dfrac{v}{v - v_s}$	$= f_s \dfrac{v - v_d}{v}$
$= f_s \dfrac{\frac{v}{v}}{\frac{v}{v} - \frac{v_s}{v}}$	$= f_s \dfrac{\frac{v}{v} - \frac{v_d}{v}}{\frac{v}{v}}$
$= f_s \dfrac{1}{1 - \frac{v_s}{v}}$	$= f_s \dfrac{1 - \frac{v_d}{v}}{1}$
	$= f_s \left(1 - \frac{v_d}{v}\right)$

THE DOPPLER EFFECT A guitar player sounds C above middle C (523 Hz) while traveling in a convertible at 24.6 m/s. If the car is coming toward you, what frequency would you hear? Assume that the temperature is 20°C.

1 ANALYZE AND SKETCH THE PROBLEM

- Sketch the situation.
- Establish a coordinate axis. Make sure that the positive direction is from the source to the detector.
- Show the velocities of the source and detector.

KNOWN

$v = +343$ m/s
$v_s = +24.6$ m/s
$v_d = 0$ m/s
$f_s = 523$ Hz

UNKNOWN

$f_d = ?$

2 SOLVE FOR THE UNKNOWN

Use $f_d = f_s \dfrac{v - v_d}{v - v_s}$ with $v_d = 0$ m/s.

$$f_d = f_s \frac{1}{1 - \dfrac{v_s}{v}}$$

$$= 523 \text{ Hz} \left(\frac{1}{1 - \dfrac{24.6 \text{ m/s}}{343 \text{ m/s}}} \right)$$ ◄ Substitute $v = +343$ m/s, $v_s = +24.6$ m/s, and $f_s = 523$ Hz.

$$= 564 \text{ Hz}$$

3 EVALUATE THE ANSWER

- **Are the units correct?** Frequency is measured in hertz.
- **Is the magnitude realistic?** The source is moving toward you, so the frequency should be increased.

PRACTICE PROBLEMS   Do additional problems. | Online Practice |

1. Repeat Example Problem 1, but with the car moving away from you. What frequency would you hear?

2. You are in an automobile, like the one in **Figure 7**, traveling toward a pole-mounted warning siren. If the siren's frequency is 365 Hz, what frequency do you hear? Use 343 m/s as the speed of sound.

3. You are in an automobile traveling at 55 mph (24.6 m/s). A second automobile is moving toward you at the same speed. Its horn is sounding at 475 Hz. What frequency do you hear? Use 343 m/s as the speed of sound.

4. A submarine is moving toward another submarine at 9.20 m/s. It emits a 3.50-MHz ultrasound. What frequency would the second sub, at rest, detect? The speed of sound in water at the depth the submarines are moving is 1482 m/s.

5. **CHALLENGE** A trumpet plays middle C (262 Hz). How fast would it have to be moving to raise the pitch to C sharp (277 Hz)? Use 343 m/s as the speed of sound.

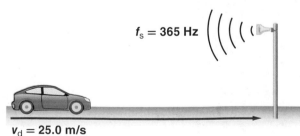

$f_s = 365$ Hz

$v_d = 25.0$ m/s

Figure 7

Applications of the Doppler effect The Doppler effect occurs in all wave motion, both mechanical and electromagnetic. It has many applications. Radar detectors use the Doppler effect to measure the speed of baseballs and automobiles. Astronomers observe light from distant galaxies and use the Doppler effect to measure their speeds. Physicians can detect the speed of the moving heart wall in a fetus by means of the Doppler effect in ultrasound.

▶ **CONNECTION TO** BIOLOGY Bats use the Doppler effect to detect and catch flying insects. When an insect is flying faster than a bat, the reflected frequency is lower, but when the bat is catching up to the insect, as in **Figure 8,** the reflected frequency is higher. Not only do bats use sound waves to navigate and locate their prey, but they often must do so in the presence of other bats. This means they must discriminate their own calls and reflections against a background of many other sounds of many frequencies. Scientists continue to study bats and their amazing abilities to use sound waves.

SECTION 1   REVIEW

Section Self-Check   Check your understanding.

6. **MAIN**IDEA What physical characteristic of a sound wave should be changed to alter the pitch of the sound? To alter the loudness?

7. **Graph** The eardrum moves back and forth in response to the pressure variations of a sound wave. Sketch a graph of the displacement of the eardrum versus time for two cycles of a 1.0-kHz tone and for two cycles of a 2.0-kHz tone.

8. **Effect of Medium** List two sound characteristics that are affected by the medium through which the sound passes and two characteristics that are not affected.

9. **Decibel Scale** How many times greater is the sound pressure level of a typical rock concert (110 dB) than a normal conversation (50 dB)?

10. **Early Detection** In the nineteenth century, people put their ears to a railroad track to get an early warning of an approaching train. Why did this work?

11. **Bats** A bat emits short pulses of high-frequency sound and detects the echoes.

   a. In what way would the echoes from large and small insects compare if they were the same distance from the bat?

   b. In what way would the echo from an insect flying toward the bat differ from that of an insect flying away from the bat?

12. **Critical Thinking** Can a trooper using a radar detector at the side of the road determine the speed of a car at the instant the car passes the trooper? Explain.

The Physics of Music

Think about all the musical instruments you have seen or heard. There are violins, flutes, drum kits, and many other instruments. How do you think the size, shape, and materials of an instrument affect the sounds it makes?

MAIN IDEA

Music consists of complex sound waves produced by vibrating objects.

Essential Questions

- What is the origin of sound?
- What are the characteristics of resonance in air columns?
- What are the characteristics of resonance on strings?
- Why are there variations in sound quality among instruments?
- How are beats produced?

Review Vocabulary

frequency the number of complete oscillations that a wave makes each second; is measured in hertz

New Vocabulary

closed-pipe resonator
open-pipe resonator
fundamental
harmonics
dissonance
consonance
beat

Sources of Sound

In the middle of the nineteenth century, German physicist Hermann Helmholtz studied sound production in musical instruments and the human voice. In the twentieth century, scientists and engineers developed electronic equipment that permits not only a detailed study of sound, but also the creation of electronic musical instruments and recording devices that allow us to listen to music whenever and wherever we wish.

Recall that sound is produced by a vibrating object. The vibrations of the object create particle motions that cause pressure oscillations in the air. A loudspeaker has a cone that is made to vibrate by electrical currents. The surface of the cone creates the sound waves that travel to your ear and allow you to hear music. Musical instruments such as gongs, cymbals, and drums are other examples of vibrating surfaces that are sources of sound.

The human voice is produced by vibrations of the vocal cords, which are two membranes located in the throat. Air from the lungs rushing through the throat starts the vocal cords vibrating. The frequency of vibration is controlled by the muscular tension placed on the vocal cords. The more tension on the vocal cords, the more rapidly they vibrate, resulting in a higher pitch sound. If the vocal cords are more relaxed, they vibrate more slowly and produce lower-pitched sounds.

In brass instruments, such as the trumpet, the tuba, and the bugle, the lips of the performer vibrate, as shown in **Figure 9.** Reed instruments, such as the clarinet and the saxophone, have a thin wooden strip called a reed that vibrates as a result of air blown across it, as shown in **Figure 9.** In flutes and organ pipes, air is forced across an opening in a pipe. Air moving past the opening sets the column of air in the instrument into vibration.

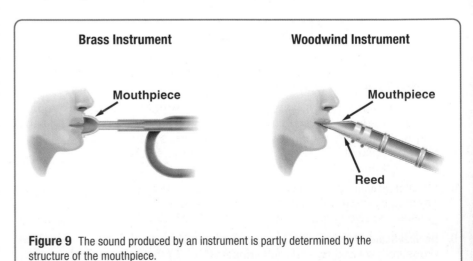

Figure 9 The sound produced by an instrument is partly determined by the structure of the mouthpiece.

Stringed instruments, such as the piano, the guitar, and the violin, work by setting wires or strings into vibration. In the piano, the wires are struck; for the guitar, they are plucked; and for the violin, the friction of the bow causes the strings to vibrate. The strings are attached to a sounding board that vibrates with the strings. The vibrations of the sounding board cause the pressure oscillations in the air that we hear as sound. Electric guitars use electronic devices to detect and amplify the vibrations of the guitar strings.

Resonance in Air Columns

If you have ever used just the mouthpiece of a brass or wind instrument, you know that while the vibration of your lips or the reed alone makes a sound, it is difficult to control the pitch. The long tube that makes up the instrument must be attached if music is to result. When the instrument is played, the air within this tube vibrates at the same frequency, or in resonance, with a particular vibration of the lips or reed. Remember that resonance increases the amplitude of a vibration by repeatedly applying a small external force to the vibrating air particles at the natural frequency of the air column. The length of the air column determines the frequencies of the vibrating air that will resonate. For wind and brass instruments, such as flutes, trumpets, and trombones, changing the length of the column of vibrating air varies the pitch of the instrument. The mouthpiece simply creates a mixture of different frequencies, and the resonating air column acts on a particular set of frequencies to amplify a single note, turning noise into music.

A tuning fork above a hollow tube can provide resonance in an air column, as shown in **Figure 10.** The tube is placed in water so that the bottom end of the tube is below the water surface. A resonating tube with one end closed to air is called a **closed-pipe resonator.** The length of the air column is changed by adjusting the height of the top of the tube above the water. If the tuning fork is struck with a rubber hammer, the sound alternately becomes louder and softer as the length of the air column is varied by moving the tube up and down in the water. The sound is loud when the air column is in resonance with the tuning fork because the resonating air column intensifies the sound of the tuning fork.

PhysicsLAB

SPEED OF SOUND
How can you determine the speed of sound?

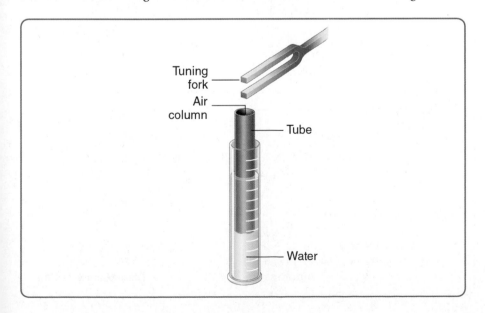

Figure 10 As the tube is raised or lowered, the length of the air column changes, which causes the sound's volume to change.

Closed Pipe

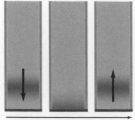

Time

Closed pipes: high pressure
reflects as high pressure

Open Pipe

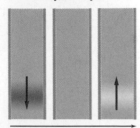

Time

Open pipes: high pressure
reflects as low pressure

Figure 11 In closed pipes, the sound wave reflects off the closed end. High pressure waves reflect as high pressure. In open pipes, the sound wave reflects off an open end. High pressure waves are reflected as low pressure.

Standing pressure wave How does resonance occur? The vibrating tuning fork produces a sound wave. This wave of alternate high- and low-pressure variations moves down the air column. When the wave hits the water surface, it is reflected back up to the tuning fork, as shown in **Figure 11**. If the reflected high-pressure wave reaches the tuning fork at the same moment that the fork produces another high-pressure wave, then the emitted and returning waves reinforce each other. This reinforcement of waves creates a standing wave, and resonance occurs.

An **open-pipe resonator** is a resonating tube with both ends open that will resonate with a sound source. In this case, the sound wave does not reflect off a closed end, but rather off an open end. If the high-pressure part of the wave strikes the open end, the rebounding wave will be low-pressure at that point, as shown in **Figure 11**.

Resonance lengths **Figure 12** shows a standing sound wave in a pipe represented by a sine wave. Sine waves can represent either the air pressure or the displacement of the air particles. Recall that standing waves have nodes and antinodes. A node is the stationary point where two equal wave pulses meet and are in the same location. An antinode is the place of largest displacement when two wave pulses meet. In the pressure graphs, the nodes are regions of mean atmospheric pressure. At the antinodes, the pressure oscillates between its maximum and minimum values. In the case of the displacement graph, the antinodes are regions of high displacement and the nodes are regions of low displacement. In both cases, two adjacent antinodes (or two nodes) are separated by one-half wavelength.

☑ **READING CHECK Explain** the difference between a node and an antinode on a displacement graph.

Figure 12 Standing waves in pipes can be represented by sine waves.

Identify Which are the areas of mean atmospheric pressure in the air pressure graphs?

View an **animation on resonance.**

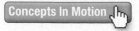

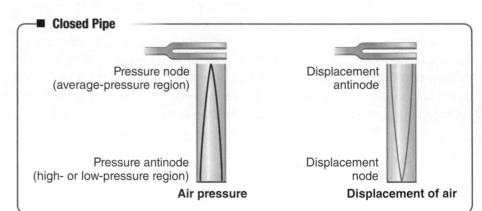

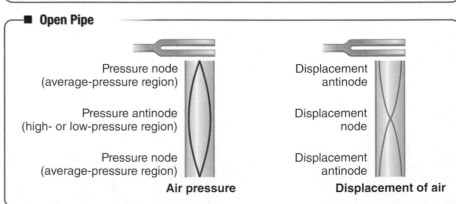

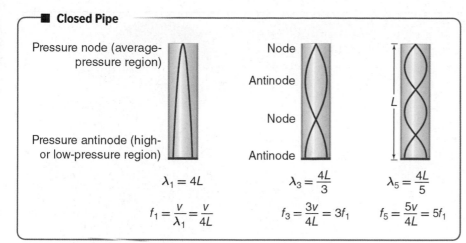

Closed Pipe

Pressure node (average-pressure region)

Pressure antinode (high- or low-pressure region)

Node

Antinode

Node

Antinode

L

$\lambda_1 = 4L$

$\lambda_3 = \dfrac{4L}{3}$

$\lambda_5 = \dfrac{4L}{5}$

$f_1 = \dfrac{v}{\lambda_1} = \dfrac{v}{4L}$

$f_3 = \dfrac{3v}{4L} = 3f_1$

$f_5 = \dfrac{5v}{4L} = 5f_1$

Figure 13 A closed pipe resonates when its length is an odd number of quarter wavelengths.

Resonance frequencies in a closed pipe If a closed end must act as a node, and an open end must act as an antinode, what is the shortest column of air that will resonate in a closed pipe? **Figure 13** shows that it must be one-fourth of a wavelength. As the frequency is increased, additional resonance lengths are found at half-wavelength intervals. Thus, columns of length $\dfrac{\lambda}{4}, \dfrac{3\lambda}{4}, \dfrac{5\lambda}{4}, \dfrac{7\lambda}{4}$, and so on will all be in resonance with a tuning fork that produces sound of wavelength λ.

In practice, the first resonance length is slightly longer than one-fourth of a wavelength. This is because the pressure variations do not drop to zero exactly at the open end of the pipe. Actually, the node is approximately 0.4 pipe diameters beyond the end. Additional resonance lengths, however, are spaced by exactly one-half of a wavelength. Measurements of the spacing between resonances can be used to find the velocity of sound in air, as in Example Problem 2.

Resonance frequencies in an open pipe The shortest column of air that can have nodes at both ends is one-half of a wavelength long, as shown in **Figure 14.** As the frequency is increased, additional resonance lengths are found at half-wavelength intervals. Thus, columns of length $\dfrac{\lambda}{2}$, $\lambda, \dfrac{3\lambda}{2}, 2\lambda$, and so on will be in resonance with a tuning fork.

If open and closed pipes of the same length are used as resonators, the wavelength of the resonant sound for the open pipe will be half as long as that for the closed pipe. Therefore, the frequency will be twice as high for the open pipe as for the closed pipe. For both pipes, resonance lengths are spaced by half-wavelength intervals.

REAL-WORLD PHYSICS
• • • • • • • • • • •

HEARING AND FREQUENCY The human auditory canal acts as a closed-pipe resonator that increases the ear's sensitivity for frequencies between 2000 and 5000 Hz, but the full range of frequencies that people hear extends from 20 to 20,000 Hz. A dog's hearing extends to frequencies as high as 45,000 Hz, and a cat's extends to frequencies as high as 100,000 Hz.

PhysicsLAB

HOW FAST DOES SOUND TRAVEL?
How can you measure sound speed using an open-pipe resonator?

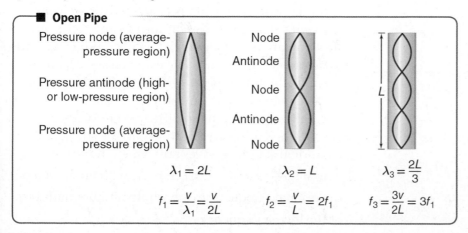

Open Pipe

Pressure node (average-pressure region)

Pressure antinode (high- or low-pressure region)

Pressure node (average-pressure region)

Node

Antinode

Node

Antinode

Node

L

$\lambda_1 = 2L$

$\lambda_2 = L$

$\lambda_3 = \dfrac{2L}{3}$

$f_1 = \dfrac{v}{\lambda_1} = \dfrac{v}{2L}$

$f_2 = \dfrac{v}{L} = 2f_1$

$f_3 = \dfrac{3v}{2L} = 3f_1$

Figure 14 An open pipe resonates when its length is an even number of quarter wavelengths.

Explain How does the length at which an open pipe resonates differ from the length at which a closed pipe resonates?

Open Pipe (Flute)

Closed Pipe (Marimba)

Closed Pipe (Seashell)

Figure 15 A flute is an example of an open-pipe resonator. The hanging pipes of a marimba and seashells are examples of closed-pipe resonators.

Figure 16 A string resonates with standing waves when its length is a whole number of half wavelengths.

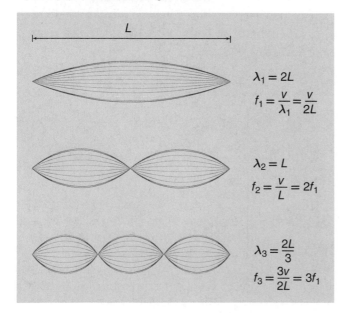

$$\lambda_1 = 2L$$
$$f_1 = \frac{v}{\lambda_1} = \frac{v}{2L}$$

$$\lambda_2 = L$$
$$f_2 = \frac{v}{L} = 2f_1$$

$$\lambda_3 = \frac{2L}{3}$$
$$f_3 = \frac{3v}{2L} = 3f_1$$

Hearing resonance Musical instruments use resonance to increase the loudness of particular notes. Open-pipe resonators include flutes, shown in **Figure 15.** Clarinets and the hanging pipes under marimbas and xylophones are examples of closed-pipe resonators. If you shout into a long tunnel, the booming sound you hear is the tunnel acting as a resonator. The seashell in **Figure 15** also acts as a closed-pipe resonator.

Resonance on Strings

Although plucking, bowing, or striking strings produces variation in waveforms, waveforms on vibrating strings have many characteristics in common with standing waves on springs and ropes. A string on an instrument is clamped at both ends, and therefore, the string must have a node at each end when it vibrates. In **Figure 16,** you can see that the first mode of vibration has an antinode at the center and is one-half a wavelength long. The next resonance occurs when one wavelength fits on the string, and additional standing waves arise when the string length is $\frac{3\lambda}{2}$, 2λ, $\frac{5\lambda}{2}$, and so on. As with an open pipe, the resonant frequencies are whole-number multiples of the lowest frequency.

Recall that the speed of a wave depends on the medium. For a string, it depends on the tension of the string, as well as its mass per unit length. The tighter the string, the faster the wave moves along it, and therefore, the higher the frequency of its standing waves. This makes it possible to tune a stringed instrument by changing the tension of its strings.

Because strings are so small in cross-sectional area, they move very little air when they vibrate. This makes it necessary to attach them to a sounding board, which transfers their vibrations to the air and produces a stronger sound wave. Unlike the strings themselves, the sounding board should not resonate at any single frequency. Its purpose is to convey the vibrations of all the strings to the air, and therefore it should vibrate well at all frequencies produced by the instrument. Because of the complicated interactions among the strings, the sounding board, and the air, the design and construction of stringed instruments are complex processes, considered by many to be as much an art as a science.

☑ **READING CHECK** **Describe** the relationship between the tension of a string and the speed of a wave as it travels along the string.

Sound Quality

A tuning fork produces a soft and uniform sound. This is because its tines vibrate like simple harmonic oscillators and produce the simple sine wave shown in the top graph in **Figure 17.** Sounds made by the human voice and musical instruments are much more complex, like the wave in the bottom graph in **Figure 17.** Both waves have the same frequency, or pitch, but they sound very different. The complex sound wave is actually a blend of several of different frequencies. The shape of the wave depends on the relative amplitudes of these frequencies. Different sources provide different combinations of frequencies. In musical terms, the difference between the waves from different instruments is called timbre (TAM bur), tone quality, or color.

MiniLAB
SOUNDS GOOD
How can you determine whether an instrument is an open-pipe resonator or a closed-pipe resonator?

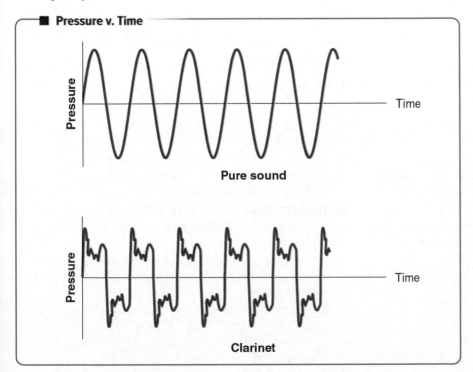

■ **Pressure v. Time**

Pressure

Time

Pure sound

Pressure

Time

Clarinet

Figure 17 The pure sound produced by a tuning fork is represented by a simple sine wave. The more complex sound produced by a clarinet is represented in the bottom graph.

FINDING THE SPEED OF SOUND USING RESONANCE

When a tuning fork with a frequency of 392 Hz is used with a closed-pipe resonator, the loudest sound is heard when the column is 21.0 cm and 65.3 cm long. What is the speed of sound in this case? Is the temperature warmer or cooler than normal room temperature, which is 20°C? Explain your answer.

1 ANALYZE AND SKETCH THE PROBLEM

- Sketch the closed-pipe resonator.
- Mark the resonance lengths.

KNOWN	UNKNOWN
$f = 392$ Hz	$v = ?$
$L_A = 21.0$ cm	
$L_B = 65.3$ cm	

Sketch showing 21.0 cm and 65.3 cm resonance lengths in the closed pipe.

2 SOLVE FOR THE UNKNOWN

Solve for the length of the wave using the length-wavelength relationship for a closed pipe.

$$L_B - L_A = \frac{1}{2}\lambda$$

$$\lambda = 2(L_B - L_A) \qquad \blacktriangleleft \text{ Rearrange the equation for } \lambda.$$

$$= 2(0.653 \text{ m} - 0.210 \text{ m}) \qquad \blacktriangleleft \text{ Substitute } L_B = 0.653 \text{ m}, L_A = 0.210 \text{ m}.$$

$$= 0.886 \text{ m}$$

Use $\lambda = \dfrac{v}{f}$

$$v = f \cdot \lambda \qquad \blacktriangleleft \text{ Rearrange the equation for } v.$$

$$= (392 \text{ Hz})(0.886 \text{ m}) \qquad \blacktriangleleft \text{ Substitute } f = 392 \text{ Hz}, \lambda = 0.886 \text{ m}.$$

$$= 347 \text{ m/s}$$

The speed is slightly greater than the speed of sound at 20°C, indicating that the temperature is slightly higher than normal room temperature.

3 EVALUATE THE ANSWER

- **Are the units correct?** $(\text{Hz})(\text{m}) = \left(\dfrac{1}{s}\right)(\text{m}) = \text{m/s}$. The answer's units are correct.

- **Is the magnitude realistic?** The speed is slightly greater than 343 m/s, which is the speed of sound at 20°C.

PRACTICE PROBLEMS

Do additional problems. | Online Practice |

13. A 440-Hz tuning fork is used with a resonating column to determine the velocity of sound in helium gas. If the spacing between resonances is 110 cm, what is the velocity of sound in helium gas?

14. The frequency of a tuning fork is unknown. A student uses an air column at 27°C and finds resonances spaced by 20.2 cm. What is the frequency of the tuning fork? Use the speed calculated in Example Problem 2 for the speed of sound in air at 27°C.

15. A 440-Hz tuning fork is held above a closed pipe. Find the spacing between the resonances when the air temperature is 20°C.

16. CHALLENGE A bugle can be thought of as an open pipe. If a bugle were straightened out, it would be 2.65-m long.

   a. If the speed of sound is 343 m/s, find the lowest frequency that is resonant for a bugle (ignoring end corrections).

   b. Find the next two resonant frequencies for the bugle.

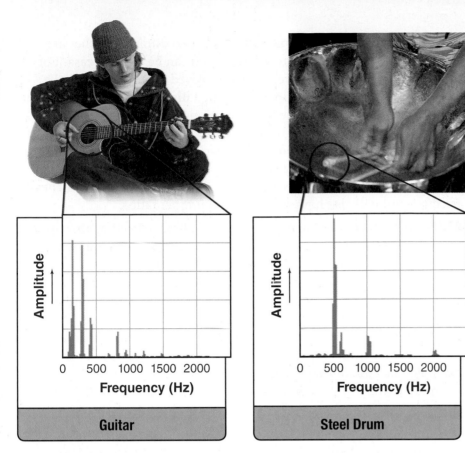

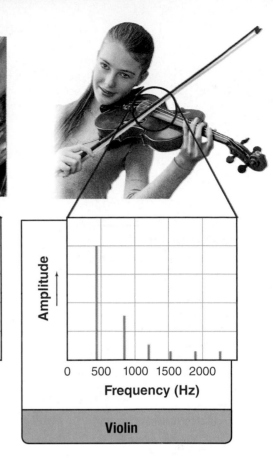

Figure 18 A guitar, a steel drum, and a violin produce characteristic sound spectra. Each spectrum is unique, as is the timbre of the instrument.

The sound spectrum: fundamental and harmonics The complex sound wave in **Figure 17** was made by a clarinet. Why does the clarinet produce such a sound wave? The air column in a clarinet acts as a closed pipe. Look back at **Figure 13,** which shows three resonant frequencies for a closed pipe. The clarinet acts as a closed pipe, so for a clarinet of length L the lowest frequency (f_1) that will be resonant is $\frac{v}{4L}$. For a musical instrument, the lowest frequency of sound that resonates is called the **fundamental.** A closed pipe also will resonate at $3f_1$, $5f_1$, and so on. These higher frequencies, which are whole-number multiples of the fundamental frequency, are called **harmonics.** It is the addition of these harmonics that gives a clarinet its distinctive timbre.

Some instruments, such as a flute, act as open-pipe resonators. Their fundamental frequency, which is also the first harmonic, is $f_1 = \frac{v}{2L}$ with subsequent harmonics at $2f_1$, $3f_1$, $4f_1$, and so on. Different combinations of these harmonics give each instrument its own unique timbre. Each harmonic on the instrument can have a different amplitude as well. A graph of the amplitude of a wave versus its frequency is called a sound spectrum. The spectra of three instruments are shown in **Figure 18.**

PHYSICS CHALLENGE

1. Determine the tension, F_T, in a violin string of mass m and length L that will play the fundamental note at the same frequency as a closed pipe also of length L. Express your answer in terms of m, L, and the speed of sound in air, v. The equation for the speed of a wave on a string is $v_{string} = \sqrt{\frac{F_T}{\mu}}$, where F_T is the tension in the string and μ is the mass per unit length of the string.

2. What is the tension in a string of mass 1.0 g and 40.0 cm long that plays the same note as a closed pipe of the same length?

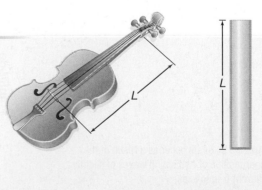

Consonance and dissonance When two different pitches are played at the same time, the resulting sound can be either pleasant or jarring. In musical terms, several pitches played together are called a chord. An unpleasant set of pitches is called **dissonance.** If the combination of pitches is pleasant, the sounds are said to be in **consonance.** What sounds pleasing varies between cultures, but most Western music is based upon the observations of Pythagoras of ancient Greece. He noted that pleasing sounds resulted when strings had lengths in small, whole-number ratios, such as 1:2, 2:3, or 3:4. This means their pitches (frequencies) will also have small, whole-number ratios.

Musical intervals Two notes with frequencies related by the ratio 1:2 are said to differ by an octave. For example, if a note has a frequency of 440 Hz, a note that is one octave higher has a frequency of 880 Hz. The fundamental and its harmonics are related by octaves; the first harmonic is one octave higher than the fundamental, the second is two octaves higher, and so on. It is the ratio of two frequencies, not the size of the interval between them, that determines the musical interval.

In other musical intervals, two pitches may be close together. For example, the ratio of frequencies for a "major third" is 4:5. An example is the notes C and E. The note C has a frequency of 262 Hz, so E has a frequency of $\left(\frac{5}{4}\right)(262 \text{ Hz}) = 327 \text{ Hz}$. In the same way, notes in a "fourth" (C and F) have a frequency ratio of 3:4, and those in a "fifth" (C and G) have a ratio of 2:3. More than two notes sounded together also can produce consonance. The three notes called do, mi, and sol make a major chord. For at least 2500 years, western music has recognized this as the sweetest of the three-note chords; it has the frequency ratio of 4:5:6.

Beats

You have seen that consonance is defined in terms of the ratio of frequencies. When the ratio becomes nearly 1:1, the frequencies become very close. Two frequencies that are nearly identical interfere to produce oscillating high and low sound levels called a **beat.** This phenomenon is illustrated in **Figure 19.** The frequency of a beat is the magnitude of difference between the frequencies of the two waves, $f_{\text{beat}} = |f_A - f_B|$. When the difference is less than 7 Hz, the ear detects this as a pulsation of loudness. Musical instruments often are tuned by sounding one against another and adjusting the frequency of one until the beat disappears.

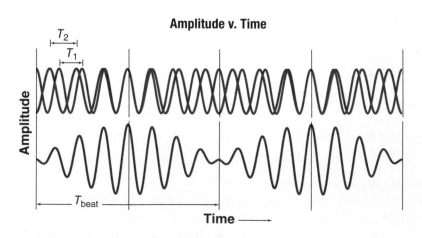

Figure 19 Beats occur as a result of the superposition of two sound waves of slightly different frequencies.

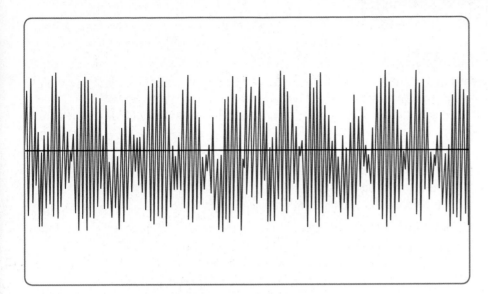

Figure 20 A noise wave consists of many different frequencies, all with about the same amplitude.

Sound Reproduction and Noise

When you listen to a live band or hear your school band practicing, you are hearing music produced directly by a human voice or musical instruments. You may want to hear live music every time you choose to listen to music. Most of the time, however, you likely listen to music that has been recorded and is played via electronic systems. To reproduce the sound faithfully, the system must accommodate all frequencies equally. A good stereo system keeps the amplitudes of all frequencies between 20 and 20,000 Hz to within a range of 3 dB.

A telephone system, on the other hand, needs only to transmit the information in spoken language. Frequencies between 300 and 3000 Hz are sufficient. Reducing the number of frequencies present helps reduce the noise. A noise wave is shown in **Figure 20.** Many frequencies are present with approximately the same amplitude. While noise is not helpful in a telephone system, some people claim that listening to white noise has a calming effect. For this reason, some dentists use noise to help their patients relax.

☑ **READING CHECK Describe** a noise wave.

SECTION 2 REVIEW

Section Self-Check ⟲ Check your understanding.

17. **MAINIDEA** What is the vibrating object that produces sounds in each of the following?

   a. a human voice

   b. a clarinet

   c. a tuba

   d. a violin

18. **Resonance in Air Columns** Why is the tube from which a tuba is made much longer than that of a cornet?

19. **Resonance in Open Tubes** How must the length of an open tube compare to the wavelength of the sound to produce the strongest resonance?

20. **Resonance on Strings** A violin sounds a note of F sharp, with a pitch of 370 Hz. What are the frequencies of the next three harmonics produced with this note?

21. **Resonance in Closed Pipes** One closed organ pipe has a length of 2.40 m.

   a. What is the frequency of the note played by this pipe?

   b. When a second pipe is played at the same time, a 1.40-Hz beat note is heard. By how much is the second pipe too long?

22. **Timbre** Why do various instruments sound different even when they play the same note?

23. **Beats** A tuning fork produces three beats per second with a second, 392-Hz tuning fork. What is the frequency of the first tuning fork?

24. **Critical Thinking** Strike a tuning fork with a rubber hammer and hold it at arm's length. Then press its handle against a desk, a door, a filing cabinet, and other objects. What do you hear? Why?

PHYSICS THAT'S ENTERTAINMENT!

SOUNDS GOOD!

Theater Acoustics

The ancient Greek theater at Epidaurus, shown in **Figure 1,** has 55 curved rows of limestone benches that can seat up to 14,000 people. Since its construction in the fourth century B.C., a great mystery has intrigued acoustic scientists: how is it possible to hear the voices of actors so clearly in the back-row seats of the theater without the aid of speakers?

It is all in the seats. Some hypothesized that music and voices were carried farther than in other theaters due to the local prevailing winds. Others thought that the Greeks used special masks that helped project the voices of actors. Many even suspected that the slope of the theater seats somehow contributed to the theater's amazing acoustics.

Modern Research It turns out that the seats do a play a role. Researchers have discovered that the shape, placement, and material of the seats all have beneficial acoustic properties. The seats in this theater absorb sound frequencies typical of audience murmurs and other background noises such as wind. At the same time, the front-row seats reflect the sounds of the actors' voices toward the back-row seats. This combination of subtracting unwanted noise and projecting desired noise means that you can enjoy a play even in the "cheap" seats.

Figure 1 Audience members in the farthest back rows of the theater at Epidaurus can still hear the voices of actors on stage—even without microphones—due to the theater's special acoustics.

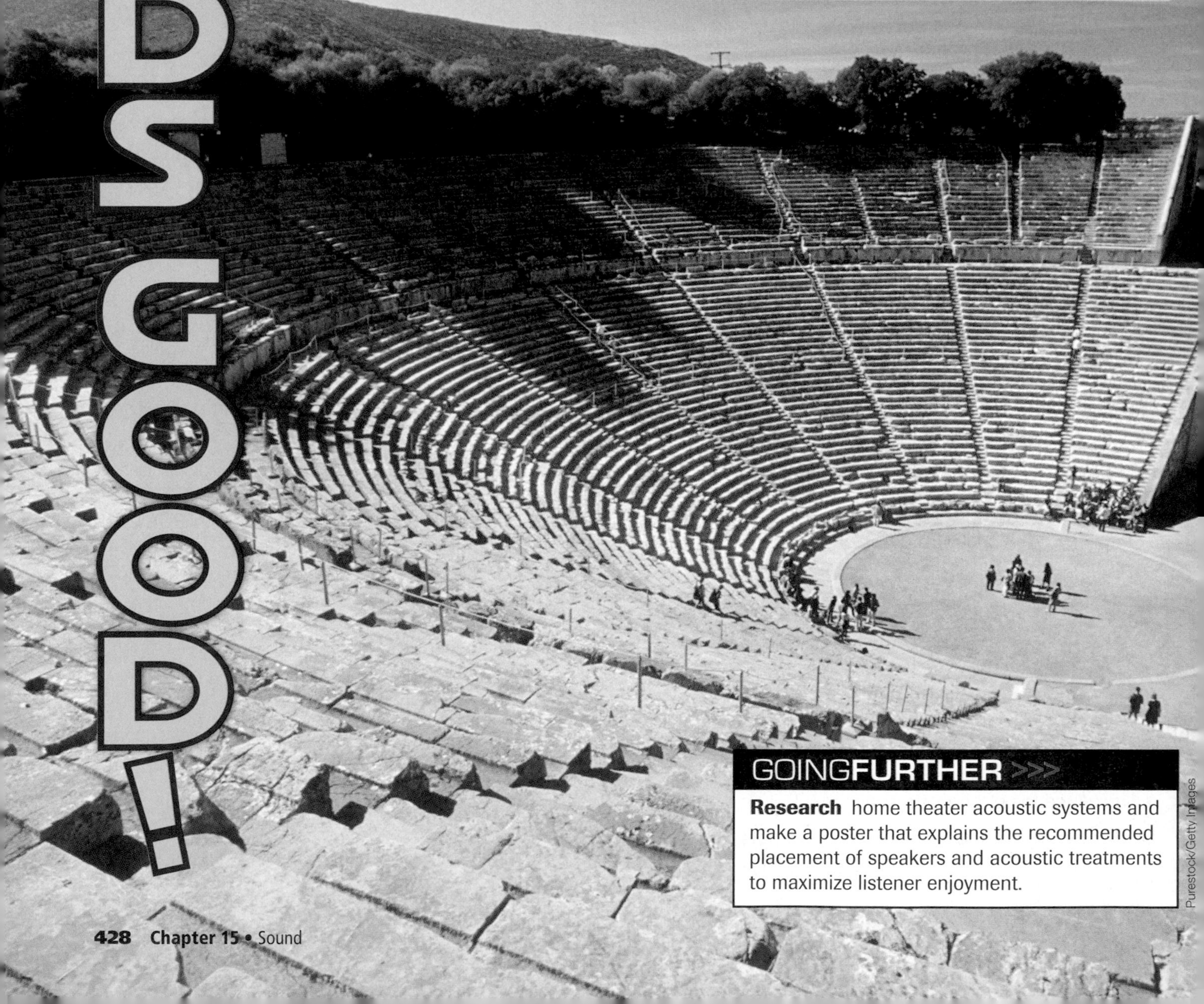

Purestock/Getty Images

GOING**FURTHER** >>>

Research home theater acoustic systems and make a poster that explains the recommended placement of speakers and acoustic treatments to maximize listener enjoyment.